Software Pioneers

Springer
Berlin
Heidelberg
New York
Barcelona
Hong Kong
London
Milan
Paris
Tokyo

Old "Bundestag", the sd&m 2001 Conference Site

Manfred Broy • Ernst Denert (Eds.)

Software Pioneers

Contributions to Software Engineering

Springer

Prof. Dr. Manfred Broy
Professor of Informatics
at Technische Universität München

- Dr. rer. nat. in Informatics, Technische Universität München
- Leibniz Award, Bundesverdienstkreuz
- Member European Academy of Sciences, IEEE Board of Governors
- Major contributions: concurrent distributed systems, systematic development of interactive systems

Prof. Dr. Ernst Denert
Chairman of the Supervisory Board
of sd&m AG

- Dr. -Ing. in Informatics, Technische Universität Berlin
- Honorary Professor at Technische Universität München
- Cofounder, former managing director and CEO of sd&m AG
- Software design and project management for several large business information systems
- Major contributions: software architecture for business information systems, author of *Software Engineering*

sd&m Conference Software Pioneers took place at the old "Bundestag" (parliament building) in Bonn, Germany, 28/29 June 2001

sd&m AG, software design & management
Consulting and Software for Custom Information Systems

sd&m develops customized software for business information systems and Internet applications supporting the customers' core business processes, and provides IT consulting. Based on its core competence – software engineering – sd&m realizes large projects for a broad variety of applications on all common system platforms. For more information on facts and figures, software engineering, projects, the team and culture, see **www.sdm.de**.

Preface

Manfred Broy

Ernst Denert

Within only a few decades software and the ability to develop high-quality software timely and at low cost became one of the most important innovations in industry. Meanwhile, for many industries, their ability to develop, use and maintain software is one of their key success factors. New companies that concentrate on software development have been created and have become substantial players in the market. sd&m AG is one example of a company that exclusively concentrates on the professional development of high-quality software. Professional software construction is impossible without carefully planned processes, professional project management and a deep understanding of the technology and its principles. Such a professionalism in the engineering of software as has been reached today would not be possible without the contributions of thousands of scientists and engineers who have developed the discipline over the years. However, as always, the major contributions to the field have not been given by large teams, but rather by a small number of individuals who have created and matured the field – the Software Pioneers.

Although software is everywhere today and most people in modern societies have to deal directly or indirectly with software, many of the major contributors to the software discipline are not known to the public nor even to younger professionals in the field. The young software engineers often do not know and do not understand the roots of our discipline. Since the progress in our field is so rapid and the technology changes so fast, one might get the impression that what was published in this field more than 30 years ago is not relevant anymore. However, this is not true at all. Over the years software engineering has developed fundamental principles, concepts and techniques that will be useful and valid independent of the advance of technology.

To make their merits more visible and to make software engineers more conscious of the fundamental contributions to the software field and also to express thanks to those individuals who helped create the field, sd&m devoted its biannual conference 2001 to the achievements of the Software Pioneers. The idea was to invite the individuals who most significantly contributed to the creation of the field of practical software engineering to an event giving them the opportunity to present their major contributions and to make visible how important these were and how much they influenced the field. The major considerations were not their contributions to informatics or to the general foundation of the discipline, but rather their influence on the practical engineering processes in a company like sd&m.

We are glad and proud that most of the Software Pioneers we invited accepted our invitation and therefore made it possible to have a conference with such an impressive list of speakers. The few that could not make it and turned down our invitation for good personal reasons are acknowledged at the end of the contribution by Ernst Denert. The conference took place in Bonn, in the old "Bundestag" in June 2001. About 1200 people were in attendance – excited and fascinated by the presentations of the Software Pioneers.

To document the conference and the everlasting contribution of the pioneers this volume was put together after the conference. For each of the pioneers, it contains a short contribution in which he looks back onto his work from the perspective of today. In addition, it contains the most important historical publications of the Software Pioneers. The volume starts with a survey paper by Ernst Denert in which he outlines the major contributions of each of the pioneers and puts their work into perspective.

The whole conference was recorded on video tapes. Therefore, we are glad to put the lectures of the Software Pioneers in full length on four DVDs together with a one hour excerpt prepared by Ernst Denert and some impressions from the conference. The DVDs supplement this volume.

It is our pleasure to thank all of those who made the conference into an overwhelming success, first of all of course the Software Pioneers who not only attended the conference, but helped us to prepare the selection of the historical papers and wrote their contribution for this volume in spite of the fact that most of them are still very busy and engaged in many activities. In particular, we thank sd&m for sponsoring the conference. We thank the staff who helped to prepare the conference, in particular Ms. Carola Lauber who did such an excellent job in making sure that everything ran smoothly both throughout the conference and afterwards in preparing this volume. We also thank the team of Springer-Verlag who was immediately ready to publish first of all a conference volume, available directly at the conference with all the historical contributions of the pioneers, and then this volume after the conference, which puts everything into perspective. Many thanks go also to those who helped to put this volume together.

We hope this volume shows very clearly the deep value of the contributions of a small number of people to an emerging field, which is changing and maturing further even today and certainly will much more tomorrow. We also hope that this volume is a contribution to the history of our young discipline and its documentation.

Manfred Broy
Ernst Denert

Munich, February 2002

The Software Pioneers with the conference organizers

Front row:
Ernst Denert
Tom DeMarco
Niklaus Wirth
John V. Guttag
Michael Jackson
Fred P. Brooks
David L. Parnas
Peter Chen
Alan Kay
Manfred Broy

Rear row:
Barry Boehm
Edsger W. Dijkstra
Michael Fagan
Friedrich L. Bauer
Erich Gamma
Ole-Johan Dahl
Kristen Nygaard
Rudolf Bayer
C.A.R. Hoare

Table of Contents

Manfred Broy
Department of Computer Science
Technische Universität München
broy@informatik.tu-muenchen.de
wwwbroy.informatik.tu-muenchen.de/~broy/

Manfred Broy

Software Engineering
From Auxiliary to Key Technology

Over the past years, computer applications have become a focus of public attention. Spectacular innovations such as the Internet, the by-now-omnipresent personal computer, as well as numerous new developments in almost all technical areas have drawn the general public's attention to information technology. A considerable portion of this attention can be attributed to the immense economic effect of information technology, including the ups and downs of the stock exchanges, the often exaggerated hopes in connection with company start-ups, and the frenzy with which a lot of companies are pursuing their orientation towards e-commerce.

Less visible and not sufficiently noted by the public is the role that software and software development issues play in this context. The reasons are obvious: software is almost intangible, generally invisible, complex, vast and difficult to comprehend. Hardly anybody, with the exception of a few experts, can truly grasp what software really is and what role it plays. Probably the best way to give a lay person an idea of that role is through numbers.

It is rather impressive that
- cellular phones nowadays are controlled by over one million lines of program code,
- a modern luxury class vehicle has more onboard computer power and software than an Apollo spacecraft.

These examples underscore the fundamental importance of software. One can also read up on the subject in a study we prepared at the end of last year for the Deutsche Forschungsministerium (German Federal Bureau of Research) (or in the by-now-famous report to former US President Clinton, the PITAC Report), which emphasizes the utmost importance of software engineering. The report states:

... that software demand exceeds the nation's capability to produce it, that we must still depend on fragile software, that technology to build reliable and secure software is inadequate and that the nation is underinvesting in fundamental software research.

Considering the extremely short period of time of only just over 40 years in which software has gained this vast importance, it is not surprising that there are still countless shortcomings and unsolved problems in the development of software. This will likely keep software engineers busy for another good few years.

On the surface, scientific disciplines are often seen as objective fields that are largely independent of human influences. This, of course, is deceiving. Scientific and especially engineering disciplines, such as software engineering, are shaped and formed by the people who created these fields. Their ideas and beliefs form the basis of and give shape to the field. Therefore, we will be able to gain a better understanding of software engineering if we get to know the people who created and pioneered this field.

Software is complex, error-prone and difficult to visualize. Thus, it is not surprising that many of the contributions of the software pioneers are centered around the objective of making software easier to visualize and understand, and to represent the phenomena encountered in software development in models that make the often implicit and intangible software engineering tasks explicit. As software makes its way into many areas of our life, these models and methods also take hold there. In addition it has become apparent that these models and procedures are not only of great interest and importance for software development, but also for a whole series of other questions in connection with complex systems. Therefore, the contributions made by the software pioneers are of general interest, even though they were usually motivated solely by problems related to software engineering. They form the basis for a new discipline of modeling processes and sequences in dynamic systems by means of logic, algebra and discrete structures.

In addition to the many still open, unsolved and exciting questions around the topic of software, one should not forget how far we have already come since the early days of computer science and software. It is well worth casting a glance back onto some major contributions made over the past

40 years. One of the motives for doing so lies in the observation that computer science – more than any other discipline – suffers from a frequent lack of scientific standards and insufficient knowledge of the scientific literature. In light of the quickly changing technology, many young computer scientists seem to be of the opinion that scientific work that was done more than five years ago can no longer be relevant today. This is a regrettable mistake.

The work of the software pioneers, who with their contributions and ideas have given decisive impulses to the development of computer science and software engineering, is as significant today as it was then, and by no means only of historical interest. Many of the ideas are as useful today as they were in the time when they emerged. But not only the results are exemplary. Possibly even more important and impressive is the attitude of the software pioneers, their personalites and the mind-set in which their contributions are rooted and explained, which offer many stimuli and inspire awe.

The contributions of the software pioneers form the basis for the work and procedures of many businesses in the software industry. In this context, however, another thought bears importance. In addition to the very clearly visible technical impact of these contributions on the practice of software engineering, their elementary scientific significance must not be overlooked. Each and every one of the software pioneers has made fundamental scientific contributions. These have not only changed our trade, they have also provided essential cultural and scientific impulses for the fundamental step from understanding the world in continuous functions to the understanding of the world in discrete, digital models, namely the models used in software engineering. The step from the number-based models of mathematics to a qualitative, logic-oriented approach is one of the most significant contributions of software engineering. We are only just starting to realize its impact on education, on science and on our view of the world.

Also remarkable are the personalities of the software pioneers. As individual as each of them is, it is incredible how many parallels one can find in their frame of minds and general attitudes towards life. They are characterized by a critical awareness, often bordering on stubbornness, an unwavering belief in the rightness of the chosen approach, a high measure of discipline and an incredible amount of energy allowing them to critically analyze the subject matter over and over again. This leads to a spirited frame of mind – critical, and sometimes inconvenient, but at the same time always creative and stimulating, too. And here's another thing we can learn from the software pioneers: individual, independent thinking, an insistence on clarity and an unwillingness to compromise, which are often a much more important starting point than the conformist pursuit of repetitively identical approaches and conventional ideas, and striving to be part of the mainstream.

The contributions of the software pioneers have led to methods that have shaped our world in a way that exceeds their own expectations. It is high time that we increase public awareness of their accomplishments and that we lay the foundation for a long-term understanding of software engineering and its pioneers in our society.

sd&m Conference on Software Pioneers, Bonn, June 28/29, 2001

Ernst Denert
sd&m AG
Thomas-Dehler-Strasse 27
81737 München
denert@sdm.de

Ernst Denert

Opening and Overview Speech

The Relevance of the Software Pioneers for sd&m

"New trends are often nothing more than the bad memory of posterity." This is not only true of fashion, but also of software. The study of history, especially the history of science and technology, is interesting in itself, but not enough so to justify the high costs of an sd&m conference, which brings together about 1200 employees and guests of our company for two days – the topic of such a conference must certainly have relevance to our daily work. So the question is, does an exploration of the software pioneers' achievements have anything to do with our daily practice at sd&m? The answer is a resounding yes. This overview demonstrates that these pioneers not only have made key academic contributions, but also have brought about enormous practical applications.

Friedrich L. Bauer – The Stack Principle

Friedrich L. Bauer invented the stack, the computer scientist's favorite data structure, in order to be able to parse and evaluate algebraic formulas using a computer. Nowadays, as the stack is hitting its mid 40s, we can hardly imagine a time when it did not exist. In today's software, the stack is omnipresent – though still used for analyzing formulas, it is primarily used for syntax analysis in compilers and also in run time systems. Another one of Bauer's epic contributions was his involvement in defining Algol 60. Though the language is no longer used, it had a massive influence on later programming languages that are still in use today, almost all of which contain the Algol 60 gene. The discovery of the block principle and the recursive procedure were also fundamental programming milestones and would have been impossible without the stack principle.

Ole-Johan Dahl – Simula 67

Nowadays, object orientation is quite rightly the epitome of design and programming methodology. Much older than generally assumed, object orientation did not first originate with C++ in the mid-1980s, or Smalltalk in the late 1970s, but almost two decades earlier with Simula 67, which was developed by Ole-Johan Dahl and his colleagues, especially Kristen Nygaard, in Norway. By lucky coincidence I came across Simula at the beginning of the 1970s. A colleague had brought a compiler from Oslo and installed it on the Computer Science Department's computer at the Technical University of Berlin. That is how I came to be one of the very early object-oriented programmers. It had a huge influence on my later project work and by extension also on sd&m. As the name suggests, Simula was designed for, but was in no way limited to, programming simulations. In terms of object-orientation, the language was fully equipped with data encapsulation and inheritance. Simula also carried the Algol gene and has passed on its object-oriented gene to all subsequent object-oriented programming languages. I still clearly remember Brad Cox, the developer of Objective C, asking me about Simula in the mid-1980s, as he was working on this rival language to C++. An understanding of Simula was an important prerequisite when designing a new object-oriented language, in order to avoid starting from scratch. Simula's strong impact on object-oriented programming is still enormously evident in our project work today.

Niklaus Wirth – Pascal

I will now turn to the school of programming that Niklaus Wirth founded, not only, but most effectively with the programming language Pascal. He wanted to create a programming language that one could use to teach good programming effectively. I would also like to add that he had already had an impact on me with his Pascal predecessor Algol W, which we also had at the TU Berlin. I used Algol W for my first lecture on data structures, Simula for the second, and finally Algol 68 in our book titled *Data*

Structures. You can write bad programs in a good language such as Pascal, but you have to work hard to program well in a bad language such as C. Wirth taught us how a good programming language makes it so much easier to write good programs – structured, clear, simple. With Pascal, he showed us how control structures and data structures work together, especially in his groundbreaking book *Algorithms and Data Structures*. Wirth also taught us "program development by stepwise refinement," from which an unfortunate and futile discussion developed concerning top-down versus bottom-up. The dogmatic programming war that ensued was certainly not Wirth's fault. Programming is the foundation of all software engineering. Wirth's school, based on Pascal, continues to have an impact on our work.

Fred Brooks – OS/360

Fred Brooks created the IBM mainframe, one of the most important computer systems of all time. Brooks first defined the architecture of the IBM/360 by specifying the command set of this machine, or to be more precise, this family of machines. The essential features of this command set can still be seen today in the /390 series. Brooks then managed the development of the OS/360 operating system for this machine series, a gigantic IBM project with more than 1000 people – a pioneering achievement with far-reaching consequences. After almost 40 years, the system created by Fred Brooks and his team is still in operation in many thousands of companies – and will remain so, even if the mainframe is often pronounced dead. Many, if not most, sd&m projects deal with this system in one way or another. Fred Brooks recorded his experiences in the wonderful book *The Mythical Man Month*, which culminates in statements such as "Adding manpower to a late project makes it later" and "Thinkers are rare, doers are rarer, thinker-doers are rarest." He came up with the last statement in order to justify why it makes sense to separate the role of the project manager and that of the designer in larger projects. This realization is one of the fundamentals on which sd&m was founded, even to the point of influencing the name sd&m.

Alan Kay – Graphic User Interface

Alan Kay is considered the inventor of personal computing, and the graphic user interface based on screen windows and the mouse. Crediting him with this achievement is in no way diminished by also recognizing Doug Engelbart, who first publicly introduced his ideas on the graphic user interface in 1968. I would like to make a confession here. In my career as a software engineer, never did I underestimate a concept more than I did the graphic user interface.

Alan Kay invented the "Dynabook," an early model of what we now call the laptop. In the 1970s, he was working at Xerox PARC (Palo Alto Research Center) where object-oriented programming was pushed by the development of Smalltalk. Smalltalk was also the impetus for creating a development environment with a graphic user interface. This was the model on which Apple based its products starting in 1982, first the Lisa and then the Mac. Not until 1990 did Microsoft with Windows enter the market, perhaps late, but hugely successfully. Nowadays, very few can imagine – myself included – using a computer without a graphic user interface; WIMP (windows, icons, menus, pointers) has become an integral part of our lives and Alan Kay is its pioneer.

Rudolf Bayer – Relational Data Model and B-Trees

Now I will touch upon databases, for they are the main elements of the business information systems that we build. We cannot accomplish anything without database systems. I cannot even talk about databases without first mentioning Ted Codd and his relational data model. Unfortunately, due to health reasons, he is no longer able to give presentations. Rudolf Bayer, who worked closely with him, devoted part of his presentation to the relational model. According to mathematical principles, a relation is a subset of the Cartesian product of n sets. At the beginning of the 1970s when I first heard that this might be something useful and involved storing data on disks, it seemed to be a strange concept. It also took a while for relational database systems to come along. The first one was the experimental System R from IBM. Commercial success did not start until the 1980s, first with Oracle, then with DB/2, Ingres, and Informix. All of a sudden, with subsets of the Cartesian product of sets, real-world, large, and finally even high performance systems were built and a lot of money has been made with them. Even the newer object-oriented DBMS have not supplanted them.

Rudolf Bayer invented the B-tree. The meaning of the letter B in the name remains ambiguous: Does it stand for balanced or for Bayer? There can be no database system without the B-tree, whether DB/2, Oracle, or any other. The relational data model could not be integrated as a high performer without the ingenious concept of B-trees – they grow very slowly, while they are also wide and even balanced as they grow. They are thus able to host huge quantities of data, which can then be accessed by the shortest routes possible. The B-tree concept was published shortly after the relational data model; nevertheless both concepts were developed separately from one another. A few years after their discovery, these two congenial ideas came together in relational database systems, one of the most important constructs in computer science.

Peter Chen – Entity/Relationship Model

In 1976, Peter Chen introduced his entity/relationship (E/R) Model, a method for modeling, specifying, and designing relational databases. From it, we learned that the world, or at least the way it is represented in information systems, consists of objects and relationships between them – such a simple concept and yet so powerful. The E/R model certainly lies to a certain extent one semantic level above that of the relational model; Chen showed us how to establish the mapping. I first attempted E/R modeling back in 1983 when sd&m was just starting out. Afterwards, I asked myself how we had modeled data previously, if at all. Of course, somehow every software designer and developer establishes a model for his system's data; since Chen, most use the E/R form. I have heard people say, "We don't have a data model." What they mean was that they do not have an E/R model, or at least none that is documented. This shows that the E/R model has become synonymous with a data model, just like Kleenex for paper tissues.

Edsger W. Dijkstra – Structured Programming

Edsger W. Dijkstra has contributed a great deal, including groundbreaking concepts, to the development of computer science. His most popular and probably the most significant contribution is structured programming, introduced in 1968 in his short, two-page article entitled "Goto statement considered harmful." This article broke the spell of the goto statement, which had been so widely used to produce spaghetti code. At the same time the article was the impetus for structured – or even better – structuring control mechanisms: sequence, case-by-case distinctions (if-then-else, case) and loops (while-do, repeat-until).

A few years earlier, Böhm and Jacopini had mathematically proven that the structured programming concept is powerful enough to program every algorithm thereby rendering goto's superfluous. When a colleague first told me that at the beginning of the 1970s, I did not want to believe him. I had just written a program for my master's thesis on the search for cycles in graphs. It was real spaghetti code of which I was incredibly proud. It took some convincing before I would accept that my masterpiece could work without goto's. Structured programming then turned into a popular trend that we are only too familiar with. It appeared in the graphic notion of the Nassi/Shneiderman diagrams, for which my former company, Softlab, invented the name "Struktogramm." I did not like this graphic notation. Many of you know my motto: Software should be written, not drawn. Nowadays, structured programming in the sense of Dijkstra is simply standard practice – hopefully. It is hard to imagine that this was once a fundamental discovery and that I was ignorant enough to believe was not going to be practical.

Tony Hoare – Program Correctness

When addressing the topic of software correctness, you could say that the name Tony Hoare is synonymous with correctness. The title of his presentation "Assertions and Program Verification" deals with exactly that topic. Like Dijkstra, he has also made substantial contributions to the progress of computer science. Among them are the Quicksort algorithm and the synchronization of parallel processes using monitors, which are a generalization of Dijkstra's basic semaphore concept. Hoare's most important concept, however, is the verification of program correctness. At first glance, such program verification seems to hold little importance for our daily work at sd&m – we will likely never formally prove that a program is correct. However, understanding the concept of program correctness is very valuable. Bill Wulf said it best when he wrote, "even if we are not going to prove a program to be correct, we should write it as if we were going to." Simply thinking about the pre- and postconditions and assertions necessary to do the proof is more likely to lead to "correct" programming from the outset. Hoare's school of thinking had a huge impact on generations of software engineers.

David Parnas – Modularization

In this presentation, it is not really appropriate to give prominence to any single pioneer. But I am going to do it anyway by acknowledging that no one had such a fundamental influence on myself and my work as David Parnas. I was lucky enough to meet him personally way back in 1972. At that time we wanted to hire him as a professor at the Technical University of Berlin. It is a well-known fact that we did not succeed. He did come to Germany, but went to Darmstadt instead. I was his guide during his visit to Berlin. At that time, he had just written two articles for *Communications of the ACM*: "A technique for software module specification with examples" as well as the famous "Criteria to be used in decomposing systems into modules." He gave lectures on these two topics in Berlin. I admit that I did not then understand their significance and believed it to be mostly superficial stuff. It was not until later, in my own development work in industry, that their significance dawned on me. When designing the START application software, I could use the principle of data abstraction and encapsulation, i.e. information hiding, to full advantage. And it has paid off to this day, not only in the START software, but above all at sd&m, where we applied this modularization principle right from the start. Modularization is the core of object orientation and one cannot imagine software engineering without it. The coincidence that led Dahl's Simula compiler and Dave Parnas to Berlin at the same time had a huge impact: it put the "d" into sd&m.

John Guttag – Abstract Data Type Specification

Dahl's Simula laid the foundation for object-oriented programming; Parnas' concept of information hiding accomplished the same for modularization; and, from John Guttag, we know how to specify abstract data types using algebraic formulas. The equation for the stack, pop(push(s,x)) = x, is very simple and easy to remember, though the same certainly cannot be said for all algebraic and other formal specifications. The hope that formal specification would create the basis for correct, provable software was never fulfilled in practice. Applying such a method to massive applications is too difficult and costly, and requires a qualification that only a few software engineers have. Additionally, if you think out and structure a system so extensively that you can specify it algebraically, you have understood it so well that the algebraic specification is no longer worth it. So is abstract data typing simply a nice theory without practical significance? The answer is no – there are applications in which the effort does pay off or is even essential, for example in safety-critical systems. However, sd&m does not build such systems. What we have learned from John Guttag and other researchers specializing in formal specifications is which questions a specification should answer: What does the module depend on? What precisely is it supposed to achieve? What are its pre- and postconditions? What knowledge does it reveal? And what information does it hide? This way of thinking is very important to ensure correctness in our work.

Michael Jackson – Structured Programming

Back when it was first introduced, Michael Jackson's structured programming (JSP) concept had often been limited and thereby trivialized to a type of graphic notation for structured control flow, a kind of alternative to the Nassi/Shneiderman diagrams. But his contribution goes much deeper than that. He assumed that data and control structures should be designed in accordance with each other or, to be more exact, that control should be derived from the data structures, because the latter are the more fundamental concept. In the end, we are talking about data processing. Data is at the forefront of information systems, defining and structuring the system more than anything else, and it is more constant than the processes. Our work is deeply affected by this understanding: object structure, the algorithms, and in some cases, even the business processes are derived from the data model.

In the 1980s, I often came across Jackson's charts at sd&m. Customers used them mostly when preparing specifications and technical concepts; however, they were scaled down to the previously mentioned control structure charts.

Michael Jackson's ideas did not gain the popularity that they might have deserved. His impact was more indirect, but not less significant for our work.

Tom DeMarco – Structured Analysis

Tom DeMarco's structured analysis had an extraordinary influence on software design. In the mid-1970s, modeling information systems using data flows had not yet hit the mainstream. Doug Ross' SADT publication came out in 1977; Tom's book *Structured Analysis and System Specification* was published in 1978. Data flow charts are one of or even the central element of structured analysis. There was now a choice of two contrary paradigms: data encapsulation, on the one hand, and data flow, on the other. It is almost like the duality between wave and particle when explaining the physics of light; both explain certain relevant physical features very well, but rule each other out as an explanation model. I have already admitted several times that I was always a believer in and a fan of data encapsulation and therefore object orientation. So, I had a fundamental problem with structured analysis and many a debate with its supporters, though never with Tom personally.

Structured analysis along with its data flow charts, supplemented by the entity/relationship model, began its climb toward success with the CASE tools in the later half of the 1980s. It lasted for about ten years. Do you still remember IEW, later called ADW, IEF, and similar tool systems? The "structured methods" were considered the standard – who knows how far they really went? They played an important role at sd&m in the form of the CASE tools that clients imposed on us. The schism between data encapsulation and data flow has since disappeared; object orientation has prevailed. Nevertheless Tom DeMarco's structured analysis not only had a far-reaching impact on our work, it also contributed greatly to explaining the methodology of software design.

Tom has always been involved with project management as well, as can be seen in his early book *Controlling Software Projects* and even more so in *Peopleware*, which helped establish the soft factors in software engineering. At the 1994 sd&m conference, he impressed us with "Beyond Peopleware," which has had a sustained impact on us even today.

Michael Fagan – Inspections

Reviews and inspections, walk-throughs, and code reading are important and well-established means for analytical quality assurance at sd&m. Even before sd&m, we had already begun working on these issues in the second half of the 1970s for the START project at Softlab. There we performed systematic reviews of module specifications and test cases, which we called interface and test conferences, respectively. Michael Fagan had provided the impetus for this with his article published in 1976 in the IBM *Systems Journal* on *Design and Code Inspections*. Reviews are easier said than done. For example: someone reviews a document, such as a specification, and finds a mistake. The author should be pleased at having one less mistake. But have you ever been pleased when someone finds a mistake in your work? Such reviews require a good amount of "egoless attitude", as Gerald Weinberg once called it.

Reviews and inspections are highly practical and very useful activities. They are therefore an integral part of sd&m's software engineering process. Nowadays, it is barely conceivable that it was not normal practice until Michael Fagan came up with the idea of inspections.

Barry Boehm – Software Economics

sd&m's stand at the 1983 Systems Trade Fair was decorated with the cover from Barry Boehm's book *Software Engineering Economics*. We presented the title page picture showing the software cost drivers. Using the statistic analyses that Boehm had performed at TRW, it demonstrated very transparently which factors heavily influence the costs of software development. The greatest influence – no surprise – was the team's qualifications. This gave us the scientific proof for what we already instinctively knew – that sd&m had to be a top team.

Barry Boehm's book introduced the concept of Cocomo, his constructive cost model, which established the relationship between the size of a software system, specified by the number of lines of code, and its development expenses, measured in man days. From his TRW analyses, Boehm derived precise formulas for his model. This was the first time a quantitative model was established to solve the practical, highly relevant problem of calculating the costs of a software project. A drawback in this model is however apparent: at the end of a project, it is easy to accurately count the number of lines of code; however, at the beginning, such a calculation is hardly possible. The function point method is an advantage here, but it is not without its own set of problems.

Planning software projects remains a difficult task: effort estimation and cost calculation, time and deadline scheduling, and team selection are all complicated to execute correctly. Such planning is, however, critical for each individual project and for sd&m overall. Barry Boehm's concept of software economics has contributed fundamental insights to project planning. I should not fail to mention the spiral model that Barry published in 1988, which had a massive impact on discussions concerning the approach to software engineering.

Erich Gamma – Design Patterns

Each designer, regardless whether he is designing a physical product or software, has a repertoire of solution patterns in his head – mostly intuitive – each of which is suited to solve a certain problem. Twenty years ago, the architect Christopher Alexander had already collected dozens of such patterns for architectural problems. He served as an example for Erich Gamma and his colleagues, who established a school concerning software design patterns in the mid-1990s. Based on object-oriented methodology, it superimposes concepts on constantly recurring patterns – on a higher level to a certain extent. In other words: if you consider the design of individual classes or modules and operations as design-in-the-small, then you can understand design patterns as a means for design-in-the-large. Such an understanding is obviously very helpful in terms of the size and complexity of the software systems that we build at sd&m. Our architecture blueprint Quasar enables us to pursue the same objective, namely reusing good, tried, and trusted concepts.

Some pioneers did not come

Not all the pioneers whom we had on our wish list could or wanted to accept our invitation. I have already mentioned Ted Codd, who could not make it for health reasons – Rudolf Bayer spoke in his place. Others I wish could have been present include John Backus, the developer of the first higher programming language, Fortran, as well as Don Knuth, who worked on L/R Parsing. It is particularly disappointing that Unix was not addressed, especially because this operating system is so significant today; all three of its developers had to decline. Finally, Tim Berners-Lee, whose report on the World Wide Web would have guided us into the newest era in software history, turned us down. Though regrettable, their absence in no way minimizes the program that we have put together.

"There is nothing more practical than a good theory."

Practitioners and managers in the IT industry tend to accuse the computer scientists at universities of being bogged down by theory, and may regard university-based research as essentially irrelevant. The Software Pioneers here have taught us that the opposite is true. To a great extent, their contributions originated from an academic-scientific environment, thus substantiating the following quote from Kant: "There is nothing more practical than a good theory."

From the Stack Principle to ALGOL

Friedrich L. Bauer
Professor of Mathematics and Informatics (Emeritus)
Technische Universität München
flb@in.tum.de

Doctorate in Mathematics, Theoretical Physics and Astronomy, University of Munich

Professor of Applied Mathematics, University of Mainz:

Founder of the "Gesellschaft für Informatik (GI)"

Creator of the Informatics exhibition at Deutsches Museum

IEEE Computer Pioneer Award

Major contributions: development of the stack principle and ALGOL 60, initiator of "Software Engineering"

Current interests: cryptology, history of computing

Friedrich L. Bauer

From the Stack Principle to ALGOL

Abstract

Certainly one of the most characteristic principles to organize computations with the help of appropriate data structures is the so-called stack principle. It is used not only when expressions with parentheses are to be evaluated, but even more importantly it is also the essential principle of organizing the data inside the memory when calling nested procedures. The stack principle also works for locally declared data within scopes and is therefore the most prominent implementation idea when dealing with high-level languages, scopes and block structures such as the languages of the ALGOL family. The paper describes the early ideas on using stacks starting from evaluating expressions and the development processes that have led to the ALGOL language family.

1. Britzlmayr and Angstl

The idea of what I later called the *stack principle* came into my mind before I became seriously acquainted with what was dubbed *program-controlled calculators* at that time. In the academic year 1948–1949 I was sitting in a class of Wilhelm Britzlmayr's, logician and honorary professor at the Ludwig-Maximilians-Universität in Munich (his regular occupation was director of a bank). One day he spoke on the syntactic aspects (but this terminology was not used at that time) of the parentheses-free notation that was advocated by Jan Lukasiewicz [1]. Something stirred my interest, and thus I was not completely lost when on June 27, 1949 in Britzlmayr's seminar a man by the name of Konrad Zuse, while giving a survey of his *Plankalkül* [4], used the checking of the well-formedness of a propositional formula written in parentheses-free notation as an example for a *Plankalkül* program (a *Rechenplan*, as he called it). The discussion between Zuse and Britzlmayr brought to light that it was an algorithm Zuse had learned from his colleague Hans Lohmeyer, a mathematician working, like Zuse, at Henschel-Flugzeug-Werke in Berlin. The algorithm originated in 1943 from the Berlin logician Karl Schröter [3], based on the 1938 work by the Viennese logician Karl Menger [2]. While Zuse wanted to sell his *Plankalkül*, Britzlmayr was interested only in the algorithm as such, and most of the discussion took place in two different jargons, which was rather confusing.

One year later, around the end of August 1950, Britzlmayr showed me a sketch, made by my classmate Helmuth Angstl, for a wooden apparatus that was intended for mechanizing the Schröter test for well-formedness (Fig. 1).

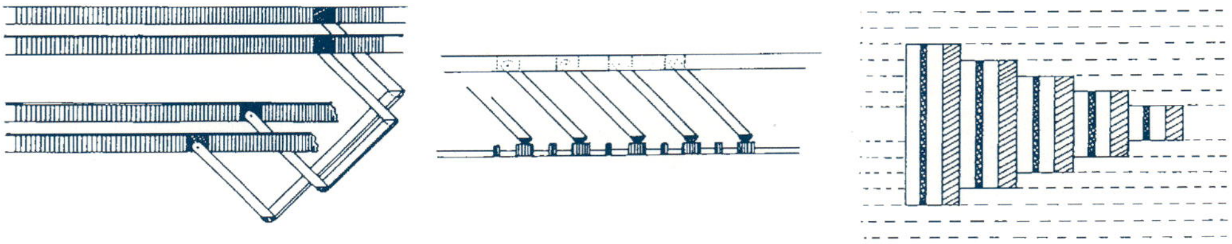

Fig. 1 Cut from the drawing of Helmut Angstl, August 1950 [30]

Angstl's Device

The way it worked can be explained briefly, say for the propositional formula $(p \to \neg q) \wedge (\neg p \to q)$, in the notation of Lukasiewicz $CIpNqINpq$. After a precise specification of the arity of the symbols involved:

$\cdot p,\ \cdot q$	one-valued, nullary (zero arguments) – variable,
$\cdot N \cdot$	one-valued, unary (one argument) – negation,
$\cdot C{:},\ \cdot I{:}$	one-valued, binary (two arguments) – conjunction, implication,

the formula reads

$$\cdot C{:}\ \ \cdot I{:}\ \ \cdot p\ \ \cdot N\cdot\ \ \cdot q\ \ \cdot I{:}\ \ \cdot N\cdot\ \ \cdot p\ \ \cdot q$$

and the outcome is a nexus such that, stepwise from right to left, the exit of a symbol is always linked with the next not-yet-occupied entry to a symbol:

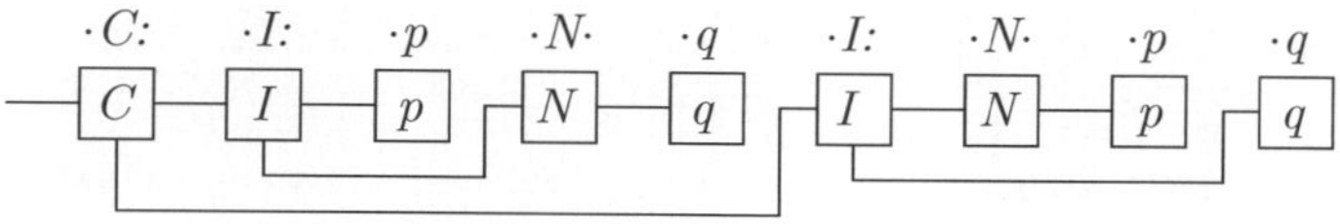

The formula is well-formed if all links can be accommodated under this rule. Angstl checked this mechanically by using for each symbol a hinge engaged in the first free position. The original drawing reflects this only unclearly.

The Algorithm

Schröter expressed it mathematically: a parentheses-free formula is well-formed, if and only if it has rank 1 and the rank of each of its right segments is positive. The rank is defined as an additive (from right to left) function over the weights of the symbol sequence, where the weight of a symbol with zero arguments (a variable) is 1 and the weight of a symbol with n arguments, $n > 0$, is $1 - n$. (This rule was rediscovered by D. C. Gerneth [5] in 1948; in 1950 Paul Rosenbloom [7] recognized its far-reaching importance and based his definition of a 'simple language' on it.) In the example of the formula $(p \to \neg q) \wedge (\neg p \to q)$ this results in

·C:	·I:	·p	·N·	·q	·I:	·N·	·p	·q	
−1	−1	1	0	1	−1	0	1	1	weights
1	2	3	2	2	1	2	2	1	rank

The rank condition is satisfied: the rank of each right segment is positive, the rank of the complete formula is 1.

A non-well-formed formula, like $CIpNqINp$, may lack a variable:

·C:	·I:	·p	·N·	·q	·I:	·N·	·p	
−1	−1	1	0	1	−1	0	1	weights
⓪	1	2	1	1	⓪	1	1	rank

or, like $IpNqINpq$, may lack an operation with two arguments:

·I:	·p	·N·	·q	·I:	·N·	·p	·q	
−1	1	0	1	−1	0	1	1	weights
②	3	2	2	1	2	2	1	rank

In both cases, the rank condition is violated. But even if the number of symbols is correct, if they are in wrong order, the rank condition will be violated:

·C:	·q	·p	·N·	·I:	·I:	·N·	·p	·q	
−1	1	1	0	−1	−1	0	1	1	weights
1	2	1	⓪	⓪	1	2	2	1	rank

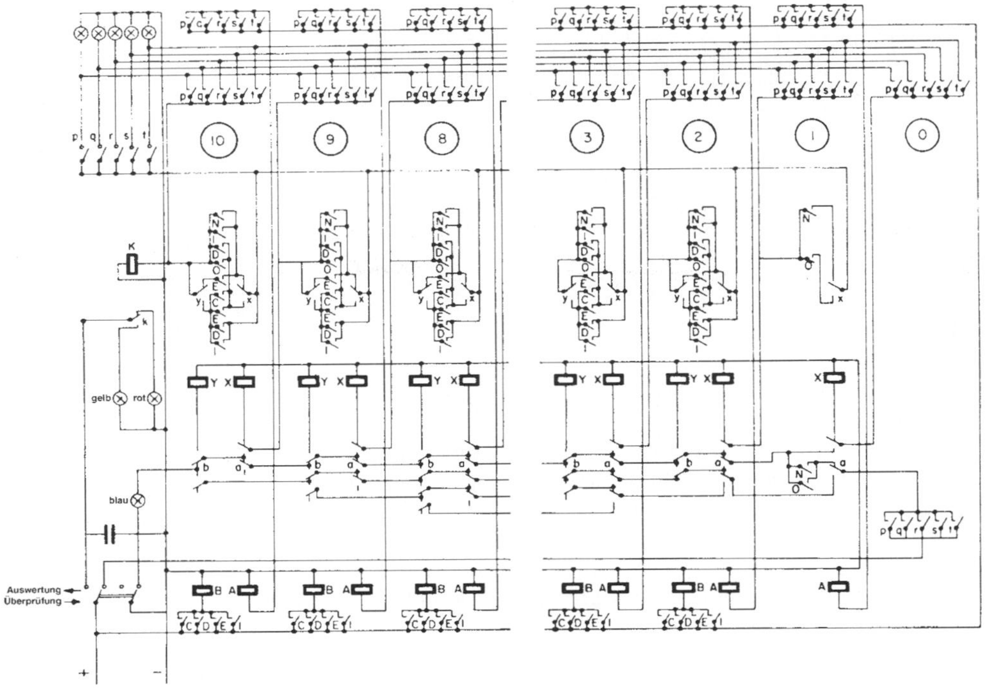

Fig. 2 Wiring design of STANISLAUS, 1950/1951 [8]

2. STANISLAUS

A Formula-Programmed Relay Calculator

Back to Angstl: I did not like his solution with the wooden apparatus, and influenced by the 1949 paper by Claude Shannon [6], *The Synthesis of Two-Terminal Switching Circuits*, I started to look for a relay realization for the formula, which was to be typed in directly. At the same time, this allowed a direct evaluation of the propositional formula for true or false instantiations of the variables; the test for well-formedness turned out to be a byproduct of such a *formula-programmed relay calculator for parentheses-free propositional formulas.*

Around the turn of the year 1950/51, during a stay in Davos, Switzerland, I made the wiring diagram for the relay calculator; in honor of the Polish school of logic I dubbed it STANISLAUS (Fig. 2). It was laid out for a formula with up to 11 symbols; thus it could handle, for example, the tautology of the transitivity of implication:

$$((p \to q) \wedge (q \to r)) \to (p \to r)\,,$$

in the notation of Lukasiewicz

$$ICIpqIqrIpr$$

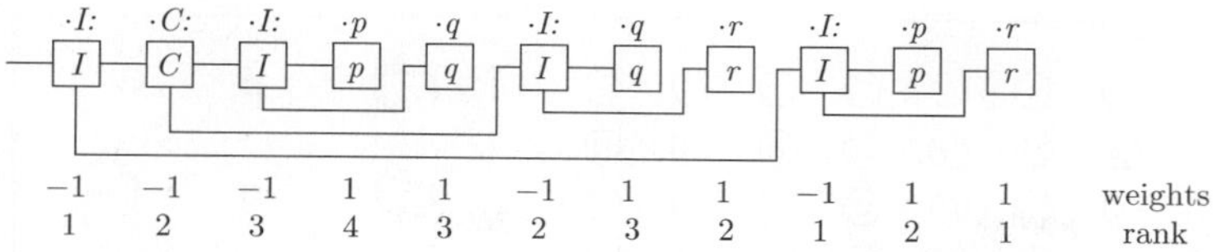

Possibilities for instantiation were provided for five variables, designated by p, q, r, s, t. The operations N (negation), C (conjunction), D (disjunction), E (equivalence), I (implication) were realized by contact switches and suitable combinations thereof, and the connections were formed by two pyramids of relay contacts facing each other, where for example the prenex formula with six variables $CCCCCpqrpqr$ needed the full depth of the rank:

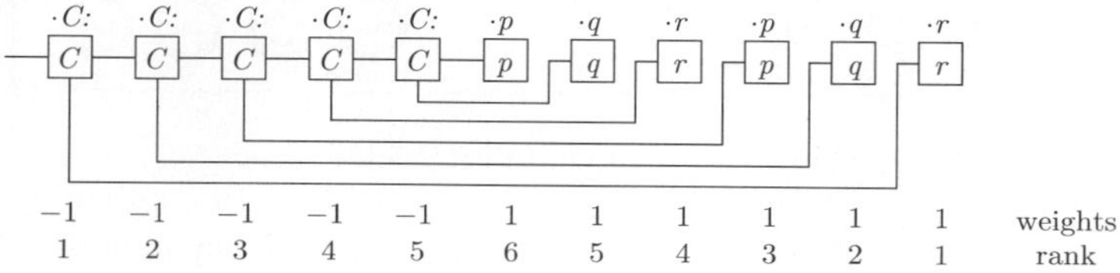

The Cellar

The staircase-like wiring for the two pyramids forming the connections between the symbols suggested calling this part of the diagram a cellar (German *Keller*). In fact, the cellar was laid out for one additional storey, the bottommost wire serving the check for well-formedness. Thus the original problem was solved as a byproduct.

Figure 3 shows a front view of the logic calculator. The formula $ICIpqIqrIpr$ is typed in. A toggle switch serves, when in the left position, to check the well-formedness, and when in the right position, to evaluate the formula. The truth values for the variables (1 for *true*, 0 for *false*) can be input as well by toggle switches. If the formula typed in is well-formed, a blue signal will appear; the evaluation will produce a yellow signal for *true*, a red one for *false*.

Having finished the wiring diagram, I started to collect material for the construction of the device. In October 1951 I entered into a consulting contract with the Siemens company in Munich; from time to time I could now help myself from the surplus bin of the Siemens-Halske AG Central Laboratory in the Munich Hoffmannstraße, where relays and keyboards were to be found. Now and again, Britzlmayr urged me to start with the actual construction, but I had too many other obligations. Towards the end of 1951, I finished my dissertation under Fritz Bopp; in October 1952 I became an assistant to Robert Sauer, which led me, among other occupations, to work in the team that under Hans Piloty and Robert Sauer constructed the PERM (*Programmgesteuerte Elektronische Rechenanlage München*, Programm-controlled computer Munich). In this respect, I could hope for some support from Piloty's mechanical workshop. It was only in 1954 that I started, under the attentive eyes of my colleague and friend Klaus Samelson and helped by Heinz Schecher, to wire the

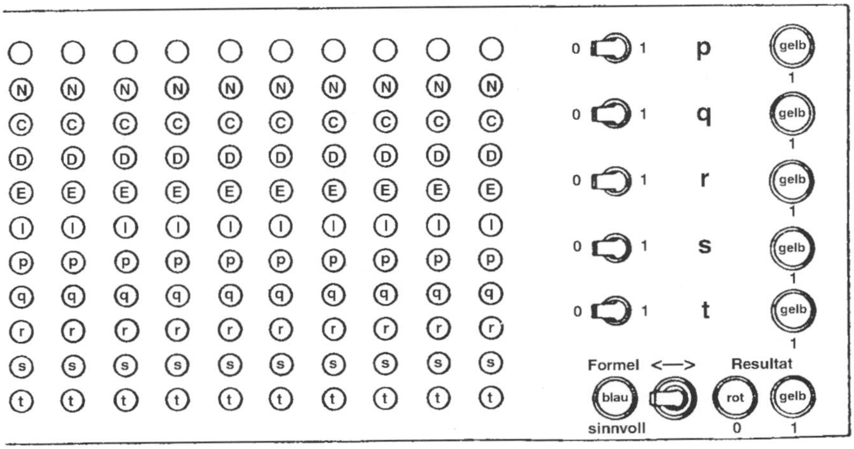

Fig. 3 Front view of STANISLAUS [8]

first segments. Then, after having seen how it would work, my interest waned, and it was only after very urgent admonition from Britzlmayr that in 1956 STANISLAUS [8] was finished. It was displayed on December 3, 1956 at a colloquium in Münster and presented on January 8, 1957 to the Munich academic public; it met with the delight of Britzlmayr. Samelson and myself did not think so highly of the apparatus: in the meantime we had greater joy with the PERM computer. STANISLAUS is now exhibited in the Deutsches Museum, Munich.

3. A Software Patent for the Stack

The Stack Principle

In 1955, Samelson and I had quite a different motivation with respect to the STANISLAUS design. In February 1952, I had visited Heinz Rutishauser in Zurich and had seen the Z4 working; since the fall of 1953, I had had close contact with him, mainly in questions of numerical mathematics, which was my main duty under Sauer and also the field where I hoped to obtain my habilitation. Naturally, I did not neglect to take notice of Rutishauser's germinative publication [9] [10] of 1951, *Über automatische Rechenplanfertigung bei programmgesteuerten Rechenmaschinen*, dealing for the first time with the translation of a mathematical formula language into machine programs by a *Superplan* in Rutishauser's terminology, a "programming program", as Andrei Ershov called it later. Samelson and I both had in mind to realize this *Formelübersetzung* for the PERM. When in mid-1955 we had time to start it and could expect a speedy finish to the construction of the PERM, it soon came to our attention that STANISLAUS solved a similar, but simplified task. Its "cellar" contained stored intermediate *yes-no* values; in the case of arithmetic formulas this would be a "number cellar". However, the classical arithmetical formula language Rutishauser had treated is not parentheses-free. Every opening parenthesis amounts to a delay in the execution of the last arithmetical operation, until the corresponding closing parenthesis is found. Thus, it was natural to provide next to the number cellar for intermediate results an "operation cellar" for delayed operation symbols. We had to rack our brains to find the interplay between these two cellars; for this task Samelson made full use of his analytical talent, a particular problem being the "invisible parentheses" caused by the precedence of certain operations like multiplication over others like addition.

The total history of the operations delayed in such a way determined the course of execution; the automaton that was sufficient for the parentheses-free notation was replaced by a *Kellerautomat* (push-down automaton) for the execution of formulas in languages with parenthetical structure. The cellar had to be programmed into what came to be called a stack. Thus the idea of the stack principle (*Kellerprinzip)* was born—free of hardware and usable for a wide class of languages—actually for context-free languages.

Σ \ χ	(	$\underline{v}$	$\times$	$+$	)	Ende
$\sigma\,\underline{v}$ τ			σ r τ	σ r τ	σ r τ	σ r τ
$\sigma\times$ $\tau\,\underline{v}$	$\sigma\times($ l $\tau\,\underline{v}$	$\sigma\times v$ l $\tau\,\underline{v}\,\underline{v}$	$\sigma\times$ l τ	$\sigma+$ r τ	σ r τ	σ r τ
$\sigma+$ $\tau\,\underline{v}$	$\sigma+($ l $\tau\,\underline{v}$	$\sigma+v$ l $\tau\,\underline{v}\,\underline{v}$	$\sigma+\times$ l $\tau\,\underline{v}$	$\sigma+$ l τ	σ r τ	σ r τ
$\sigma($ τ	$\sigma(($ l τ	$\sigma(v$ l $\tau\,\underline{v}$	$\sigma(\times$ l τ	$\sigma(+$ l τ	σv l τ	
$\alpha\rightarrow$ Φ Φ	$\Phi($ l Φ	Φv l $\Phi\,\underline{v}$				Φ
$\omega\rightarrow$ Φ $\Phi\,\underline{v}$			$\Phi\times$ l $\Phi\,\underline{v}$	$\Phi+$ l $\Phi\,\underline{v}$		Φ $\Phi\,\underline{v}$

Fig. 4 Transition matrix for the stack automaton (infix notation with $+$ and $\times$). [26]

For the concrete case of arithmetics with precedence of $\times$ over $+$ we chose in [26] a description (Fig. 4) by a matrix with an entry line for the next symbol read—where $\underline{v}$ stands for a number variable (numbers included)—and an entry column containing pairs of number cellar and operation cellar. σ stands for a sequence of operation symbols and τ for a sequence of number variables. In the matrix fields are found pairs of resulting number cellar and operation cellar contents and between them the marks r or l: r (*repetieren*) means repeat with the symbol last read, i.e., in the same column with the new pair of number cellar and operation cellar contents, and l (*lesen*) means continue reading. Crossed-out matrix fields indicate that the expression is not well-formed. α points to the starting state, ω indicates a state where the formula could end.

	1	2	3	4	5	6	7	8	9	10	11	12	13	14	15	16	17	18	19	20	21	22	23	24	25
formula	(	a	$\times$	b	$+$	c	$\times$	d	)			/	(	a	$-$	d	)		$+$	b	$\times$	c			
OC	$\boxtimes$	(	(	$\times$	$\times$	$+$	$+$	$\times$	$\times$	$+$	(	$\boxtimes$	/	(	(	$-$	$-$	(	/	$+$	$+$	$\times$	$\times$	$+$	$\boxtimes$
		$\boxtimes$	$\boxtimes$	(	(	(	(	$+$	$+$	(	$\boxtimes$		$\boxtimes$	/	/	(	(	/	$\boxtimes$	$\boxtimes$	$\boxtimes$	$+$	$+$	$\boxtimes$	
				$\boxtimes$	$\boxtimes$	$\boxtimes$	$\boxtimes$	(	(	$\boxtimes$				$\boxtimes$	$\boxtimes$	/	/	$\boxtimes$				$\boxtimes$	$\boxtimes$		
								$\boxtimes$	$\boxtimes$							$\boxtimes$	$\boxtimes$								
NC	$\boxtimes$	$\boxtimes$	a	a	b	κ_1	c	c	d	κ_2	κ_3	κ_3	κ_3	κ_3	a	a	d	κ_4	κ_4	κ_5	b	b	c	κ_6	κ_7
			$\boxtimes$	$\boxtimes$	a	$\boxtimes$	κ_1	κ_1	c	κ_1	$\boxtimes$	$\boxtimes$	$\boxtimes$	$\boxtimes$	κ_3	κ_3	a	κ_3	κ_3	$\boxtimes$	κ_5	κ_5	b	κ_5	$\boxtimes$
					$\boxtimes$		$\boxtimes$	$\boxtimes$	κ_1	$\boxtimes$					$\boxtimes$	$\boxtimes$	κ_3	$\boxtimes$	$\boxtimes$		$\boxtimes$	$\boxtimes$	κ_5	$\boxtimes$	
									$\boxtimes$								$\boxtimes$								

Fig. 5 Translation of the formula $(a \times b + c \times d)/(a - d) + b \times c$

Figure 5 [20, 21] gives an example for the translation of the formula

$$(a \times b + c \times d)/(a - d) + b \times c$$

OC is the operation cellar, NC is the number cellar, $\boxtimes$ is a testable boundary symbol for a cellar, and κ_1 to κ_7 designate intermediate results. On top of the number cellar stands initially $\boxtimes$, finally the value of the formula. In detail, the resulting computation goes as follows:

$$\begin{aligned}
5 - 6 &: \quad a \times b \Rightarrow \kappa_1 \\
9 - 10 &: \quad c \times d \Rightarrow \kappa_2 \\
10 - 11 &: \quad \kappa_1 + \kappa_2 \Rightarrow \kappa_3 \\
17 - 18 &: \quad a - d \Rightarrow \kappa_4 \\
19 - 20 &: \quad \kappa_3 / \kappa_4 \Rightarrow \kappa_5 \\
23 - 24 &: \quad b \times c \Rightarrow \kappa_6 \\
24 - 25 &: \quad \kappa_5 + \kappa_6 \Rightarrow \kappa_7
\end{aligned}$$

The translation algorithm turns out to be superior to Rutishauser's method [9] inasmuch as it avoids the Rutishauser *Springprozession*; the effort is only proportional to the length of the formula and not, as with Rutishauser, to the square of the length. In Rutishauser's terminology it amounts to the decomposition of the parenthesis mountain from the first pair of peaks in the first chain of peaks, so it was sequential. Correspondingly, in the publication the method was characterized as "sequential formula translation".

Heinz Rutishauser
(1918–1970)

Hardware Stacks

We gave a report on our results to Sauer and Piloty. Piloty remarked that the German Research Council (*Deutsche Forschungsgemeinschaft*) had a tendency to make sure that patents were obtained for the projects it supported; he urged us to examine whether this would be possible in our case. We agreed, and he offered the prospect of providing the successor machine of the PERM with a number cellar and operation cellar in hardware. This must have been in the summer or fall of 1955.

Klaus Samelson
(1918–1980)

For the patent application we had to disguise our method as hardware, and for this purpose had to become engineers. The elaboration of the patent application therefore brought a lot of work and was fun, too; on the other hand it meant that publication of our results was paralyzed. Samelson therefore reported at the Dresden meeting at the end of November 1955 [13] with great caution. Both Rutishauser and Heinz Billing in Göttingen, who was building the G3 computer, were in on the secret. The German patent application [14, in the following partly reprinted] was finally filed March 30, 1957 (Auslegeschrift 109472019, Kl. 42m), the U.S.-American one [15] March 28, 1958 within the priority time limit.

A Software Patent for the Stack

While the U.S.-American application contained an abundance of *and* and *or* gates, the German patent law allowed quite functional descriptions of methods, thus the German application stayed free of drawings for electrical circuits; it was possible to design from it immediately a program with programmed cellar structures, later called stacks. Our patent can therefore be considered as an early case of a software patent.

The actual writing of a machine program for the PERM, which in the meantime was operating, was delegated in mid-1957 to the assistants Manfred Paul and Peter Graeff; the program was ready for tests after a few months. At first, an interpreting machine program was written; then the transition to a generating translator (a compiler) meant simply, instead of immediately executing say (as above)

$$\kappa_1 + \kappa_2 \Rightarrow \kappa_3 \ ,$$

inserting into the program the corresponding instruction

$$\kappa_3 := \kappa_1 + \kappa_2 \ .$$

The hardware stacks, the building of which Piloty had suggested, were not realized in Munich, since the PERM II project was not supported by the German Research Council. Billing, however, equipped his G3 computer with the hardware for a number stack.

4. The ZMD Group and ALGOL

At a meeting in Darmstadt in October 1955, Rutishauser appealed for an attempt to unify programming languages—FORTRAN had appeared in the meantime. As a consequence, the German-Austrian-Swiss *Gesellschaft für angewandte Mathematik und Mechanik* (GaMM) established a committee for programming, which for interested people in Zurich, Munich, and Darmstadt became in 1956 the umbrella for joint work towards the development of a suitable programming language. The ZMD (Zurich-Munich-Darmstadt) group, with Rutishauser, Samelson, myself, and Hermann Bottenbruch from Darmstadt as representatives, met in the fall of 1957 in Lugano in order to pass a first proposal for the intended programming language. This proposal was based on an extension of the stack principle for arithmetics to the whole programming language and was characterized by the catch phrase: "Postpone as long as necessary, but not longer". Together with Bottenbruch, I travelled in summer 1957 through the United States and advocated the proposal to people interested: Alan Perlis, Bob Rich, Saul Gorn, and many others. John W. Carr, at that time president of the ACM, reacted with enthusiasm and arranged for a U.S.-American committee, which met several times in January 1958. Comparison with the transmitted proposals of the ZMD group showed some concord. Thus the mutual wish for a joint working meeting arose. From the European side an invitation for participation also went to the British Computer Society, but due to a mishap on the side of Maurice Wilkes no reaction came from there. Therefore a U.S.-American delegation, consisting of John Backus, Charles Katz, Alan J. Perlis, and Joe Wegstein, met from May 27 to June 2, 1958 in Zurich with the GaMM delegation Rutishauser, Samelson, Bauer, and Bottenbruch [19a, b].

The Zurich Meeting

In the summer of 1957, Bottenbruch became initiated in the Munich Sequential Formula Translator method [16], and at the Zurich meeting the ZMD group not only presented a draft [17] for the language, which at first was called *International Algebraic Language*, but also had a completed compiler design in the bag. Some U.S.-American delegates had experience with working compilers (Backus with FORTRAN, Perlis with IT, Katz with MATH-MATIC). An open discussion of the technical problems of programming language translation into machine code was left out, as there would not have been enough time. Technically speaking, the state of the art within the ZMD group was far more advanced: FORTRAN used the method of parentheses completion, introduced by P. B. Sheridan [18] and discussed as early as 1952 by Corrado Boehm [11] in his Zurich dissertation, a method that like Rutishauser's required an effort proportional to n^2; IT [12] used pure comparison of neighboring operators, enforcing an oppressive limitation to operator precedence grammars. This situation led from time to time to a paralysis of the discussion, which was basically oriented towards progress. On the whole, ALGOL, as it was called in the publication [19b], was an incarnation of the cellar principle and thus particularly well suited for efficient translation, which was a main concern of the impoverished European side. In Zurich, Samelson had particularly stressed the point that the block structure of storage allocation (*Cellar Principle of State Transition and Storage Allocation*, [30]), following so naturally from the cellar principle and becoming so typical in the later development, became the dominant organization principle of the programming language. Storage allocation with complete parentheses structure should be organized in a dynamic stack, which without further complications allowed mastery of recursive definitions. The compromise that was achieved in a struggle with the U.S.-American side did not reflect this issue in the published report; thus, later the implementation of recursive situations was reinvented by some people not present at the Zurich meeting.

The Preliminary ALGOL Report

The goals attempted by the ZMD group, to create a language *widely following mathematical notation, readable without much ado, and suitable for publications and for mechanical translation*, had been largely reached. The publication was completed in 1959 in the first issue of the new journal *Numerische Mathematik* of Springer-Verlag under the title *Report on the Algorithmic Language ALGOL*.

5. Work at Mainz University

When in 1958 I moved from Munich to Mainz, with Samelson soon following me, the ZMD group was widened to the ZMMD group. Emphasis was on finishing compilers for ALGOL 58. The common basis was the method of a stack automaton developed in Munich, which was extended without any difficulty to a full algorithmic language including statements, declarations, block structure, indexed variables, and so on. It was published in 1959 in the newly founded journal *Elektronische Rechenanlagen* ([20], in the following reprinted), and in 1960 in *Communications of the ACM* [21]. Manfred Paul, who had done most of the preparatory work, finished a compiler for the Mainz Z 22 towards the end of 1958. Soon afterwards, H. R. Schwarz and P. Läuchli followed in Zurich for the ERMETH and G. Seegmüller in Munich for the PERM.

ICIP Conference

A lot of work was caused by the preparations for ALGOL 60. At the *International Conference on Information Processing (ICIP)*, arranged in Paris, June 15–20, 1959 by UNESCO, John Backus [24] made a famous proposal for the formal description of the syntax. The Backus notation (*Backus Normal Form*), soon generally accepted, allowed one to attach in an elegant way the semantics of the programming language to the syntax of a context-free language. Manfred Paul, in his 1962 dissertation, clarified how from this description the transition matrix for the stack automaton could be derived formally.

Christopher Strachey, who—inadvertently—had not been invited to the Zurich meeting, went into the trouble of criticizing ALGOL 58 and produced not only considerable stir, but also a lot of public attention. Thus, it was not too difficult to convince the *International Federation for Information Processing*, founded in Paris, to organize a conference for the "final ALGOL", later called ALGOL 60. The preparations were this time much more intensive; a European pre-conference was held in November 1959 in Paris; it nominated seven European delegates, who met again in December 1959 in Mainz. The U.S.-American side nominated its delegates in November 1959. This time, representatives from Great Britain, France, the Netherlands, and Denmark, besides representatives from the U.S.A., Germany, and Switzerland, were invited.

Paris, 1960

Friedrich L. Bauer and Klaus Samelson at the Paris ALGOL Conference, 1960

The ALGOL conference took place in Paris, January 11–16, 1960 under the patronage of the newly founded IFIP. It led to consolidations and completions of the *Preliminary Report*. Characteristically, the introduction to the report [25a, b] says "The present report represents the union of the committee's concepts and the intersection of its agreements". In this way, contradictions could remain here and there and solutions were omitted. What made me particularly angry was that Samelson, who in 1958 regarding the question of the block structure could not win against Alan Perlis, in 1960, when acceptance of recursion was no longer an issue, was given no credit for the block structure; the editor Peter Naur, who was not present in Zurich, was not aware of this.

In the short period of six days we also did not succeed in formalizing, next to the syntax which now was formalized in BNF (Backus Normal Form), the semantics as well; it was still explained rather verbally, leading later to exegetic quarrels. Heinz Zemanek tried, with the IFIP Technical Committee 2 Working Conference *Formal Language Description Language*, held in 1964 in Baden near Vienna, to compensate for this lack. Peter Landin [29] gave a complete formal description of ALGOL 60, but it lacked the blessing of the authorities.

6. The ALCOR Group

After the ICIP Congress, June 1959 in Paris and particularly after the publication of the ALGOL 60 report, the ZMMD group decided to widen its membership and invited interested institutions in Europe and North

America to participate in the efforts for propagation of ALGOL through the construction of ALGOL compilers for various different machine configurations; the documents of the previous ZMMD group were made available for this purpose. The offer was accepted by scientific institutions (among the first were Regnecentralen Copenhagen, Bonn University, Zemanek's Mailüfterl group in Vienna, Oak Ridge National Laboratory, Neher Laboratory Leidschendam) as well as by some computer companies (Siemens and Halske AG for the 2002, Telefunken for the TR4, Standard Elektrik Lorenz AG, IBM Germany). The resulting multitude of concrete implementations was unavoidable because of the differences in the machines involved, but it was useful in its scientific aspects. For example, Albert A. Grau, Oak Ridge, introduced the concept of syntactic states and described the compiler as a system of mutually recursive subprograms [31]. Peter Lucas in Vienna went his own way [32] in generating, like Paul in Mainz [33, 34], the compiler from the syntax in BNF. Jürgen Eickel and Manfred Paul, in 1964, studied the parsing and ambiguity problem for Chomsky languages in general [39].

After my return to Munich in 1963, the build-up of computer science there became my main obligation, leaving less time to be spent on the further development of ALGOL. The way it went became more and more of a nightmare, leading to ALGOL 68 and to the ruin of the ALGOL idea. One after another, people left the IFIP Working Group 2.1 on ALGOL: Peter Naur, Niklaus Wirth, Tony Hoare, Edsger Dijkstra, Brian Randall, Gerhard Seegmüller, Wlad Turski, Mike Woodger, Hans Bekic, and Fraser Duncan.

Stacks in Hardware and Software—Worldwide

The use of cellar storage (*stacks, pushdown stores*) did not remain restricted to ALGOL compilers. Peter Deussen [35] used them for the translation of Boolean expressions in disjunctive normal form, and Walter Petry [36] for the translation of formal languages in topological structures. Conversely, familiarity with stacks did lead frequently to the systematic use of fully parenthesized language structures. Donald Knuth built his text programming system TEX in this way.

Hardware cellars, stacks, and pushdown stores have been discussed elsewhere, possibly as early as 1947 by Alan Turing, certainly in 1949 by Harry D. Huskey in connection with the ZEPHYR (SWAC) computer and in 1956 by Willem L. van der Poel in connection with the design of the MINIMA computer; in all cases presumably for the treatment of return jumps in subroutines. More generally, Charles L. Hamblin, in a December 1957 publication [37], recommended strongly the use of hardware stacks. Hardware stacks were built in the Burroughs machines, B5000 onwards, and in mini computers such as PDP-8 and PDP 11 (Fraser George Duncan [38]). Hewlett-Packard programmable pocket computers like the HP-11C, which I still have in use, work with a 4-level number stack.

In time, however, it was discovered at many places that it was preferable to implement stacks in software by "linked memory techniques", for example when A. Newell, J. C. Shaw, and H. A. Simon in 1956 programmed the "Heuristic Problem Solver" IPL-II. After 1960, chained storage structures became quite common.

Among the first to establish a connection between the theory of formal languages and our stack principle were S. Ginsburg and H. G. Rice in 1961. The connection between formal languages and stacks was also discussed in 1961 by Y. Bar-Hillel, in 1962 by S. Ginsburg, and in 1963 by B. A. Trachtenbrot. Theoretical studies on cellar automata (pushdown automata) and context-free languages were done in 1963 through an interplay of N. Chomsky and M. P. Schützenberger.

None of the early full programming languages was so vigorously influenced by the stack principle as ALGOL. Today, the stack principle is so ubiquitous in software systems of all sorts that a denotation for it is hardly needed or used.

References

[1] J. Lukasiewicz, A. Tarski: Untersuchungen über den Aussagenkalkül. C. R. Séances Soc. Sci. Lett. Varsovie, Cl. III, **23** (1930) 31–32

[2] K. Menger: Eine elementare Bemerkung über die Struktur logischer Formeln. In: K. Menger: Ergebnisse eines mathematischen Kolloquiums **3**, Deuticke, Leipzig (1935)

[3] K. Schröter: Axiomatisierung der Frege'schen Aussagekalküle. In: Forschungen zur Logik und zur Grundlegung der exakten Wissenschaften, N. F. **8**; J. Symb. Logic **9** (1943) 69

[4] K. Zuse: Über den Allgemeinen Plankalkül als Mittel zur Formulierung schematisch-kombinativer Aufgaben. Arch. Math. **1** (1948/49) 441–449

[5] D. C. Gerneth: Generalization of Menger's result on the structure of logical formulas. Bulletin of the American Mathematical Society **54** (1948) 803–804; J. Symb. Logic **13** (1948) 224

[6] C. E. Shannon: The synthesis of two-terminal switching circuits. Bell Syst. Tech. J. **28** (1949) 59–88

[7] P. C. Rosenbloom: The Elements of Mathematical Logic. Dover, New York (1950)

[8] F. L. Bauer: The Formula-Controlled Logical Computer "Stanislaus". Math. Tabl. Aids Comp. **14** (1960) 64–67

[9] H. Rutishauser: Über automatische Rechenplanfertigung bei programmgesteuerten Rechenanlagen. Z. Angew. Math. Mech. **31** (1951) 255

10] H. Rutishauser: Automatische Rechenplanfertigung bei programmgesteuerten Rechenmaschinen. Z. Angew. Math. Mech. **32** (1952) 312–313; Mitt. Inst. f. Angew. Math. der ETH Zürich, Nr. 3, 1952

11] C. Boehm: Calculatrices digitales. Du déchiffrage de formules logico-mathématiques par la machine même dans la conception du programme. Dissertation Zürich (1952); Ann. Matematica Pura Appl. Ser. 4, **37** (1954) 5–47

12] A. J. Perlis, J. W. Smith, H. R. van Zooren: Internal Translator (IT), a Compiler for the 650. Lincoln Lab. Div. 6, Document 6D-327 (1956)

13] K. Samelson: Probleme der Programmierungstechnik. Internationales Kolloquium über Probleme der Rechentechnik Dresden 1955. Berlin (1957) 61–68

[14] Deutsches Patentamt, Auslegeschrift 1 094 019, B44122IX/42m. Verfahren zur automatischen Verarbeitung von kodierten Daten und Rechenmaschine zur Ausübung des Verfahrens. Anmeldetag: 30. März 1957. Bekanntmachung der Anmeldung und Ausgabe der Auslegeschrift: 1. Dezember 1960.

Dr. Friedrich Ludwig Bauer und Dr. Klaus Samelson, München, sind als Erfinder genannt worden. Erteilt 12.8.1971, DE-PS 1 094 019

[15] United States Patent Office. Patent No. 3,047,228, Patented July 31, 1962.

Friedrich Ludwig Bauer and Klaus Samelson: Automatic Computing Machines and Method of Operation. Filed March 28, 1958, Ser. No. 724,770

[16] H. Bottenbruch: Übersetzung von algorithmischen Formelsprachen in die Programmiersprachen von Rechenmaschinen. Z. Math. Logik Grundl. Math. **4** (1958) 180–221

[17] F. L. Bauer, H. Bottenbruch, H. Rutishauser, K. Samelson: Proposal for a universal language for the description of computing processes. In: Carr, J. W. (ed.), Summer School 1958. University of Michigan (1958) 355–373

[18] P. B. Sheridan: The arithmetic translator-compiler of the IBM Fortran automatic coding system. Commun. ACM **2**, 2 (1959) 9–21

[19a] A. J. Perlis, K. Samelson (eds.): Report on the algorithmic language ALGOL by the ACM committee on programming languages and the GAMM committee on programming. Num. Math. **1** (1959) 41–60

[19b] A. J. Perlis, K. Samelson (eds.): Preliminary report – international algebraic language. Commun. ACM **1** (1959) 8–22

[20] K. Samelson, F. L. Bauer: Sequentielle Formelübersetzung.
Elektron. Rechenanlagen **1** (1959) 176–182

[21] K. Samelson, F. L. Bauer: Sequential formula translation.
Commun. ACM **3** (1960) 76–83

[22] F. L. Bauer, K. Samelson: The cellar principle for formula translation. In: Proceedings International Conference on Information Processing, UNESCO Paris 1959. Oldenburg, München (1960) 154–155

[23] F. L. Bauer, K. Samelson: The problem of a common language. In: Proceedings International Conference on Information Processing, UNESCO Paris 1959. Oldenburg, München (1960) 120–125

[24] J. W. Backus: The syntax and semantics of the proposed international algebraic language of the Zurich ACM-GAMM conference. In: Proceedings International Conference on Information Processing, UNESCO Paris 1959. Oldenburg, München (1960) 125–132

[25a] J. W. Backus, F. L. Bauer, J. Green, C. Katz, J. McCarthy, P. Naur (ed.), A. J. Perlis, H. Rutishauser, K. Samelson, B. Vauquois, J. H. Wegstein, A. van Wijngaarden, M. Woodger: Report on the algorithmic language ALGOL 60.
Num. Math. **2** (1960) 106–136

[25b] J. W. Backus, F. L. Bauer, J. Green, C. Katz, J. McCarthy, P. Naur (ed.), A. J. Perlis, H. Rutishauser, K. Samelson, B. Vauqois, J. H. Wegstein, A. van Wijngaarden, M. Woodger: Report on the algorithmic language ALGOL 60.
Commun. ACM **3** (1960) 299–314

[26] F. L. Bauer, K. Samelson: Maschinelle Verarbeitung von Programmiersprachen. In: W. Hoffmann (ed.), Digitale Informationswandler. Probleme der Informationsverarbeitung in ausgewählten Beiträgen. Vieweg, Braunschweig (1962) 227–268

[27a] R. Baumann, M. Paul: Praktische Erfahrungen im ALGOL-Betrieb. Elektron. Datenverarbeitung, Beiheft 2 (1961) 51–56

[27b] K. Samelson, F. L. Bauer: The ALCOR project. In: Symbolic Languages in Data Processing, Proceedings of the Symposium, organized and edited by the International Computation Centre Rome, March 26–31, 1962. Gordon and Breach, New York (1962) 207–217

[28a] J. W. Backus, F. L. Bauer, J. Green, C. Katz, J. McCarthy, P. Naur, A. J. Perlis, H. Rutishauser, K. Samelson, B. Vauqois, J. H. Wegstein, A. van Wijngaarden, M. Woodger: Revised report on the algorithmic language ALGOL 60. Num. Math. **4** (1963) 420–453

[28b] J. W. Backus, F. L. Bauer, J. Green, C. Katz, J. McCarthy, P. Naur, A. J. Perlis, H. Rutishauser, K. Samelson, B. Vauqois, J. H. Wegstein, A. van Wijngaarden, M. Woodger: Revised report on the algorithmic language ALGOL 60. Commun. ACM **6** (1963) 1–17

[29] P. J. Landin: A formal description of ALGOL 60. In: T. B. Steel, Jr. (ed.), Formal language description languages for computer programming, Proceedings of the IFIP Working Conference on Formal Language Description Languages, Vienna September 15–18, 1964. Amsterdam (1966) 266–294

[30] F. L. Bauer: The cellar principle of state transition and storage allocation. Ann. Hist. Comput. **12** (1990) 41–49

[31] A. A. Grau: Recursive processes and ALGOL translation. Commun. ACM **4** (1961) 10–15

[32] P. Lucas: Die Strukturanalyse von Formelübersetzern. Interner Bericht Mailüfterl Wien (1961)

[33] M. Paul: Zur Struktur formaler Sprachen. Dissertation Mainz (1962)

[34] M. Paul: A general processor for certain formal languages. In: Symbolic Languages in Data Processing, Proceedings of the Symposium, organized and edited by the International Computation Centre Rome, March 26–31, 1962. Gordon and Breach, New York (1962) 65–74

[35] P. Deussen: Bericht über ein Programm zur Übersetzung Boolescher Ausdrücke in disjunktive Normalform. Colloquium über Schaltkreis- und Schaltwerk-Theorie. Birkhäuser, Basel (1961)

[36] W. Petry: Übersetzung formaler Sprachen in topologische Strukturen. Dissertation Mainz (1962)

[37] Charles L. Hamblin: Computer languages. Austral. J. Sci. **20** (1957) 135

[38] F. G. Duncan: Stack machine development: Australia, Great Britain, and Europe. Computer **10** (1977) 50–52

[39] J. Eickel, M. Paul: The parsing and ambiguity problem for Chomsky languages. In: T. B. Steel, Jr. (ed.), Formal Language Description Languages for Computer Programming, Proceedings of the IFIP Working Conference on Formal Language Description Languages, Vienna, September 15–18, 1964. Amsterdam (1966) 52–75

Friedrich L. Bauer

K. Samelson and Friedrich L. Bauer
Sequentielle Formelübersetzung

Elektronische Rechenanlagen 1, 1959
pp. 176-182

Sequentielle Formelübersetzung

Sequential Formula Translation

von K. SAMELSON und F. L. BAUER

Universität Mainz

Elektronische Rechenanlagen 1 (1959), H. 4, S. 176—182
Manuskripteingang: 9. 9. 1959

Die Syntax einer Formelsprache wie ALGOL läßt sich als Folge von Zuständen beschreiben, die durch ein Keller genanntes Element angezeigt werden. Die Übergänge werden gesteuert durch zulässige Zustand-Zeichen-Paare, die sich in Form einer Übergangsmatrix darstellen lassen. Diese Beschreibung liefert gleichzeitig eine äußerst einfache Vorschrift zur Übersetzung der Anweisungen der Formelsprache in Maschinenprogramme. Lediglich Optimisierungsprozesse wie die rekursive Adressenfortschaltung entziehen sich der sequentiellen Behandlung.

The syntax of an algorithmic language such as ALGOL is conveniently described as a sequence of states indicated by an element called cellar. Transitions are controlled by admissable state-symbol pairings which may be represented by a transition matrix. This description at the same time furnishes an extremely simple rule for translating statements of the algorithmic language into machine programs. Sequential treatment, however, is not feasible in the case of optimizing processes such as recursive address calculation.

Verwendete Zeichen

Es gelten alle Bezeichnungen von [12], Elektronische Rechenanlagen 1 (1959), 72. Darüber hinaus oder abweichend sind verwendet:

Symbol $\chi \sim I, N$, *'go to'* etc.
Zeichen $\alpha \sim + - \times / ( )$
Ergibtzeichen $\Rightarrow$
Adresse von z $\rangle z \langle$
Inhalt einer Speicherzelle $\langle \varphi \rangle$
mit der Adresse φ
AC Inhalt des Akkumulators
æ Ende des Ausdrucks
η Inhalt der Zahlkelleradresse

φ Adresse
Φ Adressenkeller
$\varnothing$ Leersymbol beim Keller
h Zählerstand des Zahlkellers
H Zahlkeller
K Befehlsfolge
Π Programm
s Nummer des Kellersymbols
σ Kellersymbol
Σ Symbolkeller

1. Einleitung, Grund und Entwicklung der Formelübersetzung

Die schnelle Entwicklung des Baues programmgesteuerter Rechenanlagen in den letzten zehn Jahren hat dazu geführt, daß heute eine beträchtliche Anzahl verschiedener Automatentypen hergestellt wird. Alle diese Maschinentypen haben jedoch, trotz großer Unterschiede in Konstruktion und Befehlscode, zwei Charakteristika gemeinsam, die nach allgemeiner (möglicherweise nicht vorurteilsfreier) Ansicht technisch bedingt sind, nämlich

1. den in eine eindimensionale Folge von Worten fester Zeichenlänge zerlegten Speicher (Arbeitsspeicher),
2. Das entsprechend in eine Folge fester unabhängiger Elemente (der Befehle) zerlegte Programm, das von der Steuerung Befehl für Befehl abgearbeitet wird. Dies bedeutet, daß die einem ins Steuerwerk gelangenden Maschinenbefehl zukommende Operation unabhängig ist von der Befehlsvorgeschichte.

Diese beiden Merkmale stellen sich dem Benutzer der Rechenanlage, also dem Programmhersteller, als Hindernisse entgegen, insofern sie verantwortlich sind für die bekannte Unbequemlichkeit und Irrtumsanfälligkeit des Programmierens in Maschinencode. Denn sie erfordern das Operieren mit Adressen und bedingen darüber hinaus eine völlige Atomisierung des Programms. Es ist wichtig festzustellen, daß dieser Zwang unnatürlich ist: ein Problem irgendwelcher Art, das von einer Rechenanlage behandelt werden soll, entsteht in der gedanklichen Konzeption zunächst meist als Ablaufschema für gewisse größere Operationseinheiten, die durch ihren Zweck umrissen und mehr oder weniger vage durch die dem Problemkreis eigentümlichen Bezeichnungen angegeben werden. Die Ausgestaltung des Problems führt zu einer operativen Fixierung, die in möglichst rationeller Form unter Benutzung gebräuchlicher Notation geschieht, vornehmlich unter Heranziehung mathematischer Formeln und verbaler Erläuterungen. Eine Atomisierung in kleinste Einzeloperationen ist unökonomisch hinsichtlich der darauf zu verwendenden Zeit und des erforderlichen Platzes, vor allem führt sie zur Unübersichtlichkeit. Die Hinzunahme der Adressen als völlig künstlicher Elemente wiegt noch schwerer, sie erfordert umfangreiche Buchführung und überdies in rekursiven Prozessen Adressenberechnungen, die sich der eigentlichen Aufgabe überlagern. Die Verhältnisse werden geradezu paradox bei gewissen Grundaufgaben der Numerischen Mathematik: ein generell brauchbares Programm zur Lösung eines linearen Gleichungssystems enthält etwa hundert einzelne Befehle, unter denen ein einziger Additions- und ein einziger Multiplikationsbefehl der eigentlichen Aufgabe dienen. Insbesondere die mit der Einführung der Adressen verbundenen Arbeitsgänge sind weitgehend routinemäßiger Natur, und man hat daher schon frühzeitig versucht, sie wenigstens teilweise dem Rechenautomaten selbst zuzuschieben, der dabei als reiner Codeumsetzer arbeitet [1], [8], [9].

Gewisse Erleichterungen verschaffte man sich ferner durch den Gebrauch vorgefertigter Bibliotheksprogramme für standardisierte Operationseinheiten, die, mit Codeworten bezeichnet, ebenfalls vom Rechenautomaten direkt aufgerufen, d. h. in den Ablauf eingeordnet werden. Derart aufgebaute Programmierungssysteme waren ab 1954 in allgemeinem Gebrauch, wobei das Programm, das die Routinearbeiten der Programmierung („automatische Programmierung") erledigte, als Compiler bezeichnet wurde [10].

Daß man sich bei numerischen Aufgaben eine effektive Lösung des Problems der Programmierung erst erhoffen kann, wenn man bei der automatischen Programmfertigung von den in konventioneller Schreibweise geschriebenen Formeln ausgeht und alle weiteren Phasen dem Automaten überläßt, hat schon 1951 *Rutishauser* [4] erkannt. Sein Verwirklichungsvorschlag [5] sowie die daran anknüpfende Arbeit von *Böhm* [14] blieb jedoch unbeachtet, und erst 1955 wurden mit PACT [3] und FORTRAN [2] die ersten Programmierungssysteme mit Formelübersetzungscharakter aufgebaut, ohne daß jedoch etwas über die dabei verwendeten Methoden publiziert worden wäre. Etwa gleichzeitig begannen in Kenntnis der Rutishauserschen Ergebnisse ähnliche Überlegungen am Rechenzentrum der TH München, wobei die Entwicklung solcher Übersetzungsmethoden im Vordergrund stand, die auch für Anlagen von wesentlich geringerem Umfang und Leistungsfähigkeit als etwa der IBM 704 anwendbar sein sollten. Zu diesem Zwecke wurde, auf unabhängigen Vorarbeiten basierend [6], [13], eine sequentielle Übersetzungstechnik entwickelt. Die Arbeiten wurden seit 1957 im Rahmen der heutigen Arbeitsgruppe Zürich—München—Mainz—Darmstadt (ZMMD) fortgesetzt.

Inzwischen hatte sich jedoch eine prinzipielle Verschiebung der Standpunkte angebahnt: die herkömmlichen Programmierungssysteme waren noch vom Maschinencode als dem Ziel der Übersetzung her aufgebaut, und die Übersetzung selbst war schrittweise aus einer Übertragung der bisher von Menschen geleisteten Routinearbeit auf die Rechenanlage entstanden, wobei die Sprache des jeweiligen Programmierungssystems von der Struktur der Rechenanlage her immer weniger bestimmt war. Mit der Beherrschung der Technik des Übersetzungsvorganges gewann man nun auch Freiheit in der Wahl der Programmierungssprache, und die Aufstellung einer möglichst bequem handzuhabenden, übersichtlichen, selbstverständlichen Sprache trat als Aufgabe hervor, die gelöst werden mußte, bevor die Übersetzer selbst programmiert werden konnten. Insbesondere entstand die verlockende Möglichkeit, für verschiedene Rechenanlagen, zunächst innerhalb der ZMMD-Gruppe, dieselbe Programmierungssprache zu verwenden. Die Entwicklung führte 1958 zum Vorschlag einer algorithmischen Formelsprache (ALGOL) durch ein gemeinsames ACM-GAMM-Kommittee [11], [12]. In der Zwischenzeit wurde, nunmehr auf der Basis von ALGOL, die Struktur des Formelübersetzers der ZMMD-Gruppe einheitlich festgelegt und mit der Codierung für die Rechenanlagen der beteiligten Institute (ERMETH, PERM, Z 22, SIEMENS) sowie für die Rechenanlagen einiger befreundeter Institute in Deutschland, USA, Österreich und Dänemark nach diesem ALCOR (ALGOL Converter) genannten System begonnen.

Da somit dieses Projekt seiner Vollendung entgegengeht, erscheint es an der Zeit, einen Überblick über die ihm zugrunde liegenden Prinzipien der sequentiellen Übersetzung zu geben, die sowohl von dem ursprünglichen Rutishauserschen Vorschlag [5] als auch von den kürzlich veröffentlichten Methoden des FORTRAN-Systems [7] wesentlich abweichen[1]). Ausführliche Strukturpläne, die das ganze Formelübersetzungsprogramm in detaillierter Form ohne Bezugnahme auf eine spezielle Maschine beschreiben, wurden im Institut für Angewandte Mathematik der Univer-

---

[1]) Einzelne Züge des Systems finden sich bereits in der erwähnten Arbeit von *Böhm* [14], der jedoch starke Einschränkungen hinsichtlich der zulässigen Notation macht.

sität Mainz in reproduktionsfähige Form gebracht; sie bilden die Grundlage der oben erwähnten Zusammenarbeit der ALCOR-Familie.

2. Sequentielle Übersetzung und das Kellerungsprinzip

Die in einer Formelsprache wie ALGOL niedergeschriebenen Anweisungen sind eine Folge von Symbolen, die sich ihrerseits aus einem oder mehreren Charakteren zusammensetzen. Da der Aufbau von Symbolen aus Charakteren jedoch trivial (es handelt sich stets um lückenlose, eindeutig abgegrenzte Folgen) und bis zu einem gewissen Grade von technischen Gegebenheiten wie dem verwendeten Schreibgerät abhängig ist, werden wir im folgenden den Unterschied zwischen Symbolen und Charakteren unterdrücken und jedes Symbol χ als Einheit betrachten. Dies gilt insbesondere für Identifier I, Zahlen N und verbal definierte Begrenzer wie 'go to', 'if' usw.

Die Folge von Symbolen χ des Formelprogramms stellt nun (mit der üblichen Interpretation der Symbole) eine Arbeitsvorschrift dar. Dabei ist es jedoch nicht möglich, die Symbole in der angegebenen Reihenfolge in orthodoxe Maschinenoperationen zu übersetzen. Vielmehr erzwingen bereits bestimmte arithmetische Symbole, die Klammern (), und Vorrangregeln ($\times$ vor $+$) eine von der Symbolanordnung abweichende Reihenfolge der Operationen. So heißt $a \times b + c \times d$: multipliziere a mit b, multipliziere c mit d, addiere die Produkte, während die sequentielle Auswertung ergeben würde: multipliziere a mit b, addiere dazu c und multipliziere das Resultat mit d.

Es ist also bei der Abarbeitung des Formelprogramms ständig notwendig, gelesene Symbole als nicht auswertbar zu übergehen und in einem späteren, von der weiteren Symbolfolge abhängigen Zeitpunkt wiederzufinden und auszuwerten. *Rutishauser* hat mit dem „Klammergebirge" die grundsätzliche Lösung angegeben. Die von ihm vorgeschlagene Ausführung, durch Vorwärts- und Rückwärtslesen die ausführbare Operation einzukreisen, ist aber unbequem und (unnötig) zeitraubend. Daran ändert sich auch nicht viel, wenn man *Rutishausers* Methode dahingehend variiert, daß man bereits lokale Gipfel abarbeitet. Das Problem, die beim ersten Erscheinen als nicht auswertbar übergangene Information im richtigen Augenblick wieder greifbar zu haben, läßt sich aber mit Hilfe eines als Kellerung bezeichneten Prinzips weitgehend vereinfachen, das immer anwendbar ist, wenn die Struktur der Symbolfolge klammerartigen Charakter hat. Das soll heißen, daß zwei verschiedene Paare A, A' und B, B'

zusammengehöriger Elemente sich nur umfassen, aber nicht gegenseitig trennen können, daß also nur Anordnungen $ABB'A'$ und nicht $ABA'B'$ vorkommen.
Das Prinzip besagt: Man setze alle nicht sofort auswertbaren Informationen in der Reihenfolge des Einlaufens in einem besonderen Speicher, dem „Symbolkeller", ab, in dem jeweils nur das zuletzt abgesetzte, im obersten Geschoß befindliche Element interessiert und damit unmittelbar zugänglich zu sein braucht. Jedes neu gelesene Symbol wird mit dem obersten Kellersymbol verglichen. Die beiden Symbole in Konjunktion legen fest, ob das Kellersymbol in eine Operation umgesetzt werden kann, worauf es aus dem Keller entfernt wird. Je nach den Umständen wird der Vergleich mit dem nunmehr obersten Symbol des Kellers wiederholt und schließlich gegebenenfalls ein neues Zustandssymbol im Keller abgesetzt.
In der Sprechweise der Theorie der Automaten kann das Prinzip so formuliert werden: Durch die gesamte Besetzung des Kellers wird ein Zustand (des Übersetzungsvorgangs) definiert, der effektiv in jedem Augenblick nur von dem obersten Kellerzeichen abhängt, und neu gelesene Information plus Zustand bestimmen die Aktionen des Übersetzers, die aus der Abgabe von Zeichen, nämlich von Operationsanweisungen für das erzeugte Programm und der Festlegung eines neuen Zustands bestehen. Das Wesentliche ist aber die durch die Besetzung des Kellers induzierte latente Zustandsstruktur.

3. Auswertung einfacher arithmetischer Ausdrücke

Den wichtigsten Fall der Symbolfolgen mit Klammerstruktur stellen die arithmetischen Ausdrücke dar, deren Behandlung wir daher als Beispiel ausführlich besprechen wollen. Um aber den prinzipiellen Sachverhalt nicht mit relativ unwichtigen Details zu belasten, werden wir einige Vereinfachungen vornehmen.
Diese betreffen einmal die zulässigen Symbole. Wir werden Funktionen $\boldsymbol{I}$ ($\boldsymbol{P}, \ldots, \boldsymbol{P}$) und indizierte Variable $\boldsymbol{I}$ [$\boldsymbol{E}, \ldots, \boldsymbol{E}$] vorläufig ausschließen und die Additionssymbole $\pm$ nur als zweistellige Operation ($a \pm b$) und nicht als einstellige ($\pm a$) zulassen.

Weiter werden wir zur Erläuterung hinsichtlich der Rechengrößen selbst unterstellen, daß dem Rechenwerk der Maschine, für die das Programm hergestellt werden soll, ein Schnellspeicher begrenzter Kapazität zur Verfügung steht, dessen Zugriffszeit vernachlässigbar ist gegenüber der Zugriffszeit des Arbeitsspeichers, so daß für alle Zahlen, die zur Verarbeitung dem Rechenwerk zur Verfügung ge-stellt werden sollen, ein vorübergehendes Absetzen im Schnellspeicher keine Verzögerung des Ablaufs des Resultatprogramms bedeutet.

Dieser Schnellspeicher habe nun dieselbe Kellerstruktur wie der Symbolkeller, d. h., seine Plätze werden sukzessive belegt, und die jeweils zuletzt abgespeicherte (gekellerte) Zahl ist als erste abrufbar. Der Speicher werde deshalb als Zahlkeller H bezeichnet.

Jeder unter den gemachten Voraussetzungen in einem Ausdruck auftretende Identifier stellt eine Variable dar, d. h. den Decknamen für eine Zahl, und ist somit eine symbolische Adresse, die von dem Übersetzer in irgendeiner Weise auf eine echte Speicheradresse abgebildet wird, wie dies schon von allen mit symbolischen Adressen arbeitenden Compilern getan wird[2]). Zahlen N sind, gegebenenfalls nach Konvertierung, in Zellen abzusetzen und ebenfalls durch Adressen zu ersetzen, so daß wir sie weiterhin außer Betracht lassen können.

Die Auswertung eines arithmetischen Ausdrucks mit Hilfe des Kellerungsprinzips geht nun in folgender Weise vor sich:

a) Jeder auftretende Identifier I veranlaßt die Überführung des Inhalts der entsprechenden Speicherzelle in den jeweils obersten Platz des Zahlkellers H. Das Wort „veranlaßt" bedeutet hier, daß der Übersetzer die entsprechenden Befehle an den bereits aufgebauten Teil des zu erzeugenden Maschinenprogramms anfügt. Ein im Übersetzer enthaltener Zähler h hat den jeweils obersten Platz des Zahlkellers anzuzeigen und muß daher gleichzeitig eine Eins aufzählen. Bezeichnen wir den Speicher für das erzeugte Programm mit Π, die Inhalte der Plätze des Zahlkellers H mit η_h, wobei der Index h die Zählgröße darstellt, und die Maschinenbefehlsfolge $I \Rightarrow \eta_h$, die die Überführung in den Zahlkeller darstellt, mit K_I, so sind die vom Übersetzer auszuführenden Operationen:

$$I\colon h \dotplus 1 \Rightarrow h;\ K_I \Rightarrow \Pi;\ \text{lies } \chi.$$

'lies χ' bedeutet hier, daß das nächste Zeichen χ des Ausdrucks zu lesen ist.

b) Alle übrigen Symbole α, das sind $+$, $-$, $\times$, $/$, $($, $)$, werden beim Einlaufen mit dem jeweils obersten, als σ_s bezeichneten Symbol des Symbolkellers verglichen, der im Anfangs

[2]) Die einfachste Möglichkeit wäre etwa, die Zahl der zulässigen Identifier soweit zu beschränken, daß jedem Identifier ein fester oder wenigstens relativ zu dem erzeugten Programm fester Speicherplatz zugewiesen wird.

zustand das Leersymbol ∅ enthält. Jedes aus einem Kellersymbol σ_s und einem Formelzeichen α bestehende Paar veranlaßt eine bestimmte Folge von Operationen des Übersetzers entsprechend der folgenden Liste:

σ_s	α	
∅	α	$1 \Rightarrow s$; $\alpha \Rightarrow \sigma_s$; lies χ;
+ —	+ —	$K_\sigma \Rightarrow \Pi$; $\alpha \Rightarrow \sigma_s$; $h-1 \Rightarrow h$; lies χ;
× /	× /	
(	+ — × /	$s+1 \Rightarrow s$; $\alpha \Rightarrow \sigma_s$; lies χ;
+ —	× /	
+ — × /	)	
(	)	$s-1 \Rightarrow s$; lies χ;
× /	+ —	$K_\sigma \Rightarrow \Pi$; $s-1 \Rightarrow s$; $h-1 \Rightarrow h$; repetiere mit α;
+ — × /	) æ	

Die vom Übersetzer erzeugten und an den Programmspeicher Π abgegebenen Maschinenbefehlsfolgen K_σ haben dabei stets die folgende Dreiadreßform, wobei σ eines der vier Operationssymbole + — × / darstellt:

$$K_\sigma:\ \eta_{h-1}\sigma\,\eta_h \Rightarrow \eta_{h-1}.$$

Es werden also die jeweils beiden obersten Elemente des Zahlkellers η_{h-1} und η_h durch die mit σ bezeichnete Operation verknüpft und das Resultat als nunmehr oberstes Element η_{h-1} an den Zahlkeller zurückgegeben. Mit der Abgabe dieser Befehlsfolge muß daher auch der Zähler h des Zahlkellers um Eins heruntergezählt werden.

'Repetiere mit α' bedeutet, daß im nächsten Schritt mit dem gleichen Symbol α und dem neuen σ_s zu arbeiten ist.

Das Ende eines Ausdrucks muß natürlich erkennbar sein. Es ist hier mit 'æ' angedeutet und wirkt wie eine dem Anfang als öffnender Klammer zugeordnete schließende Klammer.

Die Liste von Zeichenpaaren σ_s, α läßt sich bequem durch eine Matrix darstellen, deren Zeilen den möglichen Kellersymbolen σ_s und deren Spalten den Formelzeichen α zugeordnet sind, so daß jedem Paar ein Matrixelement ent-

$$A\colon (a \times b + c \times d)/(a - d) + b \times c$$

Σ	χ (α oder I)	Π
leer	(	
(	a	$a \Rightarrow \eta_1$
(	$\times$	
($\times$	b	$b \Rightarrow \eta_2$
($\times$	+	$\eta_1 \times \eta_2 \Rightarrow \eta_1$
(+	c	$c \Rightarrow \eta_2$
(+	$\times$	
(+ $\times$	d	$d \Rightarrow \eta_3$
(+ $\times$	)	$\eta_2 \times \eta_3 \Rightarrow \eta_2$
(+		$\eta_1 + \eta_2 \Rightarrow \eta_1$
(		
leer	/	
/	(	
/(	a	$a \Rightarrow \eta_2$
/(	—	
/(—	d	$d \Rightarrow \eta_3$
/(—	)	$\eta_2 - \eta_3 \Rightarrow \eta_2$
/(		
/	+	$\eta_1/\eta_2 \Rightarrow \eta_1$
+	b	$b \Rightarrow \eta_2$
+	$\times$	
+ $\times$	c	$c \Rightarrow \eta_3$
+ $\times$	æ	$\eta_2 \times \eta_3 \Rightarrow \eta_2$
+	æ	$\eta_1 + \eta_2 \Rightarrow \eta_1$
leer		

spricht[3]). Diese Übergangsmatrix liefert eine vollständige syntaktische und operative Beschreibung aller zulässigen arithmetischen Ausdrücke.

Anfangszustand ist stets $s=0$ ($\sigma_s = \emptyset$) und $h=0$ (Zahlkeller leer), ein zulässiger Endzustand, der einem vollständigen Ausdruck entspricht, ist mit $s=0$ und $h=1$ erreicht. Der Wert eines vollständigen Ausdrucks findet sich also stets auf dem ersten Platz des Zahlkellers.

Ein einfaches Beispiel möge den Ablauf erläutern, wobei wir nur den jeweiligen Inhalt des Symbolkellers Σ, das neu ein-

---

[3]) Bei *Böhm* [14], der für eine stark eingeschränkte Formelsprache bereits eine matrixartige Übersetzungsvorschrift gibt, fehlt der Symbolkeller. *Böhm* hat jedoch bereits die Auswertung klammerfreier Ausdrücke durch Vergleich aufeinanderfolgender Operationszeichen.

laufende Zeichen χ und das in Π aufgebaute Programm angeben.

$$A : (a \times b + c \times d)/(a - d) + b \times c$$

Wie man aus der obigen Tabelle sieht, ist die Reihenfolge der Operationen im entstehenden Programm durch das Formelprogramm völlig festgelegt, und es wird kein Versuch gemacht, etwa zur Beschleunigung Umstellungen vorzunehmen. Denn die Wahl der Reihenfolge der Operationen muss völlig in der Hand des das Programm entwerfenden Mathematikers liegen. Jede Umstellung kann wegen der Ungültigkeit des assoziativen Gesetzes (wenigstens beim Rechnen mit gleitendem Komma) unerwünschte numerische Konsequenzen haben.

4. Vollständige arithmetische Ausdrücke

Wir haben nun zu diskutieren, wie das oben angegebene Schema zu variieren ist, wenn wir die angegebenen Vereinfachungen fallen lassen. Betrachten wir zunächst die Behandlung der Rechengrößen:

Das angegebene Beispiel zeigt deutlich, daß eine Anzahl unnötiger Umspeicherungen vorgenommen wird. Tatsächlich sind alle Operationen $I \Rightarrow \eta_h$ überflüssig, und Variable dürfen von ihrem Platz nur zur Ausführung von Rechenoperationen ins Rechenwerk abgerufen werden. Wenn wir uns, was weiterhin vorausgesetzt sein soll, auf den Fall der Einadreßmaschine beschränken, so fallen Ergebnisse stets im Akkumulator an. Der Zahlkeller darf nur noch dazu dienen, solche (unbenannte) Zwischenergebnisse aufzunehmen, deren Abspeicherung notwendig ist, um das Rechenwerk für die nachfolgenden Operationen freizumachen. Für diese arbeitet er in der vorher beschriebenen Weise.

Im übrigen tritt aber an die Stelle des Zahlkellers ein von dem Übersetzer auszuwertender (Variablen- oder) Adressenkeller Φ, und alle überflüssigen Transportoperationen sind durch Eintragung der entsprechenden Adressen in diesem Keller zu ersetzen, die durch den Übersetzer vorgenommen wird und die Programmerzeugung mitsteuert. Da nun auch der Akkumulator als Zahlspeicher verwendet wird, ist es zweckmäßig, auch ihm eine (identifizierbare) Pseudoadresse zuzuweisen, die in den Adressenkeller eingetragen wird. Notwendige Abspeicherungen von Zwischenresultaten ergeben sich dann daraus, daß eine öffnende Klammer auf ein arithmetisches Operationszeichen im Symbolkeller Σ

stößt, dem als oberstes Element des Adressenkellers die Adresse des Akkumulators entspricht. Eine solche Klammer wird impliziert auch durch ein einlaufendes $\times/$, das auf ein $\pm$ in Σ stößt. Da zwischen diesen Symbolen ein Identifier aufgetreten sein muß, ist in diesem Falle auch die zweithöchste Position des Adressenkellers zu kontrollieren.

Ist eine Zwischenspeicherung notwendig, so wird die Abspeicherung des AC in den gerade obersten Platz des Zahlkellers H veranlaßt, die Adresse des AC im Adreßkeller durch die Adresse von η_h ersetzt und angemerkt, daß bei Abruf der Adresse in das erzeugte Programm der Zahlkellerindex um Eins heruntergezählt werden muß.

Die vom Übersetzer in den Programmspeicher abzusetzenden Operationen K_σ erhalten jetzt im allgemeinen die Form

$$K_\sigma\colon \begin{array}{c} \langle \varphi_{f-1} \rangle \Rightarrow \mathrm{AC} \\ \mathrm{AC}\ \sigma\ \langle \varphi_f \rangle \Rightarrow \mathrm{AC}. \end{array}$$

Dabei ist jedoch stets zu prüfen, ob eine der beiden Operandenadressen φ_{f-1}, φ_f den AC darstellt. In diesem Falle fällt für σ gleich $+$ oder $\times$ der erste Befehl aus, der zweite erhält die Adresse φ_f oder φ_{f-1}, die nicht den AC darstellt. Bei σ gleich — oder / fällt der erste Befehl weg, wenn φ_{f-1} den AC darstellt. Im entgegengesetzten Falle aber, also $\varphi_f = \mathrm{AC}$, muß man für $\sigma = -$ setzen:

$$K_-\colon \begin{array}{c} -\mathrm{AC} \Rightarrow \mathrm{AC} \\ \mathrm{AC} + \langle \varphi_{f-1} \rangle \Rightarrow \mathrm{AC}, \end{array}$$

während man für σ gleich / sogar zuerst den AC sicherstellen muß.

$$K_/\colon \begin{array}{c} \mathrm{AC} \Rightarrow \eta_h \\ \langle \varphi_{f-1} \rangle \Rightarrow \mathrm{AC} \\ \mathrm{AC}/\eta_h \Rightarrow \mathrm{AC} \end{array}$$

Die beiden Fälle entsprechen Formeln vom Typ $a-(b+c)$ bzw. $a/(b+c)$, die sich bequemer mit Maschinen behandeln ließen, die „vom Speicher subtrahieren" bzw. „in den Speicher dividieren" können.

Die einstelligen Operationen $+a$, $-b$ schließlich erledigen sich begrifflich am einfachsten durch Hinzunahme eines Leerelementes im Adressenkeller, das anzeigt, daß der entsprechende Linksoperand nicht existiert.

Die Behandlung Boolescher Ausdrücke läuft offensichtlich der Behandlung arithmetischer Ausdrücke parallel.

Die Hinzunahme von Funktionen und indizierten Variablen bedeutet zunächst einmal, daß das Auftreten eines Identi-

fiers unmittelbar von einer öffnenden Klammer festgestellt werden muß, da die Kombination $I($ die Funktionen und die Kombination $I[$ die indizierten Variablen eindeutig kennzeichnet. Weiter, und das ist der wesentliche Punkt, stellen beide Symbole, Funktion und indizierte Variable, einen neuen Typ von Klammer mit besonderen Eigenschaften dar. Wenn wir uns hinsichtlich der indizierten Variablen zunächst auf den Fall beschränken, daß die durch die der Variablen zugehörige Feld-Vereinbarung (array declaration) festgelegte Speicherabbildungsfunktion (vgl. Abschnitt 7) für jedes Auftreten der Variablen vollständig ausgewertet wird, ist die Behandlung weitgehend einheitlich.
Zunächst ist der Reihe nach die Auswertung der auf den einzelnen Argument- bzw. Indexpositionen stehenden Ausdrücke zu veranlassen, wobei das trennende Komma bzw. die abschließende Klammer) oder] die Rolle des Abschlußzeichens übernimmt. Die Werte der Ausdrücke sind abzuspeichern, konsequenterweise als Zwischenergebnisse im Zahlkeller. Anschließend an die Berechnung der Argumente ist ein Sprung mit automatischer Rückkehr zu setzen, der in das durch die Funktions- bzw. Feld-Vereinbarung definierte Programm führt. Dieses endet wie üblich mit der Abgabe des ermittelten Wertes an den Akkumulator. Für indizierte Variable mit laufenden Indizes in Schleifen ist eine solche Behandlung natürlich zeitraubend und ineffektiv; sie muß durch rekursive Auswertung der Speicherabbildungsfunktion ersetzt werden, bei der innerhalb der Schleife nur Additionen auftreten, die z. B. durch Indexregister erledigt werden können. Wir kommen darauf noch zurück.

5. Anweisungen (statements)

Die Auswertung vollständiger Anweisungen verläuft nach den gleichen Prinzipien wie die der Ausdrücke, die ja den wesentlichsten Teil aller Anweisungen darstellen. Es muß nur der Symbolkeller einige weitere Symbole aufnehmen können.
Was die arithmetischen (und Booleschen) Anweisungen anbetrifft, handelt es sich hier im wesentlichen um das :=, das stets als erstes im Symbolkeller abgesetzt wird und damit an Stelle des anfänglichen Leerzustands des Kellers tritt. Als Schlußzeichen im Informationseinlauf fungiert das Anweisungstrennzeichen ; bzw. das 'end' der zusammengesetzten Anweisungen, das jeweils erst die Setzung der letzten arithmetischen Operationen des Ausdrucks auslöst und bei Koinzidenz mit dem := anzeigt, daß dieses in den abschließenden Speicherbefehl umgesetzt werden kann.

Die verbalen Klammern 'begin' und 'end' für zusammengesetzte Anweisungen werden naturgemäß ebenso behandelt wie arithmetische Klammern: 'begin' wird in den Symbolkeller abgesetzt. Ein einlaufendes 'end' dient zunächst als Schlußzeichen für die vorangegangene Anweisung und löst die Veranlassung aller im Keller anstehenden Operationen aus, bis es auf das erste 'begin' stößt, das noch gelöscht wird. Damit ist die Funktion des 'end' beendet. Ist das nächste Zeichen wieder ein 'end', so wiederholt sich der Vorgang, bis als Schlußzeichen das Trennzeichen ; eintrifft, das die Abarbeitung des Kellers bis zum nächsten gekellerten 'begin' auslöst, das nun aber natürlich unangetastet bleibt.

Ähnlich ist die Situation bei der einfachen Sprunganweisung 'go to' L. Der führende Begrenzer wird im Keller abgesetzt, anschließend die Marke L ausgewertet. Das Trennzeichen ; schließt die Auswertung ab und zeigt beim Auftreffen auf den Begrenzer im Keller, daß die zugehörige Operation „Sprung nach dem durch L bezeichneten Speicherplatz" abgesetzt werden kann.

Die Behandlung der beiden Anweisungen 'if' B und 'for' $V := l$, wo l eine Liste entweder von Ausdrücken E oder von Ausdruck-Tripeln E_i (E_s) E_e darstellt, ist zunächst ähnlich wie die der Sprunganweisung. Der Begrenzer wird im Keller abgesetzt und die anschließende Zeichenfolge B bzw. $V := l$ ausgewertet. Das abschließende Symbol ; zeigt das Ende der Auswertung an. In beiden Fällen ist jedoch die Funktion des Begrenzers noch nicht abgeschlossen.

Im Falle des 'if' kann zwar die Absetzung des an die Aussage B anschließenden bedingten Sprungbefehls durch das ; veranlaßt werden. Jedoch ist die Sprungadresse noch unbekannt. Sie liegt erst fest, wenn die nächste Anweisung voll ausgewertet ist. Daher muß das 'if' als transformiertes 'if$_1$' im Keller verbleiben, bis es auf das nächste einlaufende Trennzeichen ; oder 'end' trifft, das das Ende der bedingten Anweisung markiert. Erst damit liegt das Sprungziel im erzeugten Programm fest und kann eingetragen werden, worauf das 'if$_1$' endgültig gelöscht wird.

Der Fall des 'for' ist wesentlich komplizierter. Besteht die Liste l in 'for' $V := l$; Σ (wo Σ die qualifizierte Anweisung darstellt) aus Ausdrücken E_1 bis E_k, so ist die Anweisung unter Einführung einer zusätzlichen Indexvariablen HI und einer indizierten Variablen $V[HI]$ in die folgenden Anweisungen umzusetzen:

$V[1] := E_1$; $V[2] := E_2; \cdots$; $V[k] := E_k$;
'for' $HI := 1(1)k$;
'begin' $V := V[HI]$; Σ 'end';

Damit ist dieser Fall auf den der Progression zurückgeführt. Ähnlich hätte man vorzugehen, wenn die Elemente der Liste l selbst Progressionen E_i (E_s) E_e sind.

Einfacher ist in diesem Fall sicher, die Anweisungen 'for' $V := E_{ig}$ (E_{sg}) E_{eg} ; Σ für jedes Listenelement getrennt aufzuschreiben. In jedem Fall aber genügt die Betrachtung der einfachen Progression: 'for' $V := E_i$ (E_s) E_e ; Σ.

Nach Kellerung des 'for' kann der erste Teil der folgenden Symbolkette $V := E_i$ wie eine normale arithmetische Anweisung betrachtet werden, da ja hierdurch der erste Wert von V festgelegt wird. Als Schlußzeichen, das auf das 'for' im Keller trifft, wirkt die öffnende Klammer. Ihr Zusammentreffen mit 'for' besagt, daß sie zu ersetzen ist durch $S :=$, was zusammen mit dem folgenden E_s wieder als arithmetische Anweisung ausgewertet werden kann. Die schließende Klammer wirkt als Schlußzeichen und ist zu ersetzen durch $E :=$, worauf wieder mit der Auswertung von E_e fortgefahren werden kann. S und E sind dabei vom Übersetzer einzuführende Hilfsvariable für Schritt und Endgröße. Eine Vereinfachung ist möglich, wenn E_s oder E_k eine Zahl oder eine einzige Variable ist: In diesem Falle genügt es, wenn der Übersetzer die Hilfsvariablen S bzw. E durch die betreffenden Größenbezeichnungen ersetzt[4]).

Da in der auf das 'for' folgenden Schleife die Abschlußbedingung von dem Vorzeichen des Wertes von E_s abhängt, muß dieses noch vor dem Eintritt in die Schleife getestet werden. Ist E_s eine Zahl, so kann dies der Übersetzer übernehmen. In anderen Fällen muß eine Prüfung der Laufrichtung und eine entsprechende Festlegung der Abschlußbedingung im Programm erzeugt werden, wenn man nicht dem Übersetzer sehr unbequeme dynamische Kontrollen aufbürden will.

Die Funktion des 'for' ist mit der durch das erste Semikolon angezeigten Abarbeitung der Progressionsangaben nicht erledigt. Vielmehr muß noch die Schleifenschließung einschließlich Zählung und Prüfung veranlaßt werden. Daher ist auch das 'for' im Keller durch das erste Semikolon zu transformieren zu 'for_1'. Benützt man als Standardschleife den normalerweise effektivsten Typ mit Prüfung am Schluß und Schließung durch bedingten Sprung, so ist die Absetzung der entsprechenden Operationen bis zum Ende der auf die 'for'-Anweisung folgenden Anwei-

[4]) Auf den dubiosen Fall, daß die Anweisung $S := E_s$ in die Schleife selbst aufgenommen werden muß, weil etwa E_s von V abhängt (etwa 'for' $V := 1$ (V) N, was die Folge der ganzen Potenzen von 2 liefert), soll hier nicht weiter eingegangen werden. S ist also für die Schleife fest und in sinnvollen Fällen ungleich Null.

sung zurückzustellen. Da man aber dem Fall der leeren Schleife vom Typ 'for' $V := 1(1)0$ Rechnung tragen muß, ist vor dem Schleifenbeginn noch ein Sprung auf die Ausgangsprüfung der Schleife zu setzen. Diesem muß noch eine Marke folgen, die als Ziel für den Schleifenschließungssprung dient.

Anschließend kann die dem 'for' unterliegende einfache und zusammengesetzte Anweisung abgearbeitet werden. Das abschließende Symbol ; oder 'end' führt beim Auftreffen auf das 'for$_1$' im Keller zur Absetzung der Schließungsbefehle:

Insgesamt ist also die 'for'-Anweisung

$$\text{'for'}\ V := E_i\ (E_s)\ E_e\ ;\ \Sigma\ ;$$

vom Übersetzer wie die aufgelöste Anweisungsfolge

$$V := E_i;\quad S := E_s;\quad E := E_e;$$

$$\nparallel := \begin{cases} \geqq \text{ falls } S < 0 \\ \leqq \text{ falls } S > 0 \end{cases};\ \text{'go to'}\ L_p;\ L_B\text{: }\Sigma;$$

$$V := V + S;\ L_p\text{: 'if'}\ V \nparallel E;\ \text{'go to'}\ L_B$$

$$V := V - S;$$

zu behandeln, wobei sich bei der zweiten, dritten und vierten Anweisung die diskutierten Vereinfachungen ergeben können.

Die Prozeduranweisung schließlich ist als Aufruf eines Bibliotheksprogramms von Standardform zu behandeln, wobei die Parameter in der im Aufruf angegebenen Reihenfolge abgesetzt werden[5]). Da es sich hierbei um bekannte Techniken handelt, sind weitere Ausführungen unnötig.

Die Anweisung 'return' behandelt einen einfachen Rücksprung auf eine eingebrachte Rückkehradresse. Die Anweisung 'stop' bedeutet (unwiderrufliches) Ende des betreffenden Programmlaufs und soll die Maschine in einen Zustand versetzen, in dem sie weitere Aufträge annimmt.

6. Vereinbarungen (declarations)

Von den Vereinbarungen sollen nur die Funktions-, Prozedur- und Feld-Vereinbarungen kurz behandelt werden[6]). Funktions- und Prozedur-Vereinbarungen sowie Feld-Ver-

[5]) Insbesondere darf angenommen werden, daß Ein- und Ausgabetätigkeiten, die weithin von den Maschinencharakteristika abhängig sind, in genereller Form als Prozeduren aufgerufen werden können.

[6]) Typ-Vereinbarungen sind in selbstverständlicher Weise bei der Behandlung arithmetischer (oder Boolescher) Ausdrücke zu berücksichtigen.

einbarungen (solange man sich auf jeweils vollständige Auswertung der Speicherabbildungsfunktion, siehe 7., beschränkt) definieren jeweils Unterprogramme. Funktions- und Feld-Vereinbarungen führen zu statischen Programmen, die Prozedur-Vereinbarungen dagegen zu dynamischen[7]). Dem aus der Auswertung der Vereinbarung resultierenden Unterprogramm ist also wie bei allen Unter- bzw. Bibliotheksprogrammen ein Anschlußteil voranzustellen, der die Übernahme der Rückkehradresse und der Programmparameter durchführt, im dynamischen Fall ist ein Adaptieren zur Berechnung des benötigten Hilfsspeichers und zur Adressierung der auf den Hilfsspeicher bezüglichen Befehle (mit Hilfe eines speziellen Parameters, der den Beginn des freien Speichers angibt) hinzuzufügen. Auch hier handelt es sich um bekannte Techniken, auf die nicht näher eingegangen zu werden braucht.

7. Adressenfortschaltung bei indizierten Variablen

Wie bereits erwähnt, fügen sich die indizierten Variablen ohne Schwierigkeiten in den Rahmen der diskutierten Übersetzungstechnik, solange man die Speicherabbildungsfunktion im Programmlauf jeweils in geschlossener Form auswertet. Tatsächlich ist ja etwa die Größe $a\,[i,\,k]$ mit zugehöriger Feld-Vereinbarung 'array' ($a\,[1,1:n,m]$) als Funktion gegeben durch

$$a\,[i,k] := \langle\, k \times n + i + \,\rangle\, a\,[0,0]\,\langle\,\rangle\,.$$

Hier stellt die Variable $\rangle a \langle$ die Zahl dar, die die Adresse des die Zahl a enthaltenden Speicherplatzes angibt. Die Funktion $\langle E \rangle$ hat als Wert diejenige Zahl, die auf dem durch den (ganzzahligen) Wert von E adressierten Speicherplatz steht[8]).

Für Variable mit Laufindex in inneren Schleifen bedeutet eine solche Auswertung jedoch einen unerträglichen Zeitverlust, weshalb die Adressenberechnung stets rekursiv vorgenommen wird, insbesondere in der innersten Rekursionsstufe mit Hilfe von Indexregistern. Prinzipiell ist auch hier klar, was getan werden muß, sogar bei Zulassung allgemeinerer Speicherabbildungen:

$$a\,[i,k] := \langle\,\rangle\, a\,[0,0]\,\langle\, + P\,(i,k,p_j)\rangle$$

---

[7]) Die Prozedurvereinbarung liefert auch die Möglichkeit, in ALGOL dynamische (Bibliotheks-) Programme zu formulieren.

[8]) Beide Elemente sind in ALGOL nicht enthalten, dürften im übrigen nahezu ausreichen, um ALGOL in die oft diskutierte universal computer language UNCOL zu verwandeln.

wobei p_j weitere Parameter wie n, m darstellt und P eine beliebige Funktion ist.

Tritt nun in einer Schleife mit der Laufvariablen V die indizierte Variable $a\ [f_1\ (V),\ f_2\ (V)]$ auf, deren Indexpositionen mit Funktionen f_1 und f_2 der Variablen V besetzt sind, so hat man in der Abbildungsfunktion einzutragen

$$\begin{aligned} a\,[f_1(V), f_2(V)] &:= \langle\ \rangle\ a\,[0,0]\ \langle\ + P(f_1(V), f_2(V), p_i)\rangle \\ &= \langle\ \rangle\ a\,[0,0]\ \langle\ + Q(V, p_i)\rangle \end{aligned}$$

und die entstehende Funktion $Q\ (V,\ p_i)$ durch eine Rekursion hinsichtlich V auszudrücken, mit deren Hilfe die Abbildungsfunktion ohne Multiplikationen ausgewertet werden kann. Es muß demnach zur Adressenfortschaltung die sukzessive Bildung des vollständigen Differenzenschemas von $Q\ (V,\ p_i)$ veranlaßt werden.

In praxi bedeutet das eine ungeheure Komplikation, da der Übersetzer das Schema für die Bildung beliebiger Rekursionen in sich tragen muß. Da auch der allgemeine Fall äußerst selten vorkommt (ein nicht triviales Beispiel ist jedoch die Dreieckspeicherung dreieckiger Matrizen), ist bereits in ALGOL nur rechteckige Speicherung von Feldern, d. h. in den Indizes lineare Abbildungsfunktion, unterstellt. Ferner hat man bisher auch die auf Indexpositionen zulässigen Ausdrücke auf in dem Laufindex lineare Funktionen beschränkt. In ALGOL ist eine solche Beschränkung nicht vorgesehen, dementsprechend haben wir bei der geschlossenen Auswertung beliebige (auch indizierte) Indexausdrücke zugelassen. Bei der Adressenfortschaltung beschränken wir uns jedoch ebenfalls auf den Fall linearer Indexausdrücke. Für rekursive Auswertung der Speicherabbildungsfunktionen kommen also nur indizierte Variable der Form $a\ [i \times c_1 + E_1',\ i \times c_2 + E_2',\ \ldots,\ i \times c_k + E_k']$ in Frage, wobei i der Laufindex der betreffenden Schleife sei, die c_j seien Konstante und die E_j' Ausdrücke, die i nicht enthalten. Aus der oben angegebenen Speicherabbildungsfunktion (Fall $k = 2$) ergibt sich, wenn wir abkürzend $\rangle\, a\,[i \times c_1 + E_1', \ldots, i \times c_k + E_k']\,\langle$ durch A und $\rangle\, a\,[0,0]\,\langle$ durch A *null* ersetzen

(1) $\quad A := (E_i \times c_2 + E_2') \times n + E_i \times c_1 + E_1' + A\ \mathit{null}$

(2) $\quad A := (c_2 \times n + c_1) \times S + A$

als Rekursion für die Adressen der durch die Werte $i := E_i (S) E_e$ ausgewählten Komponenten $a\ [i \times c_1 + E_1',\ \ldots,\ i \times c_k + E_k']$ des Feldes $a\ [\,,]$. Der Wert für n ist dabei der zugehörigen Feld-Vereinbarung zu entnehmen, der Wert von A *null* ist vom Übersetzer aus der Speicherverteilung, die nach Abschluß der eigentlichen Übersetzung des Formelprogramms an Hand der Feld-Vereinbarungen in üblicher Compilertechnik auszuführen ist, zu berechnen und dem erzeugten Programm als Konstante einzuverleiben.

Das ursprüngliche Formelprogramm laute etwa

$$\text{'for'}\ i := E_i(S)E;$$
$$\ldots a\,[c_1 \times i + E_1', c_2 \times i + E_2'] \ldots;$$

Es ist vom Übersetzer zu behandeln wie: ($S>0$ vorangesetzt)

$$
\begin{aligned}
& i := E_i; \\
& A := (i \times c_2 + E_2') \times n + i \times c_1 + E_1' + A\ \mathit{null}; \\
& \mathit{delta}\ A := (c_2 \times n + c_1) \times S; \\
& \text{'go to'}\ L_p; \\
& L_B: \ldots \langle A \rangle \ldots; \\
& A := A + \mathit{delta}\ A; \\
& i := i + S; \\
& L_p: \text{'if'}\ i \leqq E;\ \text{'go to'}\ L_B
\end{aligned}
$$

Das Symbol $\langle A \rangle$ ist dabei zu interpretieren als:

man rechne mit dem Inhalt der durch die Zahl A als Adresse bezeichneten Zelle, d. h. als übliche indirekte Adresse, wenn man nicht diese Zahl A als Adresse in dem Rechenbefehl substituiert.

Stehen Indexregister für die Adressenfortschaltung zur Verfügung, so ist die Sequenz abzuändern. Wir können formal auch ein Indexregister mit einer Variablen IRK (Indexregister K) bezeichnen. Ist eine Variable mit IRK indiziert ($a\,[IRK]$), so bedeutet dies, daß der Übersetzer demjenige erzeugten Befehl, der die dieser Variablen entsprechende Adresse enthält, das Merkmal für Adressenmodifikation durch Addition des Indexregisters K anfügen muß.

Mit diesen Abkürzungen ist der oben angegebene Formelausschnitt vom Übersetzer zu behandeln wie

$$
\begin{aligned}
& i := E_i; \quad \mathit{delta}\ A := (c_2 \times n + c_1) \times S; \\
& A := E_2' \times n + E_1' + A\ \mathit{null};\ IRK := (c_2 \times n + c_1) \times i; \\
& \text{'go to'}\ L_p; \\
& L_B: \ldots \langle A \rangle\ [IRK] \ldots; \\
& IRK := IRK + \mathit{delta}\ A; \\
& i := i + S; \\
& L_p: \text{'if'}\ i \leqq E;\ \text{'go to'}\ L_B;
\end{aligned}
$$

Das allgemeine Schema zeigt bereits, daß bei der Adressenfortschaltung das Kellerungsprinzip durchbrochen wird. Denn erst das Erscheinen der indizierten Variablen im Inneren der Schleife zeigt dem Übersetzer, daß er in den bereits erzeugten Teil des Programms noch Befehlsserien einzuschieben hat. Er muß also zwei Programmteile, die Befehle der Schleife selbst und den für die Fortschaltungen notwendigen Vorbereitungsteil, gleichzeitig nebeneinander aufbauen und kann sie erst nach Abschluß der Schleife aneinanderfügen, wobei die Reihenfolge Geschmackssache ist, solange dem für das erzeugte Programm erforderlichen zeitlichen Ablauf Rechnung getragen wird.

Abgesehen davon liegt hier nun ein Fall vor, in dem die Symbolfolge nicht mehr sequentiell mit einfacher Kellerung abgearbeitet werden kann: Die auf den Indexpositionen stehenden Ausdrücke müssen in mehrere parallele Züge auseinandergefahren werden, da neben der vollständigen Auswertung zum Aufbau der Speicherabbildungsfunktion, die den Anfangswert von A bzw. A und IRK festlegt, noch zur Festlegung des Programms für die Berechnung von *delta* A vom Übersetzer die Koeffizienten der Laufvariablen i auf allen Indexpositionen zu sammeln und mit den richtigen Faktoren, die aus der Feld-Vereinbarung stammen, zu versehen sind.

Mit der Fortschaltung verknüpft sind eine Reihe von notwendigen Kontrollen und Vergleichen, die für den Übersetzer stets das Anlegen und Durchsehen von speziellen Listen bedeuten. Zunächst ist, als Voraussetzung für die Fortschaltung, vom Übersetzer die Linearität der Indexausdrücke festzustellen. Vor allem muß sichergestellt sein, daß nicht etwa die Koeffizienten eines in der Laufvariablen formal linearen Indexausdrucks Variable enthalten, die in der Schleife umgerechnet werden und damit von den Werten der Laufvariablen abhängen. Das bedeutet aber, daß alle in der betreffenden Schleife als Rechenergebnisse links von dem Symbol := in arithmetischen Anweisungen auftretenden Variablen vom Übersetzer notiert werden müssen, um gegebenenfalls mit einer in einem Indexausdruck auftretenden Variablen, die nicht als Laufvariable eines 'for'-Symbols definiert ist, verglichen werden zu können. Diese Kontrolle ist unumgänglich, da nichtlineare Indizes zugelassen sind, aber nicht fortgeschaltet werden können.

Zur richtigen Behandlung der Indexausdrücke muß der Übersetzer in jedem Augenblick die Laufvariable der gerade abgearbeiteten Schleife greifbar haben. Dies wird erreicht mit Hilfe eines neuen Kellers, des Schleifenkellers, in dem jedes Laufvariablensymbol beim Auftreten nach dem zugehörigen 'for' abgesetzt wird und aus dem es erst beim Schleifenende, das durch das Zusammentreffen des Endzeichens ; oder end mit dem transformierten 'for_1' angezeigt wird, wieder entfernt wird.

Um das erzeugte Programm so kurz und damit so effektiv wie möglich zu machen, muß der Übersetzer weiterhin eine Reihe von Identitätsprüfungen vornehmen. Zunächst ist, beim Auftreten mehrerer indizierter Variablen in einer Schleife, die mögliche Identität der zugehörigen Fortschaltgrößen *delta A* festzustellen. Zwar führen solche Identitäten, solange ohne Indexregister gearbeitet wird, nur zur Verkürzung des Vorbereitungsteils der Schleife und zur Einsparung von Hilfsspeicherzellen. Beim Einsatz von Indexregistern aber gewinnt man mehr. Denn da mit den *delta A* auch die Anfangseinstellungen der *IRK* identisch sind, kann der Übersetzer alle identischen Fortschaltungen mit einem Indexregister ausführen lassen, und man spart sowohl Fortschaltungsrechnungen in der Schleife als auch Indexregister ein. In den meisten vorkommenden Fällen erscheint die Laufvariable i einer Schleife nur in Indexausdrücken. Außerdem (und eben aus diesem Grunde) verfügen die meisten Maschinen mit Indexregistern über einen speziellen bedingten Sprung, der vom Inhalt eines Indexregisters und evtl. einer anderen Speicherzelle abhängig ist. Um dies auszunützen, muß der Übersetzer kontrollieren, ob die Laufvariablen außerhalb von Indexausdrücken vorkommt. Ist dies nicht der Fall, dann kann die laufende Berechnung der Laufvariablen ganz entfallen und durch die Indexregisterfortschaltung ersetzt werden. Entsprechend ist die Schlußbedingung auf das Indexregister umzustellen, weshalb im Vorbereitungsteil der Endwert E von i mit dem Koeffizienten von S in *delta A* zu multiplizieren ist.

Unser obiges Beispiel hat der Übersetzer dann zu behandeln wie:

$$\begin{array}{ll}
\textit{delta } A & := c_2 \times n + c_1; \\
IRK & := (E_i) \times \textit{delta } A; \\
E & := E \times \textit{delta } A; \\
\textit{delta } A & := S \times \textit{delta } A; \\
A & := E_2' \times n + E_1' + A \textit{ null}; \\
\multicolumn{2}{l}{\text{'go to' } L_p;} \\
\multicolumn{2}{l}{L_B: \ldots \langle A \rangle\, [IRK] \ldots \langle A' \rangle\, [IRK] \ldots;} \\
IRK & := IRK + \textit{delta } A; \\
\multicolumn{2}{l}{L_p: \text{'if' } IRK \leqq E; \text{'go to' } L_B;}
\end{array}$$

Die Möglichkeiten zur Vereinfachung von Schleifen sind damit natürlich noch nicht erschöpft. Jedoch ist auf das Wesentliche hingewiesen und die Diskussion soll damit abgeschlossen werden.

Wie bereits erwähnt, ist die Adressenfortschaltung von besonderer Bedeutung in den innersten Schleifen eines Programms, und an diesen Stellen sollten die Indexregister in erster Linie zur Fortschaltung eingesetzt werden. Die Tatsache, daß eine Schleife innerste Schleife ist, ist jedoch erst am Schleifenende festzustellen. Der Übersetzer muß daher die endgültige Absetzung der entsprechenden Befehle bis zum Erscheinen des die Schleife abschließenden Trennzeichens verschieben. Weiter ist im Schleifenkeller eine Anmerkung „nicht innerste Schleife" bei jedem gekellerten Laufindex, für den dies zutrifft, notwendig.

Schlußbemerkung

Die vorangegangene Darstellung zeigt, daß die Umsetzung des Formelprogramms in Maschinenoperationen durch den Übersetzer mit Ausnahme der Behandlung der Adressenfortschaltung zügig ohne Abspeicherung des Formelprogramms, also als reiner Eingabeprozeß, durchgeführt werden kann. Denn die Fernzusammenhänge im Programm beschränken sich auf Adressen, die aus während des Einleseprozesses anzulegenden Listen entnommen werden können. Dementsprechend ist es auch möglich, die Umsetzungsmethode zur sofortigen interpretativen Ausführung der im Formelprogramm angegebenen Operationen, auch mit Hilfe verdrahteter Schaltungen, anzuwenden. Eine Adressenfortschaltung macht allerdings hierbei außerordentliche Schwierigkeiten.

Literatur

[1] *H. Goldstine* and *J. von Neumann:* Planning and Coding for an Electronic Computing Instrument. Institute for Advanced Study, Princeton, N. J., 1947/48.

[2] International Business Machines Corp., FORTRAN Manual.

[3] *W. S. Melahn:* A description of a cooperative venture in the production of an automatic coding system, Journ. Assoc. Comp. Mach. 3 (1956), S. 266—271.

[4] *H. Rutishauser:* Über automatische Rechenplanfertigung bei programmgesteuerten Rechenanlagen, Z. Angew. Math. Mech. 31 (1951), 255.

[5] *H. Ruthisauser:* Automatische Rechenplanfertigung bei programmgesteuerten Rechenmaschinen, Mitt. Inst. f. Angew. Math. der ETH Zürich, Nr. 3, (1952).

[6] *K. Samelson:* Probleme der Programmierungstechnik. Intern. Kolloquium über Probleme der Rechentechnik, Dresden 1955, S. 61—68.

[7] *P. B. Sheridan:* The arithmetic translator-compiler of the IBM Fortran automatic coding system, Com. ACM Bd. 2, (1959) 2, S. 9—21.

[8] *M. V. Wilkes; D. J. Wheeler; S. Gill:* The preparation of programmes for an electronic digital computer (Cambridge, Mass., 1951).

[9] *M. V. Wilkes:* The Use of a Floating Address System for Orders in an Automatic Digital Camputer. Proc. Cambridge Philos. Soc. 49, (1953) 84—89.

[10] *Charles W. Adams* and *J. H. Laning jr.:* The M. I. T. System of Automatic Coding; Comprehensive, Summer Session and Algebraic, in: Symposium on Automatic Programming for Digital Computers, US Department of Commerce.

[11] ACM Committee on Programming Languages and GAMM Committee on Programming: Report on the Algorithmic Language ALGOL, edited by *A. J. Perlis* and *K. Samelson.* Num. Math. 1 (1959), S. 41—60.

[12] *H. Zemanek:* Die algorithmische Formelsprache ALGOL. Elektron. Rechenanl. 1 (1959), S. 72—79 und S. 140—143.

[13] *F. L. Bauer:* The Formula Controlled Logical Computer Stanislaus, erscheint in Math. Tabl. Aids Comp.

[14] *C. Böhm:* Calculatrices digitales. Du déchiffrage de formules logico-mathématiques par la machine même dans la conception du programme (Dissertation, Zürich 1952) Annali dei Matematica pura ed applicata. Ser. 4, 37 (1954), S. 5—47.

Friedrich L. Bauer

Ausschnitt aus der Patentschrift: **Verfahren zur automatischen Verarbeitung von kodierten Daten und Rechenmaschinen zur Ausübung des Verfahrens**

Auslegeschrift 1 094 019
Anmeldetag: 30. März 1957

BUNDESREPUBLIK DEUTSCHLAND

DEUTSCHES PATENTAMT

KL. 42 m 14

INTERNAT. KL. G 06 f

AUSLEGESCHRIFT 1 094 019

B 44122 IX/42m

ANMELDETAG: 30. MÄRZ 1957

BEKANNTMACHUNG DER ANMELDUNG UND AUSGABE DER AUSLEGESCHRIFT: 1. DEZEMBER 1960

1

Die Erfindung betrifft ein Betriebsverfahren für automatische mechanische, elektrische oder elektronische Rechenmaschinen und bezieht sich insbesondere auch auf den technischen und logischen Aufbau der Rechenmaschine sowie der damit in Verbindung stehenden Eingabe- und Ausgabevorrichtungen.

Die bekannten Rechenautomaten und Datenverarbeitungsanlagen erfordern im Einzelfall Anweisungen über die Art und den Ablauf der numerischen oder sonstigen informationsverarbeitenden Prozesse. Die Schreibweise, in der diese Anweisungen fixiert werden, wurde zu Beginn der Entwicklung so gewählt, daß sie gewisse als elementar erachtete technische Funktionen der Anlage beschrieb. Die so geschriebenen Anweisungen werden üblicherweise »Programm« genannt. Das Programm für einen Rechenprozeß etwa und die mathematische Formel, mit der der Mathematiker diesen Prozeß gewöhnlich beschreibt, kennzeichnen jeweils genau denselben Vorgang, allerdings in zwei grundverschiedenen Sprachen.

Die Übersetzung von der mathematischen Formelsprache ins Programm wird üblicherweise Programmierung genannt; sie hat sich in praxi als eine zeitraubende und fehleranfällige, im allgemeinen nur lästige Angelegenheit herausgestellt. Für den Mathematiker stellt die Programmierungssprache eine ungewohnte Formulierung dar, die überdies noch von Anlagentyp zu Anlagentyp wechselt. Diese bei den meisten bestehenden Maschinen jeweils verschiedene Art der Programmschreibweise zeigt bereits, wie sehr das Befehlssystem üblicher Maschinen noch von der verwendeten Technik abhängt und wie wenig die auf der ganzen Welt einheitliche mathematische Formelsprache von den Rechenautomatenbauern bisher ernst genommen wurde.

Die Mängel der üblichen Programmierung sind in der Literatur bereits vor einigen Jahren klar erkannt worden. Man ist jedoch den zunächst naheliegenden Weg gegangen, vorhandene Rechenautomaten universeller Art zu gewissen Routinearbeiten der Programmierung, die selbst Datenverarbeitungsaufgaben darstellen, heranzuziehen. Es gibt heute bereits Programme, die unter gewissen Einschränkungen die ganze Übersetzungsarbeit von einer mathematischen Formel bis zum Programm für einen üblichen Rechenautomaten erledigen.

Die Übersetzungsprogramme sind sehr kompliziert aufgebaut und dementsprechend umfangreich. Kleinere Rechenanlagen sind nicht mehr in der Lage, solche Aufgaben durchzuführen. Umfangreiche Formeln zu übersetzen, führt auch bei mittelgroßen Anlagen zu übermäßig hohem Zeitbedarf.

Demgegenüber ist es von Bedeutung, daß durch geschickte Organisation des Zusammenwirkens geeigneter Einzelkomponenten, gestützt auf grundsätzliche Studien über das Wesen von Rechnungsabläufen, unter Verwendung neuartiger Maschinenfunktionen und -steuerungsabläufe sowie Anlagenteile ein Rechenautomat gebaut werden kann, der unmittelbar durch mathematische Formeln in üblicher Schreibweise gesteuert wird, also ein formelgesteuerter Rechenautomat, der in seinem technischen Aufbau und in seiner praktischen Verwendungsmöglichkeit gegenüber den programmgesteuerten Rechenanlagen bisheriger Art einen wesentlichen Fortschrit darstellt.

Verfahren zur automatischen Verarbeitung von kodierten Daten und Rechenmaschine zur Ausübung des Verfahrens

Anmelder:

Dr. Friedrich Ludwig Bauer, München, Pörtschacherstr. 40, und Dr. Klaus Samelson, München, Hiltenspergerstr. 19

Dr. Friedrich Ludwig Bauer und Dr. Klaus Samelson, München, sind als Erfinder genannt worden

2

Eine solche Rechenmaschine muß außer den bekannten, mehr oder weniger üblichen Teilen eine Vorrichtung besitzen, die diese mathematischen Formeln in üblicher Schreibweise analysiert und eine entsprechende Folge von Steuerbefehlen löst. Dabei ergeben sich im einzelnen auch neuartige Lösungen für die Erledigung gewisser Rechenabläufe in Anpassung an diese besondere Art der Verarbeitung der mathematischen Formeln.

Die Erfindung beruht im wesentlichen auf dem Gedanken, den Komponenten einer Rechenmaschine einen Analysator beizuordnen, dem die mathematischen Formeln in üblicher Schreibweise zugeführt werden. Gemäß der Erfindung werden die den einzelnen Zeichen entsprechenden Signale in der Reihenfolge der Aufschreibung dem Analysator zugeführt und in diesem entsprechend der Reihenfolge des Eingangs geprüft, ob die Operationen sofort ausführbar sind oder ob der Eingang weiterer Signale abgewartet werden muß; in diesem letzteren Falle werden die noch nicht verarbeitbaren Zeichen in einen Speicher (Keller) eingeführt; beim Eintreffen neuer Zeichen im Analysator, die die

1 094 019

3

Ausführung einer Operation mit gespeicherten Zeichen ermöglichen, werden diese gespeicherten Zeichen in der durch die Art der Einführung festgelegten, umgekehrten Reihenfolge entnommen und verarbeitet.

Der Analysator im engeren Sinne enthält einen Formelentschlüsseler, der als Bestandteil einen Vorentschlüsseler, einen Formelumsetzer, einen Ziffernumsetzer und ein Ausgabesteuerwerk aufweist, sowie eine Hilfssteuereinrichtung zur Verarbeitung von Indizes; der Analysator im weiteren Sinne umfaßt auch einen Größenvorspeicher und einen Kennzeichenentschlüsseler.

Bei der Ausführung des Verfahrens kann auch eine weitergehende Zürückstellung der zugeführten Signale erfolgen; sie ist aber nicht notwendig und kann nur durch andere, außerhalb der Erfindung liegende Vorteile gerechtfertigt werden.

Gemäß einer weiteren Ausführungsform der Erfindung werden diejenigen Formelzeichen, welche Ziffernsymbole, also Zahlen, darstellen, von solchen Formelzeichen, welche Operationssymbole darstellen, getrennt und, sofern sie zurückgestellt werden müssen, speicherfähigen Vorrichtungen, vorzugsweise zwei verschiedenen »Kellern«, nämlich dem Zahlkeller und dem Operationskeller, zugeführt und von diesen Vorrichtungen her dem Steuerwerk zugänglich gemacht.

Dabei ist es zweckmäßig, die in dem Zahlkeller bzw. dem Operationskeller neu eintreffenden Zeichen jeweils an die Spitze der entsprechenden Sequenz zu setzen und die Entnahme eines Zeichens automatisch durch Wegnahme von der Spitze der entsprechenden Sequenz vorzunehmen.

Ein Ausführungsbeispiel für das Verfahren in seiner einfachsten Form wird als Stufe I im folgenden näher beschrieben. Es ist zur Verarbeitung einfachster Formelausdrücke geeignet. Die Formelsymbole der Arithmetik $+ - \sqrt{\ } \times : ( )$ werden dabei vorzugsweise in Form von Kodezeichen zur Auslösung von Steuerungsabläufen, d. h. von Maschinenfunktionen, benutzt. Der Zeitpunkt der endgültigen Auslösung des Steuerungsablaufes durch ein Kodezeichen hängt unter Umständen, z. B. bei der Klammer, davon ab, daß ein oder mehrere nachfolgende Kodezeichen eintreffen. Aus diesem Grund werden die Kodezeichen zunächst in dem Operationskeller zurückgestellt und erst dann, wenn der Ausführungszeitpunkt eintritt, dem Operationskeller wieder entnommen, wobei die Darbietung bei der obenerwähnten sequentiellen Aufreihung automatisch die richtige ist. In ähnlicher Weise wird durch den Zahlkeller dafür gesorgt, daß die vorzunehmenden Rechenoperationen automatisch mit den jeweils dafür in Frage kommenden Zahlen vorgenommen werden, sobald alle für die Ausführung der Operation erforderlichen Zahlenwerte vorhanden sind. Die Zahlen, die in einer Rechnung Verwendung finden sollen, können durch Ziffernsymbole und entsprechende Kodezeichen, z. B. im Dezimalsystem, dargestellt werden.

Solche Formelzeichen, welche ein Resultat verlangen, insbesondere das Gleichheitszeichen, werden einer besonderen Vorrichtung, nämlich der Ausgabesteuerung, zugeführt.

Im folgenden wird auf ein weiteres Ausführungsbeispiel des Verfahrens eingegangen, die als Stufe II bezeichnet wird.

In vielen Fällen ist es erwünscht, solche Teilergebnisse, die sich wiederholen, nur einmal zu berechnen und sie in der mathematischen Schreibweise durch besondere Symbole, z. B. Buchstaben, zu bezeichnen. Gemäß der weiteren Erfindung werden zur formelartigen Benennung von Zahlen oder Zahlsätzen, z. B. Ausgangsdaten und Teilergebnissen, besondere Zeichen als algebraische Größen-

4

symbole und zugehörige Kodezeichen benutzt, derart, daß jedes Zeichen beim erstmaligen Einlaufen in das Steuerwerk eine Reservierung von an sich beliebigen Plätzen für eine Zahl oder einen Zahlsatz in einem an sich bekannten Speicher veranlaßt, wobei eine Zuordnung zwischen diesem Platz bzw. diesen Plätzen und dem betreffenden Größensymbol bis auf Widerruf festgehalten wird. Das Größensymbol wird in Formeln vom Steuerwerk stellvertretend für die auf dem zugeordneten Platz bzw. Plätzen gespeicherte Zahl bzw. Zahlsatz behandelt.

Die Belegung eines mit einem Größensymbol bezeichneten Platzes im Zahlspeicher durch eine Zahl erfolgt durch das Ergibtsymbol ⇒ und Angabe der Größe. Es kann vorteilhaft sein, die obenerwähnte Platzreservierung erst zusammen mit der eben besprochenen Platzbelegung durchzuführen. Eine Zurückstellung des Rechenvorganges erfolgt, wenn innerhalb einer Formel ein Größensymbol ins Steuerwerk gelangt, für das noch keine Platzbelegung im Zahlspeicher vorgenommen war, wobei von außen neue Information so lange verlangt wird, bis der zu reservierende Platz nunmehr durch eine Zahl besetzt worden ist.

Durch die Einführung der Buchstaben als Zeichen ist bereits hier die Möglichkeit gegeben, die Ausgangsgrößen einer Rechnung von vornherein mit Buchstaben zu bezeichnen, so daß die mathematischen Formeln ganz oder teilweise mit algebraischen Zeichen geschrieben werden können.

Eine weitere Ausgestaltung der Rechenanlage ist durch Maßnahmen gegeben, die im folgenden beschrieben und als Stufe III bezeichnet werden. Während die bisher beschriebene Ausführung bereits eine Rechenanlage mit direkter Formelsteuerung und Niederschrift des gesamten Ablaufs, d. h. der Eingabe und Resultate, in mathematischer Schreibweise ermöglicht, ist es häufig erwünscht, die Möglichkeit der Wiederholung von Formeln ausnutzen zu können. Zu diesem Zweck wird die gesamte einlaufende, nach wie vor der direkten Formelsteuerung dienende Information gleichzeitig nebenher in einem Formelspeicher gespeichert. Zur Erschließung weiterer Möglichkeiten werden zur Numerierung von Formelgruppen besondere Zeichen als Kennzeichnungssymbole benutzt, derart, daß jedes Kennzeichnungssymbol beim erstmaligen Einlaufen in das Steuerwerk bewirkt, daß die Zuordnung zwischen dem Platz, den der Anfang der Formelgruppe in einem Formelspeicher einnimmt, und dem Kennzeichnungssymbol bis auf Widerruf festgehalten wird.

Es ist insbesondere möglich, die Anfänge von Formelgruppen zu kennzeichnen, wobei das Kennzeichnungssymbol etwa bestehen kann aus Ziffern mit Beifügung eines speziellen Zeichens, für das hinfort * benutzt wird. Eine laufende Durchnumerierung soll nicht erforderlich sein. Dabei ist es lediglich notwendig, vor dem Formelspeicher einen Vorspeicher anzuordnen, worin unter der Nummer jedes Kennzeichnungssymbols als Eingang derjenige Platz des dahinterliegenden Hauptspeichers, der mit dem betreffenden Kennzeichnungssymbol gekennzeichnet ist, festgehalten ist.

Das Verfahren kann weiterhin so ausgebildet werden, daß bereits die Angabe eines Kennzeichnungssymbols in Verbindung mit einem speziellen Zeichen, z. B. einem →, als Sprungsymbol genügt, um zu bewirken, daß die Rechnung wiederholt wird, allgemeiner, daß sie mit dem Beginn der unter dem betreffenden Kennzeichen im Formelspeicher notierten Formelgruppe fortgesetzt wird, wobei der Übergang in bekannter Weise von Bedingungen abhängig sein kann. Eine Zurückstellung des Rechenvorganges erfolgt, wenn ein Sprungsymbol auf eine noch nicht im Formelspeicher notierte Formelgruppe führt, wobei ebenfalls von außen neue Information verlangt

1 094 019

5

wird. In diesem wie auch in dem obenerwähnten Fall, daß ein Größensymbol in einer Formel erscheint, das noch keine Belegung im Speicher hat, wird die verlangte Information im Formelspeicher lediglich notiert unter Festhaltung der durch Kennzeichnungssymbole bezeichneten Plätze der Anfänge einzelner Formelgruppen im Formelspeicher. Dieser Vorgang bricht automatisch ab, wenn das Kennzeichnungssymbol der aufgerufenen Formelgruppe ausgewertet wird bzw. wenn die aufgerufene Größe mit einer Zahl belegt worden ist, wobei der Rechenvorgang an der Unterbrechungsstelle wieder einsetzt, insbesondere im letzterwähnten Fall der Sprung ausgeführt wird. Bei Ausnutzung dieses Verfahrens erfolgt eine Zurückstellung des Rechenvorganges, auch wenn das zeitlich zuletzt im Formelspeicher notierte Formelzeichen abgearbeitet ist, ohne daß es einen Sprung auf ein schon vorhandenes Kennzeichnungssymbol bewirkt. Die Meldung, daß der Formelspeicher abgearbeitet ist, bewirkt dann, daß das Steuerwerk von außen neue Information verlangt, die im Formelspeicher notiert und gleichzeitig ausgeführt wird.

An Stelle des bisher verfolgten Prinzips der baldmöglichen Ausführung aller Verarbeitungsvorgänge von Formelsymbolen kann auch wahlweise eine weitere ganze oder teilweise Zurückstellung bis zu einem geeigneten späteren Zeitpunkt vorgenommen werden.

Ein besonderes Zeichen kann als Symbol »Nicht notieren« interpretiert werden, so daß anschließend die Notierung der von außen einlaufenden Information bis auf Widerruf, z. B. durch ein auflösendes Symbol oder das nächste einlaufende Kennzeichnungssymbol für Formelgruppen, unterdrückt wird.

Die Maschine zur Ausführung des Verfahrens enthält in ihrer einfachsten Ausführungsform einen Vorentschlüßler, dem sämtliche Formelzeichen in der Reihenfolge der üblichen Schreibweise zugeführt werden und der mehrere Ausgänge aufweist, die zu einem Ziffernumsetzer, zu einem Operationsumsetzer und zu einem Ausgabesteuerwerk sowie zu einem Steuerwerk für »bedeutungslose Zeichen« führen. Die Umsetzer der Steuerwerke können mit dem Schreibwerk in Verbindung stehen.

Der Zahlkeller und der Ziffernumsetzer sind derart verbunden, daß der Zahlkeller die Zahlen in der Reihenfolge des Eintreffens von dem Umsetzer abnehmen kann und daß ferner die jeweils erste in der Sequenz stehende Zahl beim Eintreffen eines entsprechenden Befehles über das Ausgabesteuerwerk als Ergebnis dem Schreibwerk zugeführt wird.

Der Operationsumsetzer steht mit dem Operationskeller in Verbindung, so daß er in diesen die Operationssymbole in der Reihenfolge ihres Eintreffens einspeisen kann, wobei das jeweils zuletzt eingespeiste der von unten nachrückenden, früher eingespeisten Symbole bzw. das neu ankommende Symbol an das Rechenwerk abgegeben werden kann, um solche Operationen auszuführen, für die die zugehörigen Operanden an der Spitze der im Zahlkeller befindlichen Sequenz vorliegen.

Es ist zweckmäßig, daß solche Zeichen, die eine Formel abschließen, insbesondere das Gleichheitszeichen oder das Ergibtzeichen, eine Prüfung auf »sinnvolle Formel« veranlassen.

Der Operationskeller und der Zahlkeller können Einrichtungen aufweisen, die die zugeführten Zeichen dadurch sequentiell speichern, daß sie die bereitsgespeicherten Zeichen in der Reihenfolge des Eintreffens nach unten weiterschieben und eine Abgabe nur des jeweils zuletzt gespeicherten oder des obersten der von unten nachzuschiebenden Zeichen gestatten.

Bei einer anderen Ausführungsform weist der Operationskeller und/oder der Zahlkeller Einrichtungen auf,

6

die die zugeführten Zeichen dadurch sequentiell speichern, daß jedes eintreffende Zeichen auf den Platz vor dem zuletzt eingetroffenen gesetzt wird, daß dieser Platz festgehalten wird und daß ferner die Abnahme von dem zuletzt festgehaltenen Platz erfolgt.

Das Rechenwerk verarbeitet die im obersten oder in den beiden obersten Geschossen des Zahlkellers befindlichen Zahlen entsprechend den von der Operationssteuerung erhaltenen Anweisungen und gibt das Ergebnis wieder an das oberste Geschoß des Zahlkellers ab.

Das Ausgabesteuerwerk ist vorzugsweise mit dem Zahlkeller derart verbunden, daß beim Eintreffen eines Gleichheitszeichens und gegebenenfalls nachfolgender Zeichen »Ziffer verlangt« die Verbindung des Zahlkellers mit dem Schreibwerk hergestellt und die im obersten Geschoß des Zahlkellers befindliche Zahl ganz oder teilweise an das Schreibwerk abgegeben wird.

Mit der angegebenen Maschine kann auch mit Indexgrößen gerechnet werden. Die Rechnung mit Indexgrößen erfolgt dabei ganz analog wie das Rechnen mit den übrigen Rechengrößen. Zur Bezeichnung von indizierten Größen können besondere Zeichen als Indexsymbole verwendet werden, die Beginn und Ende der Indizes und die Abtrennung der einzelnen Indexstellen angeben, wobei diese Zeichen besonderen Vorrichtungen zugeführt werden, die intermediär eine Unterbrechung der laufenden Rechnung, die Auswertung der auf den Indexstellen befindlichen Ausdrücke nach dem obengenannten Verfahren und die Ansteuerung der durch die Indexauswertung festgestellten Einzelkomponente der induzierten Größe bewirken. Die Durchführung von derartigen Rechnungen wird weiter unten beispielsweise näher beschrieben, wobei diese Verfahren als Stufe IV bezeichnet sind.

Weitere Merkmale und Vorteile des Erfindungsgegenstandes gehen aus der folgenden Beschreibung von Ausführungsbeispielen hervor, die an Hand der Zeichnungen beschrieben werden.

1 094 019

20

Patentansprüche:

1. Verfahren zur automatischen Verarbeitung von kodierten Daten, z. B. arithmetischen Formeln in üblicher Schreibweise, die als kodierte Zeichen Klammern, Operationssymbole, Zahlen und Variable gemischt, Dezimalkommas, Indizes, Entscheidungssymbole sowie Formelnummern enthalten, in einer datenverarbeitenden Maschine mit einer Eingabe- und einer Ausgabevorrichtung, **dadurch gekennzeichnet,** daß die den einzelnen Zeichen entsprechenden Signale in der Reihenfolge der Aufschreibung einem Analysator (4, 28) zugeführt und in diesem entsprechend der Reihenfolge des Eingangs geprüft werden, ob die Operationen sofort ausführbar sind oder ob der Eingang weiterer Signale abgewartet werden muß, daß in diesem letzteren Fall die noch nicht verarbeitbaren Zeichen in einen Speicher (Keller) eingeführt werden und daß beim Eintreffen neuer Zeichen im Analysator (4, 28), die die Ausführung einer Operation mit gespeicherten Zeichen ermöglichen, diese gespeicherten Zeichen in der durch die Art der Einführung festgelegten umgekehrten Reihenfolge entnommen und verarbeitet werden.

2. Verfahren nach Anspruch 1, dadurch gekennzeichnet, daß im Analysator die Formelzeichen nach Ziffernsymbolen (Zahlen) und Operationssymbolen getrennt sind und, sofern sie zurückgestellt werden müssen, zwei verschiedenen speicherfähigen Vorrichtungen (11, 12), vorzugsweise zwei verschiedenen »Kellern«, nämlich dem Zahlkeller (11) und dem Operationskeller (12), zugeführt werden.

3. Verfahren nach Ansprüchen 1 und 2, dadurch gekennzeichnet, daß die in dem Zahlkeller (11) bzw. dem Operationskeller (12) neu eintreffenden Zeichen sich jeweils an die Spitze der entsprechenden Sequenz setzen und die Entnahme eines Zeichens automatisch durch Wegnahme von der Spitze der entsprechenden Sequenz erfolgt.

4. Verfahren nach Ansprüchen 1 und 2, dadurch gekennzeichnet, daß solche Formelzeichen, welche ein Resultat verlangen, insbesondere das Gleichheitszeichen, einer besonderen Vorrichtung, nämlich der Ausgabesteuerung (6), zugeführt werden, daß ferner dadurch automatisch das Endergebnis der Formelauswertung zur Ausgabe gebracht wird.

5. Verfahren nach Ansprüchen 1 bis 4, dadurch gekennzeichnet, daß zur formelartigen Benennung von Zahlen oder Zahlsätzen, z. B. Ausgangsdaten und Teilergebnissen, besondere Zeichen, z. B. Buchstaben, als algebraische Größensymbole und zugehörige Kodezeichen benutzt werden, derart, daß jedes Zeichen beim erstmaligen Einlaufen in den Analysator (4, 28) eine Reservierung von an sich beliebigen Plätzen für eine Zahl oder einen Zahlsatz in einem an sich bekannten Speicher veranlaßt, wobei eine Zuordnung zwischen diesem Platz bzw. diesen Plätzen und dem betreffenden Größensymbol im Größenvorspeicher (24) bis auf Widerruf festgehalten wird.

6. Verfahren nach Anspruch 5, dadurch gekennzeichnet, daß das Größensymbol in Formeln vom Analysator (4, 28) stellvertretend für die auf dem zugeordneten Platz bzw. Plätzen gespeicherte Zahl bzw. Zahlsätze behandelt wird.

7. Verfahren nach Ansprüchen 1 bis 6, dadurch gekennzeichnet, daß zur Bezeichnung von indizierten Größen besondere Zeichen als Indexsymbole verwendet werden, die Beginn und Ende der Indizes und die Abtrennung der einzelnen Indexstellen angeben, wobei diese Zeichen im Analysator einer Hilfssteuereinrichtung (28) zugeführt werden, die intermediär eine Unterbrechung der laufenden Rechnung, die Auswertung der auf den Indexstellen befindlichen Ausdrücke nach dem obengenannten Verfahren und die Ansteuerung der durch die Indexauswertung festgestellten Einzelkomponente der indizierten Größe bewirken.

8. Verfahren nach Ansprüchen 1, 5 und 6, dadurch gekennzeichnet, daß eine Zurückstellung des Rechenvorganges erfolgt, wenn innerhalb einer Formel ein Größensymbol in den Analysator (4, 28) gelangt, für das noch keine Platzreservierung im Zahlspeicher (22) vorgenommen war, wobei von außen neue Information so lange verlangt wird, bis der nunmehr zu reservierende Platz durch eine Zahl besetzt worden ist.

9. Verfahren nach Ansprüchen 1 bis 4, dadurch gekennzeichnet, daß zur Numerierung von Formelgruppen besondere Zeichen als Kennzeichnungssymbole und zugehörige Kodezeichen benutzt werden.

derart, daß jedes Kennzeichensymbol beim erstmaligen Einlaufen in den Analysator bewirkt, daß die Zuordnung zwischen dem Platz, den der Anfang der Formelgruppe in einem Formelspeicher (23) einnimmt, und dem Kennzeichnungssymbol bis auf Widerruf im Kennzeichenentschlüßler (25) festgehalten wird.

10. Verfahren nach Anspruch 9, dadurch gekennzeichnet, daß bereits die Angabe eines Kennzeichnungssymbols in Verbindung mit einem speziellen Sprungsymbol genügt, um zu bewirken, daß die Rechnung mit dem Beginn der unter dem betreffenden Kennzeichen im Formelspeicher (23) notierten Formelgruppe fortgesetzt wird, wobei der Sprung in bekannter Weise von Bedingungen abhängig sein kann.

11. Verfahren nach Ansprüchen 1, 9 und 10, dadurch gekennzeichnet, daß eine Zurückstellung des Rechenvorgangs erfolgt, wenn das zeitlich zuletzt im Formelspeicher (23) notierte Formelzeichen abgearbeitet ist, ohne daß es einen Sprung auf ein schon vorhandenes Kennzeichnungssymbol bewirkt, wobei von außen neue Information verlangt wird.

12. Verfahren nach Ansprüchen 1 und 9 bis 11, dadurch gekennzeichnet, daß ein Sprungsymbol, das auf eine noch nicht im Formelspeicher (23) notierte Formelgruppe führt, bzw. die Meldung, daß der Formelspeicher abgearbeitet ist bzw. daß ein Größensymbol in einer Formel erscheint, das noch keine Belegung im Zahlspeicher hat, bewirkt, daß der Analysator den Rechenvorgang abbricht und von außen neue Information verlangt; diese Information wird mindestens so lange im Formelspeicher oder, soweit es sich um eine Voreinstellungspeicherung handelt, im Zahlspeicher lediglich notiert, bis alle zur Fortsetzung der Rechnung notwendigen Angaben zur Verfügung stehen, worauf der Rechenvorgang automatisch an der Unterbrechungsstelle wieder einsetzt.

13. Verfahren nach Ansprüchen 1, 9 und 10, dadurch gekennzeichnet, daß eine Zurückstellung des Rechenvorganges erfolgt, wenn ein Sprungsymbol auf eine noch nicht im Formelspeicher (23) notierte Formelgruppe führt, wobei von außen eine neue Information verlangt wird, die im Formelspeicher lediglich notiert wird unter Festhaltung der durch Kennzeichnungssymbole bezeichneten Plätze der Anfänge einzelner Formelgruppen im Formelspeicher, und daß dieser Vorgang automatisch abbricht, wenn das Kennzeichnungssymbol der aufgerufenen Formelgruppe ausgewertet wird, wobei der Rechenvorgang wieder einsetzt.

14. Verfahren nach Ansprüchen 1 bis 13, dadurch gekennzeichnet, daß wahlweise an Stelle des bisher befolgten Prinzips der baldmöglichsten Ausführung aller Verarbeitungsvorgänge von Formelsymbolen eine weitere ganze oder teilweise Zürückstellung bis zu einem geeigneten späteren Zeitpunkt vorgenommen wird.

15. Verfahren nach Ansprüchen 1 und 9 bis 12, dadurch gekennzeichnet, daß ein besonderes Zeichen als Symbol »nicht notieren« interpretiert wird, derart, daß anschließend die Notierung der von außen einlaufenden Information bis auf Widerruf, z. B. durch ein auslösendes Symbol oder das nächste, in den Analysator (4) einlaufende Kennzeichnungssymbol für Formelgruppen, unterdrückt wird.

16. Automatische Rechenmaschine zur Ausführung des Verfahrens nach Ansprüchen 1 bis 15, dadurch gekennzeichnet, daß der Analysator einen Vorentschlüßler (5) enthält, dem sämtliche Formelzeichen in der Reihenfolge der üblichen Schreibweise zugeführt

werden, und der mehrere Ausgänge aufweist, die zu einem Ziffernumsetzer (7), zu einem Operationsumsetzer (8) und zu einem Ausgabesteuerwerk (6) sowie zu einem Steuerwerk (9) für »bedeutungslose Zeichen« führen.

17. Rechenmaschine nach Anspruch 16, dadurch gekennzeichnet, daß die Umsetzer der Steuerwerke mit dem Schreibwerk (2) in Verbindung stehen.

18. Rechenmaschine nach Ansprüchen 16 und 17, dadurch gekennzeichnet, daß der Zahlkeller (11) und der Ziffernumsetzer (7) derart verbunden sind, daß der Zahlkeller die Zahlen in der Reihenfolge des Eintreffens von dem Umsetzer abnehmen kann und daß ferner die jeweils erste in der Sequenz stehende Zahl beim Eintreffen eines entsprechenden Befehls über das Ausgangssteuerwerk als Ergebnis dem Schreibwerk (2) zugeführt wird.

19. Rechenmaschine nach Ansprüchen 16 bis 18, dadurch gekennzeichnet, daß der Operationsumsetzer (8) mit dem Operationskeller (12) in Verbindung steht, so daß er in diesen die Operationssymbole in der Reihenfolge ihres Eintreffens einspeisen kann, wobei nach der den Ablauf der direkten Formelauswertung wiedergebenden Vorschrift entweder das jeweils zuletzt eingespeiste unter gleichzeitigem Nachrücken der früher eingespeisten Symbole von unten her oder das neu ankommende Symbol an das Rechenwerk (10) abgegeben werden kann, um solche Operationen auszuführen, für die die zugehörigen Operanden an der Spitze der im Zahlkeller befindlichen Sequenz vorliegen.

20. Rechenmaschine nach Ansprüchen 16 bis 19, dadurch gekennzeichnet, daß Einrichtungen vorgesehen sind, die beim Eintreffen solcher Zeichen, die eine Formel abschließen, insbesondere des Gleichheitszeichens oder des Ergibtzeichens, eine Prüfung auf »sinnvolle Formel« veranlassen.

21. Rechenmaschine nach Ansprüchen 16 bis 20, dadurch gekennzeichnet, daß der Operationskeller (12) und der Zahlkeller (11) Einrichtungen aufweisen, die die zugeführten Zeichen dadurch sequentiell speichern, daß sie die bereits gespeicherten Zeichen in der Reihenfolge des Eintreffens nach unten weiterschieben und eine Abgabe nur des jeweils zuletzt gespeicherten oder des obersten der von unten nachzuschiebenden Zeichen gestatten.

22. Rechenmaschine nach Ansprüchen 16 bis 20, dadurch gekennzeichnet, daß der Operationskeller (12) und der Zahlkeller (11) Einrichtungen aufweisen, die die zugeführten Zeichen dadurch sequentiell speichern, daß jedes eintreffende Zeichen auf den Platz vor dem zuletzt eingetroffenen gesetzt wird, daß dieser Platz festgehalten wird und daß ferner die Abnahme von dem zuletzt festgehaltenen Platz erfolgt.

23. Rechenmaschine nach Ansprüchen 16 bis 22, dadurch gekennzeichnet, daß das Rechenwerk (10) die im obersten oder in den beiden obersten Geschossen des Zahlkellers befindlichen Zahlen entsprechend den von der Operationssteuerung erhaltenen Befehlen verarbeitet und das Ergebnis wieder an das oberste Geschoß des Zahlkellers abgibt.

24. Rechenmaschine nach Ansprüchen 16 bis 23, dadurch gekennzeichnet, daß das Ausgabesteuerwerk (16) mit dem Zahlkeller (11) derart verbunden ist, daß beim Eintreffen eines Gleichheitszeichens und gegebenenfalls nachfolgender Zeichen »Ziffer verlangt« die Verbindung des Zahlkellers mit dem Schreibwer hergestellt und die im obersten Geschoß des Zahlkellers befindliche Zahl ganz oder teilweise an das Schreib werk abgegeben wird.

1 094 019

23

25. Rechenmaschine nach Anspruch 24, dadurch gekennzeichnet, daß in dem Tastenfeld (1) eine Ergebnistaste vorgesehen ist, die bewirkt, daß die einzelnen Stellen des Ergebnisses jeweils beim Anschlag der Taste geschrieben werden, so daß jede gewünschte Anzahl von Ergebnisstellen geschrieben werden kann.

26. Rechenmaschine nach Ansprüchen 16 bis 25, dadurch gekennzeichnet, daß an den Vorentschlüßler (5) ein im Steuerwerk befindlicher Größenentschlüßler angeschlossen ist, der beim erstmaligen Eintreffen eines Größensymbols, vorzugsweise wenn es unmittelbar auf ein »Ergibtzeichen« folgt, diesem Größensymbol die Nummer eines freien Platzes im Zahlspeicher derart zuordnet, daß fernerhin dasselbe Größensymbol beim Einlaufen in den Größenspeicher unmittelbar die Ansteuerung des zugehörigen Speicherplatzes zur Aufnahme von Zahlen aus dem Zahlkeller bzw. dem Rechenwerk oder zur Abgabe von Zahlen in den Zahlkeller bzw. das Rechenwerk bewirkt.

27. Rechenmaschine nach Ansprüchen 16 bis 26, dadurch gekennzeichnet, daß an den Vorentschlüßler ein im Steuerwerk befindlicher Kennzeichenentschlüßler (25) angeschlossen ist, der beim erstmaligen Eintreffen eines Kennzeichnungssymbols für Formelgruppen diesem die Nummer desjenigen Platzes im Formelspeicher zuweist, auf den das erste Symbol der nachfolgenden Formelgruppe trifft, derart, daß fernerhin dasselbe Kennzeichnungssymbol in Verbindung mit einem Sprungsymbol unmittelbar die Ansteuerung des festgehaltenen Platzes des Anfangs der Formelgruppe im Formelspeicher bewirkt, von wo aus die Formelentschlüßlung fortgesetzt wird.

28. Rechenmaschine nach Ansprüchen 21 und 22, dadurch gekennzeichnet, daß der Zahlkeller (11) mit dem Rechenwerk (10) derart vereinigt ist, daß die üblicherweise als Multiplikanden-Divisor-Register, Akkumulator und Multiplikatorregister bezeichneten speicherfähigen Einrichtungen des Rechenwerkes ganz oder teilweise in den obersten Plätzen des Zahlkellers liegen.

29. Rechenmaschine nach Ansprüchen 16 bis 28, dadurch gekennzeichnet, daß die Plätze des Zahlkellers auch im Falle des Vorkommens von Zahlen wechselnder Länge voll ausgenutzt werden, wobei die einzelnen Zahlen gegebenenfalls durch Markierungs- oder Schlußzeichen voneinander getrennt sein können.

30. Rechenmaschine nach Anspruch 28, dadurch gekennzeichnet, daß der Zahlkeller nach oben als Appendix fortgesetzt und andererseits mit einem ringförmigen Speicher über Verschiebeeinrichtungen verbunden ist, derart, daß die Durchführung der Rechenoperationen auf synchrone Verschiebungen im Appendix und im ringförmigen Speicher unter gleichzeitiger stellenweiser Addition und Verschiebung in den Zahlkeller hinein zurückgeführt werden kann.

31. Rechenmaschine nach Anspruch 28, dadurch gekennzeichnet, daß der Zahlkeller als ringförmiger Speicher derart ausgebildet ist, daß er durch Verschiebeeinrichtungen in Verbindung mit einem weiteren ringförmigen Speicher steht, so daß die Rechenoperationen auf synchrone Verschiebungen in den beiden ringförmigen Speichern unter gleichzeitig stellenweiser Addition in den Zahlkeller hinein zurückgeführt werden können.

32. Rechenmaschine zur Ausführung des Verfahrens nach Ansprüchen 1 bis 15, dadurch gekennzeichnet, daß die Rechenregister als ansteuerbare, aber nicht notwendig verschiebbare Speicher ausgebildet sind, derart, daß über parallel ausgebildete Sucheinrichtungen die Operanden abgegriffen und dadurch die Durchführung der Rechenoperationen auf sukzessive stellenweise Additionen zurückgeführt wird, wobei das Ergebnis in einem der beiden Operandenplätze wieder aufgebaut werden kann.

24

33. Rechenmaschine nach Anspruch 32, dadurch gekennzeichnet, daß der Zahlkeller und gegebenenfalls das Multiplikanden-Divisor-Register als ansteuerbare, aber nicht notwendig verschiebbare Speicher ausgebildet sind und daß zur Durchführung der arithmetischen Operationen der Zahlkeller mitbenutzt wird.

34. Rechenmaschine nach Ansprüchen 32 und 33, dadurch gekennzeichnet, daß der Zahlkeller (11) ganz oder teilweise in Plätze des an sich vorhandenen Zahlspeichers (22), vorzugsweise in die jeweils freien Plätze, gelegt wird, wobei der jeweilige Stand des freien Speichers und der jeweilige Stand der Spitze der Zahlkellersequenz durch besondere Zählregister festgehalten werden kann.

35. Rechenmaschine zur Ausführung des Verfahrens nach Ansprüchen 1 bis 15, dadurch gekennzeichnet, daß an Stelle des Zahlkellers ein Platznummernkeller tritt, in dem anstatt der in den Zahlkeller einzufahrenden Zahlen deren Platznummern im Hauptzahlspeicher festgehalten und bei der Formelauswertung stellvertretend für die durch sie anzusteuernden Zahlen behandelt werden.

36. Rechenmaschine nach Ansprüchen 32, 33 und 34, dadurch gekennzeichnet, daß der Operand oder die beiden Operanden einer arithmetischen Operation mittels zählfähiger Register, die von den Inhalten des Platznummernspeichers her eingestellt werden, angesteuert werden und daß die Speicherplätze des Resultats von dem Register, in dem der jeweilige Stand der Spitze der Zahlkellersequenz festgehalten wird, her angesteuert werden, wobei der Stand des Registers »freier Speicher« zur Sinnvolltestung herangezogen werden kann.

37. Rechenmaschine nach Ansprüchen 35 und 36, dadurch gekennzeichnet, daß zur Kennzeichnung von Zahlen wechselnder Länge die Platznummern des Zahlenanfanges und die Stellenzahl im Größenvorspeicher (24) festgehalten werden.

38. Rechenmaschine nach Ansprüchen 29 oder 37 und 38, dadurch gekennzeichnet, daß mit Zahlen wechselnder und im Prinzip beliebiger Länge gearbeitet wird.

39. Rechenmaschine zur Ausführung des Verfahrens nach Ansprüchen 1 bis 6, dadurch gekennzeichnet, daß zur Auffindung der einzelnen Komponenten von Zahlsätzen, die indizierten Größen entsprechen, die Platznummer des Anfangs des Zahlsatzes und die Kenngrößen für den Indexlauf sowie gegebenenfalls die Zahllänge im Größenvorspeicher (24) festgehalten werden.

40. Rechenmaschine nach Anspruch 39, dadurch gekennzeichnet, daß eine Zuweisung von Speicherplätzen zu Größensymbolen, insbesondere indizierten Größensymbolen, beim ersten Auftreten einer expliziten Speicherungsvorschrift vorgenommen wird, wobei einer zählfähigen Vorrichtung der Stand des »freien Speichers« entnommen und im Vorspeicher unter dem Eingang des Größensymbols gespeichert wird und wobei ferner die Kenngrößen für den Indexlauf, gegebenenfalls einschließlich der Zahllängen, der Speicherungsvorschrift entnommen werden.

41. Rechenmaschine nach Ansprüchen 39 und 40, dadurch gekennzeichnet, daß die Indexsymbole einer Hilfssteuereinrichtung zugeführt werden, die den Übergang zur Auswertung der Formelausdrücke auf

1 094 019

25

den einzelnen Indexstellen veranlaßt und für die Einschiebung der speziellen Index-Auswertungsoperationen, die mit den Kenngrößen des Indexlaufes bzw. der Zahllänge und der Platznummer des Anfangs durchzuführen sind, sorgt und mit der so errechneten Platznummer der betreffenden Komponente der indi-

26

zierten Größe diese im Zahlspeicher aufsucht und sie der weiteren Formelauswertung zur Verfügung stellt.

In Betracht gezogene Druckschriften:
Deutsche Patentschrift Nr. 922 085;
Hollerith-Nachr., 74, 1937, S. 1022.

Hierzu 2 Blatt Zeichnungen

ZEICHNUNGEN BLATT 1 AUSGABETAG: 1. DEZEMBER 1960

DAS 1094019
KL. 42 m 14
INTERNAT. KL. G 06 f

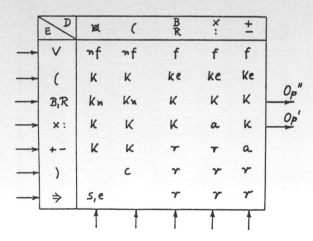

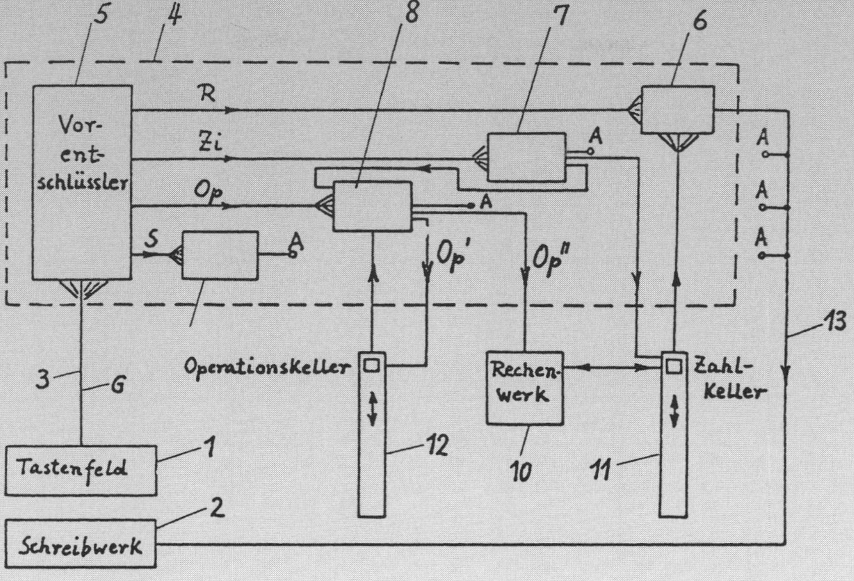

Fig.1

E \ D	⊠	(	B/R	×/:	+/−
V	nf	nf	f	f	f
(	K	K	ke	ke	Ke
B,R	Kn	Kn	K	K	K
× :	K	K	K	a	K
+ −	K	K	r	r	a
)		c	r	r	r
⇒	s,e		r	r	r

Op''
Op'

Fig.2

ZEICHNUNGEN BLATT 1 AUSGABETAG: 1. DEZEMBER 1960

DAS 1094019
KL. 42m 14
INTERNAT. KL. G 06 f

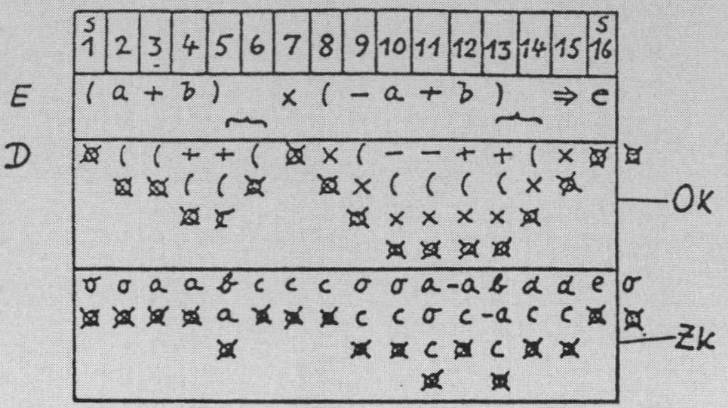

Fig. 3

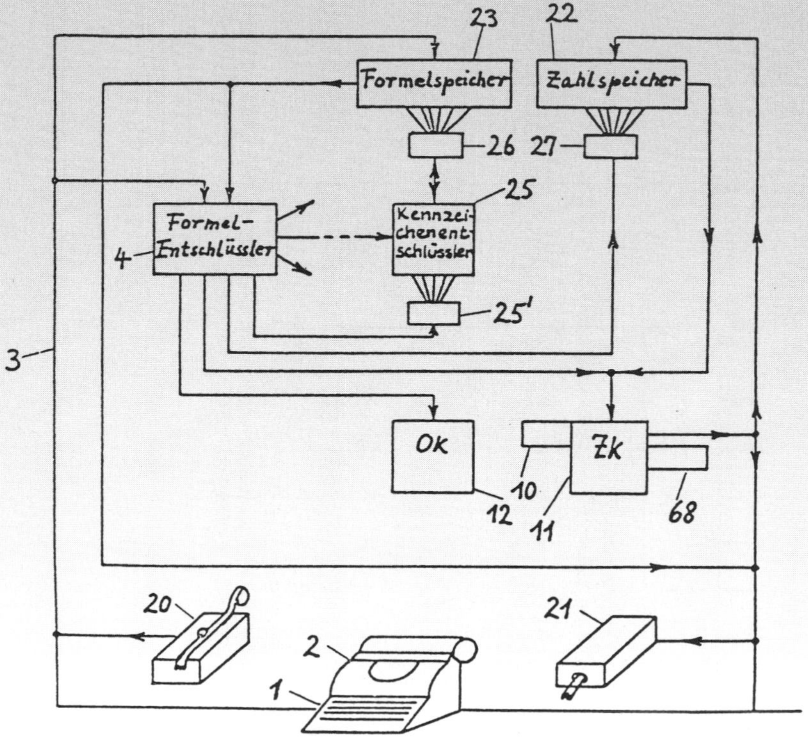

Fig. 4

009 650/226

ZEICHNUNGEN BLATT 2 AUSGABETAG: 1. DEZEMBER 1960

DAS 1094019
KL. 42 m 14
INTERNAT. KL. G 06 f

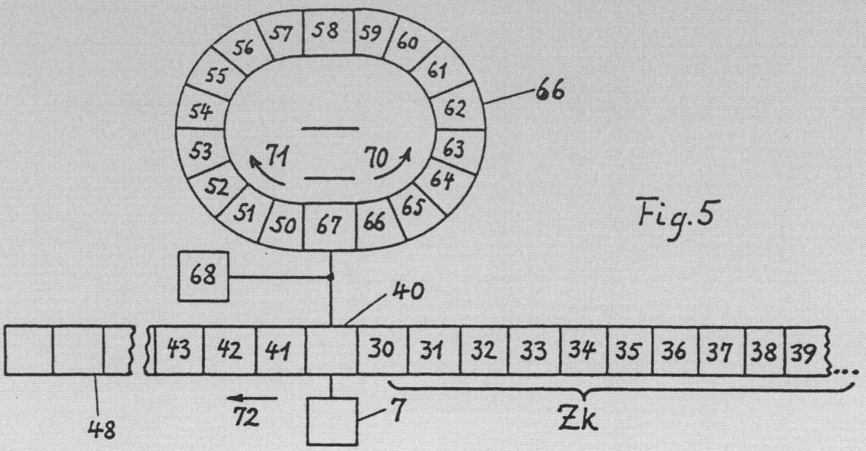

Fig. 5

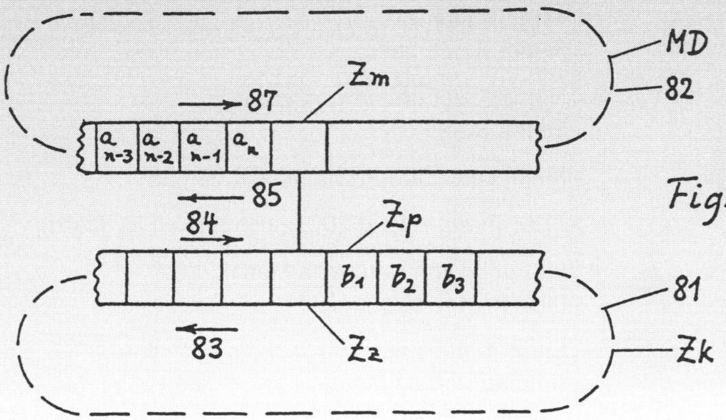

Fig. 6

The Roots of Object-Oriented Programming: Simula 67

Ole-Johan Dahl
Professor of Informatics (Emeritus)
University of Oslo
olejohan@ifi.uio.no

Norwegian Defence Research Establishment:
Senior Researcher

Norwegian Computing Center:
Senior Researcher

Major contribution: Simula 67

ACM Turing Award 2001 (also Kristen Nygaard)

IEEE John von Neumann Medal 2002 (also Kristen Nygaard)

Current interests:
simulation languages,
verifiable programming

Ole-Johan Dahl

The Roots of Object Orientation: The Simula Language

The development of the programming languages Simula I and Simula 67 is briefly described. An attempt is made also to explain the cultural impact of the languages, in particular the object-oriented aspects.

1. Introduction

In 1962 Kristen Nygaard initiated a project for the development of a discrete event simulation language, to be called Simula. At the time Nygaard was the director of research at the Norwegian Computing Center, NCC, (a semi-governmental institution). Nygaard also served as the administrative leader for the duration of the project. This required much creative manipulation in an environment that outside the NCC was largely hostile. The proper language development was a result of a close cooperation between Nygaard and myself, whereas implementation considerations were mainly my responsibility.

We were both fostered at the Norwegian Defence Research Establishment in the pioneering group headed by Jan V. Garwick, the father of computer science in Norway. But our backgrounds were nevertheless quite different. Nygaard had done Monte Carlo computations calibrating uranium rods for a nuclear reactor and later operations research on military systems. I had developed basic software together with Garwick and designed and implemented a high-level programming language. The differences in our backgrounds probably account for some of the success of the Simula project.

The present paper mainly deals with language issues, including some thoughts on their possible cultural impact, especially on later programming languages. For other aspects of the project the reader is referred to [30].

Two language versions were defined and implemented. The first one, later called Simula I, was developed under a contract by UNIVAC. (UNIVAC wanted us to provide also a Fortran-based version, but that was abandoned because the block structure turned out to be essential to our approach.) It was up and running by the end of 1964. The second version, Simula 67, was sponsored by the NCC itself. It was a generalization and refinement of the former; it was fairly ambitious and was intended mainly as a general purpose programming language, but with simulation capabilities.

2. Simula I

It was decided at an early stage that our language should be based on a well-known one. ALGOL 60 was chosen for the following main reasons:

- block structure,
- good programming security,
- European patriotism.

We realised that in order to profit from block structure in simulation models it would be necessary to break out of the strict LIFO regime of block instances in Algol. Thus, a new storage management system was developed based on a list structured free store [3]. Then a useful simulation language could be defined by adding a few special-purpose mechanisms to ALGOL 60:

- Procedure-like **activity** declaration giving rise to so-called "processes". Processes could range from record-like data structures to block-structured programs executing in a coroutine-like fashion [9, 36] over a simulated system time.

- Explicit process pointers for dynamic naming and referencing. (The pointers were indirect through list-forming "element" records.)

- A mechanism for accessing, from the outside, quantities local to the outermost block level of processes, designed so that the access security inherent in Algol would be maintained (the **inspect** construct).

- A few run time mechanisms for the scheduling and sequencing of processes in system time, such as *hold*(...), suspending the calling process for a specified amount of system time.

```
SIMULA begin activity Car;
              begin real X0,T0,V;
                    real procedure X; X := X0+V*(time-T0);
                    procedure newV(Vnew); real Vnew;
                      begin X0 := X; T0 := time; V := Vnew end;
                    Car behaviour: ......; hold(<travel time>); ......
              end of Car;
              activityPolice;
              begin ......; inspect <process> when Car do
                    if X <is within city> and V >50 then
                      begin newV(50); <give fine> end; ......
              end of Police;
              main program: <initialise>; hold(<simulation period>)
end of simulation model;
```

The skeleton example could be a small fragment of a road traffic simulation. It is taken from the Simula I manual [4], but is slightly extended. It may serve to indicate the flavour of the language.

The example shows that the idea of data objects with associated operators was under way already in 1965. According to a comment in [4] it was a pity that the variable attributes of a *Car* process could not be hidden away in a subblock. It would have required the accessing procedures to be hidden similarly.

New processes were generated explicitly. For programming security reasons, however, process deletions had to be implicit, in our implementation through reference counts and a last resort garbage collector. The bulk of the implementation effort therefore consisted in writing a new run time system for the Algol system provided by UNIVAC; the compiler extensions, on the other hand, were minimal. The "block prefix" **SIMULA** served to introduce the Simula I additions to Algol. Consequently any Algol program not containing that keyword would execute normally on our compiler. That was an important consideration in those days.

A paper on Simula I was published in CACM 1966 [5]. It was also the main topic of lectures given by myself at the NATO Summer School at Vilard-de-Lans the same year. The lectures were written up and published as a chapter of [6].

The language was used for simulation purposes as well as for teaching at several locations at home and abroad, also within the UNIVAC organization. A modified version was used for Burroughs computers. This was through the advocacy of Don Knuth and J. McNeley, the authors of SOL, another Algol-like simulation language.

3. Simula 67

In spite of the success of Simula I as a practical tool it became increasingly clear that the activity/process concepts, if stripped from all references to simulated time, would be useful for programming and system design in general. If possible the special purpose simulation facilities should be definable within the new language. Also the list processing facilities of Simula I would be useful, although we felt that the referencing mechanism should be simplified.

At the Vilard-de-Lans Summer School Tony Hoare had put forth a proposal for "record handling" with record classes and subclasses, as well as record references restricted to, or "qualified by", a given class or subclass by declaration. Attribute accessing was by dot notation, see [19], as well as [17] and [18].

We chose the terms "class" and "objects" of classes for our new Simula. The notion of subclass was especially appealing to us, since we had seen many cases of objects belonging to different classes having common properties. It would be useful to collect the common properties in a separate class, say C to be specialised differently for different purposes, possibly in different programs. The solution came with the idea of class prefixing: using C as a prefix to another class, the latter would be taken to be a subclass of C inheriting all properties of C.

Technically the subclass would be seen as a "concatenated" class in which the parameter parts, the block heads, and the block tails of the two classes were juxtaposed. (The block tail of the prefixing class could be separated into initial actions and final actions, that of the prefixed class sandwiched between them.) The attributes of a compound object would be accessible by dot notation down to the prefix level of the qualifying class of the reference used. Access to deeper levels could be achieved by class discrimination as in Simula I.

The breakthrough happened in January of 1967. An IFIP-sponsored working conference on simulation languages had been approved to take place in Oslo in May. Some hectic winter months followed, during which our new concepts were explored and tested. A paper was ready just in time for advanced distribution to the invitees [7]. The new language was to be called Simula 67 [8]. The paper appearing in the proceedings was mended by the addition of "virtual" specifications, see below.

One way of using a class, which appeared important to us, was to collect concepts in the form of classes, procedures, etc. under a single hat. The resulting construct could be understood as a kind of "application language" defined on top of the basic one. It would typically be used as a prefix to an in-line block making up the application program.

We illustrate this idea by showing a simplified version of a *SIMULATION* class defining the simulation-oriented mechanisms used in our Simula I example.

```
class SIMULATION;
begin class process;
        begin real EventTime, NextEvent; ...... end;
        ref(process) current;
        comment current points to the currently operating process.
                It is the head of the "time list" of scheduled ones,
                sorted with respect to nondecreasing EventTimes;
        real procedure time; time := current.EventTime;
        procedure hold(deltaT); real deltaT;
        begin current.EventTime := time+deltaT;
              if time ≥ current.NextEvent.EventTime then
              begin ref(process)P; P :- current; current :- P.NextEvent;
                    <move P to the right position in the time list>;
              resume(current) end end of hold;
        ......
end of SIMULATION;

SIMULATION begin
  process class Car;
  begin real X0,T0,V;
        real procedure X; X := X0+V*(time-T0);
        procedure newV(Vnew); real Vnew;
          begin X0 := X; T0 := time; V := Vnew end;
        Car behaviour: ......; hold(<travel time>); ......
  end of Car;
  process Police;
  begin ......; inspect <process> when Car do
        if X <is within city> and V > 50 then
          begin newV(50); <give fine> end; ......
  end of Police;

  main program: <initialise>; hold(<simulation period>)
end of simulation model;
```

Thus, the "block prefix" of Simula I is now an ordinary class declared within the new language, and the special purpose **activity** declarator is replaced by *process* **class**.

We chose to introduce a special set of operators for references, in order to make it clear that the item in question is a reference, not (the contents of) the referenced object. The resume operator is a basic coroutine call, defined for the whole language.

Notice that the class *SIMULATION* is completely selfcontained. If some necessary initializing operations were added, it could be separately compiled and then used repeatedly in later programs. In fact a somewhat more elaborate class is predefined in Simula 67, providing an application language for simulation modelling. That class is itself prefixed by one containing mechanisms for the management of circular lists.

It is fair to say that Simula 67 invites bottom-up program design, especially through the possibility of a separate compilation of classes. As a last minute extension, however, we introduced a top-down-oriented mechanism through a concept of "virtual procedures".

In general, attribute identifiers may be redeclared in subclasses, as is the case of inner blocks. The identity of an attribute is determined by the prefix level of the accessing occurrence, or, if the access is remote, by the class qualifying the object reference in question. In this way any ambiguity of identifier binding is resolved textually, i.e at compile time; we call it static binding.

On the other hand, if a procedure *P* is specified as **virtual** in a class *C* the binding scheme is semidynamic. Any call for *P* occurring in *C* or in any subclass of *C* will bind to that declaration of *P* which occurs at the innermost prefix level of the actual object containing such a declaration (and similarly for remote accesses). Thus, the body of the procedure *P* may, at the prefix level of *C*, be postponed to occur in any subclass of *C*. It may even be replaced by more appropriate ones in further subclasses.

This binding scheme is dynamic in the sense that it depends on the class membership of the actual object. But there is nevertheless a degree of compiler control; the access can be implemented indirectly through a table produced by the compiler for *C* and for each of its subclasses.

As a concrete example the "fine giving" operation of the above example could be formalised as a virtual procedure, as follows: Redefine the head of the prefixed block as a subclass *RoadTraffic* of *SIMULATION*. In addition to the classes *Car* and *Police* declarations introduce the following specification:

virtual procedure *Fine*(*cr*); **ref**(*Car*)*cr*;

If appropriate the *RoadTraffic* class may be separately compiled. Using that class as a block prefix at some later time, a suitable fining procedure can be defined in that block head.

There is an alternative, more implementation-oriented view of virtual procedures. As mentioned in connection with Simula I, deletion of objects would have to be implicit (in Simula 67 by garbage collector alone). But then there is a danger of flooding the memory with useless data, especially if there are implicit pointers between block instances. In Algol 60 there must be a pointer from a procedure activation back to its caller in order to implement procedure parameters and parameters "called by name". Such pointers from objects back to their generating block instance would have been destructive. So, it was decided that parameters to objects must be called by "value" (including object references). The absence of procedure parameters, however, was felt to be a nuisance. Fortunately the virtual procedure mechanism provided a solution to the dilemma: a virtual procedure can be seen as a parameter, where the actual parameter is a procedure residing safely within the object itself, at an appropriate prefix level. There is the additional advantage that the procedure has direct access to attributes of the object containing it.

Similar considerations led to forbidding class prefixing across block levels. Fortunately this would not prevent the use of separately compiled, "external" classes. Since there is no reference to nonlocal quantities in such a class, it can be called in as an external one at any block level of a user program.

4. Language Finalisation and Distribution

A couple of weeks after the IFIP Conference a private "Simula Common Base Conference" (CBC) was organised, attended by several interested persons. The objective was to agree on the definition of a common core language. We made a proposal to the CBC to extend the language by "class-like" types giving rise to permanently named objects, directly accessed, thus extending the Algol variable concept. The proposal was prudently voted down, as not sufficiently worked through. However, a Pascal-like **while** statement was added, and the virtual mechanism was slightly revised.

A "Simula Standards Group", SSG, was established and consisted of representatives from the NCC and various implementation groups. Five compilers were implemented initially. It was decided that the NCC would propose mechanisms for text handling, I/O, and file handling. Our good colleague Bjørn Myhrhaug of the NCC gave three alternatives for text handling and I/O. The ones chosen by the SSG would have required class-like types in order to be definable within the common base.

The class concept as it was formulated originally was too permissive for the purpose of developing large systems. There was no means of enforcing a programming discipline protecting local class invariants (such as those expressed verbally for the *Simulation* class example). This was pointed out by Jacob Palme of the Swedish defence research institute. He proposed hiding mechanisms for protecting variable attributes from unauthorised updating. The proposal was approved by the SSG as the last addition ever to the language. The authors toyed with the idea of class-like types for some time, but it was never implemented.

The first compilers were operative already in 1969, three for Control Data computers. Then came implementations for UNIVAC and IBM machines. The general distribution of the compilers was, however, greatly hampered by the high prices asked for the compilers by the NCC, very unwisely enforced by the NTNF (Norwegian Council for Scientific and Technical Research) stating that programming languages only had a lifetime of 3–5 years and thus had to provide profits within this time span. However, a nice compiler for the DEC 10 system, implemented by a Swedish team in the early 1970s, contributed considerably to the spreading of the language. Lectures I gave at NATO Summer Schools, as well as a chapter in [9] must have made the new concepts better known in academic circles.

The most important new concept introduced by Simula 67 was surely the idea of data structures with associated operators (and with or without own actions), called objects. There is an important difference, except in trivial cases, between

- *the inside view* of an object, understood in terms of local variables, possibly initialising operations establishing an invariant, and implemented procedures operating on the variables maintaining the invariant, and

- *the outside view*, as presented by the remotely accessible procedures, including some generating mechanism, dealing with more "abstract" entities.

This difference, as indicated by the *Car* example in Simula I and the associated comments, underlies many of our program designs from an early time on, although not usually consciously and certainly not explicitly formulated. (There is, for example, an intended invariant of any **Car** object vaguely stating that its current position X is the right one in view of the past history of the object.)

It was Tony Hoare who finally expressed mathematically the relationship of the two views in terms of an "abstraction function", see [20]. He also expressed requirements for the concrete operations to correctly represent the corresponding abstract ones. Clearly, in order to enforce the use of abstract object views, read access to variable attributes would also have to be prevented.

5. Cultural Impact

The main impact of Simula 67 has turned out to be the very wide acceptance of many of its basic concepts: objects, but usually without own actions, classes, inheritance, and virtuals, often the default or only way of binding "methods" (as well as pointers and dynamic object generation).

There is universal use of the term "object orientation", OO. Although no standard definition exists, some or all of the above ideas enter into the OO paradigm of system development. There is a large flora of OO languages around for programming and system specification. Conferences on the theory and practice of OO are held regularly. The importance of the OO paradigm today is such that one must assume something similar would have come about also without the Simula effort. The fact remains, however, that the OO principle was introduced in the mid-1960s through these languages.

Simula 67 had an immediate success as a simulation language, and was, for instance, extensively used in the design of VLSI chips, e.g. at INTEL. As a general programming language, its impact was enhanced by lectures I gave at NATO Summer Schools, materialising as a chapter in a book on structured programming [9]. The latter has influenced research on the use of abstract data types, e.g. the CLU language [30], as well as research on monitors and operating system design [21].

A major new impact area opened with the introduction of workstations and personal computers. Alan Kay and his team at Xerox PARC developed

Smalltalk [15], an interactive language building upon Simula's objects, classes, and inheritance. It is oriented towards organising the cooperation between a user and her/his personal computer.

An important step was the integration of OO with a graphical user interface, leading the way to the Macintosh Operating System, and then to Windows.

In the larger workstation field, Lisp was (and in some places still is) an important language, spawning dialects such as MACLISP [16] at MIT, and InterLisp at Xerox PARC. Both obtained OO facilities, MACLISP through ZetaLisp also introducing multiple inheritance, [2], and InterLisp through LOOPS (Lisp Object-Oriented Programming System). The object-oriented component of the merged effort, CommonLisp, is called CLOS (Common Lisp Object System) [25].

With the general acceptance of object orientation, object-oriented databases started to appear in the 1980s. The demand for software reuse also pushed OO tools, and in the 1990s OO tools for system design and development became dominant in that field. UML (Unified Modeling Language) [1] is very widely used, and CORBA (Common Object Request Broker Architecture) is a widely accepted tool for interfacing OO systems. The Microsoft Component Object Model, COM [28] is an important common basis for programming languages such as C#, as well as other tools.

A large number of OO programming languages have appeared. We list below some of the more interesting or better known languages, in addition to those mentioned above.

- BETA is a compilable language built around a single abstraction mechanism, that of patterns, which can be specialised to classes, singular objects, and types, as well as procedures. It was developed in the later 1970s by Kirsten Nygaard and colleagues in Denmark [26, 27].

- Bjarne Stroustrup extended the Unix-related language C with several Simula-inspired mechanisms. The language, called C++, has been widely used and has contributed importantly to the dissemination of the OO ideas [34]. Since C is fairly close to machine code, security aspects are not the best. As a result, complex systems may be difficult to implement correctly. C++ has been revised and extended, e.g. by multiple inheritance.

- Eiffel [29] is an OO programming language designed by Bertrand Meyer in the 1980s, well known and quite widely used. It has pre- and post-conditions and invariants.

- SELF [35] is an OO language exploring and using object cloning instead of object generation from a class declaration.

- JAVA [22] is a recent Simula-, Beta-, and C++-inspired language, owing much of its popularity to its integration with the Internet. Its syntax is unfortunately rather close to that of C++ and thus C (but with secure pointers). It contains Beta-like singular objects and nested classes, but not general block structure. Parallel, "multi-threaded", execution is introduced, but outside compiler control. As a result, much of the programming security otherwise inherent in the language is lost. The synchronisation mechanisms invite inefficient programming and do not facilitate good control of process sequencing, see [14].

Kirsten Nygaard and I believed that the use of class declarations for the definition of "application languages" as natural extensions of a basic one would be of special importance in practice. However, although various kinds of packages or modules are defined for many languages, they are not consequences of a general class declaration as in Simula 67.

The coroutine-like sequencing of Simula has not caught on as a general-purpose programming tool. A natural development, however, would have been objects as concurrent processes, e.g. as in COM.

One may fairly ask how it could happen that a team of two working in the periphery of Europe could hit on programming principles of lasting importance. No doubt a bit of good luck was involved. We were designing a language for simulation modelling, and such models are most easily conceived in terms of cooperating objects. Our approach, however, was general enough to be applicable to many aspects of system development.

Kirsten Nygaard oriented his activities for some years to trade union work, as well as system development and description, see [24]. In 1976 he turned back to programming language design, see BETA above. In [33] he introduced new constructs for OO layered distributed systems.

I was professor of Informatics at Oslo University for the period 1968-1999, developing curricula including OO programming. I explored the concept of time sequences to reason about concurrent systems [10, 11]. In [13] I apply techniques, such as Hoare logic and Guttag-Horning axiomatisation of types and subtypes [32], to the specification and proof of programs, including OO ones. See also [12].

Of the Simula authors Kirsten Nygaard especially has consistently promoted the OO paradigm for system development.

Acknowledgements

I am greatly indebted to Kristen Nygaard for his help to explain the impact of object orientation in various areas of programming and system work. Olaf Owe has contributed as well.

References

[1] G. Booch, J. Rumbaugh, I. Jacobson: The Unified Modeling Language User Guide. Addison-Wesley, 1998.

[2] H. Cannon: Flavors, A Non-hierarchical Approach to Object-Oriented Programming. Draft 1982.

[3] O.-J. Dahl: The Simula Storage Allocation Scheme. Norwegian Computing Center Doc. 162, 1963.

[4] O.-J. Dahl, K. Nygaard: SIMULA – A language for programming and description of discrete event systems. Introduction and user's manual. Norwegian Computing Center Publ. no. 11, 1965.

[5] O.-J. Dahl, K. Nygaard: SIMULA – an ALGOL-based simulation language. CACM 9(9), 671-678, 1966.

[6] O.-J. Dahl: Discrete event simulation languages. In F. Genuys, ed.: Programming Languages. Academic Press, pp 349-394, 1968.

[7] O.-J. Dahl, K. Nygaard: Class and subclass declarations. In J. Buxton, ed.: Simulation Programming Languages. Proceedings from the IFIP Working Conference in Oslo, May 1967. North Holland, 1968.

[8] O.-J. Dahl, B. Myhrhaug, K. Nygaard: SIMULA 67 Common Base Language. Norwegian Computing Center, 1968.

[9] O.-J. Dahl, C.A.R. Hoare: Hierarchical program structures. In O.-J. Dahl, E.W. Dijkstra, C.A.R. Hoare, eds.: Structured Programming. Academic Press, pp. 175-220, 1972.

[10] O.-J. Dahl: Can program proving be made practical? In M. Amirchahy, D. Neel, eds.: Les Fondements de la Programmation. IRIA, pp. 57-114, 1977.

[11] O.-J. Dahl: Time sequences as a tool for describing process behaviour. In D. Bjørner, ed.: Abstract Software Specifications, Lecture Notes in Computer Science 86, Springer, pp. 273-290, 1980.

[12] O.-J. Dahl:, O. Owe: Formal development with ABEL. In VDM91, Lecture Notes in Computer Science 552, Springer, pp. 320-363.

[13] O.-J. Dahl: Verifiable Programming. Hoare Series, Prentice Hall, 1992.

[14] O.-J. Dahl: A note on monitor versions. In Proceedings of Symposium in the Honour of C.A.R. Hoare at his resignation from the University of Oxford. Oxford University, 1999. Also available at www.ifi.uio.no/olejohan (Institute of Informatics, Oslo University).

[15] A. Goldberg, D. Robson: Smalltalk-80: The Language and Its Implementation. Addison Wesley, 1984.

[16] B.S. Greenberg: The Multics MACLISP Compiler. The Basic Hackery – A Tutorial. MIT Press, 1977, 1988, 1996.

[17] C.A.R. Hoare: Record Handling. In ALGOL Bulletin no. 21, 1965.

[18] C.A.R. Hoare: Further Thoughts on Record Handling. In ALGOL Bulletin no 23, 1966.

[19] C.A.R. Hoare: Record handling. In F. Genuys, ed.: Programming Languages. Academic Press, pp 291- 346, 1968.

[20] C.A.R. Hoare: Proof of the correctness of data representations. Acta Inf. 1, 271-281, 1972.

[21] C.A.R. Hoare: Monitors: an operating system structuring concept. Comm. ACM 17(10), pp. 549-557, 1974.

[22] J. Gosling, Bill Joy, G. Steele: The Java Language Specification. Java(tm) Series, Addison-Wesley, 1989.

[23] J. Gosling, B. Joy, G. Steele: The Java Language Specification. Java(tm) Series, Addisson-Wesley, 1996

[24] P. Handlykken, K. Nygaard: The DELTA system description language: motivation, main concepts and experience from use. In H. Hünke, ed.: Software Engineering Environments. GMD, North-Holland, 1981.

[25] S.E. Keene: Object-Oriented Programming in COMMON LISP-A Programmer's Guide to CLOS. Addison-Wesley, 1989.

[26] B.B. Kristensen, O.L. Madsen, B. Møller-Pedersen, K. Nygaard: Abstraction mechanisms in the BETA programming language. In Proceedings of the Tenth ACM Symposium on Principles of Programming Languages, Austin, Texas, 1983.

[27] O.L. Madsen, B. Møller-Pedersen, K. Nygaard: Object-Oriented Programming in the BETA Programming Language. Addison-Wesley/ACM Press, 1993.

[28] R.C. Martin: Design Principles and Design Patterns. Microsoft, www.objectmentor.com.

[29] B. Meyer: Eiffel: The Language. Prentice Hall, 1992.

[30] B. Liskov, A. Snyder, R. Atkinson, C. Schaffert: Abstraction mechanisms in CLU. Comm. ACM 20(8), 564-576, 1977.

[31] K. Nygaard, O.-J. Dahl: SIMULA session. In R. Wexelblatt, ed.: History of Programming Languages. ACM, 1981.

[32] O. Owe, O.-J. Dahl: Generator induction in order sorted algebras. Formal aspects Comput., 3:2-20, 1991.

[33] K. Nygaard: GOODS to appear on the stage. In Proceedings of the 11th European Conference on Object-Oriented Programming. Springer, 1997

[34] B. Stroustrup: The C++ Programming Language. Addison-Wesley, 1986.

[35] D. Ungar, R.B. Smith: SELF: the power of simplicity. In SIGPLAN Notices 22(12), 1987.

[36] A. Wang, O.-J. Dahl: Coroutine sequencing in a block structured environment. In BIT 11, 425-449, 1971.

Ole-Johan Dahl

Ole-Johan Dahl, Kristen Nygaard
Class and Subclass Declarations

Simulation Programming Languages,
ed. by J.N. Buxton,
North Holland, Amsterdam, 1967
pp. 158-174

CLASS AND SUBCLASS DECLARATIONS

OLE-JOHAN DAHL and KRISTEN NYGAARD
Norwegian Computing Center, Oslo, Norway

1. INTRODUCTION

A central idea of some programming languages [28, 57, 58] is to provide protection for the user against (inadvertantly) making meaningless data references. The effects of such errors are implementation dependent and can not be determined by reasoning within the programming language itself. This makes debugging difficult and impractical.

Security in this sense is particularly important in a list processing environment, where data are dynamically allocated and de-allocated, and the user has explicit access to data addresses (pointers, reference values, element values). To provide security it is necessary to have an automatic de-allocation mechanism (reference count, garbage collection). It is convenient to restrict operations on pointers to storage and retrieval. New pointer values are generated by allocation of storage space, pointing to the allocated space. The problem remains of correct interpretation of data referenced relative to user specified pointers, or checking the validity of assumptions inherent in such referencing. E.g. to speak of "*A* of *X*" is meaningful, only if there is an *A* among the data pointed to by *X*.

The record concept proposed by Hoare and Wirth [58] provides full security combined with good runtime efficiency. Most of the necessary checking can be performed at compile time. There is, however, a considerable expense in flexibility. The values of reference variables and procedures must be restricted by declaration to range over records belonging to a stated class. This is highly impractical.

The connection mechanism of SIMULA combines full security with greater flexibility at a certain expense in convenience and run time efficiency. The user is forced, by the syntax of the connection statement, to determine at run time the class of a referenced data structure (process) before access to the data is possible.

The subclass concept of Hoare [59] is an attempt to overcome the difficulties mentioned above, and to facilitate the manipulation of data structures, which are partly similar, partly distinct. This paper presents another approach to subclasses, and some applications of this approach.

2. CLASSES

The class concept introduced is a remodelling of the record class concept proposed by Hoare. The notation is an extension of the ALGOL 60 syntax. A prefix notation is introduced to define subclasses organized in a hierarchical tree structure. The members of a class are called objects. Objects belonging to the same class have similar data structures. The members of a subclass are compound objects, which have a prefix part and a main part. The prefix part of a compound object has a structure similar to objects belonging to some higher level class. It can itself be a compound object.

The figure below indicates the structure of a class hierarchy and of the corresponding objects. A capital letter denotes a class. The corresponding lower case letter denotes the data comprising the main part of an object belonging to that class.

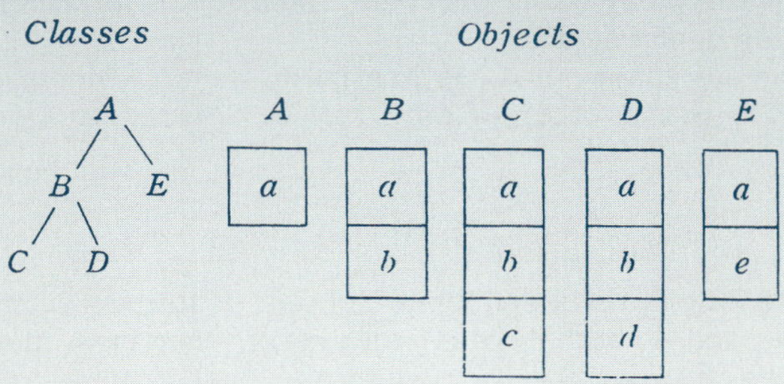

B, *C*, *D*, *E* are subclasses of *A*; *C* and *D* are sublasses of *B*.

2.1. *Syntax*

⟨class id.⟩ ::= ⟨identifier⟩
⟨prefix⟩ ::= ⟨class id.⟩
⟨class body⟩ ::= ⟨statement⟩
⟨main part⟩ ::= <u>class</u> ⟨class id.⟩ ⟨formal parameter part⟩; ⟨specification part⟩ ⟨class body⟩
⟨class declaration⟩ ::= ⟨main part⟩|⟨prefix⟩⟨main part⟩

2.2. *Semantics*

An object is an instance of a class declaration. Different instances of the same declaration are said to belong to class C, where C is the class identifier. If the class body does not take the form of an unlabelled block, it acts as if enclosed in an implicit block. The parameters and the quantities declared local to the outermost block of the class body are called the attributes of an object. The attributes can be referenced locally from within the class body, or nonlocally by a mechanism called remote acessing (5).

The parameters are transmitted by value. One possible use of the statements of the class body may be to initialize attribute values.

A prefixed class declaration represents the result of concatenating the declaration referenced by the prefix and the main part. The concatenation is recursively defined by the following rules.

1) The formal parameter lists of the former and the latter are concatenated to form one parameter list.

2) The specification parts are juxtaposed.

3) A combined class body is formed, which is a block, whose block head contains the attribute declarations of the prefix body and the main body. The block tail contains the statements of the prefix body followed by those of the main body.

The attributes of the main part are not accessible from within the prefix body, except by remote accessing. The attributes of the prefix are acessible as ordinary local quantities from within the body of the main part.

The object class represented by a prefixed class declaration is a subclass of the class denoted by the prefix. Subclasses can be nested to any depth by using prefixed class identifiers as prefixes to other class declarations.

Let A_0 be any class. If A_0 is prefixed, we will denote this prefix by A_1. The prefix of A_1 (if any) will be denoted by A_2 etc. The sequence

$$A_1, A_2, \ldots$$

will be called the "prefix sequence" of A_0. It follows from the syntax that if A_i and A_j both have A_k as prefix, they have identical prefix sequences.

It will be required that all prefix sequences are finite. (This excludes multiple occurrence of any class A_i in a prefix sequence.) Let

$$A_1, A_2 \ldots, A_n$$

be the prefix sequence of A_0. We shall say that the class A_i is "included in A_j" if $0 \leq i \leq j \leq n$.

3. OBJECT REFERENCES

Reference values in the sense of [59] are introduced, in a slightly modified form.

3.1. *Reference types*

3.1.1. *Syntax*

⟨type⟩ :: = ⟨ALGOL type⟩ | <u>ref</u> | <u>ref</u> ⟨qualification⟩
⟨qualification⟩ :: = (⟨class id.⟩)

3.1.2. *Semantics*

Associated with each object is a unique value of type ref, which is said to reference or point to the object. A reference value may, by qualifying a declaration or specification by a class identifier, be required to refer to objects belonging to either this class or any of its subclasses. In addition the value of any item of type reference is restricted to objects belonging to classes whose declarations are statically visible from the declaration or specification of the item.

The reference value none is a permissible value for any reference item, regardless of its qualification.

3.2. *Reference Expressions*

3.2.1. *Syntax*

⟨simple ref. expr.⟩ :: = none | ⟨variable⟩ | ⟨function designator⟩ | ⟨object designator⟩ | ⟨local reference⟩
⟨ref. expr.⟩ :: = ⟨simple ref. expr.⟩ | if ⟨Boolean expr.⟩ then ⟨simple ref. expr.⟩ else ⟨ref. expr.⟩
⟨object designator⟩ :: = ⟨class id.⟩ ⟨actual parameter part⟩
⟨local reference⟩ :: = this ⟨class id.⟩

3.2.2. *Semantics*

A reference expression is a rule for computing a reference value. Thereby reference is made to an object, except if the value is none, which is a reference to "no object".

i) *Qualification.* A variable or function designator is qualified according to its declaration or specification. An object designator or local reference is qualified by the stated class identifier. The expression none is not qualified.

No qualification will be regarded as qualification by a universal class, which includes all declared classes.

ii) *Object generation.* As the result of evaluating an object designator an object of the stated class is generated. The class body is executed. The value of the object designator is a reference to the generated object. The life span of the object is limited by that of its reference value.

iii) *Local reference.* A local reference "this *C*" is a meaningful expression within the class body of the class *C* or of any subclass of *C*. Its value is a reference to the current instance of the class declaration (object).

Within a connection block (5.2) connecting an object of class *C* or a subclass of *C* the expression "this *C*" is a reference to the connected object.

The general rule is that a local reference refers to the object, whose attributes are local to the smallest enclosing block, and which belongs to a class included in the one specified. If there is no such object, the expression is illegal.

4. REFERENCE OPERATIONS

4.1. *Assignment*

4.1.1. *Syntax*

⟨reference assignment⟩ :: = ⟨variable⟩: =⟨reference expr.⟩|
⟨variable⟩: =⟨reference assignment⟩

4.1.2. *Semantics*

Let the left and right hand sides be qualified by *Cl* and *Cr*, respectively, and let the value of the right hand side be a reference to an object of class *Cv*. The legality and effect of the statement depends on the relations that hold between these classes.

Case 1. *Cl* includes *Cr*: The statement is legal, and the assignment is carried out.

Case 2. *Cl* is a subclass of *Cr*: The statement is legal, and the assignment is carried out if *Cl* includes *Cv*, or if the value is <u>none</u>. If *Cl* does not include *Cv*, the effect of the statement is undefined (cf. 6.1).

Case 3. *Cl* and *Cr* satisfy neither of the above relations: The statement is illegal.

The following additional rule is considered: The statement is legal only if the declaration of the left hand item (variable, array or ⟨type⟩ procedure) is within the scope of the class identifier *Cr* and all its subclasses. (The scope is in this case defined after having effected all concatenations implied by prefixes.)

This rule would have the following consequences.

1) Accessible reference values are limited to pointers to objects, whose attributes are accessible by remote referencing (5).

2) Classes represented by declarations local to different instances of the same block are kept separate.

3) Certain security problems are simplified.

4.2. *Relations*

4.2.1. *Syntax*

⟨relation⟩:: = ⟨ALGOL relation⟩|
⟨reference expr.⟩ = ⟨reference expr.⟩|
⟨reference expr.⟩ ≠ ⟨reference expr.⟩|
⟨reference expr.⟩ <u>is</u> ⟨class id.⟩

4.2.2. *Semantics*

Two reference values are said to be equal if the point to the same object, or if both are <u>none</u>. A relation "*X* <u>is</u> *C*" is true if the object referenced by *X* belongs to the class *C* or to any of its subclasses.

4.3. *For statements*

4.3.1. *Syntax*

⟨for list element⟩:: = ⟨ALGOL for list element⟩ | ⟨reference expr.⟩ |
⟨reference expr.⟩ while ⟨Boolean expr.⟩

4.3.2. *Semantics*

The extended for statement will facilitate the scanning of list structures.

5. ATTRIBUTE REFERENCING

An attribute of an object is identified completely by the following items of information:

1) the value of a ⟨reference expr.⟩ identifying an object,

2) a ⟨class id.⟩ specifying a class, which includes that of the object, and

3) the ⟨identifier⟩ of an attribute declared for objects of the stated class.

The class identification, item 2, is implicit at run time in a reference value, however, in order to obtain runtime efficiency it is necessary that this information is available to the compiler.

For a local reference to an attribute, i.e. a reference from within the class body, items 1 and 2 are defined implicitly. Item 1 is a reference to the current instance (i.e. object), and item 2 is the class identifier of the class declaration.

Non-local (remote) referencing is either through remote identifiers or through connection. The former is an adaptation of the technique proposed in [57], the latter corresponds to the connection mechanism of SIMULA [28].

5.1. *Remote Identifiers*

5.1.1. *Syntax*

⟨remote identifier⟩ :: = ⟨reference expr.⟩ . ⟨identifier⟩
⟨identifier 1⟩ :: = ⟨identifier⟩ | ⟨remote identifier⟩

Replace the meta-variable ⟨identifier⟩ by ⟨identifier 1⟩ at appropriate places of the ALGOL syntax.

5.1.2. *Semantics*

A remote identifier identifies an attribute of an individual object. Item 2 above is defined by the qualification of the reference expression. If the latter has the value none, the meaning of the remote identifier is undefined (cf. 6.2).

5.2. *Connection*

5.2.1. *Syntax*

```
⟨connection block 1⟩ :: = ⟨statement⟩
⟨connection block 2⟩ :: = ⟨statement⟩
⟨connection clause⟩ :: = when ⟨class id.⟩ do ⟨connection block 1⟩
⟨otherwise clause⟩ :: = ⟨empty⟩ | otherwise ⟨connection block 2⟩
⟨connection part⟩ :: = ⟨connection clause⟩ |
                       ⟨connection part⟩ ⟨connection clause⟩
⟨connection statement⟩ :: = inspect ⟨reference expr.⟩ do
                            ⟨connection block 2⟩ |
                            inspect ⟨reference expr.⟩
                            ⟨connection part⟩ ⟨otherwise clause⟩
```

5.2.2. *Semantics*

The connection mechanism serves a double purpose:

1) To define item 1 above implicitly for attribute references within connection blocks. The reference expression of a connection statement is evaluated once and its value is stored. Within a connection block this value is said to reference the connected object. It can itself be accessed through a ⟨local reference⟩ (see section 3.2.2).

2) To discriminate on class membership at run time, thereby defining item 2 implicitly for attribute references within alternative connection blocks. Within a ⟨connection block 1⟩ item 2 is defined by the class identifier of the connection clause. Within a ⟨connection block 2⟩ it is defined by the qualification of the reference expression of the connection statement.

Attributes of a connected object are thus immediately accessible through their respective identifiers, as declared in the class declaration corresponding to item 2. These identifiers act as if they were declared local to the connection block. The meaning of such an identifier is undefined, if the corresponding ⟨local reference⟩ has the value none. This can only happen within a ⟨connection block 2⟩.

6. UNDEFINED CASES

In defining the semantics of a programming language the term "undefined" is a convenient stratagem for postponing difficult decisions concerning special cases for which no obvious interpretation exists. The most difficult ones are concerned with cases, which can only be recognized by runtime checking.

One choice is to forbid offending special cases. The user must arrange his program in such a way that they do not occur, if necessary by explicit checking. For security the compiled program must contain implicit checks, which to some extent will duplicate the former. Failure of a check results in program termination and an error message. The implicit checking thus represents a useful debugging aid, and, subject to the implementor's foresight, it can be turned off for a "bugfree" program (if such a thing exists).

Another choice is to define ad hoc, but "reasonable" standard behaviours in difficult special cases. This can make the language much more easy to use. The programmer need not test explicitly for special cases, provided that the given ad hoc rule is appropriate in each situation. However, the language then has no implicit debugging aid for locating unforeseen special cases (for which the standard rules are not appropriate).

In the preceding sections the term undefined has been used three times in connection with two essentially different special cases.

6.1. *Conflicting reference assignment*

Section 4.1.2, case 2, *Cl* does not include *Cv*: The suggested standard behaviour is to assign the value none.

6.2. *Non-existing attributes*

Sections 5.1.2 and 5.2.2: The evaluation of an attribute reference, whose item 1 is equal to none, should cause an error printout and program termination. Notice that this trap will ultimately catch most unforeseen instances of case 6.1.

7. EXAMPLES

The class and subclass concepts are intended to be general aids to data structuring and referencing. However, certain widely used classes might well be included as specialized features of the programming language.

As an example the classes defined below may serve to manipulate circular lists of objects by standard procedures. The objects of a list may have different data structures. The "element" and "set" concepts of SIMULA will be available as special cases in a slightly modified form.

```
class linkage; begin ref (linkage) suc, pred; end linkage;
linkage class link; begin
   procedure out; if suc ≠ none then
      begin pred. suc: = suc; suc. pred: = pred;
            suc: = pred: = none end out;
   procedure into (L); ref (list) L;
      begin if suc ≠ none then out;
            suc: = L; pred: = suc. pred;
            suc. pred: = pred. suc: = this linkage end into;
   end link;
linkage class list;
   begin suc: = pred: = this linkage end list;
```

Any object prefixed by "link" can go in and out of circular lists. If *X* is a reference expression qualified by link or a subclass of link, whose value is different from none, the statements

X. into (L)   and   X. out

are meaningful, where L is a reference to a list.

Examples of user defined subclasses are:

```
link class car (license number, weight);
   integer license number: real weight; . . . ;
car class truck (load); ref (list) load; . . . ;
car class bus (capacity); integer capacity;
   begin ref (person) array passenger [1 : capacity] . . . end;
list class bridge; begin real load: . . . end;
```

Multiple list memberships may be implemented by means of auxiliary objects.

```
link class element (X); ref X;;
```

A circular list of element objects is analogous to a "set" in SIMULA. The declaration "set S" of SIMULA is imitated by "ref (list) S" followed by the statement "S: = list".

The following are examples of procedures closely similar to the corresponding ones of SIMULA.

```
procedure include (X, S); value X; ref X; ref (list) S;
if X ≠ none then element (X). into (S);
ref (linkage) procedure suc (X): value X; ref (linkage) X;
   suc: = if X ≠ none then X. suc else none;
ref (link) procedure first (S); ref (list) S;
   first: = S. suc;
Boolean procedure empty (S); value S; ref (list) S;
   empty: = S. suc = S;
```

Notice that for an empty list S "suc (S)" is equal to S, whereas "first (S)" is equal to none. This is a result of rule 6.1 and the fact that the two functions have different qualifications.

8. EXTENSIONS

8.1. *Prefixed Blocks*

8.1.1. *Syntax*

```
⟨prefixed block⟩:: = ⟨block prefix⟩ ⟨main block⟩
⟨block prefix⟩:: = ⟨object designator⟩
⟨main block :: = ⟨unlabelled block⟩
⟨block⟩ :: = ⟨ALGOL block⟩ | ⟨prefixed block⟩
              ⟨label⟩:⟨prefixed block⟩
```

8.1.2. *Semantics*

A prefixed block is the result of concatenating (2.2) an instance of a class declaration and the main block. The formal parameters of the former are given initial values as specified by the actual parameters of the block prefix. The latter are evaluated at entry into the prefixed block.

8.2. *Concatenation*

The following extensions of the concepts of class body and concatenation give increased flexibility.

8.2.1. *Syntax*

⟨class body⟩ :: = ⟨statement⟩ | ⟨split body⟩
⟨split body⟩ :: = ⟨block head⟩;⟨part 1⟩ <u>inner</u>; ⟨part 2⟩
⟨part 1⟩ :: = ⟨empty⟩ | ⟨statement⟩;⟨part 1⟩
⟨part 2⟩ :: = ⟨compound tail⟩

8.2.2. *Semantics*

If the class body of a prefix is a split body, concatenation is defined as follows: the compound tail of the resulting class body consists of part 1 of the prefix body, followed by the statements of the main body, followed by part 2 of the prefix body. If the main body is a split body, the result of the concatenation is itself a split body.

For an object, whose class body is a split body, the symbol inner represents a dummy statement. A class body must not be a prefixed block.

8.3. *Virtual quantities*

The parameters to a class declaration are called by value. Call by name is difficult to implement with full security and good efficiency. The main difficulty is concerned with the definition of the dynamic scope of the actual parameter corresponding to the formal name parameter. It is felt that the cost of an unrestricted call by name mechanism would in general be out of proportion to its gain.

The virtual quantities described below represent another approach to call by name in class declarations. The mechanism provides access at one prefix level of the prefix sequence of an object to quantities declared local to the object at lower prefix levels.

8.3.1. *Syntax*

⟨class declaration⟩ :: = ⟨prefix⟩⟨class declarator⟩⟨class id.⟩
⟨formal parameter part⟩;
⟨specification part⟩⟨virtual part⟩
⟨class body⟩
⟨virtual part⟩ :: = ⟨empty⟩ | <u>virtual</u>: ⟨specification part⟩

8.3.2. *Semantics*

The identifiers of a ⟨virtual part⟩ should not otherwise occur in the heading or in the block head of the class body. Let $A_1, \ldots, A_n$ be the prefix sequence of A_0 and let X be an identifier occurring in the ⟨virtual part⟩ of A_i. If X identifies a parameter of A_j or a quantity declared local to the body of A_j, $j < i$, then for an object of class A_0 identity is established between the virtual quantity X and the quantity X local to A_j.

If there is no A_j, $j < i$, for which X is local, a reference to the virtual quantity X of the object constitutes a run time error (in analogy with 6.2).

8.3.3. *Example*

```
class A; virtual: real X, Y,Z;. . . ;
A class B(X,Y); real X,Y;. . . ;
A class C(Y,Z); real Y,Z;. . . ;
A class D(Z,X); real Z,X;. . . ;
ref (A) Q;
```

The attribute reference $Q.X$ is meaningful if Q refers to an object of class B or D. Notice that all three subclasses contain objects with only two attributes.

8.4. *Example*

As an example on the use of the extended class concept we shall define some aspects of the SIMULA concepts "process", "main program", and "SIMULA block".

Quasi-parallel sequencing is defined in terms of three basic procedures, which operate on a system variable SV. SV is an implied and hidden attribute of every object, and may informally be characterized as a variable of "type label". Its value is either null or a program point [5]. SV of a class object initially contains the "exit" information which refers back to the object designator. SV of a prefixed block has the initial value null. The three basic procedures are:

1) detach. The value of SV is recorded, and a new value, called a reactivation point, is assigned referring to the next statement in sequence. Control proceeds to the point referenced by the old value of SV. The effect is undefined if the latter is null.

2) resume(X); ref X. A new value is assigned to SV referring to the next statement in sequence. Control proceeds to the point referenced by SV of the object X. The effect is undefined if $X.SV$ is null or if X is none. null is assigned to $X.SV$.

3) goto(X); ref X. Control proceeds to the point referenced by SV of the object X. The effect is undefined if $X.SV$ is null or if X is none. null is assigned to $X.SV$.

```
class SIMULA; begin
   ref(process)current;
   class process; begin ref(process)nextev; real evtime;
      detach; inner; current: =nextev; goto(nextev)end;
      procedure schedule(X,T); ref(process)X; real T;
                begin X.evtime: = T; ------------ end;
   process class main program; begin
                L: resume(this SIMULA); go to L end;
   schedule(main program, 0)end SIMULA;
```

The "sequencing set" of SIMULA is here represented by a simple chain of processes, starting at "current", and linked by the attribute "nextev". The "schedule" procedure will insert the referenced process at the correct position in the chain, according to the assigned time value. The details have been omitted here.

The "main program" object is used to represent the SIMULA object within its own sequencing set.

Most of the sequencing mechanisms of SIMULA can, except for the special syntax, be declared as procedures local to the SIMULA class body.

Examples:

```
procedure passivate; begin current: =current. nextev;
                                    resume(current)end;
procedure activate(X); ref X; inspect X when process do
    if nextev = none then
    begin nextev: =current; evtime: =current. evtime;
          current: =this process; resume(current)end;
procedure hold(T); real T; inspect current do
    begin current: =nextev; schedule(this process, evtime+T);
          resume(current)end;
```

Notice that the construction "process class" can be regarded as a definition of the symbol "activity" of SIMULA. This definition is not entirely satisfactory, because one would like to apply the prefix mechanism to the activity declarations themselves.

9. CONCLUSION

The authors have for some time been working on a new version of the SIMULA language, tentatively named SIMULA 67. A compiler for this language is now being programmed and others are planned. The first compiler should be working by the end of this year.

As a part of this work the class concept and the prefix mechanism have been developed and explored. The original purpose was to create classes and subclasses of data structures and processes. Another useful possibility is to use the class concept to protect whole families of data, procedures, and subordinate classes. Such families can be called in by prefixes. Thereby language "dialects" oriented towards special problem areas can be defined in a convenient way. The administrative problems in making user defined classes generally available are important and should not be overlooked.

Some areas of application of the class concept have been illustrated in the preceding sections, others have not yet been explored. An interesting area is input/output. In ALGOL the procedure is the only means for handling I/0. However, a procedure instance is generated by the call, and does not survive this call. Continued existence, and existence in parallel versions is wanted for buffers and data defining external layout, etc. System classes, which include the declarations of local I/0 procedures, may prove useful.

The SIMULA 67 will be frozen in June this year, and the current plan is to include the class and reference mechanisms described in sections 2-6. Class prefixes should be permitted for activity declarations. The "element" and "set" concepts of SIMULA will be replaced by appropriate system defined classes. Additional standard classes may be included.

SIMULA is a true extension of ALGOL 60. This property will very probably be preserved in SIMULA 67.

DISCUSSION

Garwick:

This language has been designed with a very specific line of thought just as GPL has been designed with a very specific line. Dahl's line is different from mine. His overriding consideration has been security. My effort has always been security but not to the same degree. I think that Dahl has gone too far in this respect and thereby lost quite a number of facilities, especially a thing like the "call by name". He can of course use a reference to a variable; this corresponds very closely to the FORTRAN type of "call by address", as opposed to the call by name in ALGOL and so for instance he can not use Jensens device. As you know in GPL, I use pointers. A pointer is not the same as a reference; it is a more general concept. So I think the loss of facilities here is a little too much to take for the sake of security.

The "virtuals" seem to be very closely corresponding to the "externals" in FORTRAN or assembly languages. But you see first of all you can only access things which belong to the same complex structure and secondly it seems to me that it is pretty hard to get type declarations for these procedures. You have to have declared the type of the value of the procedure and the type of parameters. In the example given the procedures seem to be parameterless and they do not deliver any value for the function. So I would like to know how Dahl would take care of that.

Dahl:

We think of SIMULA as an extension of ALGOL 60. We therefore provide exactly the same kind of specification for a virtual quantity as you would do for a formal parameter. You can write <u>procedure</u> *P*; <u>real</u> <u>procedure</u> *Q*; <u>array</u> *A*; and so forth.

I would much have preferred to specify the formal parameters of *P* within the virtual specification of *P* itself. Then, of course, alternative actual declarations in subclasses could have been simplified by omitting much of the procedure heading. This would have made it possible to check at compile time the actual parameters of a call for a virtual procedure. But in order to be consistent with ALGOL 60, we decided not to do it in this way.

The virtual quantities are in many ways similar to ALGOL's name parameters, but not quite as powerful. It turns out that there is no analogy to Jensen's device. This, I feel, is a good thing, because I hate to implement Jensen's device. It is awful.

If you specify a virtual real X, then you have the option to provide an actual declaration real X in a subclass. But you cannot declare a real expression for X. So, if you specify a quantity which looks like a variable, you can only provide an actual quantity which is a variable. This concept seems more clean to me than the call by name of ALGOL.

To begin with, the whole concept of virtual variables seemed to be superfluous because there was nothing more to say about a virtual variable than what had already been said in the specification. But there is: you can say whether or not it actually exists. A virtual variable X takes no space in the data record of an object if there is no actual declaration of X at any subclass level of the object. Therefore you can use the device for saving space, or for increasing the flexibility in attribute referencing without wasting space. If you access any virtual quantity out of turn, the implementation can catch you and give a run time error message. It is a problem similar to the "null" problem.

Strachey:

Supposing you had classes C and D, could you then define procedures P in both and if so, if you defined one in C and one in D, both being called P, which one would win? Do the scopes go the reverse way from the ordinary scopes or do they go the same way?

Dahl:

Thank you for reminding me of the problem which exists here. The concatenation rule states that declarations given at different prefix levels are brought together into a single block head. Name conflicts in a concatenated block head are regarded as errors of the same kind as redeclarations in an ordinary ALGOL block head. However, if there is a "name conflict" between a declared quantity and a virtual one, identity is established between the two, if the declaration and specification "match".

Strachey:

The other thing I was going to ask about is whether you have thought about the question of achieving security, not by making it impossible to refer to any thing which has gone away but by making it impossible to cause anything which is referred to, to go away. That is to say, by keeping an account of the number of pointers or references to each record, which is one of the methods of garbage collection and only letting it go away when this count reaches zero. The curious thing is this is generally faster than garbage collection.

Dahl:

We have made some experiments on that recently which suggest that it may not be faster.

Strachey:

Anyway, have you thought of this as an alternative method for providing security?

Dahl:

Evidently an actual parameter called by name is represented at run-time by a pointer of some kind, and you could achieve security by instructing the garbage collector to follow such pointers in addition to stored reference values. But then the price you pay for the call by name is much higher than for instance in ALGOL, where data referenced by any parameter has to be retained for other reasons. In my view, a call by name mechanism for classes would be a convenient device which would invite a programmer to entirely misuse the computer - by writing programs where no data can ever be de-allocated and without realizing it.

Petrone:

My first question was covered by Strachey but I now have another question which has arisen from his question. I am asking you whether the call by name mechanism was already present in the old SIMULA in the array case. And did you use it in garbage collection on arrays?

Dahl:

That is quite correct. There is a pointer from the object to the array, and the garbage collector will follow it. The reason why we did that is that an array is usually a big thing, which it is reasonable to regard as a separate object.

It is not reasonable to give a small thing like a real variable an independent existence, because that may cause very severe fragmentation of the store. Fragmentation is a disaster if you do not have a compacting scheme, and if you have one the fragmentation will tend to increase the time for each garbage collection and also the frequency of calling for it.

Petrone:

Your concatenation mechanism expresses the possibility of generating families of activity declarations - I am speaking now in terms of your old SIMULA - and the virtual mechanism seems to be a restricted call by name of quantities declared within such a family. Maybe it would be better to restrict the call by name to within an activity block, so that an activity block is equivalent to an ALGOL program with the full call by name mechanism available for procedures.

Dahl:

SIMULA in new and old versions has the complete call by name mechanism for parameters to procedures. You could also have name parameters to classes at no extra cost if you restricted any actual parameter called by name to be computable within the block enclosing the referenced class declaration. That is, it must only reference quantities which are local to that block or to outer blocks. But this is a rather unpleasant restriction considering that an actual parameter may be part of a generating expression occurring deep down in a block hierarchy.

Teaching Programming Principles: Pascal

Niklaus Wirth
Professor of Informatics (Emeritus)
ETH Zürich
wirth@inf.ethz.ch

Electronics Engineer, ETH Zürich,
Ph.D., University of California, Berkeley

Stanford University: Assistant Professor

IEEE Emanuel Piore Prize,
IEEE Computer Pioneer Award,
ACM Turing Award

Major contributions:
PL360, ALGOL W, Pascal,
Modula-2, Oberon

Current interests:
circuit design with programmable
devices, language Lola

Niklaus Wirth

Pascal and Its Successors

The programming language Pascal was designed in 1969 in the spirit of Algol 60 with a concisely defined syntax representing the paradigm of structured programming. Seven years later, with the advent of the micro-computer, it became widely known and was adopted in many schools and universities. In 1979 it was followed by Modula-2 which catered to the needs of modular programming in teams. This was achieved by the module construct and the separate compilation facility. In an effort to reduce language complexity and to accommodate object-oriented programming, Oberon was designed in 1988. Here we present some aspects of the evolution of this family of programming languages.

1. Introduction

Many times I have been asked how one "invents" a programming language. One cannot really tell, but it certainly is a matter of experience with the subject of programming, and of careful deliberation. Sometimes I answered: "Like one designs an airplane. One must identify a number of necessary building blocks and materials, and then assemble and combine them properly to a functioning whole." This answer may not be entirely satisfactory, but at least in both cases the result either flies or crashes.

Programming languages were one of the first topics that established computing science as a discipline with its own identity. The topic belonged neither to mathematics nor to electrical engineering. It was ALGOL 60 that introduced rigor and precision to the subject through its formal definition of syntax. A flurry of activities began in academia to investigate language properties, to find faults and inconsistencies, to devise powerful algorithms of syntax analysis, and to cope with the challenges of compilers. Soon the range of application of Algol was felt to be too narrow. A new, better language was required, perhaps a successor to Algol. Committees were established and hot controversies raged, some protagonists dreaming of grandiose formal systems, some thinking more modestly of a practical improvement. It was this environment that bred Pascal.

2. Structured Programming and Pascal

Pascal was born in 1969 out of an act of liberation [1] – in more than one sense. Confronted with the duty to teach programming, I had been faced with the dire options of Fortran and Algol. The former did not appeal to my taste as a scientist, the latter not to those of the practical engineer. I liberated myself from this jail by designing Pascal, convinced that an elegant style and an effective implementation were not mutually exclusive. I felt strongly – and still do – that a language used in teaching must display some style, elegance, consistency, while at the same time also reflecting the needs (but not necessarily bad habits) of practice. I wanted to design a language for both my classroom and my "software factory".

The second liberation alluded to the design constraint imposed by committee work. In 1966, Algol W [2] was a compromise bowing to many divergent opinions and requests from both an Algol committee and an Algol community. Surely, many of them were inspiring and beneficial, but others were incompatible and hindering. Some members had high ambitions of creating a language with novel features whose consequences were to be the subject of further research, whereas I, having brought up as an engineer, felt uneasy with proposals whose realization was still the subject of speculation. I wanted to have at least a concrete idea of how a construct was to be represented on available computers, and these, let me add, were rather ill-suited for any feature not already present in Fortran.

The general idea dominating the design of Pascal was to provide a language appealing to systematic thinking, mirroring conventional mathematical notation, satisfying the needs of practical programming, and encouraging a structured approach. The rules governing the language should be intuitive and simple, and freely combinable. For example, if x+y stands for an expression, x+y should be usable as a subexpression, in assignments, as a procedure parameter, or as an index. For example, if a widely established convention interprets x-y-z to mean (x-y)-z, we should not redefine this to denote x-(y-z). Or if x=y has been used for centuries to denote equality of x and y, we should refrain from the arrogance of replacing it with x==y. Clearly, Pascal was to build upon the notational grounds established by mathematics and Algol. Pascal and its successors were therefore called Algol-like.

Today, it is hard to imagine the circumstances prevailing in the 1960s. We must recall that the computing community was strictly split into two professional camps. The scientists and engineers used Fortran for their programming large-scale, word-oriented, binary computers, whereas the business community used Cobol for their smaller, character-oriented, decimal machines. System programmers were labouring within computer companies using proprietary machine-code assemblers. There were attempts to unite the two worlds, such as the highly innovative Burroughs B-5000 computer, or IBM's programming language PL/I. Both were ill-fated and devoured considerable budgets. Pascal was another such attempt, although less ambitious and without budget or industrial support. It applied the idea of recursively defined structures not only to executable statements, but also to data types. As elements, it adopted arrays (vectors, matrices) from Fortran and Algol, as well as records and files from Cobol. It allowed them to be freely combined and nested.

The other fact about the 1960s that is difficult to imagine today is the scarcity of computing resources. Computers with more than 8K of memory words and less than 10μs for the execution of an instruction were called super-computers. No wonder it was mandatory for the compiler of a new language to generate at least equally dense and efficient code as its Fortran competitor. Every instruction counted, and, for example, generating sophisticated subroutine calls catering to rarely used recursion was considered an academic pastime. Index checking at run-time was judged to be a superfluous luxury. In this context, it was hard if not hopeless to compete against highly optimized Fortran compilers.

Yet, computing power grew with each year, and with it the demands on software and on programmers. Repeated failures and blunders of industrial products revealed the inherent difficulties of intellectually mastering the ever increasing complexity of the new artefacts. The only solution lay in structuring programs, to let the programmer ignore the internal details of the pieces when assembling them into a larger whole. This school of thought was called *structured programming* [3], and Pascal was designed explicitly to support this discipline. Its foundations reached far deeper than simply "programming without go to statements" as some people believed. It is more closely related to the top-down approach to problem solving.

Besides structured statements, the concept of data types characterized Pascal profoundly. It implies that every object, be it a constant, a variable, a function, or a parameter has a type. Data typing introduces redundancy, and this redundancy can be used to detect inconsistencies, that is, errors. If the type of all objects can be determined by merely reading the program text, that is, without executing the program, then the type is called *static*, and checking can be performed by the compiler. Surely errors detected by the compiler are harmless and cheap compared to those detected during program execution in the field by the customer. Thus static typing became an important concept in software engineering, the discipline emerging in the 1970s to cope with the construction of large software complexes.

A particularly successful concept was the integration of pointers into static typing as suggested by Hoare [4] and adopted in Pascal. The simple idea is to attribute a fixed type not only with every data object, but also with every pointer, such that a pointer variable can at any time only refer to an object of the type to which it is bound (or to none at all). Programming with pointers, then called *list processing*, notoriously fraught with pitfalls, now became as safe as programming without pointers.

Yet, Pascal also suffered from certain deficiencies, more or less significant depending on personal perception and application. One of them had its origin in a too-dogmatic interpretation of static typing, requiring that the type of every procedure parameter be known at compile-time. Since this included index bounds in the case of array types, the frequently convenient dynamic arrays were excluded. In hindsight, this rigidity was silly and kept many Algolites from adopting Pascal. Arrays are typically passed by a reference, and for dynamic arrays only the array bounds must be added to this information. The limited additional complexity of the compiler would certainly have been outweighed by the gained language flexibility.

Certain other deficiencies were due to the author's lack of courage to throw some rules inherited from Algol overboard, in fear of antagonizing influential Algol programmers. The prime entry in this list is the famed go to statement, retained although, in principle, always replaceable by an if, while, or repeat construct. Another retained mistake was the lack of full type specification of parameters of formal procedures, through which, in principle, the entire type system could be undermined. This is illustrated by the following condensed, artificial example. Incidentally, it may also serve as an example of programming puzzles popular at the time.

```
PROCEDURE P(b: BOOLEAN; q: PROCEDURE);
    VAR i: INTEGER;

    PROCEDURE Q; BEGIN i := i+1 END Q;

BEGIN i := 0;
    IF b THEN P(FALSE, Q) ELSE q;
    Print(i)

END P
```

The puzzle: Which sequence of numbers will be printed by the call P(TRUE, P)? Note that no parameter types need be specified for q!

We are here confronted with a case where a certain combination of concepts leads to difficulties in interpretation, although each concept in isolation is harmless and well-defined. Here it is the combination of nested procedures, local scopes, and recursion that causes the problem. It is one of the outstanding challenges of language design to exclude unexpected effects.

Last but not least, Pascal adopted from Algol a few syntactic ambiguities – a deadly sin. I refer to the lack of an explicit closing symbol for nestable constructs. The prime example is the conditional statement. As a consequence, the nested if statement

```
IF b0 THEN IF b1 THEN S0 ELSE S1
```

can be interpreted as

```
IF b0 THEN [IF b1 THEN S0 ELSE S1]
```

or alternatively as

```
IF b0 THEN [IF b1 THEN S0] ELSE S1
```

This case demonstrates that one should not commit a mistake simply because everybody else does, particularly if there exists a known, elegant solution to eliminate the mistake. A thorough account of the development of Pascal is contained in [5].

3. Modular Programming and Modula-2

With various defects clearly identified and new challenges in programming emerging, the time seemed ripe for a fresh start, for a successor language. The two foremost novel challenges were multiprogramming and information hiding. For me personally, a third, quite practical challenge became an ambition: to create a language adequate for describing entire systems, from storage allocator to document editor, from process manager to compiler, and from display driver to graphics editor. I perceived that many problems in software development stemmed from the mixing of parts written in different languages. The challenge became real within our project to design and build the workstation Lilith in 1978 [7]. Its precursor was Xerox PARC's pioneering workstation Alto [6]. Alto's software was mostly written in Mesa; Lilith's software entirely in Modula-2. It would have been prohibitive to implement more than one language. Evidently, Modula was born out of an act of necessity [8].

The cornerstone of Modula-2 was the module construct. Whereas Pascal had served to build monolithic programs, Modula-2 was suitable for systems consisting of a hierarchy of units with properly defined interfaces. Such a unit was called *module*, and later *package* in Ada. In short, a module is like a Pascal program with the addition of an explicit interface specification to other modules. This is obtained as follows: Modules are described by two, distinct texts, a *definition part* and an *implementation* part. In the former all objects are defined which are visible by other modules, typically types and procedure signatures. They are said to be exported. The latter part contains all local, hidden objects, and the bodies of procedures, i.e. their implementations. Hence the term *information hiding*. The heading contains lists of identifiers to be imported from other modules. A small example follows:

```
DEFINITION MODULE Files;
    TYPE File;  (*opaque type*)
        Rider = RECORD eof: BOOLEAN END ;  (*other fields hidden*)

    PROCEDURE Set(VAR r: Rider; f: File; pos: INTEGER);
    PROCEDURE Read(VAR r: Rider; VAR ch: CHAR);
    PROCEDURE Write(VAR r: Rider; ch: CHAR);
    PROCEDURE Length(f: File): INTEGER;
END Files.
```

This key feature catered to the urgent demands of programming in teams. Now it became possible to determine jointly a modular decomposition of the task and to agree on the interfaces of the planned system. Thereafter, the team members could proceed independently in implementing the parts assigned to them. This style is called *modular programming*. The concept of module arose earlier in work by Parnas and, in conjunction with multiprogramming by Hoare and Brinch Hansen, where the module construct was called a *monitor* [9, 10]. The module was also present in a concrete form in Mesa, which in Modula was simplified and generalized.

The module construct would, however, have remained of mostly academic interest only, were it not for the technique of *separate compilation*, which was from its inception combined with the module. By separate compilation we understand that (1) full type checking is performed by the compiler not only within a module, but also across module interfaces, and (2) that compatibility (or version) checking between modules to be joined is achieved by a simple key comparison when the modules are linked and loaded. We refrain from expounding technical details, but emphasize that this is a crucial requirement for the design of complex systems, yet still poorly handled by most systems of commercial provenance.

Besides the successful feature of the module with separate compilation, the language also had some drawbacks. Surely, the evident deficiencies of Pascal had been mended. The syntax was now unambiguous, type specifications were complete, and the set of basic data types were adequately comprehensive. But as a result, the language, and with it the compiler, had become relatively large and bulky, although still orders of magnitude less than comparable commercial ventures. The goal of making the language powerful enough to describe entire systems was achieved by introducing certain low-level features, mostly for accessing particular machine resources (such as I/O device registers) and for breaching or overriding the rigid type system. Such facilities, like type casting, are inherently contrary to the notion of abstraction by high-level language, and should be avoided. They were called *loopholes*, because they allow breaking the rules imposed by the abstraction. But sometimes these rules appear too rigid, and use of a loophole becomes unavoidable. The dilemma was resolved through the module facility which would allow one to confine the use of such "naughty" tricks to specific, low-level server modules. It turned out that this was a naive view of the nature of programmers. The lesson: If you introduce a feature that can be abused, then it will be abused, and frequently so!

4. Object-Oriented Programming and Oberon

The advent of the personal computer around 1980 dramatically changed the way in which computers were used. Direct, immediate interaction replaced remote access and batch processing. User interfaces became an important issue. They were shaped by the novel mouse and the high-resolution display, replacing the 24-lines-by-80-characters screens. They established the paradigm of windows and multitasking. This paradigm was pioneered by the Xerox Alto workstation, and in particular the Smalltalk project [11]. Along with it came the *object-oriented* style of programming. Object-oriented design emerged from the specialized subject of discrete event simulation and its language Simula [12], whose authors Dahl and Nygaard had realized that its concepts had a scope far beyond simulation. Some of the proponents of object-oriented programming even suggested that all of programming should be converted to the new view of the world.

We felt that a revolution was undesirable to cure the lamented ills of the software profession, and we considered evolution as the wiser approach. Tempted to design a version of Modula stripped down to essentials, we also wanted to identify those features that were indispensable to encompass object orientation. Our findings were revealing: a single feature would suffice, all other ingredients were already present in Modula. The one feature to be added had to allow the construction of data type hierarchies, called *subclassing* in Smalltalk. Our own term was *type extension*: The new type *adds* attributes to those of the old type. Type extension had the welcome side effect of practically eliminating all needs for loopholes.

The absence of loopholes is the acid test for the quality of a language. After all, a language constitues an abstraction, a formal system, determined by a set of consistent rules and axioms. Loopholes serve to break these rules and can be understood only in terms of another, underlying system, an implementation. The principal purpose of a language, however, is to shield the programmer from implementation details, and to let him think exclusively in terms of the higher-level abstraction. Hence, a language should be fully defined without reference to any implementation or computer.

The language Oberon was born out of the ambition to simplify language to the essentials. The result turned out to be a considerably more powerful and more elegant language than its predecessors. The defining report of

Pascal required 30 pages, that of Modula grew to 45 pages, Oberon's could do with 16 [13]. Not surprisingly, implementations profited substantially from this achievement after 25 years of experience in language design – from a continuous evolution.

One of the simplifications was the reunification of the definition and implementation parts of modules. An Oberon module is again defined by a single text. Its heading contains a single list of server modules (rather than of individual, imported objects). Declarations of objects that are to be accessible in client modules are specially marked. Unmarked, local objects remain hidden. From a didactic point of view this reunification may be regrettable, because ideally definition parts are designed among team members and form contracts between them, whereas implementation parts can thereafter be designed by individual members without regard for the others, as long as the definition part remains untouched. However, the proliferation of files and the burden of keeping corresponding parts consistent was considered a drawback. Moreover, reunification eliminated the compiler's duty to check for consistency between definition and implementation parts. Also, a definition part can readily be extracted from the module text by a simple tool.

5. Conclusions and Outlook

In his article about Oberon, M. Franz wrote in [14]: "The world will remember Niklaus Wirth primarily as 'that language guy' and associate his name with Pascal." His observation is accurate; also the invitation to speak at this conference hinted that I should concentrate on Pascal. My disobedient digression stems from my conviction that its successors Modula and Oberon are much more mature and refined designs than Pascal. They form a family, and each descendant profited from experiences with its ancestors. At the end, the time span was 25 years.

Why, then, did Pascal capture all the attention, and Modula and Oberon got so little? Again I quote Franz: "This was, of course, partially of Wirth's own making." He continues: "He refrained from ... names such as Pascal-2, Pascal+, Pascal 2000, but instead opted for Modula and Oberon." Again Franz is right. To my defense I can plead that Pascal-2 and Pascal+ had already been taken by others for their own extensions of Pascal, and that I felt that these names would have been misleading for languages that were, although similar, syntactically distinct from Pascal. I emphasized progress rather than continuity, evidently a poor marketing strategy.

But of course the naming is by far not the whole story. For one thing, we were not sufficiently active – today we would say aggressive – in making our developments widely known. Instead of asking what went wrong with Modula and Oberon, however, let us rather ask what went right with Pascal. In my own perception, the following factors were decisive:

1. Pascal, incorporating the concepts of structured programming, was sufficiently different and progressive from Fortran to make a switch worth while. Particularly so in America, where Algol had remained virtually unknown.

2. In the early 1970s, an organization of users (Pascal User Group PUG) was formed and helped to make Pascal widely known and available. It also published a newsletter.

3. Pascal was ported just in time for the first micro computers (UCSD) [15], and thereby reached a large population of newcomers unconstrained by engrained habits and legacy code.

4. Pascal was picked up by start-up companies. Borland's Pascal implementation was the first compiler to be available for less than $50, turning it into a household article.

5. UCSD as well as Borland properly integrated the compiler into a complete development tool, including a program editor, a file system, and a debugging aid. This made Pascal highly attractive to schools and to beginners. It changed the manner in which programs were "written". A fast write-compile-test-correct cycle and interactivity were the new attractions.

6. Shortly after the initial spread, an ever growing number of text books on programming in Pascal appeared. They were as important as the compiler for Pascal's popularity in schools and universities. Text books and software entered a symbiotic partnership.

Perhaps it is worth observing that this chain reaction started around 1977, fully seven years after Pascal had been published and implemented on a CDC mainframe computer. Meanwhile, Pascal had been ported to numerous other large computers, but remained largely within universities. This porting effort was significantly facilitated by our project resulting in the Pascal P-compiler generating P-code, the predecessor of the later M-code (for Modula) and Java byte-code.

In contrast to Pascal, Modula and Oberon did not appear at a time when computing reached new segments of the population. The module concept was not perceived in teaching as sufficiently significant to warrant a change to a new, albeit similar language. Text books had been selected, investments in learning had been made, time was not ripe for a change. Industry did not exactly embrace Modula either, with a few exceptions, mainly in Britain. A more palatable solution was to extend Pascal, retaining upward compatibility and old shortcomings. And competition appeared in the form of C++ and Ada with powerful industrial backing.

Oberon fared even worse. It was virtually ignored by industry. This is astounding, as not only the elegant and powerful language was presented in 1988, but also a compact and fast compiler in 1990, along with a modern, flexible development environment for workstations, complete with window system, network, document and graphics editor, neatly fitting into about 200 Kbytes of memory, the compiler alone taking less than 50 Kbytes. The entire system was fully described in a single, comprehensive book, including its source code [16]. It carried the proof that such a system could be built using only a small fraction of manpower typically allotted to such endeavors, if proper methods and tools were employed.

One is tempted to rationalize this history with recent, deep-set changes in the computing field. Computing resources are no longer at a premium, memory is counted in megabytes rather than kilobytes as was done 15 years ago, instruction cycles in nanoseconds instead of microseconds, and clock frequencies in gigahertz instead of megahertz. The incentive to economize has dwindled alarmingly. The only scarce resource is manpower, competent manpower. Industry has difficulties even in finding good C programmers, those of finding Oberon programmers are insurmountable. So how could one reasonably expect companies to adopt Oberon?

We recognize a deadlock: the adoption of new tools and methods is impracticable, yet the retention of the old ones implies stagnation. Are we therefore condemned to eternally produce an ever growing mountain of software of ever growing complexity, software that nobody fully understands, although everybody is well aware of its defects and deficiencies? In order to avert this highly unpleasant vision of the future, some fresh starts will have to be undertaken now and then. They require the courage to discard and abandon, to select simplicity and transparency as design goals rather than complexity and obscure sophistication.

In the field of programming languages two fresh starts have been attempted recently. I refer to Java and C#. Both tend to restrict rather than augment the number of features, trying to suggest a certain programming discipline. Hence they move in the right direction. This is encouraging. But there still remains a long way to reach Pascal, and even longer to Oberon.

References

[1] N. Wirth. The programming language Pascal. *Acta Inf.*, 1, 35 – 63 (1971).

[2] N. Wirth and C.A.R. Hoare. A contribution to the development of ALGOL. Comm. ACM 9, 6, 413–432 (June 1966).

[3] E. W. Dijkstra. Notes on structured programming. In O.-J. Dahl, E. W. Dijkstra and C.A.R. Hoare, eds., *Structured Programming*. Academic Press, 1972.

[4] C.A.R. Hoare. Notes on data structuring. In O.-J. Dahl, E. W. Dijkstra and C.A.R. Hoare, eds., *Structured Programming*. Academic Press, 1972; and C.A.R. Hoare. Record handling. In F. Genuys, ed., *Programming Languages*. Academic Press, 1968.

[5] N. Wirth. Recollections about the development of Pascal. In T.J. Bergin, R.G. Gibson, eds., *History of Programming Languages II*. Addison-Wesley, 1996, ISBN 0-201-89502-1.

[6] C.P. Thacker et al. Alto: a personal computer. Tech. Report, Xerox PARC, CSL-79-11.

[7] N. Wirth. The personal computer Lilith. In *Proc. 5th International Conf. on Software Engineering*, IEEE Computer Society Press, 1981.

[8] N. Wirth. *Programming in Modula-2*. Springer, 1974, ISBN 0-387-50150-9.

[9] C.A.R. Hoare. An operating system structuring concept. *Comm. ACM*, 548–557 (Oct. 1974).

[10] P. Brinch Hansen. *Operating System Principles*. Prentice-Hall, 1973, ISBN 0-13-637843-9.

[11] A. Goldberg and A. Kay, eds. Smalltalk-72 Instruction Manual. Tech. Report, Xerox PARC, March 1976.

[12] O.-J. Dahl and C.A.R. Hoare. Hierarchical program structures. In O.-J. Dahl, E. W. Dijkstra and C.A.R. Hoare, eds., *Structured Programming*. Academic Press, 1972.

[13] N. Wirth. The programming language Oberon. *Software – Practice and Experience*, 18, 7, 671–690. (July 1988).

[14] M. Franz. Oberon: the overlooked jewel. In L. Böszörményi, J. Gutknecht and G. Pomberger, eds., *The School of Niklaus Wirth*. D-Punkt Verlag, 2000, ISBN 3-932588-85-1, and Morgan Kaufman Pub., 2000, ISBN 1-55860-723-4 .

[15] UCSD Pascal Users Manual. SofTech, 1981.

[16] N. Wirth and J. Gutknecht. *Project Oberon*. Addison-Wesley, 1992, ISBN 0-201-54428-8.

Niklaus Wirth

The Programming Language Pascal

Acta Informatica, Vol. 1, Fasc. 1, 1971
pp. 35-63

Acta Informatica 1, 35–63 (1971)

The Programming Language Pascal

N. Wirth*

Received October 30, 1970

Summary. A programming language called Pascal is described which was developed on the basis of Algol 60. Compared to Algol 60, its range of applicability is considerably increased due to a variety of data structuring facilities. In view of its intended usage both as a convenient basis to teach programming and as an efficient tool to write large programs, emphasis was placed on keeping the number of fundamental concepts reasonably small, on a simple and systematic language structure, and on efficient implementability. A one-pass compiler has been constructed for the CDC 6000 computer family; it is expressed entirely in terms of Pascal itself.

1. Introduction

The development of the language *Pascal* is based on two principal aims. The first is to make available a language suitable to teach programming as a systematic discipline based on certain fundamental concepts clearly and naturally reflected by the language. The second is to develop implementations of this language which are both reliable and efficient on presently available computers, dispelling the commonly accepted notion that useful languages must be either slow to compile or slow to execute, and the belief that any nontrivial system is bound to contain mistakes forever.

There is of course plenty of reason to be cautious with the introduction of yet another programming language, and the objection against teaching programming in a language which is not widely used and accepted has undoubtedly some justification—at least based on short-term commercial reasoning. However, the choice of a language for teaching based on its widespread acceptance and availability, together with the fact that the language most widely taught is thereafter going to be the one most widely used, forms the safest recipe for stagnation in a subject of such profound paedagogical influence. I consider it therefore well worth-while to make an effort to break this vicious circle.

Of course a new language should not be developed just for the sake of novelty; existing languages should be used as a basis for development wherever they meet the chosen objectives, such as a systematic structure, flexibility of program and data structuring, and efficient implementability. In that sense Algol 60 was used as a basis for Pascal, since it meets most of these demands to a much higher degree than any other standard language [1]. Thus the principles of structuring, and in fact the form of expressions, are copied from Algol 60. It was, however, not deemed appropriate to adopt Algol 60 as a subset of Pascal; certain construction principles, particularly those of declarations, would have been incom-

* Fachgruppe Computer-Wissenschaften, Eidg. Technische Hochschule, Zürich, Schweiz.

patible with those allowing a natural and convenient representation of the additional features of Pascal. However, conversion of ALGOL 60 programs to Pascal can be considered as a negligible effort of transcription, particularly if they obey the rules of the IFIP ALGOL Subset [2].

The main extensions relative to ALGOL 60 lie in the domain of data structuring facilities, since their lack in ALGOL 60 was considered as the prime cause for its relatively narrow range of applicability. The introduction of record and file structures should make it possible to solve commercial type problems with Pascal, or at least to employ it successfully to demonstrate such problems in a programming course. This should help erase the mystical belief in the segregation between scientific and commercial programming methods. A first step in extending the data definition facilities of ALGOL 60 was undertaken in an effort to define a successor to ALGOL in 1965 [3]. This language is a direct predecessor of Pascal, and was the source of many features such as e.g. the while and case statements and of record structures.

Pascal has been implemented on the CDC 6000 computers. The compiler is written in Pascal itself as a one-pass system which will be the subject of a subsequent report. The "dialect" processed by this implementation is described by a few amendments to the general description of Pascal. They are included here as a separate chapter to demonstrate the brevity of a manual necessary to characterise a particular implementation. Moreover, they show how facilities are introduced into this high-level, machine independent programming language, which permit the programmer to take advantage of the characteristics of a particular machine.

The syntax of Pascal has been kept as simple as possible. Most statements and declarations begin with a unique key word. This property facilitates both the understanding of programs by human readers and the processing by computers. In fact, the syntax has been devised so that Pascal texts can be scanned by the simplest techniques of syntactic analysis. This textual simplicity is particularly desirable, if the compiler is required to possess the capability to detect and diagnose errors and to proceed thereafter in a sensible manner.

2. Summary of the Language

An algorithm or computer program consists of two essential parts, a description of *actions* which are to be performed, and a description of the *data* which are manipulated by these actions. Actions are described in Pascal by so-called *statements*, and data are described by so-called *declarations* and *definitions*.

The data are represented by values of *variables*. Every variable occuring in a statement must be introduced by a *variable declaration* which associates an identifier and a data type with that variable. The *data type* essentially defines the set of values which may be assumed by that variable. A data type may in Pascal be either directly described in the variable declaration, or it may be referenced by a type identifier, in which case this identifier must be described by an explicit *type definition*.

The basic data types are the *scalar* types. Their definition indicates an ordered set of values, i.e. introduces an identifier as a constant standing for each value in the set. Apart from the definable scalar types, there exist in Pascal four *standard scalar types* whose values are not denoted by identifiers, but instead by numbers and quotations respectively, which are syntactically distinct from identifiers. These types are: *integer*, *real*, *char*, and *alfa*.

The set of values of type *char* is the character set available on the printers of a particular installation. *Alfa* type values consist of sequences of characters whose length again is implementation dependent, i.e. is the number of characters packed per word. Individual characters are not directly accessible, but *alfa* quantities can be unpacked into a character array (and vice-versa) by a standard procedure.

A scalar type may also be defined as a *subrange* of another scalar type by indicating the smallest and the largest value of the subrange.

Structured types are defined by describing the types of their components and by indicating a *structuring method*. The various structuring methods differ in the selection mechanism serving to select the components of a variable of the structured type. In Pascal, there are five structuring methods available: array structure, record structure, powerset structure, file structure, and class structure.

In an *array structure*, all components are of the same type. A component is selected by an array selector, or computable *index*, whose type is indicated in the array type definition and which must be scalar. It is usually a programmer-defined scalar type, or a subrange of the type *integer*.

In a *record structure*, the components (called *fields*) are not necessarily of the same type. In order that the type of a selected component be evident from the program text (without executing the program), a record selector does not contain a computable value, but instead consists of an identifier uniquely denoting the component to be selected. These component identifiers are defined in the record type definition.

A record type may be specified as consisting of several *variants*. This implies that different variables, although said to be of the same type, may assume structures which differ in a certain manner. The difference may consist of a different number and different types of components. The variant which is assumed by the current value of a record variable is indicated by a component field which is common to all variants and is called the *tag field*. Usually, the part common to all variants will consist of several components, including the tag field.

A *powerset structure* defines a set of values which is the powerset of its base type, i.e. the set of all subsets of values of the base type. The base type must be a scalar type, and will usually be a programmer-defined scalar type or a subrange of the type *integer*.

A *file structure* is a sequence of components of the same type. A natural ordering of the components is defined through the sequence. At any instance, only one component is directly accessible. The other components are made accessible through execution of standard file positioning procedures. A file is at any time in one of the three modes called *input*, *output*, and *neutral*. According to the mode, a file can be read sequentially, or it can be written by appending components to the existing sequence of components. File positioning procedures may influence the mode. The file type definition does not determine the number of components, and this number is variable during execution of the program.

The *class structure* defines a class of components of the same type whose number may alter during execution of a program. Each declaration of a variable with class structure introduces a set of variables of the component type. The set is initially empty. Every activation of the standard procedure *alloc* (with the class as implied parameter) will generate (or allocate) a new component variable in the class and yield a value through which this new component variable may be accessed. This value is called a *pointer*, and may be assigned to variables of type pointer. Every pointer variable, however, is through its declaration bound to a fixed class variable, and because of this *binding* may only assume values pointing to components of that class. There exists a pointer value **nil** which points to no component whatsoever, and may be assumed by any pointer variable irrespective of its binding. Through the use of class structures it is possible to construct data corresponding to any finite graph with pointers representing edges and component variables representing nodes.

The most fundamental statement is the *assignment statement*. It specifies that a newly computed value be assigned to a variable (or component of a variable). The value is obtained by evaluating an *expression*. Pascal defines a fixed set of operators, each of which can be regarded as describing a mapping from the operand types into the result type. The set of operators is subdivided into groups of

1. arithmetic operators of addition, subtraction, sign inversion, multiplication, division, and computing the remainder. The operand and result types are the types *integer* and *real*, or subrange types of *integer*.

2. Boolean operators of negation, union (or), and conjunction (and). The operand and result types are *Boolean* (which is a standard type).

3. set operators of union, intersection, and difference. The operands and results are of any powerset type.

4. relational operators of equality, inequality, ordering and set membership. The result of relational operations is of type *Boolean*. Any two operands may be compared for equality as long as they are of the same type. The ordering relations apply only to scalar types.

The assignment statement is a so-called *simple statement*, since it does not contain any other statement within itself. Another kind of simple statement is the *procedure statement*, which causes the execution of the designated procedure (see below). Simple statements are the components or building blocks of *structured statements*, which specify sequential, selective, or repeated execution of their components. Sequential execution of statements is specified by the *compound statement*, conditional or selective execution by the *if statement* and the *case statement*, and repeated execution by the *repeat statement*, the *while statement*, and the *for statement*. The if statement serves to make the execution of a statement dependent on the value of a *Boolean* expression, and the case statement allows for the selection among many statements according to the value of a selector. The for statement is used when the number of iterations is known beforehand, and the repeat and while statements are used otherwise.

A statement can be given a name (identifier), and be referenced through that identifier. The statement is then called a *procedure*, and its declaration a *procedure declaration*. Such a declaration may additionally contain a set of variable declarations, type definitions and further procedure declarations. The variables, types and procedures thus defined can be referenced only within the procedure itself, and are therefore called *local* to the procedure. Their identifiers have significance only within the program text which constitutes the procedure declaration and which is called the *scope* of these identifiers. Since procedures may be declared local to other procedures, scopes may be nested.

A procedure may have a fixed number of parameters, which are classified into constant-, variable-, procedure-, and function parameters. In the case of a variable parameter, its type has to be specified in the declaration of the formal parameter. If the actual variable parameter contains a (computable) selector, this selector is evaluated before the procedure is activated in order to designate the selected component variable.

Functions are declared analogously to procedures. In order to eliminate side-effects, assignments to non-local variables are not allowed to occur within the function.

3. Notation, Terminology, and Vocabulary

According to traditional Backus-Naur form, syntactic constructs are denoted by English words enclosed between the angular brackets ⟨and⟩. These words also describe the nature or meaning of the construct, and are used in the accompanying description of semantics. Possible repetition of a construct is indicated by an asterisk (0 or more repetitions) or a circled plus sign (1 or more repetitions). If a sequence of constructs to be repeated consists of more than one element, it is enclosed by the meta-brackets {and}.

The basic *vocabulary* consists of basic symbols classified into letters, digits, and special symbols.

```
⟨letter⟩ ::= A|B|C|D|E|F|G|H|I|J|K|L|M|N|O|P|Q|R|S|T|U|V|W|X|Y|Z|
             a|b|c|d|e|f|g|h|i|j|k|l|m|n|o|p|q|r|s|t|u|v|w|x|y|z
⟨digit⟩ ::= 0|1|2|3|4|5|6|7|8|9
⟨special symbol⟩ ::= +|−|*|/|∨|∧|¬|=|≠|<|>|≦|≧|(|)|[|]|{|}|:=|
             ₁₀|.|,|;|:|'|↑|div|mod|nil|in|
             if|then|else|case|of|repeat|until|while|do|
             for|to|downto|begin|end|with|goto|
             var|type|array|record|powerset|file|class|
             function|procedure|const
```

The construct

{⟨any sequence of symbols not containing "}"⟩}

may be inserted between any two identifiers, numbers (cf. 4), or special symbols. It is called a *comment* and may be removed from the program text without altering its meaning.

4. Identifiers and Numbers

Identifiers serve to denote constants, types, variables, procedures and functions. Their association must be unique within their scope of validity, i.e. within the procedure or function in which they are declared (cf. 10 and 11).

⟨identifier⟩ ::= ⟨letter⟩ ⟨letter or digit⟩*

⟨letter or digit⟩ ::= ⟨letter⟩ | ⟨digit⟩

The decimal notation is used for numbers, which are the constants of the data types *integer* and *real*. The symbol $_{10}$ preceding the scale factor is pronounced as "times 10 to the power of".

⟨number⟩ ::= ⟨integer⟩ | ⟨real number⟩
⟨integer⟩ ::= ⟨digit⟩$^{\oplus}$
⟨real number⟩ ::= ⟨digit⟩$^{\oplus}$. ⟨digit⟩$^{\oplus}$ |
    ⟨digit⟩$^{\oplus}$. ⟨digit⟩$^{\oplus}$$_{10}$ ⟨scale factor⟩ | ⟨integer⟩$_{10}$ ⟨scale factor⟩
⟨scale factor⟩ ::= ⟨digit⟩$^{\oplus}$ | ⟨sign⟩ ⟨digit⟩$^{\oplus}$
⟨sign⟩ ::= + | −

Examples:

1   100   0.1 $5_{10}-3$ $87.35_{10}+8$

5. Constant Definitions

A constant definition introduces an identifier as a synonym to a constant.

⟨unsigned constant⟩ ::= ⟨number⟩ | '⟨character⟩$^{\oplus}$' | ⟨identifier⟩ | **nil**
⟨constant⟩ ::= ⟨unsigned constant⟩ | ⟨sign⟩ ⟨number⟩
⟨constant definition⟩ ::= ⟨identifier⟩ = ⟨constant⟩

6. Data Type Definitions

A data type determines the set of values which variables of that type may assume and associates an identifier with the type. In the case of structured types, it also defines their structuring method.

⟨type⟩ ::= ⟨scalar type⟩ | ⟨subrange type⟩ | ⟨array type⟩ | ⟨record type⟩ |
    ⟨powerset type⟩ | ⟨file type⟩ | ⟨class type⟩ | ⟨pointer type⟩ |
    ⟨type identifier⟩
⟨type identifier⟩ ::= ⟨identifier⟩
⟨type definition⟩ ::= ⟨identifier⟩ = ⟨type⟩

6.1. Scalar Types

A scalar type defines an ordered set of values by enumeration of the identifiers which denote these values.

⟨scalar type⟩ ::= (⟨identifier⟩ {, ⟨identifier⟩}*)

Examples:

(*red, orange, yellow, green, blue*)
(*club, diamond, heart, spade*)
(*Monday, Tuesday, Wednesday, Thursday, Friday, Saturday, Sunday*)

Functions applying to all scalar types are:

succ the succeeding value (in the enumeration)
pred the preceding value (in the enumeration)

6.1.1. Standard Scalar Types

The following types are standard in Pascal, i.e. the identifier denoting them is predefined:

integer the values are the integers within a range depending on the particular implementation. The values are denoted by integers (cf. 4) and not by identifiers.

real the values are a subset of the real numbers depending on the particular implementation. The values are denoted by real numbers as defined in paragraph 4.

Boolean (*false, true*)

char the values are a set of characters depending on a particular implementation. They are denoted by the characters themselves enclosed within quotes.

alfa the values are sequences of n characters, where n is an implementation dependent parameter. If α and β are values of type alfa

$$\alpha = a_1 \dots a_k \dots a_n$$
$$\beta = b_1 \dots b_k \dots b_n,$$

then

$\alpha = \beta$, if and only if $a_i = b_i$ for $i = 1 \dots n$,
$\alpha < \beta$, if and only if $a_i = b_i$ for $i = 1 \dots k-1$ and $a_k < b_k$,
$\alpha > \beta$, if and only if $a_i = b_i$ for $i = 1 \dots k-1$ and $a_k > b_k$.

Alfa values are denoted by sequences of (at most) n characters enclosed in quotes. Trailing blanks may be omitted. Alfa quantities may be regarded as a packed representation of short character arrays (cf. also 10.1.3.).

6.1.2. Subrange Types

A type may be defined as a subrange of another scalar type by indication of the least and the highest value in the subrange. The first constant specifies the lower bound, and must not be greater than the upper bound.

⟨subrange type⟩ ::= ⟨constant⟩ .. ⟨constant⟩

Examples:

1..100
−10..+10
Monday..Friday

6.2. Structured Types

6.2.1. Array Types

An array type is a structure consisting of a fixed number of components which are all of the same type, called the *component type*. The elements of the array are designated by indices, values belonging to the so-called *index type*. The array type definition specifies the component type as well as the index type.

⟨array type⟩ ::= **array** [⟨index type⟩ **{**, ⟨index type⟩**}***] **of** ⟨component type⟩
⟨index type⟩ ::= ⟨scalar type⟩ | ⟨subrange type⟩ | ⟨type identifier⟩
⟨component type⟩ ::= ⟨type⟩

If n index types are specified, the array type is called *n-dimensional*, and a component is designated by n indices.

Examples:

array [1..100] **of** *real*
array [1..10, 1..20] **of** 0..99
array [−10..+10] **of** *Boolean*
array [*Boolean*] **of** *Color*

6.2.2. Record Types

A record type is a structure consisting of a fixed number of components, possibly of different types. The record type definition specifies for each component, called *field*, its type and an identifier which denotes it. The scope of these so-called *field identifiers* is the record definition itself, and they are also accessible within a field designator (cf. 7.2) refering to a record variable of this type.

A record type may have several *variants*, in which case a certain field is designated as the *tag field*, whose value indicates which variant is assumed by the record variable at a given time. Each variant structure is identified by a case label which is a constant of the type of the tag field.

⟨record type⟩ ::= **record** ⟨field list⟩ **end**
⟨field list⟩ ::= ⟨fixed part⟩ | ⟨fixed part⟩; ⟨variant part⟩ | ⟨variant part⟩
⟨fixed part⟩ ::= ⟨record section⟩ **{**; ⟨record section⟩**}***
⟨record section⟩ ::= ⟨field identifier⟩ **{**, ⟨field identifier⟩**}***: ⟨type⟩
⟨variant part⟩ ::= **case** ⟨tag field⟩ : ⟨type identifier⟩ **of** ⟨variant⟩ **{**; ⟨variant⟩**}***
⟨variant⟩ ::= **{**⟨case label⟩ :**}**$^{\oplus}$ (⟨field list⟩) | **{**⟨case label⟩**}**$^{\oplus}$
⟨case label⟩ ::= ⟨unsigned constant⟩
⟨tag field⟩ ::= ⟨identifier⟩

Examples:

```
record day: 1..31;
       month: 1..12;
       year: 0..2000
end

record name, firstname: alfa;
       age: 0..99;
end

record x, y: real;
       area: real;
case s: Shape of
triangle:  (side: real;
            inclination, angle1, angle2: Angle);
rectangle: (side1, side2: real;
            skew, angle3: Angle);
circle:    (diameter: real)
end
```

6.2.3. Powerset Types

A powerset type defines a range of values as the powerset of another scalar type, the so-called *base type*. Operators applicable to all powerset types are:

∨ union

∧ intersection

— set difference

in membership

⟨powerset type⟩ ::= **powerset** ⟨type identifier⟩ | **powerset** ⟨subrange type⟩

6.2.4. File Types

A file type definition specifies a structure consisting of a sequence of components, all of the same type. The number of components, called the *length* of the file, is not fixed by the file type definition, i.e. each variable of that type may have a value with a different, varying length.

Associated with each variable of file type is a *file position* or *file pointer* denoting a specific element. The file position or the file pointer can be moved by certain standard procedures, some of which are only applicable when the file is in one of the three *modes*: input (being read), output (being written), or neutral (passive). Initially, a file variable is in the neutral mode.

⟨file type⟩ ::= **file of** ⟨type⟩

6.2.5. Class Types

A class type definition specifies a structure consisting of a class of components, all of the same type. The number of components is variable; the initial number upon declaration of a variable of class type is zero. Components are created (allocated) during execution of the program through the standard procedure *alloc*. The maximum number of components which can thus be created, however, is specified in the type definition.

⟨class type⟩ ::= **class** ⟨maxnum⟩ **of** ⟨type⟩

⟨maxnum⟩ ::= ⟨integer⟩

6.2.6. Pointer Types

A pointer type is associated with every variable of class type. Its values are the potential pointers to the components of that class variable (cf. 7.5), and the pointer constant **nil**, designating no component. A pointer type is said to be *bound* to its class variable.

⟨pointer type⟩ ::= ↑⟨class variable⟩

⟨class variable⟩ ::= ⟨variable⟩

Examples of type definitions:

Color = (*red*, *yellow*, *green*, *blue*)
Sex = (*male*, *female*)
Charfile = **file of** *char*
Shape = (*triangle*, *rectangle*, *circle*)
Card = **array** [1..80] **of** *char*
Complex = **record** *realpart*, *imagpart*: *real* **end**
Person = **record** *name*, *firstname*: *alfa*;
age: *integer*;
married: *Boolean*;
father, *youngestchild*, *eldersibling*: ↑*family*;
case *s*: *Sex* **of**
male: (*enlisted*, *bold*: *Boolean*);
female: (*pregnant*: *Boolean*;
size: **array** [1..3] **of** *integer*)
end

7. Declarations and Denotations of Variables

Variable declarations consist of a list of identifiers denoting the new variables, followed by their type.

⟨variable declaration⟩ ::= ⟨identifier⟩ **{**, ⟨identifier⟩**}***: ⟨type⟩

Two *standard file variables* can be assumed to be predeclared as

input, *output*: **file of** *char*

The file *input* is restricted to input mode (reading only), and the file *output* is restricted to output mode (writing only). A Pascal program should be regarded as a procedure with these two variables as formal parameters. The corresponding actual parameters are expected either to be the standard input and output media of the computer installation, or to be specifyable in the system command activating the Pascal system.

Examples:

```
x, y, z: real
u, v: Complex
i, j: integer
k: 0..9
p, q: Boolean
operator: (plus, times, absval)
a: array[0..63] of real
b: array[Color, Boolean] of
            record occurrence: integer;
                   appeal: real
            end
c: Color
f: file of Card
hue1, hue2: powerset Color
family: class 100 of Person
p1, p2: ↑family
```

Denotations of variables either denote an entire variable or a component of a variable.

⟨variable⟩ ::= ⟨entire variable⟩ | ⟨component variable⟩

7.1. Entire Variables

An entire variable is denoted by its identifier.

⟨entire variable⟩ ::= ⟨variable identifier⟩

⟨variable identifier⟩ ::= ⟨identifier⟩

7.2. Component Variables

A component of a variable is denoted by the denotation for the variable followed by a selector specifying the component. The form of the selector depends on the structuring type of the variable.

⟨component variable⟩ ::= ⟨indexed variable⟩ | ⟨field designator⟩ |
⟨current file component⟩ | ⟨referenced component⟩

7.2.1. Indexed Variables

A component of an *n*-dimensional array variable is denoted by the denotation of the variable followed by *n* index expressions.

⟨indexed variable⟩ ::= ⟨array variable⟩ [⟨expression⟩ **{**, ⟨expression⟩**}***]

⟨array variable⟩ ::= ⟨variable⟩

The types of the index expressions must correspond with the index types declared in the definition of the array type.

Examples:

a [12]
a [*i* + *j*]
b [*red*, *true*]
b [*succ* (*c*), *p* ∧ *q*]
f ↑ [1]

7.2.2. Field Designators

A component of a record variable is denoted by the denotation of the record variable followed by the field identifier of the component.

⟨field designator⟩ ::= ⟨record variable⟩ . ⟨field identifier⟩
⟨record variable⟩ ::= ⟨variable⟩
⟨field identifier⟩ ::= ⟨identifier⟩

Examples:

u . *realpart*
v . *realpart*
b [*red*, *true*] . *appeal*
p2 ↑ . *size*

7.2.3. Current File Components

At any time, only the one component determined by the current file position (or file pointer) is directly accessible.

⟨current file component⟩ ::= ⟨file variable⟩ ↑
⟨file variable⟩ ::= ⟨variable⟩

7.2.4. Referenced Components

Components of class variables are referenced by pointers.

⟨referenced component⟩ ::= ⟨pointer variable⟩ ↑
⟨pointer variable⟩ ::= ⟨variable⟩

Thus, if *p1* is a pointer variable which is bound to a class variable *v*, *p1* denotes that variable and its pointer value, whereas *p1* ↑ denotes the component of *v* referenced by *p1*.

Examples:

p1 ↑ . *father*
p1 ↑ . *eldersibling* ↑ . *youngestchild*

8. Expressions

Expressions are constructs denoting rules of computation for obtaining values of variables and generating new values by the application of operators. Expressions consist of operands, i.e. variables and constants, operators, and functions.

The rules of composition specify operator *precedences* according to four classes of operators. The operator ¬ has the highest precedence, followed by the so-called multiplying operators, then the so-called adding operators, and finally, with the lowest precedence, the relational operators. Sequences of operators of the same precedence are executed from left to right. These rules of precedence are reflected by the following syntax:

⟨factor⟩ ::= ⟨variable⟩ | ⟨unsigned constant⟩ | ⟨function designator⟩ |
⟨set⟩ | (⟨expression⟩) | ¬ ⟨factor⟩

⟨set⟩ ::= [⟨expression⟩ **{**,⟨expression⟩**}***] | [ ]

⟨term⟩ ::= ⟨factor⟩ | ⟨term⟩ ⟨multiplying operator⟩ ⟨factor⟩

⟨simple expression⟩ ::= ⟨term⟩ |
⟨simple expression⟩ ⟨adding operator⟩ ⟨term⟩ |
⟨adding operator⟩ ⟨term⟩

⟨expression⟩ ::= ⟨simple expression⟩ |
⟨simple expression⟩ ⟨relational operator⟩
⟨simple expression⟩

Expressions which are members of a set must all be of the same type, which is the base type of the set. [ ] denotes the empty set.

Examples:

Factors:

x

15

$(x+y+z)$

$sin(x+y)$

$[red, c, green]$

$\neg p$

Terms:

$x * y$

$i/(1-i)$

$p \wedge q$

$(x \leqq y) \wedge (y < z)$

Simple expressions:

$x+y$

$-x$

$hue1 \vee hue2$

$i * j + 1$

9.1.1. Assignment Statements

The assignment statement serves to replace the current value of a variable by a new value indicated by an expression. The assignment operator symbol is :=, pronounced as "becomes".

⟨assignment statement⟩ ::= ⟨variable⟩ := ⟨expression⟩ |
⟨function identifier⟩ := ⟨expression⟩

The variable (or the function) and the expression must be must be of identical type (but neither class nor file type), with the following exceptions permitted:

1. the type of the variable is *real*, and the type of the expression is *integer* or a subrange thereof.

2. the type of the expression is a subrange of the type of the variable.

Examples:

$x := y + 2.5$
$p := (1 \leqq i) \wedge (i < 100)$
$i := sqr(k) - (i * j)$
$hue := [blue, succ(c)]$

9.1.2. Procedure Statements

A procedure statement serves to execute the procedure denoted by the procedure identifier. The procedure statement may contain a list of *actual parameters* which are substituted in place of their corresponding *formal parameters* defined in the procedure declaration (cf. 10). The correspondence is established by the positions of the parameters in the lists of actual and formal parameters respectively. There exist four kinds of parameters: variable-, constant-, procedure parameters (the actual parameter is a procedure identifier), and function parameters (the actual parameter is a function identifier).

In the case of variable parameters, the actual parameter must be a variable. If it is a variable denoting a component of a structured variable, the selector is evaluated when the substitution takes place, i.e. before the execution of the procedure. If the parameter is a constant parameter, then the corresponding actual parameter must be an expression.

⟨procedure statement⟩ ::= ⟨procedure identifier⟩ |
⟨procedure identifier⟩ (⟨actual parameter⟩
{, ⟨actual parameter⟩}*)
⟨procedure identifier⟩ ::= ⟨identifier⟩
⟨actual parameter⟩ ::= ⟨expression⟩ | ⟨variable⟩ |
⟨procedure identifier⟩ | ⟨function identifier⟩

Examples:

next
Transpose (a, n, m)
Bisect $(sin, -1, +2, x, q)$

9.1.3. Goto Statements

A goto statement serves to indicate that further processing should continue at another part of the program text, namely at the place of the label. Labels can be placed in front of statements being part of a compound statement (cf. 9.2.1.).

⟨goto statement⟩ ::= **goto** ⟨label⟩

⟨label⟩ ::= ⟨integer⟩

The following restriction holds concerning the applicability of labels:

The scope (cf. 10) of a label is the procedure declaration within which it is defined. It is therefore not possible to jump into a procedure.

9.2. Structured Statements

Structured statements are constructs composed of other statements which have to be executed either in sequence (compound statement), conditionally (conditional statements), or repeatedly (repetitive statements).

⟨structured statement⟩ ::= ⟨compound statement⟩ |
⟨conditional statement⟩ | ⟨repetitive statement⟩|
⟨with statement⟩

9.2.1. Compound Statements

The compound statement specifies that its component statements are to be executed in the same sequence as they are written. Each statement may be preceded by a label which can be referenced by a goto statement (cf. 9.1.3.).

⟨compound statement⟩ ::=
begin ⟨component statement⟩ **{**;⟨component statement⟩**}*** **end**

⟨component statement⟩ ::=
⟨statement⟩ | ⟨label definition⟩⟨statement⟩

⟨label definition⟩ ::= ⟨label⟩ :

Example:

begin *z* := *x*; *x* := *y*; *y* := *z* **end**

9.2.2. Conditional Statements

A conditional statement selects for execution a single one of its component statements.

⟨conditional statement⟩ ::= ⟨if statement⟩ | ⟨case statement⟩

9.2.2.1. If Statements

The if statement specifies that a statement be executed only if a certain condition (*Boolean* expression) is *true*. If it is *false*, then either no statement is to be executed, or the statement following the symbol **else** is to be executed.

⟨if statement⟩ ::= **if** ⟨expression⟩ **then** ⟨statement⟩ |
if ⟨expression⟩ **then** ⟨statement⟩ **else** ⟨statement⟩

The expression between the symbols **if** and **then** must be of type *Boolean*.

Note: The syntactic ambiguity arising from the construct

if ⟨expression—1⟩ **then if** ⟨expression—2⟩ **then** ⟨statement—1⟩
  else ⟨statement—2⟩

is resolved by interpreting the construct as equivalent to

if ⟨expression—1⟩ **then**
  begin if ⟨expression—2⟩ **then** ⟨statement—1⟩ **else** ⟨statement—2⟩
  end

Examples:

if $x<1.5$ **then** $z := x+y$ **else** $z := 1.5$
if $p \neq$ **nil then** $p := p\uparrow.father$

9.2.2.2. Case Statements

The case statement consists of an expression (the selector) and a list of statements, each being labeled by a constant of the type of the selector. It specifies that the one statement be executed whose label is equal to the current value of the selector.

⟨case statement⟩ ::= **case** ⟨expression⟩ **of**
  ⟨case list element⟩ **{**;⟨case list element⟩**}*** **end**
⟨case list element⟩ ::= **{**⟨case label⟩:**}**$^{\oplus}$ ⟨statement⟩ | **{**⟨case label⟩:**}**$^{\oplus}$

Example:

case *operator* **of**
plus: $x := x+y$;
times: $x := x*y$;
absval: **if** $x<0$ **then** $x := -x$
end

9.2.3. Repetitive Statements

Repetitive statements specify that certain statements are to be executed repeatedly. If the number of repetitions is known beforehand, i.e. before the repetitions are started, the for statement is the appropriate construct to express this situation; otherwise the while or repeat statement should be used.

⟨repetitive statement⟩ ::= ⟨while statement⟩ |
  ⟨repeat statement⟩ | ⟨for statement⟩

9.2.3.1. While Statements

⟨while statement⟩ ::= **while** ⟨expression⟩ **do** ⟨statement⟩

The expression controlling repetition must be of type *Boolean*. The statement is repeatedly executed until the expression becomes *false*. If its value is *false* at the beginning, the statement is not executed at all. The while statement

while e **do** S

is equivalent to

```
if e then
   begin S;
      while e do S
   end
```

Examples:

```
while (a[i] ≠ x) ∧ (i < n) do i := i + 1
while i > 0 do
begin if odd(i) then z := z * x;
      i := i div 2;
      x := sqr(x)
end
```

9.2.3.2. Repeat Statements

⟨repeat statement⟩ ::=
  repeat ⟨statement⟩ **{**;⟨statement⟩**}*** **until** ⟨expression⟩

The expression controlling repetition must be of type *Boolean*. The sequence of statements between the symbols **repeat** and **until** is repeatedly (and at least once) executed until the expression becomes *true*. The repeat statement

```
repeat S until e
```

is equivalent to

```
begin S;
   if ¬e then
      repeat S until e
end
```

Examples:

```
repeat k := i mod j;
       i := j;
       j := k
until j = 0

repeat get(f)
until (f↑ = a) ∨ eof(f)
```

9.2.3.3. For Statements

The for statement indicates that a statement is to be repeatedly executed while a progression of values is assigned to a variable which is called the *control variable* of the for statement.

⟨for statement⟩ ::= **for** ⟨control variable⟩ := ⟨for list⟩ **do** ⟨statement⟩

⟨for list⟩ ::= ⟨initial value⟩ **to** ⟨final value⟩ |
        ⟨initial value⟩ **downto** ⟨final value⟩

⟨control variable⟩ ::= ⟨identifier⟩

⟨initial value⟩ ::= ⟨expression⟩

⟨final value⟩ ::= ⟨expression⟩

The control variable, the initial value, and the final value must be of the same scalar type (or subrange thereof).

A for statement of the form

```
for v := e1 to e2 do S
```

is equivalent to the statement

```
if e1 ≦ e2 then
begin v := e1; S;
   for v := succ(v) to e2 do S
end
```

and a for statement of the form

```
for v := e1 downto e2 do S
```

is equivalent to the statement

```
if e1 ≧ e2 then
begin v := e1; S;
   for v := pred(v) downto e2 do S
end
```

Note: The repeated statement S must alter neither the value of the control variable nor the final value.

Examples:

```
for i := 2 to 100 do if a[i] > max then max := a[i]
for i := 1 to n do
for j := 1 to n do
begin x := 0;
   for k := 1 to n do x := x + a[i, k]* b[k, j];
   c[i, j] := x
end
for c := red to blue do try(c)
```

9.2.4. With Statements

⟨with statement⟩ ::= **with** ⟨record variable⟩ **do** ⟨statement⟩

Within the component statement of the with statement, the components (fields) of the record variable specified by the with clause can be denoted by their field identifier only, i.e. without preceding them with the denotation of the entire record variable. The with clause effectively opens the scope containing the field identifiers of the specified record variable, so that the field identifiers may occur as variable identifiers.

Example:

```
with date do
begin
   if month = 12 then
   begin month := 1; year := year + 1
   end else month := month + 1
end
```

This statement is equivalent to

```
begin
   if date.month = 12 then
   begin date.month := 1; date.year := date.year + 1
   end else date.month := date.month + 1
end
```

10. Procedure Declarations

Procedure declarations serve to define parts of programs and to associate identifiers with them so that they can be activated by procedure statements. A procedure declaration consists of the following parts, any of which, except the first and the last, may be empty:

```
⟨procedure declaration⟩ ::=
   ⟨procedure heading⟩
   ⟨constant definition part⟩ ⟨type definition part⟩
   ⟨variable declaration part⟩
   ⟨procedure and function declaration part⟩ ⟨statement part⟩
```

The procedure heading specifies the identifier naming the procedure and the formal parameter identifiers (if any). The parameters are either constant-, variable, procedure-, or function parameters (cf. also 9.1.2.).

```
⟨procedure heading⟩ ::= procedure ⟨identifier⟩; |
   procedure ⟨identifier⟩ (⟨formal parameter section⟩
    {;⟨formal parameter section⟩}*);
⟨formal parameter section⟩ ::=
   ⟨parameter group⟩ |
   const ⟨parameter group⟩ {;⟨parameter group⟩}* |
   var ⟨parameter group⟩ {;⟨parameter group⟩}* |
   function ⟨parameter group⟩ |
   procedure ⟨identifier⟩ {,⟨identifier⟩}*
⟨parameter group⟩ ::= ⟨identifier⟩ {,⟨identifier⟩}*: ⟨type identifier⟩
```

A parameter group without preceding specifier implies constant parameters.

The constant definition part contains all constant synonym definitions local to the procedure.

```
⟨constant definition part⟩ ::= ⟨empty⟩ |
   const ⟨constant definition⟩ {,⟨constant definition⟩}*;
```

The type definition part contains all type definitions which are local to the procedure declaration.

⟨type definition part⟩ ::= ⟨empty⟩ |
type ⟨type definition⟩ **{**;⟨type definition⟩**}***;

The variable declaration part contains all variable declarations local to the procedure declaration.

⟨variable declaration part⟩ ::= ⟨empty⟩ |
var ⟨variable declaration⟩ **{**;⟨variable declaration⟩**}***;

The procedure and function declaration part contains all procedure and function declarations local to the procedure declaration.

⟨procedure and function declaration part⟩ ::=
{⟨procedure or function declaration⟩ ;**}***
⟨procedure or function declaration⟩ ::=
⟨procedure declaration⟩ | ⟨function declaration⟩

The statement part specifies the algorithmic actions to be executed upon an activation of the procedure by a procedure statement.

⟨statement part⟩ ::= ⟨compound statement⟩

All identifiers introduced in the formal parameter part, the constant definition part, the type definition part, the variable-, procedure or function declaration parts are *local* to the procedure declaration which is called the *scope* of these identifiers. They are not known outside their scope. In the case of local variables, their values are undefined at the beginning of the statement part.

The use of the procedure identifier in a procedure statement within its declaration implies recursive execution of the procedure.

Examples of procedure declarations:

```
procedure readinteger (var x: integer);
   var i, j: integer;
begin i := 0;
   while (input↑ ≧ '0') ∧ (input↑ ≦ '9') do
   begin j := int(input↑) — int('0');
         i := i * 10 + j;
         get(input)
   end;
   x := i
end

procedure Bisect (function f: real; const low, high: real;
   var, zero: real; p: Boolean);
   var a, b, m: real;
begin a := low; b := high;
   if (f(a) ≧ 0) ∨ (f(b) ≦ 0) then p := false else
```

```
        begin p := true;
            while abs(a - b) > eps do
            begin m := (a + b)/2;
                if f(m) > 0 then b := m else a := m
            end;
            zero := a
        end
    end

    procedure GCD(m, n: integer; var, x, y, z: integer); {m ≧ 0, n > 0}
    var a1, a2, b1, b2, c, d, q, r: integer;
    begin {Greatest Common Divisor x of m and n,
            Extended Euclid's Algorithm, cf. [4], p. 14}
        c := m; d := n;
        a1 := 0;  a2 := 1;  b1 := 1;  b2 := 0;
        while d ≠ 0 do
        begin {a1*m + b1*n = d, a2*m + n2*n = c,
                gcd(c, d) = gcd(m, n)}
                q := c div d; r := c mod d;
                {c = q*d + r, gcd(d, r) = gcd(m, n)}
                a2 := a2 - q*a1; b2 := b2 - q*b1;
                {a2*m + b2*n = r, a1*m + b1*n = d}
                c := d;  d := r;
                r := a1;  a1 := a2;  a2 := r;
                r := b1;  b1 := b2;  b2 := r;
                {a1*m + b1*n = d, a2*m + b2*n = c,
                gcd(c, d) = gcd(m, n)}
        end;
        {gcd(c, 0) = c = gcd(m, n)}
        x := c;  y := a2;  z := b2
        {x = gcd(m, n), y*m + z*n = gcd(m, n)}
    end
```

10.1. Standard Procedures

Standard procedures are supposed to be predeclared in every implementation of Pascal. Any implementation may feature additional predeclared procedures. Since they are, as all standard quantities, assumed as declared in a scope surrounding the Pascal program, no conflict arises form a declaration redefining the same identifier within the program. The standard procedures are listed and explained below.

10.1.1. File Positioning Procedures

put(f) advances the file pointer of file *f* to the next file component. It is only applicable, if the file is either in the output or in the neutral mode. The file is put into the output mode.

get (*f*) advances the file pointer of file *f* to the next file component. It is only applicable, if the file is either in the input or in the neutral mode. If there does not exist a next file component, the end-of-file condition arises, the value of the variable denoted by $f\uparrow$ becomes undefined, and the file is put into the neutral mode.

reset (*f*) the file pointer of file *f* is reset to its beginning, and the file is put into the neutral mode.

10.1.2. Class Component Allocation Procedure

alloc (*p*) allocates a new component in the class to which the pointer variable *p* is bound, and assigns the pointer designating the new component to *p*. If the component type is a record type with variants, the form

alloc (*p*, *t*) can be used to allocate a component of the variant whose tag field value is *t*. However, this allocation does not imply an assignment to the tag field. If the class is already compleately allocated, the value **nil** will be assigned to *p*.

10.1.3. Data Transfer Procedures

Assuming that *a* is a character array variable, *z* is an alfa variable, and *i* is an integer expression, then

pack (*a*, *i*, *z*) packs the *n* characters $a[i] \ldots a[i+n-1]$ into the alfa variable *z* (for *n* cf. 6.1.1.), and

unpack (*z*, *a*, *i*) unpacks the alfa value *z* into the variables $a[i] \ldots a[i+n-1]$.

11. Function Declarations

Function declarations serve to define parts of the program which compute a scalar value or a pointer value. Functions are activated by the evaluation of a function designator (cf. 8.2) which is a constituent of an expression. A function declaration consists of the following parts, any of which, except the first and the last, may be empty (cf. also 10.).

⟨function declaration⟩ ::=
  ⟨function heading⟩
  ⟨constant definition part⟩ ⟨type definition part⟩
  ⟨variable declaration part⟩
  ⟨procedure and function declaration part⟩ ⟨statement part⟩

The function heading specifies the identifier naming the function, the formal parameters of the function (note that there must be at least one parameter), and the type of the (result of the) function.

⟨function heading ::= **function** ⟨identifier⟩ (⟨formal parameter section⟩
  {; ⟨formal parameter section⟩}*) : ⟨result type⟩ ;

⟨result type⟩ ::= ⟨type identifier⟩

The type of the function must be a scalar or a subrange type or a pointer type. Within the function declaration there must be at least one assignment statement assigning a value to the function identifier. This assignment determines the result of the function. Occurrence of the function identifier in a function designator within its declaration implies recursive execution of the function. Within the statement part no assignment must occur to any variable which is not local to the function. This rule also excludes assignments to parameters.

Examples:

```
function Sqrt(x: real): real;
    var x0, x1: real;
begin x1 := x;  {x>1, Newton's method}
    repeat x0 := x1; x1 := (x0 + x/x0)*0.5
        {x0² − 2*x1*x0 + x = 0}
    until abs(x1 − x0) ≦ eps;
    {(x0 − eps) ≦ x1 ≦ (x0 + eps),
    (x − 2*eps*x0) ≦ x0² ≦ (x + 2*eps*x0)}
    Sqrt := x0
end

function Max(a: vector; n: integer): real;
    var x: real;  i: integer;
begin x := a[1];
    for i := 2 to n do
    begin {x = max(a1 ... ai−1)}
        if x < a[i] then x := a[i]
        {x = max(a1 ... ai)}
    end;
    {x = max(a1 ... an)}
    Max := x
end

function GCD(m, n: integer): integer;
begin if n = 0 then GCD := m else GCD := GCD(n, m mod n)
end

function Power(x: real; y: integer): real; {y ≧ 0}
    var w, z: real;  i: integer;
begin w := x;  z := 1;  i := y;
    while i ≠ 0 do
    begin {z*w^i = x^y}
        if odd(i) then z := z*w;
        i := i div 2;  {z*w^2i = x^y}
        w := sqr(w)  {z*w^i = x^y}
    end;
    {i = 0, z = x^y}
    Power := z
end
```

11.1. Standard Functions

Standard functions are supposed to be predeclared in every implementation of Pascal. Any implementation may feature additional predeclared functions (cf. also 10.1.).

The standard functions are listed and explained below:

11.1.1. Arithmetic Functions

abs (x) computes the absolute value of x. The type of x must be either *real* or *integer*, and the type of the result is the type of x.

sqr (x) computes x^2. The type of x must be either *real* or *integer*, and the type of the result is the type of x.

sin (x)
cos (x)
exp (x)
ln (x)
sqrt (x)
arctan (x)
} the type of x must be either *real* or *integer*, and the type of the result is *real*

11.1.2. Predicates

odd (x) the type of x must be *integer*, and the result is x **mod** $2=1$

eof (f) indicates, whether the file f is in the end-of-file status.

11.1.3. Transfer Functions

trunc (x) x must be of type *real*, and the result is of type *integer*, such that $abs(x)-1 < trunc(abs(x)) \leqq abs(x)$

int (x) x must be of type *char*, and the result (of type *integer*) is the ordinal number of the character x in the defined character set.

chr (x) x must be of type *integer*, and the result (of *type char*) is the character whose ordinal number is x.

11.1.4. Further Standard Functions

succ (x) x is of any scalar or subrange type, and the result is the successor value of x (if it exists).

pred (x) x is of any scalar or subrange type, and the result is the predecessor value of x (if it exists).

12. Programs

A Pascal program has the form of a procedure declaration without heading (cf. also 7.4.).

⟨program⟩ ::= ⟨constant definition part⟩⟨type definition part⟩
⟨variable declaration part⟩
⟨procedure and function declaration part⟩⟨statement part⟩.

13. Pascal 6000

The version of the language Pascal which is processed by its implementation on the CDC 6000 series of computers is described by a number of amendments to the preceding Pascal language definition. The amendments specify extensions and restrictions and give precise definitions of certain standard data types. The section numbers used hereafter refer to the corresponding sections of the language definition.

3. Vocabulary

Only capital letters are available in the basic vocabulary of symbols. The symbol **eol** is added to the vocabulary. Symbols which consist of a sequence of underlined letters are called *word-delimiters*. They are written in Pascal 6000 without underlining and without any surrounding escape characters. Blanks or end-of-lines may be inserted anywhere except within :=, word-delimiters, identifiers, and numbers. The symbol $_{10}$ is written as '.

4. Identifiers

Only the 10 first symbols of an identifier are significant. Identifiers not differing in the 10 first symbols are considered as equal. Word-delimiters must not be used as identifiers. At least one blank space must be inserted between any two word-delimiters or between a word-delimiter and an adjacent identifier.

6. Data Types

6.1.1. Standard Scalar Types

integer is defined as
type *integer* $= -2^{48} + 1 \,..\, 2^{48} - 1$

real is defined according to the CDC 6000 floating point format specifications. Arithmetic operations on real type values imply rounding.

char is defined by the CDC 6000 display code character set. This set is incremented by the character denoted by **eol**, signifying end-of-line.

The ordered set is:

eol	*A*	*B*	*C*	*D*	*E*	*F*	*G*	*H*	*I*
J	*K*	*L*	*M*	*N*	*O*	*P*	*Q*	*R*	*S*
T	*U*	*V*	*W*	*X*	*Y*	*Z*	0	1	2
3	4	5	6	7	8	9	+	−	*
/	(	)	$	=	␣	,	.	'	[
]	:	≠	{	∨	∧	↑	}	<	>
≤	≥	¬	;						

(Note that the characters ' { } are special features on the printers of the ETH installation, and correspond to the characters ≡ ⌈ ↓ at standard CDC systems.)

alfa the number *n* of characters packed into an alfa value is 10 (cf. 6.1.1.).

6.2.3. Powerset Types

The base type of a powerset type must be either

1. a scalar type with less than 60 values, or

2. a subrange of the type *integer*, with a minimum element $min(T) \geqq 0$ and a maximum element $max(T) < 59$, or

3. a subrange of the type char with the maximum element $max(T) <$ '>'.

6.2.4. and 6.2.5. File and Class Types

No component of any structured type can be of a file type or of a class type.

7. Variable Declarations

File variables declared in the main program may be restricted to either input or output mode by appending the specifiers

[*in*] or [*out*]

to the file identifier in its declaration. Files restricted to input mode (input files) are expected to be Permanent Files attached to the job by the SCOPE Attach command, and files restricted to output mode may be catalogued as Permanent Files by the SCOPE Catalog command. In both commands, the file identifier is to be used as the Logical File Name [5].

10. and 11. Procedure and Function Declarations

A procedure or a function which contains local file declarations must not be activated recursively.

14. Glossary

actual parameter	9.1.2.
adding operator	8.1.3.
array type	6.2.1.
array variable	7.2.1.
assignment statement	9.1.1.
case label	6.2.2.
case list element	9.2.2.2.
case statement	9.2.2.2.
class type	6.2.5.
class variable	6.2.6.
component statement	9.2.1.
component type	6.2.1.
component variable	7.2.
compound statement	9.2.1.
conditional statement	9.2.2.
constant	5.
constant definition	5.
constant definition part	10.
control variable	9.2.3.3.
current file component	7.2.3.
digit	3.
entire variable	7.1.
expression	8.
factor	8.
field designator	7.2.2.
field identifier	7.2.2.
field list	6.2.2.
file type	6.2.4.
file variable	7.2.3.
final value	9.2.3.3.
fixed part	6.2.2.
for list	9.2.3.3.
for statement	9.2.3.3.
formal parameter section	10.
function declaration	11.
function designator	8.2.
function heading	11.
function identifier	8.2.
goto statement	9.1.3.
identifier	4.
if statement	9.2.2.1.
index type	6.2.1.
indexed variable	7.2.1.
initial value	9.2.3.3.
integer	4.
label	9.1.3.
label definition	9.2.1.
letter	3.
letter or digit	4.

The author gratefully acknowledges his indeptedness to C. A. R. Hoare for his many valuable suggestions concerning overall design strategy as well as details, and for his critical scrutiny of this paper.

References

1. Naur, P.: Report on the algorithmic language ALGOL 60. Comm ACM 3, 299–314 (1960).
2. Report on Subset ALGOL 60 (IFIP): Comm. ACM 7, 626–628 (1964).
3. Wirth, N., Hoare, C. A. R.: A contribution to the development of ALGOL. Comm. ACM 9, 413–432 (1966).
4. Knuth, D. E.: The art of computer programming, Vol. 1. Addison-Wesley 1968.
5. Control Data 6000 Computer Systems, SCOPE Reference Manual, Pub. No. 60189400.

Prof. Dr. N. Wirth
Eidgenössische Technische Hochschule
Fachgruppe Computer-Wissenschaften
Clausiusstraße 55
CH-8006 Zürich
Schweiz

Niklaus Wirth

Program Development by Stepwise Refinement

Communications of the ACM,
Vol. 14 (4), 1971
pp. 221-227

Program Development by Stepwise Refinement

Niklaus Wirth
Eidgenössische Technische Hochschule
Zürich, Switzerland

The creative activity of programming—to be distinguished from coding—is usually taught by examples serving to exhibit certain techniques. It is here considered as a sequence of design decisions concerning the decomposition of tasks into subtasks and of data into data structures. The process of successive refinement of specifications is illustrated by a short but nontrivial example, from which a number of conclusions are drawn regarding the art and the instruction of programming.

Key Words and Phrases: education in programming, programming techniques, stepwise program construction
CR Categories: 1.50, 4.0

1. Introduction

Programming is usually taught by examples. Experience shows that the success of a programming course critically depends on the choice of these examples. Unfortunately, they are too often selected with the prime intent to demonstrate what a computer can do. Instead, a main criterion for selection should be their suitability to exhibit certain widely applicable *techniques*. Furthermore, examples of programs are commonly presented as finished "products" followed by explanations of their purpose and their linguistic details. But active programming consists of the design of *new* programs, rather than contemplation of old programs. As a consequence of these teaching methods, the student obtains the impression that programming consists mainly of mastering a language (with all the peculiarities and intricacies so abundant in modern PL's) and relying on one's intuition to somehow transform ideas into finished programs. Clearly, programming courses should teach methods of design and construction, and the selected examples should be such that a gradual *development* can be nicely demonstrated.

This paper deals with a single example chosen with these two purposes in mind. Some well-known techniques are briefly demonstrated and motivated (strategy of preselection, stepwise construction of trial solutions, introduction of auxiliary data, recursion), and the program is gradually developed in a sequence of *refinement steps*.

In each step, one or several instructions of the given program are decomposed into more detailed instructions. This successive decomposition or refinement of specifications terminates when all instructions are expressed in terms of an underlying computer or programming language, and must therefore be guided by the facilities available on that computer or language. The result of the execution of a program is expressed in terms of data, and it may be necessary to introduce further data for communication between the obtained subtasks or instructions. As tasks are refined, so the data may have to be refined, decomposed, or structured, and it is natural to *refine program and data specifications in parallel*.

Every refinement step implies some design decisions. It is important that these decision be made explicit, and that the programmer be aware of the underlying criteria and of the existence of alternative solutions. The possible solutions to a given problem emerge as the leaves of a tree, each node representing a point of deliberation and decision. Subtrees may be considered as *families of solutions* with certain common characteristics and structures. The notion of such a tree may be particularly helpful in the situation of changing purpose and environment to which a program may sometime have to be adapted.

A guideline in the process of stepwise refinement should be the principle to decompose decisions as much as possible, to untangle aspects which are only seemingly interdependent, and to defer those decisions which concern details of representation as long as possible. This will result in programs which are easier to adapt to different environments (languages and computers), where different representations may be required.

The chosen sample problem is formulated at the beginning of section 3. The reader is strongly urged to try to find a solution by himself before embarking on the paper which—of course—presents only one of many possible solutions.

2. Notation

For the description of programs, a slightly augmented *Algol 60* notation will be used. In order to express repetition of statements in a more lucid way than by use of labels and jumps, a statement of the form

repeat ⟨statement sequence⟩
until ⟨Boolean expression⟩

is introduced, meaning that the statement sequence is to be repeated until the Boolean expression has obtained the value **true**.

error. Typically, there exists a set A of candidates for solutions, among which one is to be selected which satisfies a certain condition p. Thus a solution is characterized as an x such that $(x \in A) \wedge p(x)$.

A straightforward program to find a solution is:

repeat Generate the next element of A and call it x
until $p(x) \vee$ (no more elements in A);
if $p(x)$ **then** x = solution

The difficulty with this sort of problem usually is the sheer size of A, which forbids an exhaustive generation of candidates on the grounds of efficiency considerations. In the present example, A consists of $64!/(56! \times 8!) \doteq 2^{32}$ elements (board configurations). Under the assumption that generation and test of each configuration consumes 100 μs, it would roughly take 7 hours to find a solution. It is obviously necessary to invent a "shortcut," a method which eliminates a large number of "obviously" disqualified contenders. This *strategy of preselection* is characterized as follows: Find a representation of p in the form $p = q \wedge r$. Then let $B_r = \{x \mid (x \in A) \wedge r(x)\}$. Obviously $B_r \subseteq A$. Instead of generating elements of A, only elements of B are produced and tested on condition q instead of p. Suitable candidates for a condition r are those which satisfy the following requirements:

---

[1] This problem was investigated by C. F. Gauss in 1850.

...ing a power-
...posal, one might easily be content with this gain in performance. If one is less fortunate and is forced to, say, solve the problem by hand, it would take 280 hours of generating and testing configurations at the rate of one per second. In this case it might pay to spend some time finding further shortcuts. Instead of applying the same method as before, another one is advocated here which is characterized as follows: Find a representation of trial solutions x of the form $[x_1, x_2, \cdots, x_n]$, such that every trial solution can be generated in steps which produce $[x_1]$, $[x_1, x_2]$, $\cdots$, $[x_1, x_2, \cdots, x_n]$ respectively. The decomposition must be such that:

1. Every step (generating x_j) must be considerably simpler to compute than the entire candidate x.
2. $q(x) \supset q(x_1 \cdots x_j)$ for all $j \leq n$.

Thus a full solution can never be obtained by extending a partial trial solution which does not satisfy the predicate q. On the other hand, however, a partial trial solution satisfying q may not be extensible into a complete solution. This method of *stepwise construction of trial solutions* therefore requires that trial solutions failing at step j may have to be "shortened" again in order to try different extensions. This technique is called *backtracking* and may generally be characterized by the program:

```
j := 1;
repeat trystep j;
  if successful then advance else regress
until (j < 1) ∨ (j > n)
```

In the 8-queens example, a solution can be constructed by positioning queens in successive columns starting with column 1 and adding a queen in the next column in each step. Obviously, a partial configuration not satisfying the mutual nonaggression condition may never be extended by this method into a full solution. Also, since during the jth step only j queens have to be considered and tested for mutual nonaggression, finding a partial solution at step j requires less effort of inspection than finding a complete solution under the condition that all 8 queens are on the board all the time. Both stated criteria are therefore satisfied by the decomposition in which step j consists of finding a safe position for the queen in the jth column.

The program subsequently to be developed is based on this method; it generates and tests 876 partial configurations before finding a complete solution. Assuming again that each generation and test (which is now more easily accomplished than before) consumes one second, the solution is found in 15 minutes, and with the computer taking 100 μs per step, in 0.09 seconds.

4. Development of the Program

We now formulate the stepwise generation of partial solutions to the 8-queens problem by the following first version of a program:

```
variable board, pointer, safe;
considerfirstcolumn;
repeat trycolumn;
   if safe then
   begin setqueen; considernextcolumn
   end else regress
until lastcoldone ∨ regressoutoffirstcol
```

This program is composed of a set of more primitive instructions (or procedures) whose actions may be described as follows:

considerfirstcolumn. The problem essentially consists of inspecting the safety of squares. A pointer variable designates the currently inspected square. The column in which this square lies is called the currently inspected column. This procedure initializes the pointer to denote the first column.

trycolumn. Starting at the current square of inspection in the currently considered column, move down the column either until a safe square is found, in which case the Boolean variable *safe* is set to **true**, or until the last square is reached and is also unsafe, in which case the variable *safe* is set to **false.**

setqueen. A queen is positioned onto the last inspected square.

considernextcolumn. Advance to the next column and initialize its pointer of inspection.

regress. Regress to a column where it is possible to move the positioned queen further down, and remove the queens positioned in the columns over which regression takes place. (Note that we may have to regress over at most two columns. Why?)

The next step of program development was chosen to refine the descriptions of the instructions *trycolumn* and *regress* as follows:

```
procedure trycolumn;
repeat advancepointer; testsquare
until safe ∨ lastsquare

procedure regress;
   begin reconsiderpriorcolumn
```

3. The 8-Queens Problem and an Approach to Its Solution[1]

Given are an 8 $\times$ 8 chessboard and 8 queens which are hostile to each other. Find a position for each queen (a configuration) such that no queen may be taken by any other queen (i.e. such that every row, column, and diagonal contains at most one queen).

This problem is characteristic for the rather frequent situation where an analytical solution is not known, and where one has to resort to the method of trial and error. Typically, there exists a set A of candidates for solutions, among which one is to be selected which satisfies a certain condition p. Thus a solution is characterized as an x such that $(x \in A) \wedge p(x)$.

A straightforward program to find a solution is:

repeat Generate the next element of A and call it x
until $p(x) \vee$ (no more elements in A);
if $p(x)$ **then** x = solution

The difficulty with this sort of problem usually is the sheer size of A, which forbids an exhaustive generation of candidates on the grounds of efficiency considerations. In the present example, A consists of $64!/(56! \times 8!) \doteq 2^{32}$ elements (board configurations). Under the assumption that generation and test of each configuration consumes 100 μs, it would roughly take 7 hours to find a solution. It is obviously necessary to invent a "shortcut," a method which eliminates a large number of "obviously" disqualified contenders. This *strategy of preselection* is characterized as follows: Find a representation of p in the form $p = q \wedge r$. Then let $B_r = \{x \mid (x \in A) \wedge r(x)\}$. Obviously $B_r \subseteq A$. Instead of generating elements of A, only elements of B are produced and tested on condition q instead of p. Suitable candidates for a condition r are those which satisfy the following requirements:

---

[1] This problem was investigated by C. F. Gauss in 1850.

1. B_r is much smaller than A.
2. Elements of B_r are easily generated.
3. Condition q is easier to test than condition p.

The corresponding program then is:

repeat Generate the next element of B and call it x
until $q(x) \vee$ (no more elements in B);
if $q(x)$ **then** x = solution

A suitable condition r in the 8-queens problem is the rule that in every column of the board there must be exactly one queen. Condition q then merely specifies that there be at most one queen in every row and in every diagonal, which is evidently somewhat easier to test than p. The set B_r (configurations with one queen in every column) contains "only" $8^8 = 2^{24}$ elements. They are generated by restricting the movement of queens to columns. Thus all of the above conditions are satisfied.

Assuming again a time of 100 μs for the generation and test of a potential solution, finding a solution would now consume only 100 seconds. Having a powerful computer at one's disposal, one might easily be content with this gain in performance. If one is less fortunate and is forced to, say, solve the problem by hand, it would take 280 hours of generating and testing configurations at the rate of one per second. In this case it might pay to spend some time finding further shortcuts. Instead of applying the same method as before, another one is advocated here which is characterized as follows: Find a representation of trial solutions x of the form $[x_1, x_2, \cdots, x_n]$, such that every trial solution can be generated in steps which produce $[x_1]$, $[x_1, x_2]$, $\cdots$, $[x_1, x_2, \cdots, x_n]$ respectively. The decomposition must be such that:

1. Every step (generating x_j) must be considerably simpler to compute than the entire candidate x.
2. $q(x) \supset q(x_1 \cdots x_j)$ for all $j \leq n$.

Thus a full solution can never be obtained by extending a partial trial solution which does not satisfy the predicate q. On the other hand, however, a partial trial solution satisfying q may not be extensible into a complete solution. This method of *stepwise construction of trial solutions* therefore requires that trial solutions failing at step j may have to be "shortened" again in order to try different extensions. This technique is called *backtracking* and may generally be characterized by the program:

```
j := 1;
repeat trystep j;
  if successful then advance else regress
until (j < 1) ∨ (j > n)
```

In the 8-queens example, a solution can be constructed by positioning queens in successive columns starting with column 1 and adding a queen in the next column in each step. Obviously, a partial configuration not satisfying the mutual nonaggression condition may never be extended by this method into a full solution. Also, since during the jth step only j queens have to be considered and tested for mutual nonaggression, finding a partial solution at step j requires less effort of inspection than finding a complete solution under the condition that all 8 queens are on the board all the time. Both stated criteria are therefore satisfied by the decomposition in which step j consists of finding a safe position for the queen in the jth column.

The program subsequently to be developed is based on this method; it generates and tests 876 partial configurations before finding a complete solution. Assuming again that each generation and test (which is now more easily accomplished than before) consumes one second, the solution is found in 15 minutes, and with the computer taking 100 μs per step, in 0.09 seconds.

4. Development of the Program

We now formulate the stepwise generation of partial solutions to the 8-queens problem by the following first version of a program:

```
variable board, pointer, safe;
considerfirstcolumn;
repeat trycolumn;
  if safe then
  begin setqueen; considernextcolumn
  end else regress
until lastcoldone ∨ regressoutoffirstcol
```

This program is composed of a set of more primitive instructions (or procedures) whose actions may be described as follows:

considerfirstcolumn. The problem essentially consists of inspecting the safety of squares. A pointer variable designates the currently inspected square. The column in which this square lies is called the currently inspected column. This procedure initializes the pointer to denote the first column.

trycolumn. Starting at the current square of inspection in the currently considered column, move down the column either until a safe square is found, in which case the Boolean variable *safe* is set to **true**, or until the last square is reached and is also unsafe, in which case the variable *safe* is set to **false.**

setqueen. A queen is positioned onto the last inspected square.

considernextcolumn. Advance to the next column and initialize its pointer of inspection.

regress. Regress to a column where it is possible to move the positioned queen further down, and remove the queens positioned in the columns over which regression takes place. (Note that we may have to regress over at most two columns. Why?)

The next step of program development was chosen to refine the descriptions of the instructions *trycolumn* and *regress* as follows:

```
procedure trycolumn;
repeat advancepointer; testsquare
until safe ∨ lastsquare

procedure regress;
  begin reconsiderpriorcolumn
```

```
    if – regressoutoffirstcol then
    begin removequeen;
       if lastsquare then
       begin reconsiderpriorcolumn;
          if ¬ regressoutoffirstcol then
             removequeen
       end
    end
 end
```

The program is expressed in terms of the instructions:

considerfirstcolumn
considernextcolumn
reconsiderpriorcolumn
advancepointer
testsquare (sets the variable *safe*)
setqueen
removequeen

and of the predicates:

lastsquare
lastcoldone
regressoutoffirstcol

In order to refine these instructions and predicates further in the direction of instructions and predicates available in common programming languages, it becomes necessary to express them in terms of data representable in those languages. A decision on how to represent the relevant facts in terms of data can therefore no longer be postponed. First priority in decision making is given to the problem of how to represent the positions of the queens and of the square being currently inspected.

The most straightforward solution (i.e. the one most closely reflecting a wooden chessboard occupied by marble pieces) is to introduce a Boolean square matrix with $B[i, j]$ = **true** denoting that square (i, j) is occupied. The success of an algorithm, however, depends almost always on a suitable choice of its data representation in the light of the ease in which this representation allows the necessary operations to be expressed. Apart from this, consideration regarding storage requirements may be of prime importance (although hardly in this case). A common difficulty in program design lies in the unfortunate fact that at the stage where decisions about data representations have to be made, it often is still difficult to foresee the details of the necessary instructions operating on the data, and often quite impossible to estimate the advantages of one possible representation over another. In general, it is therefore advisable to delay decisions about data representation as long as possible (but not until it becomes obvious that no realizable solution will suit the chosen algorithm).

In the problem presented here, it is fairly evident even at this stage that the following choice is more suitable than a Boolean matrix in terms of simplicity of later instructions as well as of storage economy. j is the index of the currently inspected column; (x_j, j) is the coordinate of the last inspected square; and the position of the queen in column $k < j$ is given by the coordinate pair (x_k, k) of the board. Now the variable declarations for pointer and board are refined into:

```
integer j   (0 ≤ j ≤ 9)
integer array x[1:8]   (0 ≤ xᵢ ≤ 8)
```

and the further refinements of some of the above instructions and predicates are expressed as:

```
procedure considerfirstcolumn;
  begin  j := 1;  x[1] := 0 end
procedure considernextcolumn;
  begin— j := j+1;  x[j] := 0 end
procedure reconsiderpriorcolumn; j := j−1
procedure advancepointer;
  x[j] := x[j] + 1
Boolean procedure lastsquare;
  lastsquare := x[j] = 8
Boolean procedure lastcoldone;
  lastcoldone := j > 8
```

```
Boolean procedure regressoutoffirstcol;
  regressoutoffirstcol := j < 1
```

At this stage, the program is expressed in terms of the instructions:

testsquare
setqueen
removequeen

As a matter of fact, the instructions *setqueen* and *removequeen* may be regarded as vacuous, if we decide that the procedure *testsquare* is to determine the value of the variable *safe* solely on the grounds of the values $x_1 \cdots x_{j-1}$ which completely represent the positions of the $j - 1$ queens so far on the board. But unfortunately the instruction *testsquare* is the one most frequently executed, and it is therefore the one instruction where considerations of efficiency are not only justified but essential for a good solution of the problem. Evidently a version of *testsquare* expressed only in terms of $x_1 \cdots x_{j-1}$ is inefficient at best. It should be obvious that *testsquare* is executed far more often than *setqueen* and *removequeen*. The latter procedures are executed whenever the column (j) is changed (say m times), the former whenever a move to the next square is undertaken (i.e. x_j is changed, say n times). However, *setqueen* and *removequeen* are the only procedures which affect the chessboard. Efficiency may therefore be gained by the method of *introducing auxiliary variables* $V(x_1 \cdots x_j)$ such that:

1. Whether a square is safe can be computed more easily from $V(x)$ than from x directly (say in u units of computation instead of ku units of computation).
2. The computation of $V(x)$ from x (whenever x changes) is not too complicated (say of v units of computation).

The introduction of V is advantageous (apart from considerations of storage economy), if

$$n(k-1)u > mu \qquad \text{or} \qquad \frac{n}{m}(k-1) > \frac{v}{u},$$

i.e. if the gain is greater than the loss in computation units.

A most straightforward solution to obtain a simple version of *testsquare* is to introduce a Boolean matrix B such that $B[i, j]$ = **true** signifies that square (i, j) is not taken by another queen. But unfortunately, its recomputation whenever a new queen is removed (v) is prohibitive (why?) and will more than outweigh the gain.

The realization that the relevant condition for safety of a square is that the square must lie neither in a row nor in a diagonal already occupied by another queen, leads to a much more economic choice of V. We introduce Boolean arrays a, b, c with the meanings:

a_k = **true** : no queen is positioned in row k
b_k = **true** : no queen is positioned in the /-diagonal k
c_k = **true** : no queen is positioned in the \\-diagonal k

The choice of the index ranges of these arrays is made in view of the fact that squares with equal sum of their coordinates lie on the same /-diagonal, and those with equal difference lie on the same \\-diagonal. With row and column indices from 1 to 8, we obtain:

```
Boolean array a[1:8], b[2:16], c[-7:7]
```

Upon every introduction of auxiliary data, care has to be taken of their *correct initialization*. Since our algorithm starts with an empty chessboard, this fact must be represented by initially assigning the value **true** to all components of the arrays a, b, and c. We can now write:

```
procedure testsquare;
  safe := a[x[j]] ∧ b[j+x[j]] ∧ c[j−x[j]]
procedure setqueen;
  a[x[j]] := b[j+x[j]] := x[j−x[j]] := false
procedure removequeen;
  a[x[j]] := b[j+x[j]] := c[j−x[j]] := true
```

The correctness of the latter procedure is based on the fact that each queen currently on the board had been positioned on a safe square, and that all queens positioned after the one to be removed now had already been removed. Thus the square to be vacated becomes safe again.

A critical examination of the program obtained so far reveals that the variable $x[j]$ occurs very often, and is not taken by another queen. But unfortunately, its recomputation whenever a new queen is removed (v) is prohibitive (why?) and will more than outweigh the gain.

The realization that the relevant condition for safety of a square is that the square must lie neither in a row nor in a diagonal already occupied by another queen, leads to a much more economiç choice of V. We introduce Boolean arrays a, b, c with the meanings:

a_k = **true** : no queen is positioned in row k
b_k = **true** : no queen is positioned in the /-diagonal k
c_k = **true** : no queen is positioned in the \-diagonal k

The choice of the index ranges of these arrays is made in view of the fact that squares with equal sum of their coordinates lie on the same /-diagonal, and those with equal difference lie on the same \-diagonal. With row and column indices from 1 to 8, we obtain:

```
Boolean array a[1:8], b[2:16], c[-7:7]
```

Upon every introduction of auxiliary data, care has to be taken of their *correct initialization*. Since our algorithm starts with an empty chessboard, this fact must be represented by initially assigning the value **true** to all components of the arrays a, b, and c. We can now write:

```
procedure testsquare;
  safe := a[x[j]] ∧ b[j+x[j]] ∧ c[j−x[j]]
procedure setqueen;
  a[x[j]] := b[j+x[j]] := x[j−x[j]] := false
```

```
procedure removequeen;
  a[x[j]] := b[j+x[j]] := c[j−x[j]] := true
```

The correctness of the latter procedure is based on the fact that each queen currently on the board had been positioned on a safe square, and that all queens positioned after the one to be removed now had already been removed. Thus the square to be vacated becomes safe again.

A critical examination of the program obtained so far reveals that the variable $x[j]$ occurs very often, and in particular at those places of the program which are also executed most often. Moreover, examination of $x[j]$ occurs much more frequently than reassignment of values to j. As a consequence, the principle of introduction of auxiliary data can again be applied to increase efficiency: a new variable

```
integer i
```

is used to represent the value so far denoted by $x[j]$. Consequently $x[j] := i$ must always be executed before j is increased, and $i := x[j]$ after j is decreased. This final step of program development leads to the reformulation of some of the above procedures as follows:

```
procedure testsquare;
  safe := a[i] ∧ b[i+j] ∧ c[i−j]
procedure setqueen;
  a[i] := b[i+j] := c[i−j] := false
procedure removequeen;
  a[i] := b[i+j] := c[i−j] := true
procedure considerfirstcolumn;
  begin j := 1;   i := 0 end
procedure advancepointer;   i := i+1
procedure considernextcolumn;
  begin x[j] := i;   j := j+1;   i := 0 end
Boolean procedure lastsquare;
  lastsquare := i = 8
```

The final program, using the procedures

```
testsquare
setqueen
regress
removequeen
```

and with the other procedures directly substituted, now has the form

```
j := 1;  i := 0;
repeat
  repeat i := i+1;  testsquare
  until safe ∨ (i=8);
  if safe then
  begin setqueen;  x[j] := i;  j := j+1;  i := 0
  end else regress
until (j > 8) ∨ (j < 1);
if j > 8 then PRINT(x) else FAILURE
```

It is noteworthy that this program still displays the structure of the version designed in the first step. Naturally other, equally valid solutions can be suggested and be developed by the same method of stepwise program refinement. It is particularly essential to demonstrate this fact to students. One alternative solution was suggested to the author by E. W. Dijkstra. It is based on the view that the problem consists of a stepwise extension of the board by one column containing a safely positioned queen, starting with a null-board and terminating with 8 columns. The process of extending the board is formulated as a procedure, and the natural method to obtain a complete board is by *recursion* of this procedure. It can easily be composed of the same set of more primitive instructions which were used in the first solution.

```
procedure Trycolumn(j);
  begin integer i;  i := 0;
    repeat i := i+1;  testsquare;
      if safe then
      begin setqueen;  x[j] := i;
        if j < 8 then Trycolumn (j+1);
        if ¬ safe then removequeen
      end
    until safe ∨ (i=8)
  end
```

The program using this procedure then is

Trycolumn(1);
if *safe* **then** PRINT(x) **else** FAILURE

(Note that due to the introduction of the variable *i* local to the recursive procedure, every column has its own pointer of inspection *i*. As a consequence, the procedures

testsquare
setqueen
removequeen

must be declared locally within *Trycolumn* too, because they refer to the *i* designating the scanned square in the *current* column.)

5. The Generalized 8-Queens Problem

In the practical world of computing, it is rather uncommon that a program, once it performs correctly and satisfactorily, remains unchanged forever. Usually its users discover sooner or later that their program does not deliver all the desired results, or worse, that the results requested were not the ones really needed. Then either an extension or a change of the program is called for, and it is in this case where the method of stepwise program design and systematic structuring is most valuable and advantageous. If the structure and the program components were well chosen, then often many of the constituent instructions can be adopted unchanged. Thereby the effort of redesign and reverification may be drastically reduced. As a matter of fact, the *adaptability* of a program to changes in its objectives (often called maintainability) and to changes in its environment (nowadays called portability) can be measured primarily in terms of the degree to which it is neatly structured.

It is the purpose of the subsequent section to demonstrate this advantage in view of a generalization of the original 8-queens problem and its solution through an extension of the program components introduced before.

The generalized problem is formulated as follows:

> Find *all* possible configurations of 8 hostile queens on an 8 × 8 chessboard, such that no queen may be taken by any other queen.

The new problem essentially consists of two parts:

1. Finding a method to generate further solutions.
2. Determining whether all solutions were generated or not.

It is evidently necessary to generate and test candidates for solutions in some *systematic manner*. A common technique is to find an *ordering of candidates* and a condition to identify the last candidate. If an ordering is found, the solutions can be mapped onto the integers. A condition limiting the numeric values associated with the solutions then yields a criterion for termination of the algorithm, if the chosen method generates solutions strictly in increasing order.

It is easy to find orderings of solutions for the present problem. We choose for convenience the mapping

$$M(x) = \sum_{j=1}^{8} x_j 10^{j-1}$$

An upper bound for possible solutions is then

$$M(x_{\max}) = 88888888$$

and the "convenience" lies in the circumstance that our earlier program generating one solution generates the minimum solution which can be regarded as the starting point from which to proceed to the next solution. This is due to the chosen method of testing squares strictly proceeding in increasing order of $M(x)$ starting with 00000000. The method for generating further solutions must now be chosen such that starting with the configuration of a given solution, scanning proceeds in the same order of increasing M, until either the next higher solution is found or the limit is reached.

6. The Extended Program

The technique of extending the two given programs finding a solution to the simple 8-queens problem is based on the idea of modification of the global structure only, and of using the same building blocks. The global structure must be changed such that upon finding a solution the algorithm will produce an appropriate indication—e.g. by printing the solution—and then proceed to find the next solution until it is found or the limit is reached. A simple condition for reaching the limit is the event when the first queen is moved beyond row 8, in which case regression out of the first column will take place. These deliberations lead to the following modified version of the nonrecursive program:

```
considerfirstcolumn;
  repeat trycolumn;
    if safe then
    begin setqueen;  considernextcolumn;
      if lastcoldone then
      begin PRINT(x); regress
      end
    end else regress
  until regressoutoffirstcol
```

Indication of a solution being found by printing it now occurs directly at the level of detection, i.e. before leaving the repetition clause. Then the algorithm proceeds to find a next solution whereby a shortcut is used by directly regressing to the prior column; since a solution places one queen in each row, there is no point in further moving the last queen within the eighth column.

The recursive program is extended with even greater ease following the same considerations:

```
procedure Trycolumn(j);
begin integer i;
   ⟨declarations of procedures testsquare, advancequeen,
   setqueen, removequeen, lastsquare⟩
   i := 0;
   repeat advancequeen;   testsquare;
      if safe then
      begin setqueen;   x[j] := i;
         if ¬ lastcoldone then Trycolumn(j+1) else PRINT(x);
         removequeen
      end
   until lastsquare
end
```

The main program starting the algorithm then consists (apart from initialization of *a*, *b*, and *c*) of the single statement *Trycolumn*(1).

In concluding, it should be noted that both programs represent the same algorithm. Both determine 92 solutions in the *same* order by testing squares 15720 times. This yields an average of 171 tests per solution; the maximum is 876 tests for finding a next solution (the first one), and the minimum is 8. (Both programs coded in the language Pascal were executed by a CDC 6400 computer in less than one second.)

7. Conclusions

The lessons which the described example was supposed to illustrate can be summarized by the following points.

1. Program construction consists of a sequence of *refinement steps*. In each step a given task is broken up into a number of subtasks. Each refinement in the description of a task may be accompanied by a refinement of the description of the data which constitute the means of communication between the subtasks. Refinement of the description of program and data structures should proceed in parallel.

2. The degree of *modularity* obtained in this way will determine the ease or difficulty with which a program can be adapted to changes or extensions of the purpose or changes in the environment (language, computer) in which it is executed.

3. During the process of stepwise refinement, a *notation* which is natural to the problem in hand should be used as long as possible. The direction in which the notation develops during the process of refinement is determined by the language in which the program must ultimately be specified, i.e. with which the notation ultimately becomes identical. This language should therefore allow us to express as naturally and clearly as possible the structures of program and data which emerge during the design process. At the same time, it must give guidance in the refinement process by exhibiting those basic features and structuring principles which are natural to the machine by which programs are supposed to be executed. It is remarkable that it would be difficult to find a language that would meet these important requirements to a lesser degree than the one language still used most widely in teaching programming: Fortran.

4. Each refinement implies a number of *design decisions* based upon a set of design criteria. Among these criteria are efficiency, storage economy, clarity, and regularity of structure. Students must be taught to be conscious of the involved decisions and to critically examine and to reject solutions, sometimes even if they are correct as far as the result is concerned; they must learn to weigh the various aspects of design alternatives in the light of these criteria. In particular, they must be taught to revoke earlier decisions, and to back up, if necessary even to the top. Relatively short sample problems will often suffice to illustrate this important point; it is not necessary to construct an operating system for this purpose.

5. The detailed elaborations on the development of even a short program form a long story, indicating that

careful programming is not a trivial subject. If this paper has helped to dispel the widespread belief that programming is easy as long as the programming language is powerful enough and the available computer is fast enough, then it has achieved one of its purposes.

Acknowledgments. The author gratefully acknowledges the helpful and stimulating influence of many discussions with C.A.R. Hoare and E. W. Dijkstra.

References

The following articles are listed for further reference on the subject of programming.

1. Dijkstra, E. W. A constructive approach to the problem of program correctness. *BIT 8* (1968), 174–186.
2. Dijkstra, E. W. Notes on structured programming. EWD 249, Technical U. Eindhoven, The Netherlands, 1969.
3. Naur, P. Programming by action clusters. *BIT 9* (1969) 250–258.
4. Wirth, N. Programming and programming languages. Proc. Internat. Comput. Symp., Bonn, Germany, May 1970.

The /360 Architecture and Its Operating System

Frederick P. Brooks
Kenan Professor of Computer Science
University of North Carolina at Chapel Hill
brooks@cs.unc.edu

Ph.D. in Computer Science, Harvard

IBM: Corporate Project Manager

University of North Carolina: Chair of Department of Computer Science

National Medal of Technology, IEEE John von Neumann Medal, ACM Turing Award, Bower Science Award

Member of National Academy of Engineering

Fellow of the American Academy of Arts and Sciences

Major contributions: architecture of the IBM Stretch and Harvest computers and the System/360 hardware, development of the OS/360 operating system

Current interests: computer architecture, software engineering, virtual reality

Frederick P. Brooks

The IBM Operating System/360

The System/360 Computer Family

From its inception in 1961 until mid-1965, I had the once-in-a-lifetime opportunity to manage the IBM System/360 Computer Family Project – first the hardware and then the Operating System/360 software. This computer family, announced on April 7, 1964, and first shipped in February, 1965, defined "third-generation" computers and introduced (semi-)integrated-circuit technology into off-the-shelf computer products.

Context

Since many of my audience weren't born then, let me briefly set the context and, in particular, IBM's computer product situation as of 1961. IBM's first-generation (vacuum-tube technology) computers – the 701 large scientific machine, the 702 large commercial machine, and the 650 small universal machine had each evolved a whole family of successors, by 1961 all second-generation (discrete-transistor-technology): 701→7094, 702→7080 III, 650→7074, 650→1620. These had been joined by new second-generation sibling families: the 1401 small commercial and fast

I/O machine and successors, 1401-1410-7010, and the Stretch (7030) supercomputer. Each of these families had reached an end-of-the-line architecturally, chiefly because the instruction formats could not address as much memory as users needed and could afford.

Therefore IBM decided in 1961 to define and build a single new product family intended ultimately to replace all six of the extant families. The family would be the carrier of a new and substantially cheaper circuit technology – wafer-fabricated transistors and diodes. Seven simultaneously engineered models, ranging from the cheapest and slowest to a competitive supercomputer, were undertaken, along with a host of new I/O devices.

A radical conceptual innovation was that all of these models (except the cheapest, Model 20) would be logically identical, upward-*and-downward*-compatible implementations of a single architecture. In software engineering terms, a computer architecture is an abstract data type. It defines the set of valid data, their abstract representations, and the syntax and semantics of the set of operations proper to that type. Each implementation then is an instance of that type. In practice, our first implementations ranged from 8 to 64 bits wide and had various memory and circuit speeds.

Since a computer architecture defines exactly what programs will run and what results they will produce, the product family's strict compatibility enabled us to design a single software support package that would support the whole product family, and could be cost-shared across the entire family, with its large combined market forecast. This in turn enabled building a software support package of unprecedented richness and completeness. In those days, manufacturers gave away operating systems and compilers to stimulate the sale and use of hardware. Within this context, the attached set of slides gives a great amount of detail about this undertaking. Rather than restate all this in prose, I provide elaborating comments keyed to the slides.

IBM Operating System/360
Announced 1964; Shipped 1965
Fred Brooks
System/360 Manager, 1961-65
OS/360 Manager, 1964-65
Now: University of North Carolina
at Chapel Hill
brooks@cs.unc.edu
1 © FPB – 6/16/2001

This figure is essentially the table of contents for what follows. The concepts marked with an asterisk are the most important and novel.

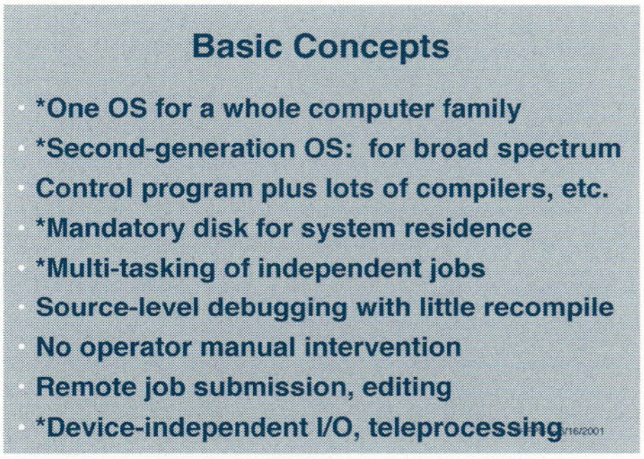
Basic Concepts
*One OS for a whole computer family
*Second-generation OS: for broad spectrum
Control program plus lots of compilers, etc.
*Mandatory disk for system residence
*Multi-tasking of independent jobs
Source-level debugging with little recompile
No operator manual intervention
Remote job submission, editing
*Device-independent I/O, teleprocessing

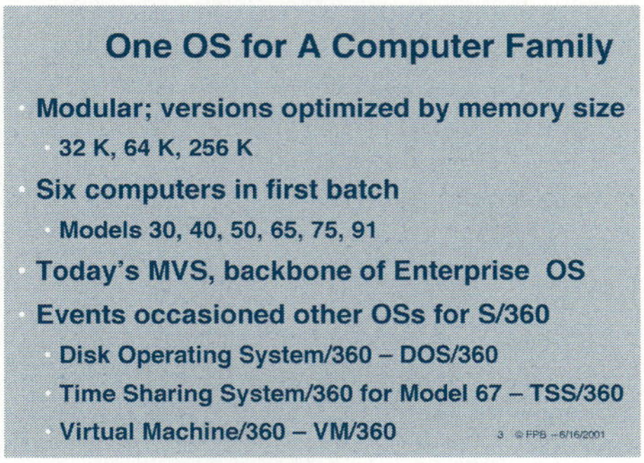
One OS for A Computer Family
Modular; versions optimized by memory size
32 K, 64 K, 256 K
Six computers in first batch
Models 30, 40, 50, 65, 75, 91
Today's MVS, backbone of Enterprise OS
Events occasioned other OSs for S/360
Disk Operating System/360 – DOS/360
Time Sharing System/360 for Model 67 – TSS/360
Virtual Machine/360 – VM/360
3 © FPB – 6/16/2001

Just as the first-generation operating systems were developed for second-generation computers, so OS/360 is the first of the second-generation operating systems, and it was developed for the first of the third-generation computers. First-generation operating systems were sharply differentiated by the application areas; OS/360 was designed to cover the entire spectrum of applications.

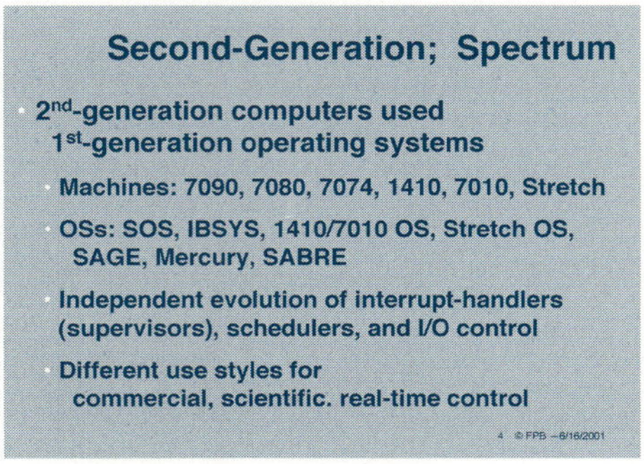
Second-Generation; Spectrum
2nd-generation computers used
1st-generation operating systems
Machines: 7090, 7080, 7074, 1410, 7010, Stretch
OSs: SOS, IBSYS, 1410/7010 OS, Stretch OS,
SAGE, Mercury, SABRE
Independent evolution of interrupt-handlers
(supervisors), schedulers, and I/O control
Different use styles for
commercial, scientific. real-time control
4 © FPB – 6/16/2001

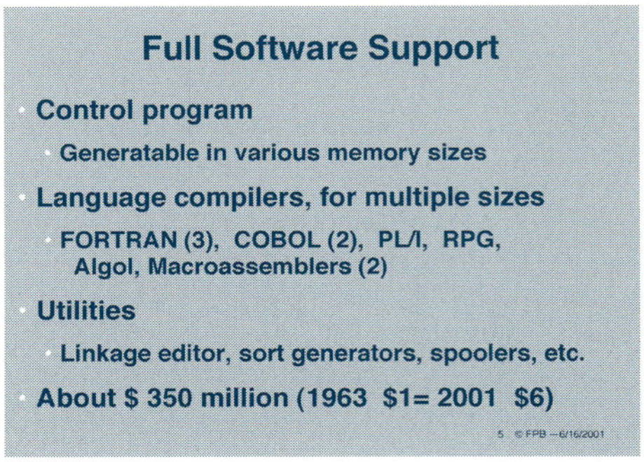
Full Software Support
Control program
Generatable in various memory sizes
Language compilers, for multiple sizes
FORTRAN (3), COBOL (2), PL/I, RPG, Algol, Macroassemblers (2)
Utilities
Linkage editor, sort generators, spoolers, etc.
About $ 350 million (1963 $1= 2001 $6)
5 ©FPB –6/16/2001

The name *Operating System/360* is sometimes used to describe the entire software support set. The same name is also used, more properly, to describe the control program alone. We shall use it hereafter only in this more restrictive sense. I like to think of the entire set as one big peach (the control program) and a bowl full of independent cherries (the compilers and utilities). OS/360 is modular; at system generation one can configure it to contain various functions, and to fit various resident memory sizes and resident disk sizes.

Several Fortran and Cobol compilers, optimized for different memory sizes, were built concurrently. Fortran H (for 256K configurations) was an exceptionally fine optimizing compiler, setting new standards for the quality of global optimization. The team was coherent and super-experienced– I think this was their fourth Fortran compiler.

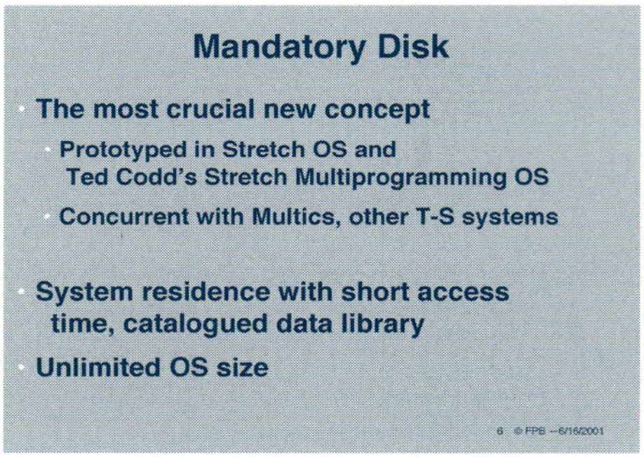
Mandatory Disk
The most crucial new concept
Prototyped in Stretch OS and Ted Codd's Stretch Multiprogramming OS
Concurrent with Multics, other T-S systems
System residence with short access time, catalogued data library
Unlimited OS size
6 ©FPB –6/16/2001

The new availability of an inexpensive disk drive, the IBM 2311, with its then-immense capacity of 7 MB, meant that we could design the operating system to assume operating system residence on a "random-access" device rather than on magnetic tape, as in most first-generation operating systems. This made the biggest single difference in the design concepts.

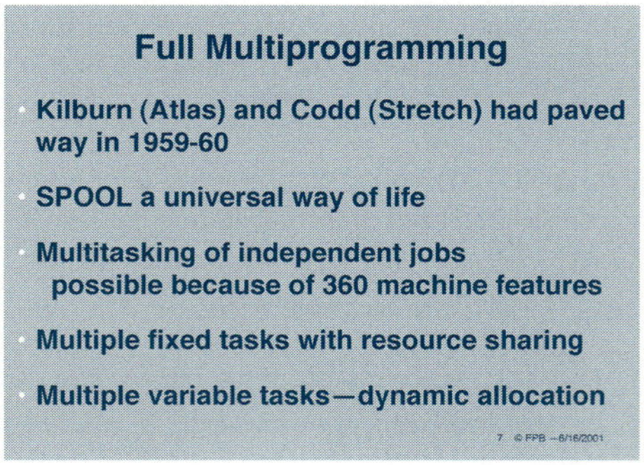
Full Multiprogramming
Kilburn (Atlas) and Codd (Stretch) had paved way in 1959-60
SPOOL a universal way of life
Multitasking of independent jobs possible because of 360 machine features
Multiple fixed tasks with resource sharing
Multiple variable tasks—dynamic allocation
7 ©FPB –6/16/2001

Late-first-generation IBM operating systems provided for Simultaneous Peripheral Operation On Line (SPOOL), so that at any given time a second-generation computer could be executing one main application and several card-to-tape, tape-to-card, tape-to-printer utilities. The latter were "trusted programs," carefully written so as not to corrupt or intrude on the main application, which usually ran tape-to-tape. OS/360 made the big leap to concurrent operation of independent, untrusted programs – a leap made possible by the architectural supervisory capabilities of the System/360 computers. Early OS/360 versions supported multiple tasks of fixed sizes for which memory allocation was straightforward. Within two years the MVS version supported multiprogramming in full generality.

A key new concept, now routine, is that the OS, not the operator, *controls* the computer. As late as 1987, some supercomputers, such as the CDC-ETA 10, were still running under operator manual control. A corollary of OS control, pioneered in Stretch and routine today, is that the keyboard or console is just another I/O device, with very few buttons (e.g. Power, Restart) that directly *do* anything.

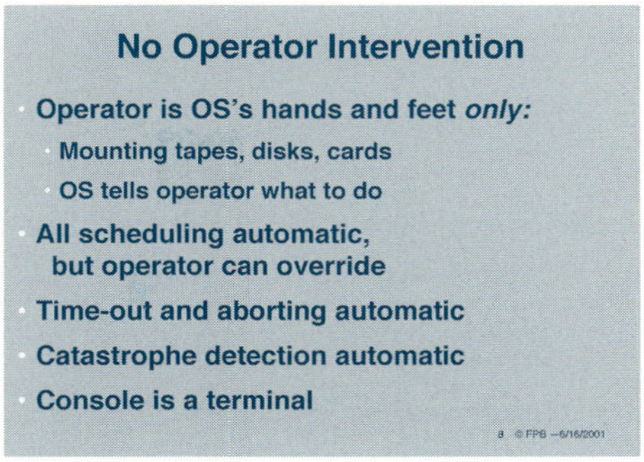
No Operator Intervention
Operator is OS's hands and feet *only:*
Mounting tapes, disks, cards
OS tells operator what to do
All scheduling automatic, but operator can override
Time-out and aborting automatic
Catastrophe detection automatic
Console is a terminal

OS/360 was designed from the ground up as a teleprocessing system, but not really a terminal-based time-sharing system. This concept contrasts with that of the contemporary MIT Multics System. OS/360 was designed for industrial-strength high-volume scientific and data processing applications; Multics was designed as an exploratory system, primarily for program development.

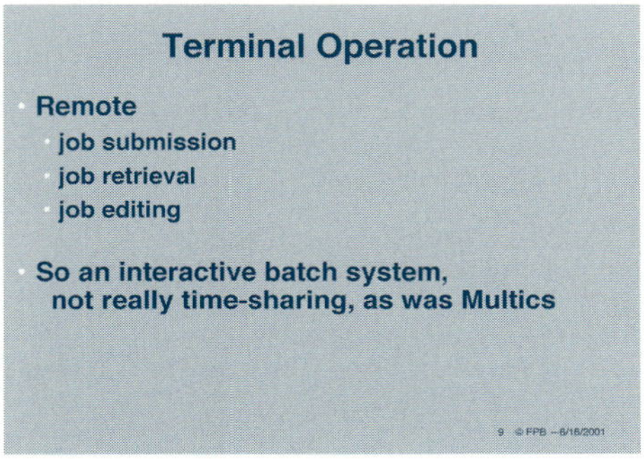
Terminal Operation
Remote
job submission
job retrieval
job editing
So an interactive batch system, not really time-sharing, as was Multics

Although strict program compatibility was the most distinctive new *concept* of the System/360 computer family, the rich set of I/O devices was its most important system attribute in terms of market breadth and performance enhancement. The single standard I/O interface radically reduced the engineering cost for new I/O devices, radically simplified system configuration, and radically eased configuration growth and change.

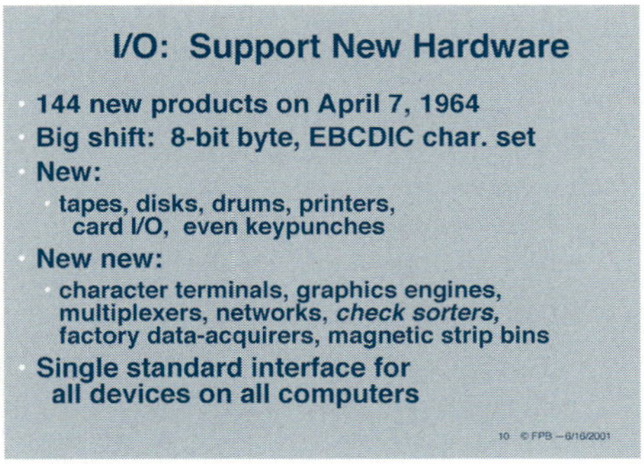
I/O: Support New Hardware
144 new products on April 7, 1964
Big shift: 8-bit byte, EBCDIC char. set
New:
tapes, disks, drums, printers, card I/O, even keypunches
New new:
character terminals, graphics engines, multiplexers, networks, *check sorters,* factory data-acquirers, magnetic strip bins
Single standard interface for all devices on all computers

The check sorters, curiously enough, posed the most rigid constraint on operating system performance, since they alone of all I/O devices rigidly demanded a quick response – the time of flight between the reading head and the pocket gate is fixed.

Device-Independent Data Mgt

- OS/360's major innovation
  - Across installations and configuration updates
- Data management is *40%* of control program
  - Includes device drivers, error-handlers
- All datasets, mounted or not, are known to OS — hierarchical catalog
- Automatic blocking and buffering

11 © FPB –6/16/2001

The crucial software innovation corresponding to the standard hardware I/O interface was a single system of I/O control and data management for all kinds of I/O devices. This was a radical departure from earlier operating systems. I consider it the most important innovation in OS/360.

Deferred Dataset Binding

- Program's dataset names not bound to actual datasets, devices until scheduling time
- Job Control Language (JCL) and Data Definition (DD) statements/cards
- SPOOLing a trivial special case (Simultaneous peripheral operation on-line)

12 © FPB –6/16/2001

OS/360, more explicitly than any of its predecessors, recognized scheduling time as a binding occasion distinct from compile time and run time. Not only were program modules bound to each other at schedule time, dataset names were bound to particular datasets on particular devices at scheduling time. This binding was specified by the Job Control Language, which was executed by the scheduler.

Data Access Methods

- Optimized for disk performance
- Sequential A-M (tape-like, buffered)
  - e.g., for sorting
- Direct A-M (pure random access)
  - e.g., airline reservations
- Partitioned A-M (fast block transfer)
  - e.g., systems residence
- Indexed sequential A-M (sequential, buffered, with query)
  - e.g., utility billing

13 © FPB –6/16/2001

Four access methods were designed especially for exploitation of the fleet of new disk types, across the range of disk applications. Two other access methods were designed especially to provide full flexibility and ease of use for telecommunications.

Overall Structure of OS/360

- Supervisor:
  - handles interrupts
  - allocates memory dynamically
  - allocates cycles among tasks by priority
- Scheduler:
  - Sequences and parallels jobs by priority
  - Allocates I/O devices and prescribes mounting
  - Commands operator
- Data Management

14 © FPB –6/16/2001

In OS/360, three independent streams of control program evolution come together. The system structure mirrors this diverse ancestry. The Supervisor evolved from early interrupt-handling routines; the Data Management System from earlier packages of I/O subroutines; the Scheduler from earlier tape-based job schedulers.

OS/360 Should Have Been Different

- Job Control Language—wrong concept
- *But not* time sharing—Not our market
- Fixed-length blocks on all disks, etc.
- One channel, control unit per device
- One sequential access method
- Only interactive debugging
- Streamline function ruthlessly
- Sysgen-optimized standard packages

15 © FPB – 6/18/2001

The Job Control Language is the worst programming language ever designed by anybody anywhere – it was designed under my management. The very concept is wrong; we did not see it as a programming language but as "a few control cards to precede the job." I have elaborated on its flaws in my Turing lecture [2].

We should have cut loose from the key-count-data variable-length block structure established for IBM's earlier disks and designed for one or two sizes of fixed-length blocks on all random-access devices.

The I/O device-control unit-channel attachment tree is unnecessarily complex [3]. We should have specified one (perhaps virtual) channel and one (perhaps virtual) control unit per device. I believe we could have invented one sequential disk access method that would have combined the optimizations of the three: SAM, PAM, ISAM.

Two radically different debugging systems are provided, one conceived for interactive use from terminals with rapid recompilation, the other conceived for batch operation. It was the best batch debugging system ever designed, yet totally obsolete from birth.

OS/360 is too rich. Systems residence on a disk removed the size constraint that had disciplined earlier OS designers – we put in many functional goodies of marginal usefulness [4, Chap. 5]. Featuritis is even now not yet dead in the software community.

The system-generation process for OS/360 is wondrously flexible and wondrously onerous. We should have configured a small number of standard packages to meet the needs of most users, and offered these to supplement the fully flexible configuration process.

Process Should Have Been Different

- Rigid architectural control over *all* interfaces
  - *Include*d external variable declarations
- Built in PL/I, not PLS-Assembler
  - Trained: "How to write *good* PL/I"
- Hidden the control blocks
  - but we were 7 years before Parnas
- Surgical teams
- Maintained a performance model

16 © FPB – 6/18/2001

I have treated this topic at length in *The Mythical Man-Month* [4]. Here I will only elaborate a little. I am firmly convinced that if we had built the whole thing in PL/I, the best high-level language available at the time, the OS would have been just as fast, far cleaner and more reliable, and built more swiftly. In fact, it was built in PLS, a syntactically sugared assembler language. Using PL/I would have required careful training of our force in how to write *good* PL/I code, code that would compile to fast run-time code.

We should have maintained rigid architectural control over all the interfaces, insisting that all declarations of external variables be *included* from libraries, not crafted anew in each instance.

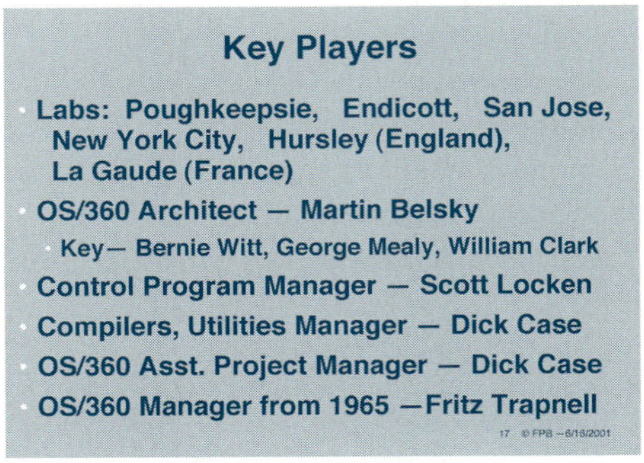
Key Players
Labs: Poughkeepsie, Endicott, San Jose, New York City, Hursley (England), La Gaude (France)
OS/360 Architect — Martin Belsky
Key— Bernie Witt, George Mealy, William Clark
Control Program Manager — Scott Locken
Compilers, Utilities Manager — Dick Case
OS/360 Asst. Project Manager — Dick Case
OS/360 Manager from 1965 —Fritz Trapnell
17 © FPB –6/18/2001

About 1000 people worked on the entire OS/360 software package. Here I identify both the teams and those individuals who contributed most to the conceptual structure.

References

[1] IBM Systems Journal, 5, 1 (1966)

[2] Brooks, F.P. Jr., "The Design of Design." ACM 1999 Turing Award Lecture, Communications of the ACM, to appear.

[3] Blaauw, G.A., and F.P. Brooks, Jr., 1997. Computer Architecture: Concepts and Evolution. Addison-Wesley, Boston, Section 8.22

[4] Brooks, F.P. Jr., 1995. The Mythical Man-Month: Essays on Software Engineering, 20th Anniversary Edition. Addison-Wesley, Boston.

Frederick P. Brooks

G.H. Mealy, B.I. Witt, W.A. Clark

The Functional Structure of OS/360

IBM Systems Journal,
Vol. 5 (1), 1966
pp. 2-51

In the design of OS/360*, a modular operating system being implemented for a range of* SYSTEM/360 *configurations, the fundamental objective has been to bring a variety of application classes under the domain of one coherent system. The conceptual framework of the system as a whole, as well as the most distinctive structural features of the control program, are heavily influenced by this objective.*

The purpose of this paper is to present the planned system in a unified perspective by discussing design objectives, historical background, and structural concepts and functions. The scope of the system is surveyed in Part I, whereas the rest of the paper is devoted to the control program. Design features relevant to job scheduling and task management are treated in Part II. The third part discusses the principal activities involved in cataloging, storing, and retrieving data and programs.

The functional structure of OS/360

Part I — **Introductory survey**
by G. H. Mealy

Part II — **Job and task management**
by B. I. Witt

Part III — **Data management**
by W. A. Clark

Individual acknowledgements cannot feasibly be given. Contributing to OS/360 are several IBM programming centers in America and Europe. The authors participated in the design of the control program.

A brief outline of the structural elements of OS/360 *is given in preparation for the subsequent sections on control-program functions.*

Emphasis is placed on the functional scope of the system, on the motivating objectives and basic design concepts, and on the design approach to modularity.

The functional structure of OS/360

Part I Introductory survey

by G. H. Mealy

The environment that may confront an operating system has lately undergone great change. For example, in its several compatible models, SYSTEM/360 spans an entire spectrum of applications and offers an unprecedented range of optional devices.[1] It need come as no surprise, therefore, that OS/360—the Operating System for SYSTEM/360—evinces more novelty in its scope than in its functional objectives.

In a concrete sense, OS/360 consists of a library of programs. In an abstract sense, however, the term OS/360 refers to one articulated response to a composite set of needs. With integrated vocabularies, conventions, and modular capabilities, OS/360 is designed to answer the needs of a SYSTEM/360 configuration with a standard instruction set and thirty-two thousand or more bytes of main storage.[2]

The main purpose of this introductory survey is to establish the scope of OS/360 by viewing the subject in a number of different perspectives: the historical background, the design objectives, and the functional types of program packages that are provided. An effort is made to mention problems and design compromises, i.e., to comment on the forces that shaped the system as a whole.

Basic objectives

The notion of an operating system dates back at least to 1953 and

MIT's Summer Session Computer and Utility System.[3] Then, as now, the operating system aimed at non-stop operation over a span of many jobs and provided a computer-accessible library of utility programs. A number of operating systems came into use during the last half of the decade.[4] In that all were oriented toward overlapped setup in a sequentially executed job batch, they may be termed "first generation" operating systems.

A significant characteristic of batched-job operation has been that each job has, more or less, the entire machine to itself, save for the part of the system permanently resident in main storage. During the above-mentioned period of time, a number of large systems—typified by SAGE, MERCURY, and SABRE—were developed along other lines; these required total dedication of machine resources to the requirements of one "real-time" application. It is interesting that one of the earliest operating systems, the Utility Control Program developed by the Lincoln Laboratory, was developed solely for the checkout of portions of the SAGE system. By and large, however, these real-time systems bore little resemblance to the first generation of operating systems, either from the point of view of intended application or system structure.

Because the basic structure of OS/360 is equally applicable to batched-job and real-time applications, it may be viewed as one of the first instances of a "second-generation" operating system. The new objective of such a system is to accommodate an environment of diverse applications and operating modes. Although not to be discounted in importance, various other objectives are not new—they have been recognized to some degree in prior systems. Foremost among these secondary objectives are:

- Increased throughput
- Lowered response time
- Increased programmer productivity
- Adaptability (of programs to changing resources)
- Expandability

throughput

OS/360 seeks to provide an effective level of machine throughput in three ways. First, in handling a stream of jobs, it assists the operator in accomplishing setup operations for a given job while previously scheduled jobs are being processed. Second, it permits tasks from a number of different jobs to concurrently use the resources of the system in a multiprogramming mode, thus helping to ensure that resources are kept busy. Also, recognizing that the productivity of a shop is not solely a function of machine utilization, heavy emphasis is placed on the variety and appropriateness in source languages, on debugging facilities, and on input convenience.

response time

Response time is the lapse of time from a request to completion of the requested action. In a batch processing context, response time (often called "turn-around time") is relatively long: the user gives a deck to the computing center and later obtains printed results. In a mixed environment, however, we find a whole spectrum of response times. Batch turn-around time is at the "red" end of the spectrum, whereas real-time requirements fall at the "violet" end. For example, some real-time applications need response times in the order of milliseconds or lower. Intermediate in the spectrum are the times for simple actions such as line entry from a keyboard where a response time of the order of one or two seconds is desirable. Faced with a mixed environment in terms of applications and response times, OS/360 is designed to lend itself to the whole spectrum of response times by means of control-program options and priority conventions.

productivity

For the sake of programmer productivity and convenience, OS/360 aims to provide a novel degree of versatility through a relatively large set of source languages. It also provides macroinstruction capabilities for its assembler language, as well as a concise job-control language for assistance in job submission.

adaptability

A second-generation operating system must be geared to change and diversity. SYSTEM/360 itself can exist in an almost unlimited variety of machine configurations: different installations will typically have different configurations as well as different applications. Moreover, the configuration at a given installation may change frequently. If we look at application and configuration as the environment of an operating system, we see that the operating system must cope with an unprecedented number of environments. All of this puts a premium on system modularity and flexibility.

Adaptability is also served in OS/360 by the high degree to which programs can be device-independent. By writing programs that are relatively insensitive to the actual complement of input/output devices, an installation can reduce or circumvent the problems historically associated with device substitutions.

expandability

As constructed, OS/360 is "open-ended"; it can support new hardware, applications, and programs as they come along. It can readily handle diverse currency conventions and character sets. It can be tailored to communicate with operators and programmers in languages other than English. Whenever so dictated by changing circumstances, the operating system itself can be expanded in its functional capabilities.

Design concepts

In the notion of an "extended machine," a computing system is viewed as being composed of a number of layers, like an onion.[5,6] Few programmers deal with the innermost layer, which is that provided by the hardware itself. A FORTRAN programmer, for instance, deals with an outer layer defined by the FORTRAN language. To a large extent, he acts as though he were dealing with hardware that accepted and executed FORTRAN statements directly. The SYSTEM/360 instruction set represents two inner layers, one when operating in the supervisor state, another when operating in the problem state.

The supervisor state is employed by OS/360 for the *supervisor* portion of the control program. Because all other programs operate in the problem state and must rely upon unprivileged instructions, they use *system macroinstructions* for invoking the supervisor. These macroinstructions gain the attention of the supervisor by means of SVC, the supervisor-call instruction.

All OS/360 programs with the exception of the supervisor operate in the problem state. In fact, one of the fundamental design tenets is that these programs (compilers, sorts, or the like) are, to all intents and purposes, problem programs and must be treated as such by the supervisor. Precisely the same set of facilities is offered to system and problem programs. At any point in time, the system consists of its given supervisor plus all programs that are available in on-line storage. Inasmuch as an installation may introduce new compilers, payroll programs, etc., the extended machine may grow.

In designing a method of control for a second-generation system, two opposing viewpoints must be reconciled. In the first-generation operating systems, the point of view was that the machine executed an incoming stream of programs; each program and its associated input data corresponded to one application or problem. In the first-generation real-time systems, on the other hand, the point of view was that incoming pieces of data were routed to one of a number of processing programs. These attitudes led to quite different system structures; it was not recognized that these points of view were matters of degree rather than kind. The basic consideration, however, is one of emphasis: programs are used to process data in both cases. Because it is the combination of program and data that marks a unit of work for control purposes, OS/360 takes such a combination as the distinguishing property of a *task*. As an example, consider a transaction processing program and two input transactions, A and B. To process A and B, two tasks are introduced into the system, one consisting of A plus the program, the second consisting of B plus the program. Here, the two tasks use the same program but different sets of input data. As a further illustration, consider a master file and two programs, X and Y, that yield different reports from the master file. Again, two tasks are introduced into the system, the first consisting of the master file plus X, and the second of the master file plus Y. Here the same input data join with two different programs to form two different tasks.

In laying down conceptual groundwork, the OS/360 designers have employed the notion of multitask operation wherein, at any time, a number of tasks may contend for and employ system resources. The term *multiprogramming* is ordinarily used for the case in which one CPU is shared by a number of tasks, the term *multiprocessing*, for the case in which a separate task is assigned to each of several CPU's. Multitask operation, as a concept, gives recognition to both terms. If its work is structured entirely in the form of tasks, a job may lend itself without change to either environment.

In os/360, any named collection of data is termed a *data set.* A data set may be an accounting file, a statistical array, a source program, an object program, a set of job control statements, or the like. The system provides for a cataloged library of data sets. The library is very useful in program preparation as well as in production activities; a programmer can store, modify, recompile, link, and execute programs with minimal handling of card decks.

System elements

As seen by a user, os/360 will consist of a set of language translators, a set of service programs, and a control program. Moreover, from the viewpoint of system management, a SYSTEM/360 installation may look upon its own application programs as an integral part of the operating system.

translators

A variety of translators are being provided for FORTRAN, COBOL, and RPGL (a Report Program Generator Language). Also to be provided is a translator for PL/I, a new generalized language.[7] The programmer who chooses to employ the assembler language can take advantage of macroinstructions; the assembler program is supplemented by a macro generator that produces a suitable set of assembly language statements for each macroinstruction in the source program.

service programs

Groups of individually translated programs can be combined into a single executable program by a linkage editor. The linkage editor makes it possible to change a program without re-translating more than the affected segment of the program. Where a program is too large for the available main-storage area, the function of handling program segments and overlays falls to the linkage editor.

The sort/merge is a generalized program that can arrange the fixed- or variable-length records of a data set into ascending or descending order. The process can employ either magnetic-tape or direct-access storage devices for input, output, and intermediate storage. The program is adaptable in the sense that it takes advantage of all the input/output resources allocated to it by the control program. The sort/merge can be used independently of other programs or can be invoked by them directly; it can also be used via COBOL and PL/I.

Included in the service programs are routines for editing, arranging, and updating the contents of the library; revising the index structure of the library catalog; printing an inventory list of the catalog; and moving and editing data from one storage medium to another.

the control program

Roughly speaking, the control program subdivides into master scheduler, job scheduler, and supervisor. Central control lodges in the supervisor, which has responsibility for the storage allocation, task sequencing, and input/output monitoring functions. The master scheduler handles all communications to and from the operator, whereas the job scheduler is primarily concerned with job-stream analysis, input/output device allocation and setup, and job initiation and termination.

supervisor

Among the activities performed by the supervisor are the following:

- Allocating main storage
- Loading programs into main storage
- Controlling the concurrent execution of tasks
- Providing clocking services
- Attempting recoveries from exceptional conditions
- Logging errors
- Providing summary information on facility usage
- Issuing and monitoring input/output operations

The supervisor ordinarily gains control of the central processing unit by way of an interruption. Such an interruption may stem from an explicit request for services, or it may be implicit in SYSTEM/360 conventions, such as in the case of an interruption that occurs at the completion of an input/output operation. Normally, a number of data-access routines required by the data management function are coordinated with the supervisor. The access routines available at any given time are determined by the requirements of the user's program, the structure of the given data sets, and the types of input/output devices in use.

job scheduler

As the basic independent unit of work, a job consists of one or more steps. Inasmuch as each job step results in the execution of a major program, the system formalizes each job step as a task, which may then be inserted into the task queue by the initiator-terminator (a functional element of the job scheduler). In some cases, the output of one step is passed on as the input to another. For example, three successive job steps might involve file maintenance, output sorting, and report tabulation.

The primary activities of the job scheduler are as follows:

- Reading job definitions from source inputs
- Allocating input/output devices
- Initiating program execution for each job step
- Writing job outputs

In its most general form, the job scheduler allows more than one job to be processed concurrently. On the basis of job priorities and resource availabilities, the job scheduler can modify the order in which jobs are processed. Jobs can be read from several input devices and results can be recorded on several output devices—the reading and recording being performed concurrently with internal processing.

master scheduler

The master scheduler serves as a communication control link between the operator and the system. By command, the operator can alert the system to a change in the status of an input/output unit, alter the operation of the system, and request status information. The master scheduler is also used by the operator to alert the job scheduler of job sources and to initiate the reading or processing of jobs.

The control program as a whole performs three main functions: job management, task management, and data management. Since Part II of this paper discusses job and task management, and Part III is devoted entirely to data management, we do not further pursue these functions here.

System modularity

Two distinguishable, but by no means independent, design problems arise in creating a system such as OS/360. The first one is to prescribe the range of functional capabilities to be provided; essentially, this amounts to defining two operating systems, one of maximum capability and the other a nucleus of minimum capability. The second problem is to ascertain a set of building blocks that will answer reasonably well to the two predefined operating systems as well as to the diverse needs bounded by the two. In resolving the second problem, which brings us to the subject of modularity, no single consideration is more compelling than the need for efficient utilization of main storage.

As stated earlier, the tangible OS/360 consists of a library of program modules. These modules are the blocks from which actual operating systems can be erected. The OS/360 design exploits three basic principles in designing blocks that provide the desired degree of modularity. Here, these well-known principles are termed parametric generality, functional redundancy, and functional optionality.

parametric generality

The degree of generality required by varying numbers of input/output devices, control units, and channels can be handled to a large extent by writing programs that lend themselves to variations in parameters. This has long been practiced in sorting and merging programs, for example, as well as in other generalized routines. In OS/360, this principle also finds frequent application in the process that generates a specific control program.

functional redundancy

In the effort to optimize performance in the face of two or more conflicting objectives, the most practical solution (at least at the present state of the art) is often to write two or more programs that exploit dissimilar programming techniques. This principle is most relevant to the program translation function, which is especially sensitive to conflicting performance measures. The same installation may desire to effect one compilation with minimum use of main storage (even at some expense of other objectives) and another compilation with maximum efficacy in terms of object-program running time (again at the expense of other objectives). Where conflicting objectives could not be reconciled by other means, the OS/360 designers have provided more than one program for the same general translation or service function. For the COBOL language, for example, there are two translation programs.

For the nucleus of the control program that resides in main storage, the demand for efficient storage utilization is especially

functional optionality

pressing. Hence, each functional capability that is likely to be unused in some installations is treated as a separable option. When a control program is generated, each omitted option yields a net saving in the main-storage requirement of the control program.

The most significant control program options are those required to support various job scheduling and multitask modes of operation. These modes carry with them needs for optional functions of the following kinds:

- Task synchronization
- Job-input and job-output queues
- Distinctive methods of main-storage allocation
- Main-storage protection
- Priority-governed selection among jobs

In the absence of any options, the control program is capable of ordinary stacked-job operation. The activities of the central processing unit and the input/output channels are overlapped. Many error checking and recovery functions are provided, interruptions are handled automatically, and the standard data-management and service functions are included. Job steps are processed sequentially through single task operations.

The span of operating modes permitted by options in the control program can be suggested by citing two limiting cases of multitask operation. The first and least complicated permits a scheduled job step to be processed concurrently with an initial-input task, say A, and a result-output task, say B. Because A and B are governed by the control program, they do not correspond to job steps in the usual sense. The major purpose of this configuration is to reduce delays between the processing of successive job steps: tasks A and B are devoted entirely to input/output functions.

In the other limiting case, up to n jobs may be in execution on a concurrent basis, the parameter n being fixed at the time the control program is generated. Contending tasks may arise from different jobs, and a given task can dynamically define other tasks (see the description of the ATTACH macroinstruction in Part II) and assign task priorities. Provision is made for removal of an entire job step (from the job of lowest priority) to auxiliary storage in the event that main storage is exhausted. The affected job step is resumed as soon as the previously occupied main-storage area becomes available again.

In selecting the options to be included in a control program, the user is expected to avail himself of detailed descriptions and accompanying estimates of storage requirements.

system generation

To obtain a desired operating system, the user documents his machine configuration, requests a complement of translators and service programs, and indicates desired control-program options—all via a set of macroinstructions provided for the purpose. Once this has been done, the fabrication of a specific operating system from the OS/360 systems library reduces to a process of two stages.

First, the macroinstructions are analyzed by a special program and formulated into a job stream. In the second stage, the assembler program, the linkage editor, and the catalog service programs join in the creation of a resident control program and a desired set of translators and service programs.

Summary comment

Intended to serve a wide variety of computer applications and to support a broad range of hardware configurations, OS/360 is a modular operating system. The system is not only open-ended for the class of functions discussed in this paper, but is based on a conceptual framework that is designed to lend itself to additional functions whenever warranted by cumulative experience.

The ultimate purpose of an operating system is to increase the productivity of an entire computer installation; personnel productivity must be considered as well as machine productivity. Although many avenues to increased productivity are reflected in OS/360, each of these avenues typically involves a marginal investment on the part of an installation. The investment may take the form of additional personnel training, storage requirements, or processing time. It repays few installations to seek added productivity through every possible avenue; for most, the economies of installation management dictate a well-chosen balance between investment and return. Much of the modularity in OS/360 represents a design attempt to permit each installation to strike its own economic balance.

CITED REFERENCES AND FOOTNOTES

1. For an introduction to SYSTEM/360, see G. A. Blaauw and F. P. Brooks, Jr., "The structure of SYSTEM/360, Part I, outline of the logical structure," *IBM Systems Journal* **3**, No. 2, 119–135 (1964).
2. The restrictions exclude MODEL 44, as well as MODEL 20. The specialized operating systems that support these excluded models are not discussed here.
3. C. W. Adams and J. H. Laning, Jr., "The MIT systems of automatic coding: Comprehensive, Summer Session, Algebraic," *Symposium on Automatic Coding for Digital Computers*, Office of Naval Research, Department of the Navy (May 1954).
4. In the case of the IBM 709 and 704 computers, the earliest developments were largely due to the individual and group efforts of SHARE installations. The first operating systems developed jointly by IBM and SHARE were the SHARE Operating System (SOS) and the FORTRAN Monitor System (FMS).
5. G. F. Leonard and J. R. Goodroe, "An environment for an operating system," *Proceedings of the 19th National ACM Conference* (August 1964).
6. A. W. Holt and W. J. Turanski, "Man to machine communication and automatic code translation," *Proceedings Western Joint Computer Conference* (1960).
7. G. Radin and H. P. Rogoway, "NPL: highlights of a new programming language," *Communications of the ACM* **8**, No. 1, 9–17 (January 1965).

This part of the paper discusses the control-program functions most closely related to job and task management.

Emphasized are design features that facilitate diversity in application environments as well as those that support multitask operation.

The functional structure of OS/360

Part II Job and task management

by B. I. Witt

One of the basic objectives in the development of OS/360 has been to produce a general-purpose monitor that can jointly serve the needs of real-time environments, multiprogramming for peripheral operations, and traditional job-shop operations. In view of this objective, the designers found it necessary to develop a more generalized framework than that of previously reported systems. After reviewing salient aspects of the design setting, we will discuss those elements of OS/360 most important to an understanding of job and task management.

Background

Although the conceptual roots of OS/360 task management are numerous and tangled, the basic notion of a task owes much to the systems that have pioneered the use of on-line terminals for inventory problems. This being the case, the relevant characteristics of an on-line inventory problem are worthy of review. We may take the airline seat-reservation application as an example: a reservation request reduces the inventory of available seats, whereas a cancellation adds to the inventory. Because a reply to a ticket agent must be sent within a matter of seconds, there is no opportunity to collect messages for later processing. In the contrasting environment where files are updated and reports made on a daily or weekly basis, it suffices to collect and sort transactions before posting them against a master file.

Three significant consequences of the on-line environment can be recognized:

- Each message must be processed as an independent task
- Because there is no opportunity to batch related requests, each task expends a relatively large amount of time in references to the master file
- Many new messages may be received by the system before the task of processing an older message is completed

What is called for, then, is a control program that can recognize the existence of a number of concurrent tasks and ensure that whenever one task cannot use the CPU, because of input/output delays, another task be allowed to use it. Hence, the CPU is considered a resource that is allocated to a task.

Another major consideration in on-line processing is the size and complexity of the required programs. Indeed, the quantity of code needed to process a transaction can conceivably exceed main storage. Furthermore, subprogram selection and sequence depend upon the content of an input message. Lastly, subprograms brought into main storage on behalf of one transaction may be precisely those needed to process a subsequent transaction. These considerations dictate that subprograms be callable by name at execution time and relocatable at load time (so that they may be placed in any available storage area); they also urge that a single copy of a subprogram be usable by more than one transaction.

The underlying theme is that a task—the work required to process a message—should be an identifiable, controllable element. To perform a task, a variety of system resources are required: the CPU itself, subprograms, space in main and auxiliary storage, data paths to auxiliary storage (e.g., a channel and a control unit), interval timer and others.

Since a number of tasks may be competing for a resource, an essential control program function is to manage the system's resources, i.e., to recognize requests, resolve conflicting demands, and allocate resources as appropriate. In this vein, the general purpose multitask philosophy of the OS/360 control program design has been strongly influenced by task-management ideas that have already been tested in on-line systems.[1] But there is no reason to limit the definition of "task" to the context of real-time inventory transactions. The notion of a task may be extended to any unit of work required of a computing system, such as the execution of a compiler, a payroll program, or a data-conversion operation.

Basic definitions

In the interests of completeness, this section briefly redefines terms introduced in Part I. Familiarity with the general structure of SYSTEM/360 is assumed.[2]

From the standpoint of installation accounting and machine

room operations, the basic unit of work is the *job*. The essential characteristic of a job is its independence from other jobs. There is no way for one job to abort another. There is also no way for the programmer to declare that one job must be contingent upon the output or the satisfactory completion of another job. Job requirements are specified by control statements (usually punched in cards), and may be grouped to form an input job stream. For the sake of convenience, the job stream may include input data, but the main purpose of the job stream is to define and characterize jobs. Because jobs are independent, the way is open for their concurrent execution.

job step

By providing suitable control statements, the user can divide a job into *job steps*. Thus, a job is the sum of all the work associated with its component job steps. In the current OS/360, the steps of a given job are necessarily sequential: only one step of a job can be processed at a time. Furthermore, a step may be conditional upon the successful completion of one or more preceding steps; if the specified condition is not met, the step in question can be bypassed.

task

Whenever the control program recognizes a job step (as the result of a job control statement), it formally designates the step as a *task*. The task consists, in part or in whole, of the work to be accomplished under the direction of the program named by the job step. This program is free to invoke other programs in two ways, first within the confines of the original task, and second within the confines of additionally created tasks. A task is created (except in the special case of initial program loading) as a consequence of an ATTACH macroinstruction. At the initiation of a job step, ATTACH is issued by the control program; during the course of a job step, ATTACH's may be issued by the user's programs.

From the viewpoint of the control system, all tasks are independent in the sense that they may be performed concurrently. But in tasks that stem from one given job (which implies that they are from the same job step), dependency relationships may be inherent because of program logic. To meet this possibility, the system provides means by which tasks from the same job can be synchronized and confined within a hierarchical relationship. As a consequence, one task can await a designated point in the execution of another task. Similarly, a task can wait for completion of a subtask (a task lower in the hierarchy). Also, a task can abort a subtask.

Although a job stream may designate many jobs, each of which consists of many job steps and, in turn, leads to many tasks, a number of quite reasonable degenerate cases may be imagined; e.g., in an on-line inventory environment, the entire computing facility may be dedicated to a single job that consists of a single job step. At any one time, this job step may be comprised of many tasks, one for each terminal transaction. On the other hand, in many installations, it is quite reasonable to expect almost all jobs to consist of several steps (e.g., compile/link-edit/execute) with no step consisting of more than one task.

In most jobs, the executable programs and the data to be processed are not new to the system—they are carried over from earlier jobs. They therefore need not be resubmitted for the new job; it is sufficient that they be identified in the control statements submitted in their place as part of a job stream. A job stream consists of such control statements, and optionally of data that is new to the system (e.g., unprocessed keypunched information). Control statements are of six types; the three kinds of interest here are *job*, *execute*, and *data definition* statements.

control statements

The first statement of each job is a job statement. Such a statement can provide a job name, an account number, and a programmer's name. It can place the job in one of fifteen priority classes; it can specify various conditions which, if not met at the completion of each job step, inform the system to bypass the remaining steps.

The first statement of each job step is an execute statement. This statement typically identifies a program to be executed, although it can be used to call a previously cataloged procedure into the job stream. The first statement can designate accounting modes, conditional tests that the step must meet with respect to prior steps, permissable execution times, and miscellaneous operating modes.

A data definition statement permits the user to identify a data set, to state needs for input/output devices, to specify the desired channel relationships among data sets, to specify that an output data set be passed to a subsequent job step, to specify the final disposition of a data set, and to incorporate other operating details.

program structure

In OS/360, a ready-for-execution program consists of one or more subprograms called *load modules*; the first load module to be executed is the one that is named in the execute control statement. At the option of the programmer, a program can take one of the following four structures:

Simple structure. One load module, loaded into main storage as an entity, contains the entire program.

Planned overlay structure. The program exists in the library as a single load module, but the programmer has identified program segments that need not be in main storage at the same given time. As a consequence, one area of storage can be used and reused by the different segments. The OS/360 treatment of this structure follows the guide lines previously laid down by Heising and Larner.[3] A planned overlay structure can make very effective use of main storage. Because the control system intervenes only once to find a load module, and linkages from segment to segment are aided by symbol resolution in advance of execution, this structure also serves the interest of execution efficiency.

Dynamic serial structure. The advantages of planned overlay tend to diminish as job complexity increases, particularly if the selection

Figure 1

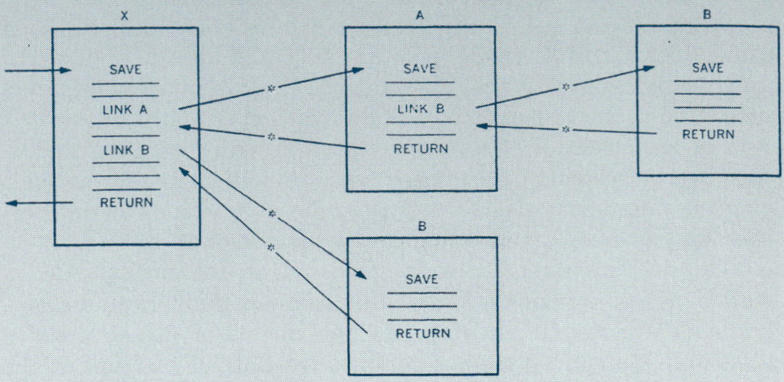

of segments is data dependent (as is the case in most on-line inventory problems). For this situation, OS/360 provides means for calling load modules dynamically, i.e., when they are named during the execution of other load modules. This capability is feasible because main storage is allocated as requests arise, and the conventions permit any load module to be executed as a subroutine. It is consistent with the philosophy that tasks are the central element of control, and that all resources required by a task for its successful performance—the CPU, storage, *and programs*—may be requested whenever the need is detected. In the dynamic serial structure, more than one load module is called upon during the course of program execution. Following standard linkage conventions, the control system acts as intermediary in establishing subroutine entry and return. Three macroinstructions are provided whereby one load module can invoke another: LINK, XCTL (transfer control), and LOAD.

The action of LINK is illustrated in Figure 1. Of the three programs (i.e., load modules) involved, X is the only one named at task-creation time. One of the instructions generated by LINK is a supervisor call (SVC), and the program name (such as A or B in the figure) is a linkage parameter. When the appropriate program of the control system is called, it finds, allocates space for, fetches, and branches to the desired load module. Upon return from the module (effected by the macroinstruction RETURN), the occupied space is liberated but not reused unless necessary. Thus, in the example, if program B is still intact in main storage at the second call, it will not be fetched again (assuming that the user is operating under "reusable" programming conventions, as discussed below).

As suggested by Figure 2, XCTL can be used to pass control to successive phases of a program. Standard linkage conventions are observed, parameters are passed explicitly, and the supervisor

Figure 2

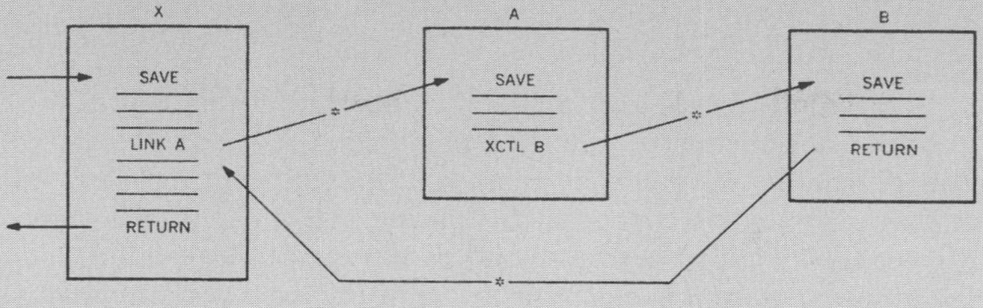

* SUPERVISORY ACTION

Figure 3a, 3b

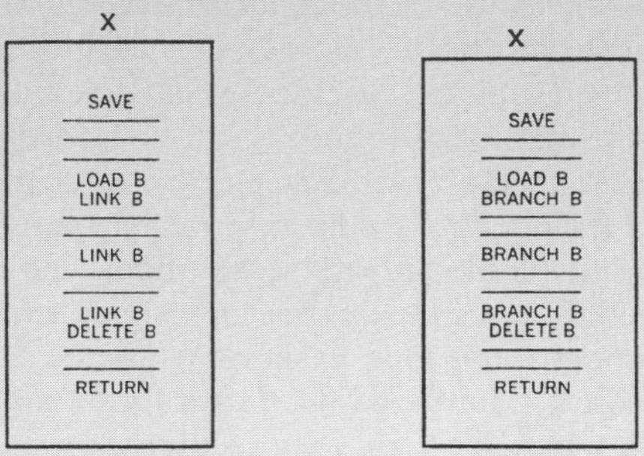

3a 3b

functions are similar to those needed for LINK. However, a program transferring via XCTL is assumed to have completed its work, and all its allocated areas are immediately liberated for reuse.

The LOAD macroinstruction is designed primarily for those cases in which tasks make frequent use of a load module, and reusable conventions are followed. LOAD tells the supervisor to bring in a load module and to preserve the module until liberated by a DELETE macroinstruction (or automatically upon task termination). Control can be passed to the module by a LINK, as in Figure 3a, or by branch instructions, as in Figure 3b.

Dynamic parallel structure. In the three foregoing structures, execution is serial. The ATTACH macroinstruction, on the other hand, creates a task that can proceed in parallel with other tasks, as permitted by availability of resources. In other respects, ATTACH is much like LINK. But since ATTACH leads to the creation of a new task, it requires more supervisor time than LINK and should not be used unless a significant degree of overlapped operation is assured.

program usability

Load modules in the library are of three kinds (as specified by the programmer at link-edit time): *not reusable, serially reusable,* and *reenterable.* Programs in the first category are fetched directly from the library whenever needed. This is required because such programs may alter themselves during execution in a way that prevents the version in main storage from being executed more than once.

A serially reusable load module, on the other hand, is designed to be self-initializing; any portion modified in the course of execution is restored before it is reused. The same copy of the load module may be used repeatedly during performance of a task. Moreover, the copy may be shared by different tasks created from the same job step; if the copy is in use by one task at the time it is requested by another task, the latter task is placed in a queue to wait for the load module to become available.

A reenterable program, by design, does not modify itself during execution. Because reenterable load modules are normally loaded in storage areas protected by the storage key used for the supervisor, they are protected against accidental modification from other programs. A reenterable load module can be loaded once and used freely by any task in the system at any time. (A reenterable load module fetched from a private library, rather than from the main library, is made available only to tasks originating from the same job step.) Indeed, it can be used concurrently by two or more tasks in multitask operations. One task may use it, and before the module execution is completed, an interruption may give control to a second task which, in turn, may reenter the module. This in no way interferes with the first task resuming its execution of the module at a later time.

In a multitask environment, concurrent use of a load module by two or more tasks is considered normal operation. Such use is important in minimizing main storage requirements and program reloading time. Many OS/360 control routines are written in reenterable form.

A reenterable program uses machine registers as much as possible; moreover, it can use temporary storage areas that "belong" to the task and are protected with the aid of the task's storage key. Temporary areas of this sort can be assigned to the reenterable program by the calling program, which uses a linkage parameter as a pointer to the area. They can also be obtained dynamically with the aid of the GETMAIN macroinstruction in the reenterable program itself. GETMAIN requests the supervisor to allocate additional main storage to the task and to point out the location of the area to the requesting program. Note that the storage obtained is assigned to the task, and not to the program that requested the space. If another task requiring the same program should be given control of the CPU before the first task finishes its use of the program, a *different* block of working storage is obtained and allocated to the second task.

Whenever a reenterable program (or for that matter any program) is interrupted, register contents and program status word are saved by the supervisor in an area associated with the interrupted task. The supervisor also keeps all storage belonging to the task intact—in particular, the working storage being used by the reenterable program. No matter how many intervening tasks use the program, the original task can be allowed to resume its use of the program by merely restoring the saved registers and program status word. The reenterable program is itself unaware of which task is using it at any instant. It is only concerned with the contents of the machine registers and the working storage areas pointed to by designated registers.

Job management

The primary functions of job management are

- Allocation of input/output devices
- Analysis of the job stream
- Overall scheduling
- Direction of setup activities

In the interests of efficiency, job management is also empowered to transcribe input data onto, and user output from, a direct-access device.

In discussing the functions of OS/360, a distinction must be made between job management and task management. Job management turns each job step over to task management as a formal task, and then has no further control over the job step until completion or abnormal termination. Job management primes the pump by defining work for task management; task management controls the flow of work. The functions of task management (and to some degree of data management) consist of the fetching of required load modules; the dynamic allocation of CPU, storage space, channels, and control units on behalf of competing tasks; the services of the interval timer; and the synchronization of related tasks.

schedulers

Job management functions are accomplished by a *job scheduler* and a *master scheduler*. The job scheduler consists mainly of control programs with three types of functions: read/interpret, initiate/terminate, and write. The master scheduler is limited in function to the handling of operator commands and messages to the console operator.

In its most general form, the job scheduler permits priority scheduling as well as sequential scheduling. The *sequential scheduling system* is suggested by Figure 4. A reader/interpreter scans the control statements for one job step at a time. The initiator allocates input/output devices, notifies the operator of the physical volumes (tape reels, removable disks, or the like) to be mounted, and then turns the job step over to task management.

In a *priority scheduling system*, as suggested by Figure 5, jobs are not necessarily executed as encountered in an input job stream. Instead, control information associated with each job enters an input work queue, which is held on a direct-access device. Use of this queue, which can be fed by more than one input job stream, permits the system to react to job priorities and delays caused by the mounting and demounting of input/output volumes. The initiator/terminator can look ahead to future job steps (in a given job) and issue volume-mounting instructions to the operator in advance.

Some versions of the system have the capability of processing jobs in which control information is submitted from remote on-line terminals. A reader/interpreter task is attached to handle the job control statements, and control information is placed in the input work queue and handled as in the case of locally submitted jobs. Output data sets from remote jobs are routed to the originating terminal.

For each step of a selected job, the initiator ensures that all necessary input/output devices are allocated, that direct-access storage space is allocated as required, and that the operator has mounted any necessary tape and direct-access volumes. Finally, the initiator requests that the supervisor lend control to the program named in the job step. At job step completion, the terminator removes the work description from control program tables, freeing input/output devices, and disposing of data sets.

multijob initiation

One version of the initiator/terminator, optional for larger systems where it is practical to have more than one job from the input work queue under way, permits *multijob initiation*. When the system is generated, the maximum number of jobs that are allowed to be executed concurrently can be specified. Although each selected job is run one step at a time, jobs are selected from the queue and initiated as long as (1) the number of jobs specified

Figure 4

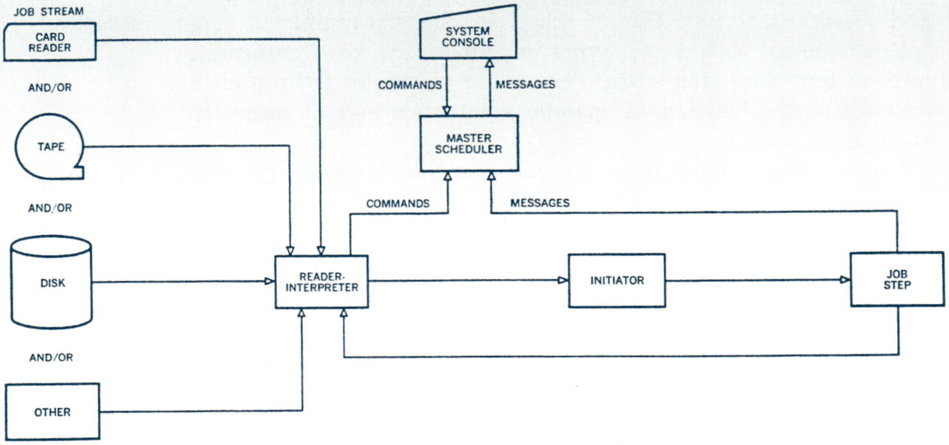

by the user is not exceeded; (2) enough input/output devices are available; (3) enough main storage is available; (4) jobs are in the input work queue ready for execution; and (5) the initiator has not been detached by the operator.

Multijob initiation may be used to advantage where a series of local jobs is to run simultaneously with an independent job requiring input from remote terminals. Typically, telecommunication jobs have periods of inactivity, due either to periods of low traffic or to delays for direct-access seeks. During such delays, the locally available jobs may be executed.

During execution, output data sets may be stored on a direct-access storage device. Later, an output writer can transcribe the data to a system output device (normally a printer or punch). Each system output device is controlled by an output writer task. Moreover, output devices can be grouped into usage classes. For example, a single printer might be designated as a class for high-priority, low-volume printed output, and two other printers as a class for high-volume printing. The data description statement allows output data sets to be directed to a class of devices; it also allows a specification that places a reference to the data on the output work queue. Because the queue is maintained in priority sequence, the output writers can select data sets on a priority basis.

In systems with input and output work queues, the output writer is the final link in a chain of control routines designed to ensure low turn-around time, i.e., time from entry of the work

Figure 5

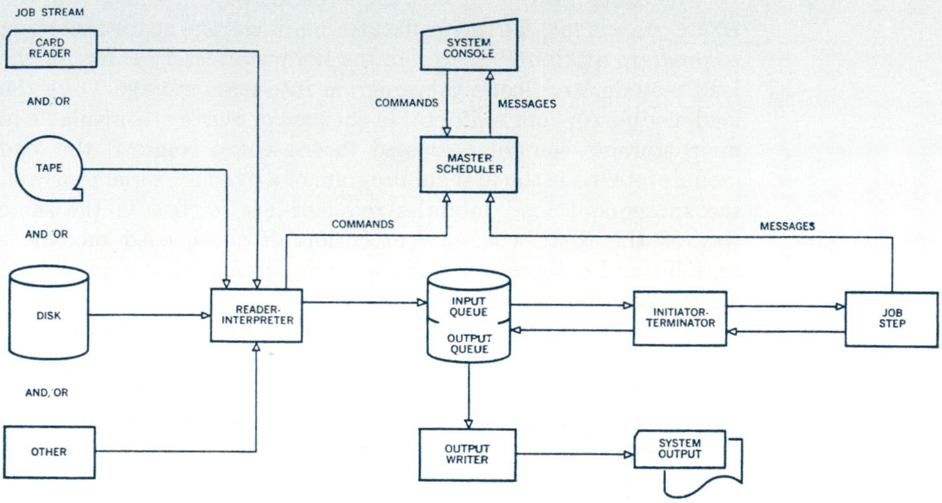

statement to a usable output. At two intermediate stages of the work flow, data are accessible as soon as prepared, without any requirement for batching; and at each of these stages, priorities are used to place important work ahead of less important work that may have been previously prepared. These stages occur when the job has just entered the input work queue, and when the job is completed with its output noted in the output work queue.

Note that a typical priority scheduling system, even one that handles only a single job at a time, may require multitask facilities to manage the concurrent execution of a reader, master scheduler, and a single user's job.

Task management

As stated earlier, job management turns job steps over to task management, which is implemented in a number of supervisory routines. All work submitted for processing must be formalized as a task. (Thus, a program is treated as data until named as an element of a task.) A task may be performed in either a *single-task* or *multitask* environment. In the single task environment, only one task can exist at any given time. In the multitask environment, several tasks may compete for available resources on a priority basis. A program that is written for the single-task environment and follows normal conventions will work equally well in the multitask environment.

single-task operation

In a single-task environment, the job scheduler operates as a task that entered the system when the system was initialized. Each job step is executed as part of this task, which, as the only task in the system, can have all available resources. Programs can have a simple, overlay, or dynamic serial structure.

The control program first finds the load module named in the EXEC statement. Then it allocates main storage space according to program attributes stated in the library directory entry for the load module, and loads the program into main storage. Once the load module (or root segment, in the case of overlay) is available in main storage, control is passed to the entry point. If the load module fetched is the first subprogram of a dynamic serial program, the subsequent load modules required are fetched in the same way as the first, with one exception: if the needed module is reusable and a copy is already in main storage, that copy is used for the new requirement.

When the job step is completed, the supervisor informs the job scheduler, noting whether completion was normal or abnormal.

By clearly distinguishing among tasks, the control system can allow tasks to share facilities where advantageous to do so.

multitask operation

Although the resource allocation function is not absent in a single-task system, it comes to the fore in a multitask system. The system must assign resources to tasks, keep track of all assignments, and ensure that resources are appropriately freed upon task completion. If several tasks are waiting for the same resource, queuing of requests is required.

Each kind of resource is managed by a separate part of the control system. The CPU manager, called the *task dispatcher*, is part of the supervisor; the queue on the CPU is called the *task queue*. The task queue consists of task control blocks ordered by priority. There is one task control block for each task in the system. Its function is to contain or point to all control information associated with a task, such as register and program-status-word contents following an interrupt, locations of storage areas allocated to the task, and the like. A task is *ready* if it can use the CPU, and *waiting* if some event must occur before the task again needs the CPU.

A task can enter the waiting state directly via the WAIT macroinstruction, or it may lapse into a waiting state as a result of other macroinstructions. An indirect wait may occur, for example, as a result of a GET macroinstruction, which requests the next input record. If the record is already in a main storage buffer area, the control program is not invoked and no waiting occurs; otherwise, a WAIT is issued by the GET routine and the task delayed until the record becomes available.

Whenever the task dispatcher gains control, it issues the Load Program Status Word instruction that passes control to the ready task of highest priority. If none of the tasks are ready, the task dispatcher then instructs the CPU to enter the hardware waiting condition.

By convention, the completed use of a resource is always signaled by an interruption, whereupon the appropriate resource manager seizes control.

Let *subtask* denote a task attached by an existing task within a job step. Subtasks can share some of the resources allocated to the attaching task—notably the storage protection key, main storage areas, serially reusable programs (if not already in use), reenterable programs, and data sets (as well as the devices on which they reside). Data sets for a job step are initially presented to the job scheduler by data definition statements. When the job scheduler creates a task for the job step, these data sets become available to all load modules operating under the task, with no restriction other than that data-set boundaries be heeded. When the task attaches a subtask, it may pass on the location of any data control block: using this, the subtask gains access to the data set.

synchronized events

We have mentioned the ways by which an active task can enter a waiting state in anticipation of some specific event. After the event has occurred, the required notification is effected with the aid of the POST macroinstruction. If the event is governed by the control program, as in the instance of a read operation, the supervisor issues the POST; for events unknown to the supervisor, a user's program (obviously not part of the waiting task) must issue a POST.

A task program may issue several requests and then await the completion of a given number of them. For example, a task may specify by READ, WRITE, and ATTACH macroinstructions that three asynchronous activities be performed, but that as soon as two have been completed, the task be placed in the ready condition. When each of these requests is initially made to the control program, the location of a one-word *event control block* is stated. The event control block provides communication between the task (which issued the original request and the subsequent WAIT) and the posting agency—in this case, the control program. When the WAIT macroinstruction is issued, its parameters supply the addresses of the relevant event control blocks. Also supplied is a *wait count* that specifies how many of the events must occur before the task is ready to continue.

When an event occurs, a *complete* flag in the appropriate event control block is set by the POST macroinstruction, and the number of complete flags is tested against the wait count. If they match, the task is placed in the ready condition. A *post code* specified in the POST macroinstruction is also placed in the event control block; this code gives information regarding the manner in which completion occurred. After the task again gains control, the user program can determine which events occurred and in what manner.

Requests for services may result in waits of no direct concern to the programmer, as, for example, in the case of the GET macroinstruction previously mentioned. In all such instances, event control blocks and wait specifications are handled entirely by the supervisor.

Another form of synchronization allows cooperating tasks to share certain resources in a "serially reusable" way. The idea (already invoked in the discussion of programs) may be applied to any shared facility. For example, the facility may be a table that has to be updated by many tasks. In order to produce the desired result, each task must complete its use of the table before another task gains access to it (just as each task had to complete its use of a self-initiating program before another task was allowed to use the program). To control access to such a facility, the programmer may create a queue of all tasks requiring access, and limit access to one task at a time. Queuing capabilities are provided by two macroinstructions: enqueue (ENQ) and dequeue (DEQ). The nature of the facility, known only to the tasks that require it, is of no concern to the operating system so long as a queue control block associated with the facility is provided by the programmer. ENQ causes a request to be placed in a queue associated with the queue control block. If the *busy indicator* in the control block is on, the task issuing the ENQ is placed in the wait condition pending its turn at the facility. If the busy indicator is off, the issuing task becomes first in the queue, the busy indicator is turned on, and control is returned to the task. When finished with the facility, a task liberates the facility and posts the next task on the queue by issuing DEQ.

In a multitask operation, competing requests for service or resources must be resolved. In some cases, choices are made by considering hardware optimization, as, for example, servicing requests for access to a disk in a fashion that minimizes disk seeking time. In most cases, however, the system relies upon a priority number provided by the user. The reason for this is that the user can best select priority criteria. He may reconcile such factors as the identification of the job requestor, response-time requirements, the amount of time already allocated to a task, or the length of time that a job has been in the system without being processed. The net result is stated in a priority number ranging from 0 to 14 in order of increasing importance.

task priority

Initial priorities, specified in job statements, affect the sequence in which jobs are selected for execution. The operator is free to modify such priorities up to the time that the job is actually selected. Changes to priorities may be made dynamically by the change priority (CHAP) macroinstruction, which allows a program to modify the priority of either the active task or of any of its subtasks. Means are provided whereby unauthorized modification can be prevented.

When the job scheduler initiates a job step, the current priority of the job is used to establish a *dispatch priority* and a *limit priority*. The former is used by the resource managers, where applicable, to resolve contention for a resource. The limit priority, on the other hand, serves to control dynamic priority assignments. CHAP permits each task to change its dispatching priority to any point in the range between zero and its limit. Furthermore, when a task attaches a subtask, it is free to set the subtask's dispatching and limit priorities at any point in the range between zero and the limit of the attacher; the subtask's dispatching priority can however exceed that of the attacher. For example, were task A, with limit and dispatching priorities both equal to 10, to attach subtask B with a higher relative dispatching priority than itself, it could use CHAP to lower its own dispatching priority to 7 and attach B with limit and dispatching priorities set to 8.

It is expected that most installations will ordinarily use three levels of priority for batch-processing jobs. Normal work will automatically be assigned a median priority. A higher number will be used for urgent jobs, and a lower one for low-priority work.

task termination

Normally, programs are expected to signal completion of their execution by RETURN or XCTL. If the program at the highest control level within the task executes a RETURN, the supervisor treats it as an end-of-task signal. Whenever RETURN is used, one of the general registers is used to transmit a return code to the caller. The return code at task termination may be inspected by the attaching task, and is used by the job scheduler to evaluate the condition parameters in job control statements. It may, for example, find that all remaining steps are to be skipped.

In addition, any program operating on behalf of a task can execute a macroinstruction to discontinue task execution abnormally. The control program then takes appropriate action to liberate resources, dispose of data sets, and remove the task control block. Although abnormal termination of a task causes abnormal termination of all subtasks, it is possible for abnormal subtasks to terminate without causing termination of the attaching task.

main storage allocation

The supervisor is designed to allocate main storage dynamically, when space is demanded by a task or the control program itself. An *implicit* request is generated internally within the control program, on behalf of some other control program service. An example is LINK, in which the supervisor finds a program, allocates space, and fetches it. To make *explicit* requests for additional main storage areas, a user program may employ the GETMAIN or GETPOOL macroinstructions.

Also provided are means for dynamic release of main storage areas. Implicit release may take place when a program is no longer in use, as signaled by RETURN, XCTL, or DELETE. Explicit release is requested by the FREEMAIN or FREEPOOL macroinstructions.

Explicit allocation by GETMAIN can be for fixed or variable areas, and can be conditional or unconditional:

- *Fixed area.* The amount of storage requested is explicitly given.
- *Variable area.* A minimum acceptable amount of storage is specified, as well as a larger amount preferred. If the larger amount is not available, the supervisor responds to the request with the largest available block that equals or exceeds the stated minimum.
- *Conditional.* Space is requested if available, but the program can proceed without it.
- *Unconditional.* The task cannot proceed without the requested space.

storage protection

The operating system uses the SYSTEM/360 storage protection feature to protect storage areas controlled by the supervisor from damage by user jobs and to protect user jobs from each other. This is done by assigning different protection keys to each of the job steps selected for concurrent execution. However, if multiple tasks result from a single job step (by use of the ATTACH macroinstruction), all such tasks are given the same protection key to allow them to write in common communication areas.

Each job step is assigned two logically different pools, each consisting of one or more storage blocks. The first of these is used to store non-reusable and serially-reusable programs from any source, and reenterable programs from sources other than the main library. This pool is not designated by number. The second pool, numbered 00, is used for any task work areas obtained by the supervisor and for filling all GETMAIN or GETPOOL requests—unless a non-zero pool number is specified.

When the highest-level task of a job step is terminated, all storage pools are released for reassignment. However, when a task attaches a subtask, and makes storage areas available to the subtask, it may suit the purposes of the task not to have the storage areas released upon completion of the subtask. To provide for this possibility, programs may call for the creation of pools numbered 01 or higher. Such a pool may be made available to a subtask in either of two ways—that is, by *passing* or *sharing*. If a pool created by a task is passed to a subtask, termination of the subtask results in release of the pool. On the other hand, subtask termination does not result in the release of a shared pool. In both cases, the subtask that receives a pool may add to the pool, delete from it, or release it in the same way as the originating task.

passing and sharing

Because Pool 00 refers to the same set of storage blocks for all tasks in a job step, it need not be passed or shared, and is not released until the job step is completed.

roll-out and roll-in

If two or more job steps are being executed, and one requires more additional main storage than is available for allocation, the control program intervenes. First, the supervisor attempts to free space occupied by a program that is neither in use nor reserved by a task. Failing that, it may suspend the execution of one or more tasks by storing the associated information in auxiliary storage. The storage and retrieval operations occasioned by competing demands for main-storage space are termed *roll-out* and *roll-in*.

The decision to roll out one or more tasks is made on the basis of task priorities. A main storage demand by a task can cause as many lower-priority tasks to be rolled out as necessary to satisfy the demand. If the lowest-priority task in the system needs more space to continue, it is placed in a wait state pending main storage availability.

During roll-out, all tasks operating under a single job step are removed as a group. Input/output operations under way at the time of the roll-out are allowed to reach completion.

Roll-in takes place automatically as soon as the original space is again available, and execution continues where it left off. Since its task control block remains in a wait status and its input/output units are not altered, a task may still be considered in the system after roll-out.

Significance of multitask operations

It may be expected that multitask operations will not only provide powerful capabilities for many existing environments, but will also serve as a foundation for more complex environments for some time to come.

Fast turnaround in job-shop operations is achieved by allowing concurrent operation of input readers, output writers, and user's programs. It is possible to handle a wide variety of telecommunication activities, each of which is characterized by many tasks (most of them in wait conditions). Also, complex problems can be programmed in segments that concurrently share system resources and hence optimize the use of those resources. With some versions of the job scheduler, multitask operations permit job steps from several different jobs to be established as concurrent tasks. To serve such current multitask needs, the structure of the control system consists of two primary classes of elements: (1) queues representing unsatisfied requirements of tasks for certain resources, and (2) tables identifying available resources. Some of the control information is in main storage; some is in auxiliary storage. This structure facilitates dynamic configuration changes, such as addition or removal of programs in main storage, and attached input/output devices.

Perhaps more important for future systems, the structure may prove adaptable in the management of additional CPU's. For example, if multiple CPU's were given access to the job queue (now stored on a disk), each CPU could queue new jobs as well as initiate jobs already on the queue. Similarly, if multiple CPU's were given access to main storage, each CPU could add tasks to the task queue and dispatch tasks already on the task queue. That is, a system could be designed wherein, by executing the task-dispatcher control routine (which itself is in the shared main storage), any CPU could be assigned a ranking task on the queue; and while executing a task, any CPU could add new tasks to the queue by means of the ATTACH macroinstruction.

Summary

In OS/360, for which the basic unit of work is the task, resources are allocated only to tasks. In general, resources are allocated dynamically to permit easier planning on the part of the programmer, achieve more efficient utilization of storage space, and open the way for concurrent execution of a number of tasks.

Users notify the system of work requirements by defining each job as a sequence of job-control statements. The number of tasks entailed by a job depends upon the nature of the job. The system permits job definitions to be cataloged, thereby simplifying the job resubmittal process. Reading of job specifications and source data, printing of job results, and job execution can occur simultaneously for different jobs. Job inputs and outputs may be queued in direct-access storage, thereby avoiding the need for external batching and permitting priority-governed job execution. In its multijob-initiation mode, the system can process a number of jobs concurrently.

CITED REFERENCES AND FOOTNOTE

1. An historic review of operating systems, with emphasis on I/O control and job scheduling, appears in Reference 4. Operating systems that provided for multiprogramming are described in References 5, 6, and 7. One on-line inventory application is described in Reference 8, and some indication of techniques used in its solution are given in References 9 and 10.
2. G. A. Blaauw and F. P. Brooks, Jr., "The structure of SYSTEM/360: Part I—outline of the logical structure," *IBM Systems Journal* **3**, No. 2, 119–135 (1964).
3. W. P. Heising and R. A. Larner, "A semi-automatic storage allocation

system at loading time," *Communications of the ACM* **4**, No. 10, 446–449 (October 1961).

4. T. B. Steel, Jr., "Operating systems," *Datamation* **10**, No. 5, 26–28 (May 1964).
5. E. S. McDonough, "STRETCH experiment in multiprogramming," *Digest of Technical Papers*, ACM 62 National Conference, 28 (1962).
6. E. F. Codd, "Multi-programming," *Advances in Computers*, Volume 3, Edited by Franz L. Alt and Morris Rubinoff.
7. G. F. Leonard, "Control techniques in the CL-II Programming System," *Digest of Technical Papers*, ACM 62 National Conference, 29 (1962).
8. M. N. Perry and W. R. Plugge, "American Airlines SABRE electronic reservation system," WJCC *Proceedings*, 593–601 (May 1961).
9. M. N. Perry, "Handling of very large programs," *Proceedings of IFIP Congress 65*, Volume 1, 243–247 (1965).
10. W. B. Elmore and G. J. Evans, Jr., "Dynamic control of core memory in a real-time system," *Proceedings of IFIP Congress 65*, Volume 1, 261–266 (1965).

Concepts underlying the data-management capabilities of OS/360 are introduced; distinctive features of the access methods, catalog, and relevant system macroinstructions are discussed.

To illustrate the way in which the control program adapts to actual input/output requirements, a read operation is examined in considerable detail.

The functional structure of OS/360

Part III Data management

by W. A. Clark

The typical computer installation is confronted today with an imposing mass of data and programs. Moreover, with the applicable technologies developing at a rapid pace, the current trend is toward increasing diversity and change in input/output and auxiliary storage devices. Together, these factors dictate that the so-called "input/output" process be viewed in new perspective. Whereas the support provided by a conventional input/output control system is usually limited to data transfer and label processing, the current need is for a data management system that encompasses identification, storage, survey, and retrieval needs—for programs as well as data. Not only should the system employ the capabilities of both direct-access and serial-access devices, but ideally should be able to satisfy a storage or input/output requirement with any storage device that meets the functional specifications of the given requirement.

Our purpose here is to discuss the main structural aspects of OS/360 from the standpoint of data management. In identifying, storing, and retrieving programs and data via OS/360, a programmer normally reckons with device classes rather than specific devices. Because actual devices are not assigned until job-step execution time, a novel degree of device independence is achieved. Moreover, as befits a system intended for a wide range of applications, OS/360 provides for several data organizations and search schemes. Various buffering and transmittal options are provided.

Background

Although the data management services provided by OS/360 are deliberately similar to those provided by predecessor systems, the system breaks with the past in the manner in which it adapts to specific needs.

compiled I/O routines

For mobilizing the input/output routines needed in a given job step, one well-known scheme places these routines into the user's program during the compilation process. No post-assembly program fetching or editing is then required; a complete, executable program results. This scheme has significant disadvantages. It requires that a fairly complete description of device types and intended modes of operation be stated in the source program. Compilation is made more difficult by having to concern itself with details of the input/output function, and compiled programs can be made obsolete by environmental changes that affect the input/output function.

interpretive I/O routines

These disadvantages led the designers of some prior operating systems, for example, IBSYS/IBJOB, to circumvent the inclusion of input/output routines in assembled programs by providing a set of input/output "packages" that could be mobilized at program-loading time.[1] Designed to operate interpretively, these optional packages permitted a source program to be less specific about devices and operating modes; moreover, they permitted change in the input/output environment without program reassembly. On the other hand, interpretive execution tends to reduce the efficiency of packages and limit the feasible degree of system complexity and expandability.

generated I/O routines

Faced with unprecedented diversity in storage devices and potential applications in addition to the complexities of multitask operation, the OS/360 designers have carried the IBSYS/IBJOB philosophy further, but with a number of significant tactical differences. Data-management control facilities are not obtained at program-loading time; instead, they are tailored to current needs during the very course of program execution (wherever the programmer uses an OPEN macroinstruction). The data-access routines are reenterable, and different tasks with similar needs may share the same routines. Because routines do not act interpretively, they can be highly specialized as well as economical of space. A program chooses one of the available access methods and requests input/output operations using appropriate macroinstructions. Device types, buffering techniques, channel affinities, and data attributes are later specified via data-definition statements in the job stream. In fact, the OS/360 job stream permits final specification of nearly every data or processing attribute that does not require re-resolution of main-storage addresses in an assembled program. These attributes include blocking factors, buffering techniques, error checks, number of buffers, and the like.

System definitions

An operating system deals with many different categories of information. Examples from a number of categories are a source program, an assembled program, a set of related subroutines, a message queue, a statistical table, and an accounting file. Each of these examples consists of a collection of related data items. In the OS/360 context, such a collection is known as a *data set*. In the operational sense, a data set is defined by a data-set label that contains a name, boundaries in physical storage, and other parameters descriptive of the data set. The data-set label is normally stored with the data set itself.

A standard unit of auxiliary storage is called a *volume*. Each direct-access volume (disk pack, data cell, drum, or disk area served by one access mechanism) is identified by a volume label. This label always contains a volume serial number; in the case of direct-access devices, it also includes the location of a *volume table of contents* (VTOC) that contains the labels of each data set stored in the corresponding volume. A label to describe the VTOC and an additional label to account for unused space are created. Before being used in the system, each direct-access volume is initialized by a utility program that generates the volume label and, for direct-access devices, constructs the table of contents. This table is designed to hold labels for the data sets to be written on the volume.

volume

Given the volume serial number and data-set name, the control program can obtain enough information from the label to access the data set itself.

A job step can place a data set in direct-access storage via a data definition (DD) statement that requests space, specifies the kind of volume, and gives the data-set name. At job-step initiation, the system allocates space and creates a label for each area requested by a DD statement. Finally, during job-step execution, the label is completed and updated via OPEN and CLOSE macro-instructions.

Each reel of magnetic tape is considered a volume. In view of the serial properties of tape, the method used for identifying volumes and data sets departs somewhat from the method used for direct-access devices. The standard procedure still employs volume labels and data-set labels; but each data-set label exists in two parts: a header label preceding its data set, and a trailer label that follows it. The location of a data set in a tape volume is represented by a sequence number that facilitates tape searching.

Although the system includes a generalized labeling procedure, it permits a user to employ his own tape-label conventions and label-checking routines if so desired. Unlabeled tapes may be used, in which case the responsibility for mounting the right volumes reverts to the operator.

To free the programmer of the need to maintain inventories of his data sets, the system provides a *data-set catalog*. Held in direct-access storage, this catalog consists of a tree-organized set of indexes. To best serve the needs of individual installations, the organization of the tree structure is left to the user. Each qualifier of a data-set name corresponds to an additional level in the tree. For example, the data set PAYROLL.MASTER.SEGMENT1 is found by searching a master index for PAYROLL, a second-level index for MASTER, and a third-level index for SEGMENT1. Stored with the latter argument are entries that identify the volume containing the data set and the device type; in the case of serial-access devices, a sequence number is also stored.

catalog

A volume containing all or part of the catalog is called a *control volume*. Normally, the operating system resides in a control volume known as the *system residence volume*. The use of a distinctive control volume for a group of related data sets makes it convenient to move the portion of the catalog that is relevant to the group. This is particularly important in planning for the possibility that groups of data sets may be moved from one computer to another.

control volume

A data-set search starts in a system residence volume and continues, level by level, until a volume identification number is obtained. If the required volume is not already mounted, a message is issued to the operator. Then, if the data set is stored in a direct-access device, the search for the data-set location resumes with the volume label of the indicated volume, continues in the volume table of contents, and proceeds from there to the data set's starting location. On the other hand, if the data set is held on a serial-access device, the search continues using a sequence number as an argument.

To simplify DD (data definition) statements for recurrent updating jobs, data sets related by name and cataloging sequence can be identified as a *generation group*. In applications that regularly use the n most prior generations of a group to produce a new generation, the new generation may be named (and later referred to) relative to the most recent generation. Thus, the DD statement need not be changed from run to run. When the index for the generation group is established, the programmer specifies n. As each new generation is cataloged, the oldest generation is deleted from the catalog. Provision is also made for the special case in which n varies systematically, starting at 1 and increasing by 1 until it reaches a user-specified number N, at which time it starts over at 1.

generation group

To safeguard sensitive data, any data set may be flagged in its label as "protected." This protection flag is tested as a consequence of the OPEN instruction; if the flag is on, a correct *password* must be entered from the console. The data set name and appended password serve as an argument for searching a control table. The OPEN routine is not permitted to continue unless a matching entry is found in the table.

password

Because the control table has its own security flag and *master password*, it can be reached only by the control program and those programmers privileged to know the master password.

record and block

In discussing the internal structure and disposition of a data set, it is necessary to distinguish between the *record*, an application-defined entity, and the *block*, which has hardware-defined boundaries and is governed by operational considerations. Let b denote block length (in bytes) and B a maximum block length. Although OS/360 requires that B be specified for each given data set, conventions permit three block-format categories: unspecified, variable, and fixed. The first category requires that $b \leq B$ for all blocks. The second is similar to the first, except that each b is stored in a count field at the beginning of its block. The third category dictates that all blocks be of length B.

A fixed or variable block may contain multiple records. A fixed block contains records of fixed size. In the variable block, records may vary in size, and each record is preceded by a field that records its size. For storage devices that employ interblock gaps, it is well known that record blocking can increase effective data rates, conserve storage, and reduce the needed number of input/output operations in processing a data set. For data sets of unspecified block format, the system makes no distinction between block and record; any applicable blocking and deblocking must be done by the user's program. The unspecified format is intended for use with peripheral equipment, such as transmission devices, address label printers, and the like.

buffer

A *buffer* is a main storage area allocated to the input/output function. The portion of a buffer that holds one record is called a *buffer segment*. A group of buffers in an area of storage formatted by the system is called a *buffer pool*; a data set associated with a buffer pool is assigned buffers from the pool. Unless a programmer assigns a buffer pool to a data set, OS/360 does so; unless buffer size is specified by the programmer, OS/360 sets the size to B.

In processing records from magnetic tape, it is customary to read and process records from one or more data sets, and to create one or more new data sets. A number of buffering considerations come into play. It may suffice to process a record within a buffer; it may be preferable to move the input record to a work area and the updated record from the work area into an output buffer; other possibilities may suggest themselves. Moreover, in processing records from direct-access storage, the same data set may be accessed for input and output.

transmittal modes

To allow flexibility in buffer usage, the OS/360 record-transfer routines invoked by the GET and PUT macroinstructions permit three *transmittal modes*. In the "move" mode, each record is moved from an input buffer to a work area and finally to an output buffer. In the "locate" mode, a record is never actually moved, but a pointer to the record's buffer segment is made available to the application program. In the "substitute" mode, which also uses pointers, the application program provides a work area equal in size to a record, and the buffer segment and work area effectively change roles.

To supplement the transmittal modes in special cases, three

methods of buffer allocation are defined. *Simple buffering*, the most general method, allocates one or more buffers to each data set. *Exchange buffering*, used with fixed-length records, utilizes data-chaining facilities to effect record gather and record scatter operations. Buffer segments from an input data set are exchanged with buffer segments of an output data set or work area. Not only can each buffer segment be treated, in turn, as an input area, work area, and output area, but chaining allows noncontiguous segments to simulate a block. Exchange buffering is particularly useful in updating sequential files, merging, and array manipulation.

buffer allocation

Chained-segment buffering is designed for messages of variable size. Segments are established dynamically, with chaining being used to relate physically separate segments. This method is designed to circumvent the need for a static allocation of space to a remote terminal: of the many terminals that can be present in a system, only a fraction are ordinarily in use at a given time.

Access principles

To fall within the OS/360 data-management framework, a data set must belong to one of five organizational categories. As will be seen, the classification is based mainly on search considerations.

Data sets consisting of records held in serial-access storage media (such as magnetic tapes, paper tapes, card decks, or printed listings) are said to possess *sequential* organization. If so desired, a data set held in a direct-access device may also be organized sequentially.

data-set categories

Three of the five categories apply solely to direct-access devices. The *indexed sequential* organization stores records in sequence on a key (record-contained identifier). Because the system maintains an index table that contains the locations of selected records in the sequence, records can be accessed randomly as well as sequentially. A *direct* organization is similar, but dispenses with the index table and leaves record addressing entirely up to the programmer. A *partitioned* organization divides a sequentially organized data set into *members*; member names and locations are held in a directory for the data set. A member consists simply of one or more blocks. Included primarily for data sets consisting of programs or subroutines, this organization is suitable for any data set of randomly retrieved sequences of blocks.

A *telecommunications* organization is provided for queues of messages received from or enroute to remote on-line terminals. Provision is made for forming message queues and for retrieving messages from queues. Queues may be held in direct-access storage as well as in main storage.

A broad distinction is made between two classes of data-management languages. Designed for programming simplicity, the *queued access* languages apply only to organizations with sequential properties. The programmer typically uses the macro-

language categories

Table 1

Organization	Language category	
	Queued	Basic
Sequential	QSAM	BSAM
Indexed Sequential	QISAM	BISAM
Direct		BDAM
Partitioned		BPAM
Telecommunication	QTAM	BTAM

instructions GET and PUT to retrieve and store records, and buffers are managed automatically by the system. On the other hand, the *basic access* languages provide for automatic device control, but not for automatic buffering and blocking. Typically, the READ and WRITE macroinstructions are used to retrieve and store blocks of data. Because the programmer retains control over device-dependent operations (such as card reader or punch-stacker selection, tape backspacing, and the like), he may use any desired searching, buffering, or blocking methods.

access methods

Of the ten possible combinations of data-set and language categories, eight are recognized by the system as *access methods*. These eight methods bear the mnemonic names given in Table 1: QSAM denotes "queued sequential access method," and so on. For each access method, a vocabulary of suitable macroinstructions is provided.

To employ a given access method, a programmer resorts to the vocabulary of macroinstructions provided for that method. Vocabularies for six of the methods are summarized in Table 2. Although six macroinstructions are common to all of these methods, the parameters to be specified in a macroinstruction may vary from method to method. If so desired for specialized applications, a programmer can circumvent the system-supported access methods and employ the *execute channel program* (EXCP) macroinstruction in fashioning his own access method. In this case, he must prepare his own channel program (sequence of channel command words).

A few words on each vocabulary element of Table 2 help to clarify access principles. At assembly time, the DCB macroinstruction reserves space for a *data control block* and fills in control block fields that designate the desired access method, name a relevant DD statement, and select some of the possible options. The application programmer is expected to provide symbolic addresses of any applicable supplementary routines, as for example, special label-processing routines.

The programmer issues an OPEN macroinstruction for each data control block. At execution time, OPEN supplies information not declared in the DCB macroinstruction, selects access routines and establishes linkages, issues volume mounting messages to the operator, verifies labels, allocates buffer pools, and positions

Table 2 Access-method vocabularies

Macro-instruction	QSAM	QISAM	BSAM	BPAM	BISAM	BDAM	*Macroinstruction function in brief*
DCB	•	•	•	•	•	•	Generate a data control block
OPEN	•	•	•	•	•	•	Open a data control block
CLOSE	•	•	•	•	•	•	Close a data control block
BUILD	•	•	•	•	•	•	Structure named area as a buffer pool
GETPOOL	•	•	•	•	•	•	Allocate space to and format buffer pool
FREEPOOL	•	•	•	•	•	•	Liberate buffer-pool space
GET	•	•					Obtain a record from an input data set
PUT	•	•					Include a record in an output data set
PUTX	•	•					Include an input record in an output data set
RELSE	•	•					Force end of input block
TRUNC	•						Force end of output block
FEOV	•		•				Force end of volume
CNTRL	•		•				Control reader or printer operation
PRTOV	•		•				Test for printer carriage overflow
SETL		•					Set lower limit for scan
ESETL		•					Postpone fetching during scan
CHECK			•	•			Wait for I/O completion and verify proper operation
NOTE			•	•			Note where a block is read or written
POINT			•	•			Point to a designated block
FIND				•			Obtain the address of a named member
BLDL				•			Build a special directory in main store
STOW				•			Update the directory
RELEX					•	•	Release exclusive control of a block
FREEDBUF					•	•	Free dynamically obtained buffer
GETBUF			•	•	•	•	Assign a buffer from the pool
FREEBUF			•	•	•	•	Return a buffer to the pool
WAIT			•	•	•	•	Wait for I/O completion
READ			•	•	•	•	Read a block
WRITE			•	•	•	•	Write a block

volumes. The programmer may free a data control block and return associated buffers to the pool by the CLOSE macroinstruction; if he omits CLOSE, the system performs the corresponding functions at task termination.

The programmer can request the system to allocate and format a buffer pool at execution time by issuing a GETPOOL macroinstruction, which specifies the address of the data control block, the buffer length, and the desired number of buffers. When a pool area is no longer needed, it can be returned to the system by FREEPOOL.

Where the programmer's knowledge permits him to allocate space more wisely than the control program, he may choose to designate the area to be set aside for a buffer pool. The area may, for example, supplant a subroutine that is no longer needed. By issuing a BUILD macroinstruction, he can request the system to employ the reserved area as a buffer pool, the details being similar to GETPOOL. With subsequent BUILD's, moreover, he can restructure the area again and again.

QSAM

QSAM corresponds closely to the schemes most favored in previous input/output systems. QSAM yields a great deal of service to the programmer for a minimum investment in programming effort. Retrieval is afforded by GET, which supplies one record to the program; disposition of an output record is afforded by PUT or PUTX. PUT transfers a record from a work area or buffer to a data set; PUTX transfers a record from one data set to another. In consequence, PUT involves one data control block, whereas PUTX involves two.

To aid the programmer in creating short blocks and in disposing of a block before all records therein have been processed, two macroinstructions permit intervention in buffer control. RELSE requests the system to release the remaining buffer segments in an input buffer, i.e., to view the buffer as empty. Analogously, TRUNC asks the system to view an output buffer as full, and to go on to another buffer.

FEOV requests the system to force an end-of-volume status for a designated data set, and thereupon to undertake the normal volume-switching procedure. CNTRL provides for specialized card-reader, printer, or tape control functions.

QISAM

The QISAM scheme is closely akin to QSAM, but the macroinstructions provide the additional functions required of indexed sequential data sets and direct-access devices. Records are arranged in logical sequence on the key, a field that is part of each record. Record keys are related to physical addresses by indexes. For a record with a given data key, a *cylinder index* yields cylinder address, and a *track index* yields track-within-cylinder address. To facilitate in-channel searches, the key of the last record in each block is placed in a hardware-defined control field.

In the initial creation of a data set, PUT's are used in the "load" mode to store records and generate indexes. Successive GET's in the "scan" mode retrieve records sequentially; SETL (set lower limit) may be issued to designate the first record obtained. Unless a SETL is issued, retrieval starts from the first record of the data set. In scan mode, PUTX may follow a GET to return an updated record to the data set. ESETL (end of scan) halts any anticipatory buffering on the part of the system until issuance of a subsequent GET.

BISAM

BISAM applies to the same sequential data organization as QISAM, but selective reading and writing is permitted through the READ and WRITE macroinstructions. Using BISAM, new records can be inserted without destroying sequence. If an insertion does not fit into the intended track, the system moves one or more records from the track to an overflow area and then reflects this overflow status in the appropriate indexes. (The existence of overflows does not alter the ability of QISAM to scan records in logical sequence.)

To permit other operations to be synchronized with BISAM input/output operations, a WAIT macroinstruction supplements READ and WRITE. (Because WAIT serves a general function in synchronizing tasks, it is discussed in Part II.)

In a multitask environment, it is possible that one task may want to use or update a record while the record is being updated by another task. To forestall confusion in the order that updating operations are accomplished, READ can request exclusive control of the record during updating. For a record being updated in place, WRITE releases exclusive control. If the record is not updated in place, the RELEX macroinstruction can be used to release control.

Because record insertions may lead to overflows, and overflows tend to reduce input/output performance, the system is designed to provide statistics that can help a programmer in determining when data-set reorganization is desirable. Held in the data control block are the number of unused tracks in an independent overflow area and, optionally, the number of full cylinder areas, as well as the number of accesses to overflow records not appearing at the head of overflow chains. Reorganization can be accomplished via the QISAM load mode, using the existing data set as input.

As implied by the above discussion, QISAM and BISAM complement one another and may be used together where the user needs to access a data set randomly as well as sequentially. For the sake of convenience, a data control block for an indexed sequential data set can be opened jointly for QISAM and BISAM.

BSAM

BSAM assumes a sequentially organized data set and deals with blocks rather than records. A block is called into a specified buffer by READ. Unless program execution is deliberately suspended during the retrieval period by a CHECK macroinstruction, the program may continue during reading. Similarly, after an output operation is initiated, CHECK can be used to postpone further processing until the operation is completed.[2] Following a READ or WRITE, the macroinstruction NOTE saves the applicable block address in a standard register; subsequently, the preserved address may be helpful in logically repositioning the volume by POINT.

BDAM

Of the access methods for direct-access devices, BDAM offers the greatest variety of access possibilities. Using WRITE and READ, the programmer can store or retrieve a block from a data set by specifying a track address and block number. Optionally, he may specify a number relative to the data set itself, either (1) a relative track number at which a search should start for a given key or (2) a relative block number. The relative numbers, which help to isolate application programs from device peculiarities, are converted to actual track addresses and block numbers by the system. GETBUF and FREEBUF are the means by which buffers can be explicitly requested and released. A dynamic buffer option, requested in the DCB macroinstruction, enables the programmer to obtain automatic buffer management (BUILD and GETPOOL are not used in conjunction with the option). The FREEDBUF macroinstruction permits release of a buffer under the dynamic option.

BPAM

BPAM is designed for storing and retrieving members of a partitioned data set held on a direct-access device. Associated with the data set is a *directory* that relates member name to track address. To prepare for access, a FIND performs the directory search. A located member can be retrieved using one or more READ's, as required by the number of blocks in the member. New members can be placed by one or more WRITE's, followed by a STOW that enters the member's name and location in the directory. CHECK again serves to synchronize the program with data-transmission operations.

A summary of the main characteristics of the eight access-methods appear in Table 3.

Control elements and system operation

data control block

With general definitions and access methods in mind, we turn to the internal structure of OS/360 as it pertains to data management.

Associated with each data set of a problem program is a data control block (DCB), which must be opened before any data transfer takes place. However, some data sets, e.g., the catalog data set, are opened automatically by the control program, and may be indirectly referred to or used in a problem program without additional opening or closing. Data-access macroinstructions, such as GET and PUT, logically refer to a data set, but actual reference is always via a data control block.

The data control block is generated and partially filled when the DCB macroinstruction is encountered at compilation time. The routine called at execution time by OPEN completes the data control block with information gained principally from a job-stream DD statement or cataloged procedure. For input data sets, a final source of such information is the data-set label. In the case of an output data set where the label has yet to be created, the final source can be the label of another data set or another DD statement.

In addition to completing the data control block, the OPEN routine ensures that needed access routines are loaded and address relations are completed. The routine prepares buffer areas and generates channel command word lists; it initializes data sets by reading or writing labels and performs a number of other housekeeping operations.

The selection of access routines is governed by choices in data organization, buffering technique, access language, input/output unit characteristics, and other factors. The selection is relayed to the supervisor, which allocates main storage space and performs the loading.

In operation, some access routines are treated as part of the user's program and are entered directly rather than through a supervisor-call interruption. These routines block and deblock

Table 3 Access-method summary

Characteristic	QSAM	QTAM	QISAM	BSAM	BTAM	BPAM	BISAM	BDAM
Data set organization	Sequential or member of partitioned	Telecom	Indexed sequential	Sequential	Telecom	Partitioned	Indexed Sequential	Direct
Basic element of data set	Record	Message or message segment	Record	Record	Message	Member	Record	Record
Basic concern of access method	Record	Message	Record	Block	Block	Block	Block	Block
Primary input and output macroinstructions	GET PUT PUTX	GET PUT RETRIEVE	Scan SETL Scan GET Load PUT Scan PUTX	READ WRITE	READ WRITE	FIND READ WRITE STOW	READ WRITE	READ WRITE
Buffer pool acquisition	BUILD GETPOOL Automatic	BUFFER	BUILD GETPOOL Automatic	BUILD GETPOOL Automatic	BUILD GETPOOL Automatic	BUILD GETPOOL Automatic	BUILD GETPOOL Automatic	BUILD GETPOOL Automatic
Buffer management for a data set	Automatic (simple or exchange)	Automatic (chained-segment)	Automatic (simple)	GETBUF FREEBUF	GETBUF FREEBUF	GETBUF FREEBUF	Dynamic FREEDBUF GETBUF FREEBUF	Dynamic FREEDBUF GETBUF FREEBUF
Transmittal mode	Move Locate Substitute	Move	Move Locate					
Synchronization	Automatic	Automatic	Automatic	CHECK WAIT	WAIT	CHECK WAIT	WAIT	WAIT
Record/block format*	F, V record	U message	F, V record	F, V record	U block	F, V, U block	F, V record	F, V, U block
Special provisions for data-set search		Sequence number; relative addressing	Cylinder & track indexes			Directory of members	Cylinder & track indexes	Can use relative record or track number

* F, V, and U denote fixed, variable, and unspecified lengths.

records, control the buffers, and call the input/output supervisor when a request for data input or output is needed. Other routines, treated as part of the I/O supervisor and therefore executed in the privileged mode, perform error checks, prepare user-oriented completion codes, post interruptions, and bridge discontinuities in the storage areas assigned to a data set.

I/O supervisor

The input/output supervisor performs all actual device control (as it must if contending programs are not to conflict in device usage); it accepts input/output requests, queues the requests if necessary, and issues instructions when a path to the desired input/output unit becomes available. The I/O supervisor also ensures that input/output requests do not exceed the storage areas allocated to a data set. The completion of each input/output operation is posted and, where necessary, standard input/output error-recovery procedures are performed. EXCP, the execute channel program macroinstruction, is employed in all communication between access routines and the input/output supervisor.

To portray the mechanics of data management, let us consider one job step and the data-management operations that support a READ macroinstruction for a cataloged data set in the BSAM context.

To begin with, we observe the state of the system just before the job is introduced; of interest at this point are the devices, control blocks, programs, and catalog elements that exist prior to job entry. Next to be considered are the data-management activities involved in DD-statement processing, and in establishment by the job scheduler of a task for the given job step. Third, we consider the activities governed by the OPEN macroinstruction; these activities tailor the system to the requirements of the job step. Finally, operation of the READ macroinstruction is considered, with special attention to the use of the EXCP macroinstruction. Essential to the four stages of the discussion are four cumulative displays. Frequent reference to numbered points within the figures is made by means of parenthetical superscripts in the text. The description refers more often to the objects generated and manipulated by the system than to the functional programs that implement the system.

catalog organization

The basic aspects of catalog implementation become apparent when we consider the manner in which the system finds a volume containing a cataloged data set. Recall that each direct-access volume contains a volume label that locates its VTOC (volume table of contents) and that the VTOC contains a data-set label for each volume-contained data set. Identified by data-set name, the data-set label holds attributes (such as record length) and specifies the location in the volume of the data set.

Search for a data set begins (see Figure 1) in the VTOC of the system residence volume, where a data-set label identifying the portion of the catalog in this volume(1) appears. This part of the catalog is itself organized as a partitioned data set whose directory is the highest level (most significant) index of the catalog. For

Figure 1 Control elements: before job entry

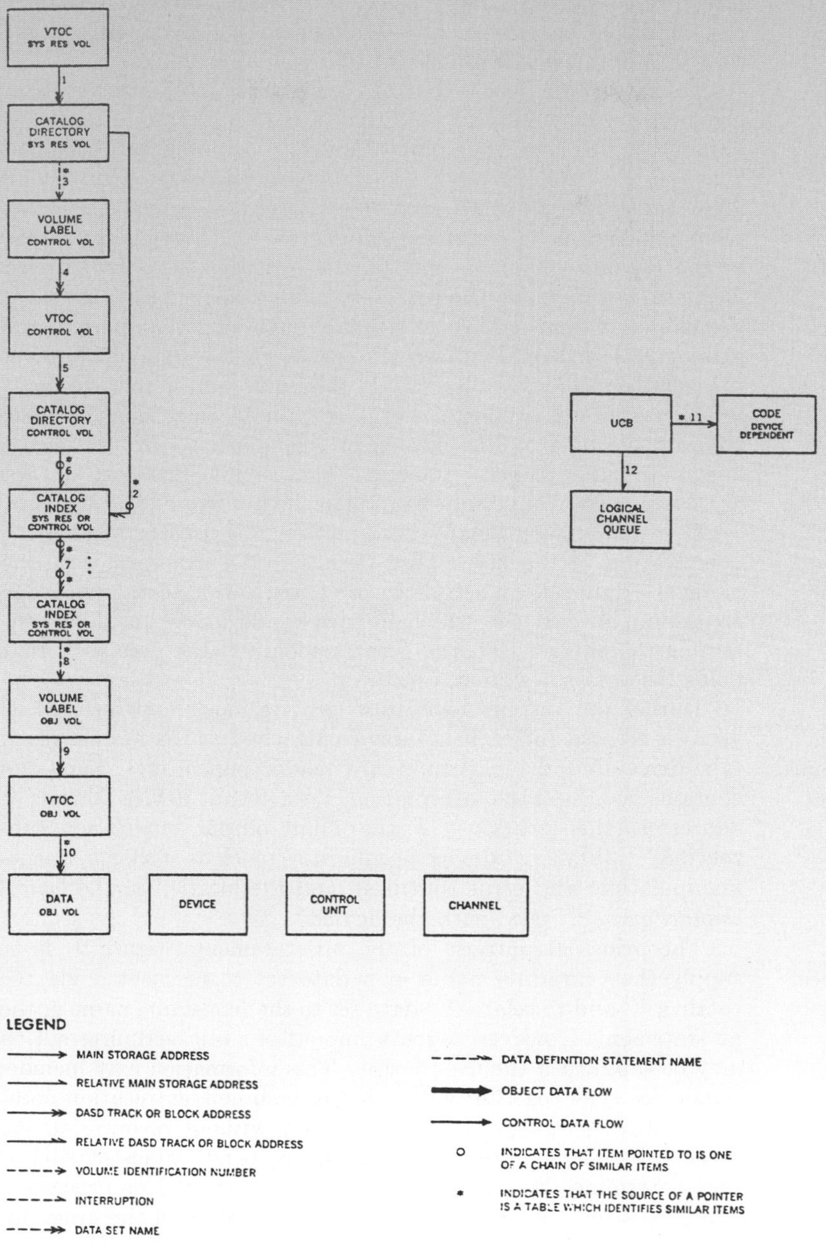

data sets cataloged on the system residence volume, entries in this directory contain the addresses of lower-level indexes;[(2)] for data sets cataloged on other control volumes,[(3)] directory entries contain the appropriate volume identification numbers.

Assume for the moment that the search is for a data set cataloged on control volume *V* and that *V* is not the system residence volume. In this case, the volume label of *V* contains the location of *V*'s VTOC.[(4)] (Volume label and VTOC are recorded separately to allow for device peculiarities.) One of the data-set labels in this VTOC identifies the part of the catalog on *V*;[(5)] just as in the case of the residence volume, this part is organized as a partitioned data set. Inasmuch as the directory of this partitioned data set is the subset of the highest-level index governing that part of the catalog recorded on *V*, directory entries contain the addresses of the next-level indexes on *V*.[(6)] It should be added that all index levels needed to catalog a data set appear on a single control volume; the part of the catalog on any given control volume is known to other control volumes, because the directory entries of the given control volume appear in the directories of the others.

Each index level below the directory[(7)] is used to resolve one qualification in the name of a data set. For example, were the name of a data set A.B.C, a directory entry A would locate an index containing an entry B, which in turn would locate an index containing the entry C. This last entry identifies the volume[(8)] that holds the data set named A.B.C.[3]

unit control block

During the system generation process, one *unit control block* (UCB) is created for each I/O device attached to the system (each tape drive, disk drive, drum, card reader/punch, etc). Each UCB contains device-status information, the relevant device address or addresses, the locations of the input/output supervisor subroutines[(11)] that treat device peculiarities (such as start-I/O, queue-manipulation, and error routines), and the location of the *logical channel queue*[(12)] used with the device.[4]

DD-statement processing

The principal purpose of the DD statement (Figure 2) is to supply the (variable) name of a data set to be located via the catalog,[(13)] and to relate the data set to the (constant) name of the DD statement. However, a great amount of additional information may be supplied if the user desires. This information may include: the device type together with a list of volume identification numbers which serve to locate the data set without recourse to the catalog;[(14,15)] label information used to create new labels; attributes that determine the nature of the data set created or processed; and processing options that modify the operation of the program. After being encoded by the job scheduler, most of this information is included in a *job file control block* (JFCB)[(16)] that is used in lieu of the original DD statement.

As was suggested above, a data set can be located either by an explicit list of volume identification numbers and an indication of the device type (if this information is given on the DD statement), or by data-set name alone. In the latter case, a list of volume

Figure 2 Control elements: job scheduling—hexagonal blocks denote elements of first concern at time job is scheduled

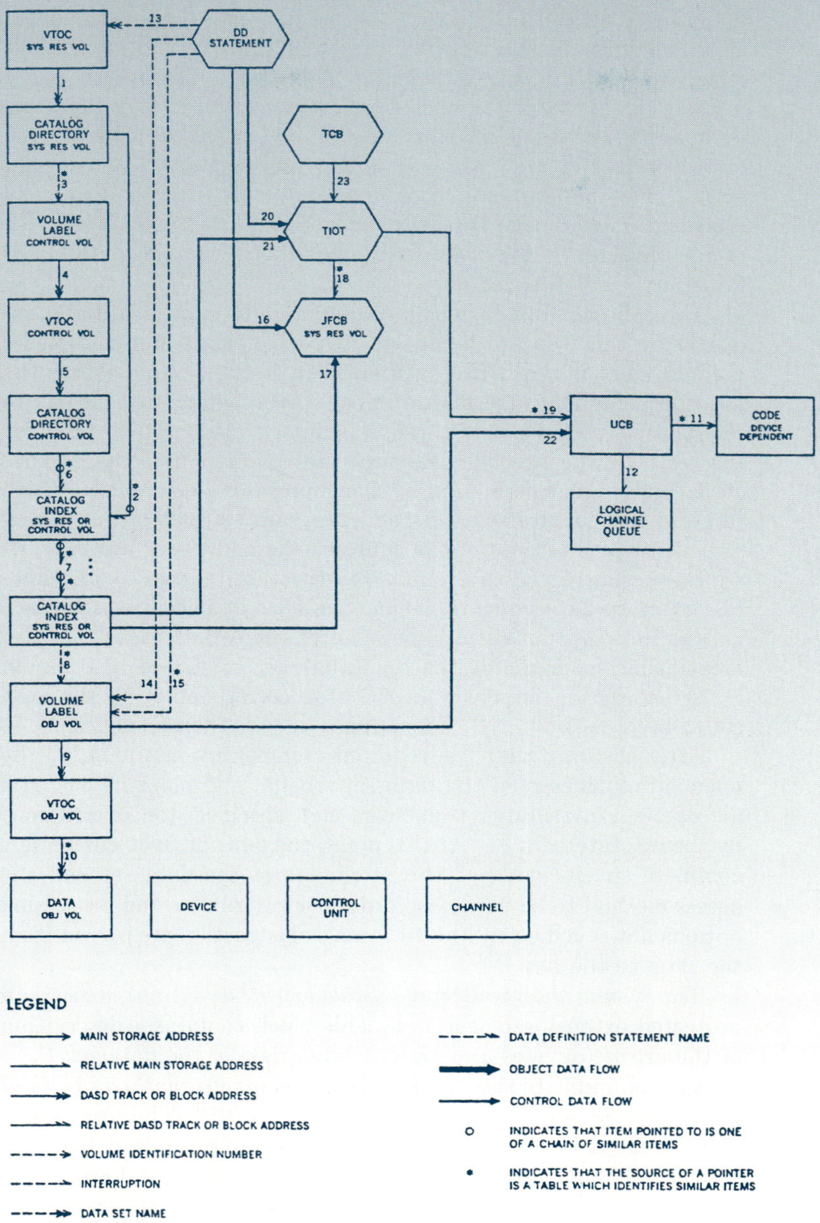

identification numbers is extracted from the catalog and placed in the JFCB.[17]

Prior to establishing a task for the job step, the job scheduler assigns devices to the step. To represent this assignment, the job scheduler constructs a *task input/output table* (TIOT). An entry is made in this table for each DD statement supplied by the user; each entry relates a DD-statement name to the location of the corresponding JFCB[18] and the unit or units assigned to the data set.[19] The assignment of a specific device derives from the specification of device type supplied through the DD statement[20] or the catalog,[21] together with a table of available units maintained by the job scheduler.

The job scheduler then assures that all volumes initially required by the step are mounted. As each volume is mounted, its volume label is read; the volume identification number and the location of its VTOC are placed in the corresponding UCB for future reference.[22] Finally, the job scheduler "attaches" a task for the step. In the process, the supervisor constructs a *task control block* (TCB). The TCB is used by the supervisor as an area in which to store the general registers and program status word of a task at a point of interruption; it contains the address of the TIOT.[23]

OPEN

Execution of the OPEN macroinstruction (Figure 3) identifies one or more data control blocks (DCB's) to be initialized:[24] since an SVC interruption results, the TCB of the calling task[25] is also identified. The name of the DD statement, contained in the DCB, is used to locate the entry in the TIOT corresponding to the data set to be processed.[23,26] The related JFCB is then retrieved.[18]

After assuring that the required volumes are mounted,[19] the open subroutines read the data-set label(s) and place in the JFCB all data-set attributes that were not specified (or overridden) by the DD statement.[27] At this point, the DCB and JFCB comprise a complete specification of the attributes of the data set and the access method to be used. Next, data-set attributes and processing options not specified by the DCB macroinstruction are passed from the JFCB to the DCB.[28]

The system then constructs a *data extent block* (DEB), logically a protected extension of the DCB. This block contains a description of the *extent* (devices and track boundaries) of the data set,[29,30] flags which indicate the set of channel commands that may be used with the data set,[37] and a priority indicator.[31] The DEB is normally located via the DCB;[32] but in order to purge a failing task or close the DCB upon task termination, it may be located via the TCB.[33] If the data set is to be retrieved sequentially, the address of the first block of the data set is moved to the DCB.[34]

Next, the access-method routines are selected and loaded. The addresses of these routines are placed in the DCB.[35] If privileged interrupt-handling or error routines are required, they are loaded and their addresses recorded in the DEB.[36] Finally, the channel programs which will later be used to access the data set are generated. For each channel program, an *input/output block* (IOB) is

Figure 3 Control elements: OPEN macroinstruction—oblate blocks denote elements of first concern during execution of OPEN macroinstruction

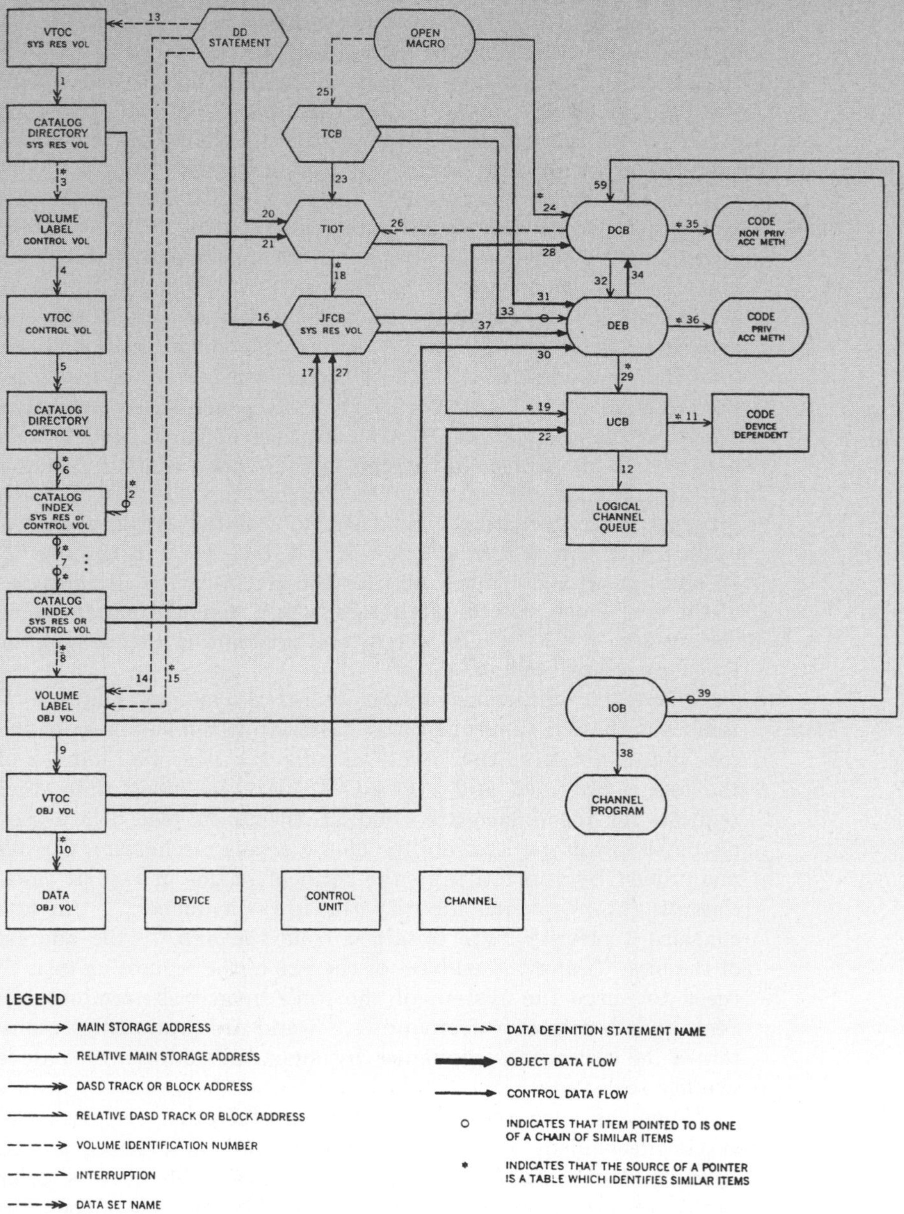

created.[38] The IOB is the major interface between the problem program (or the access-method routines) and the I/O supervisor. It contains flags that govern the channel program, the location of the DCB,[59] the location of an event control block used with the channel program, the location of the channel program itself, the "seek address," and an area into which the I/O supervisor can move the channel status word at the completion of the channel program. IOB's are linked in a chain originating at the DCB.[39]

READ

The READ macroinstruction (see Figure 4) identifies a parameter list, called the *data event control block* (DECB),[40] that is prepared either by the user or the READ macroinstruction. This block contains the address of a buffer,[41] the length of the block to be read (or the length of the buffer), the address of the DCB associated with the data set,[43] an event control block, and the like. Buffer address and block or buffer length are obtained from the DCB if not supplied by the user.[44] Using an address previously placed in the DCB,[35] the READ macroinstruction branches to an access-method routine that assigns an IOB and a channel program to the DECB. Subsequently, the routine modifies the channel program to reflect the block length and the location of the buffer;[42] it then records the address of the DECB in the IOB.[45] In addition, the routine computes the track and block addresses of the next block and updates the IOB and channel program using the results.[42,46,47,48] The access method routine then issues the EXCP macroinstruction.

EXCP

The EXCP macroinstruction causes an SVC interruption[49] that calls the I/O supervisor and passes to it the addresses of the IOB and, indirectly, the DCB.[59] Using the DCB, the address of the DEB is obtained and verified.[32] Next, assuming that other requests for the device are pending, the IOB is placed in a seek queue to await the availability of the access mechanism. Queues maintained by the IOS take the form of chains of *request queue elements* (RQE's) which identify the IOB's in queues.[51] An RQE contains a priority byte obtained from the DEB,[52] the address of the DEB,[53] and the address of the TCB of the requesting task[54] (used to purge the system of the IOB's upon task termination). Seek queues originate from UCB's,[50] and are (optionally) maintained in ascending sequence by cylinder address to reduce average seek time.

When, as a result of the completion of other requests, the access mechanism becomes available to the current IOB, a seek operation is initiated using the track address in the IOB. Just prior to this, the track address is verified (using the contents of the DEB) to ensure that the seek address lies within the extent of the data set. Assuming that the seek operation was not immediately completed, seek commands to other devices are issued; the channel is then used for other operations if possible. At the completion of the relevant seek operation,[56] the RQE is removed from the top of the seek queue and placed in the appropriate logical channel queue[55] in priority sequence. For the performance of all of these functions,

Figure 4 Control elements: READ and EXCP macroinstructions—elliptical blocks denote elements of first concern during execution of READ or EXCP macroinstruction

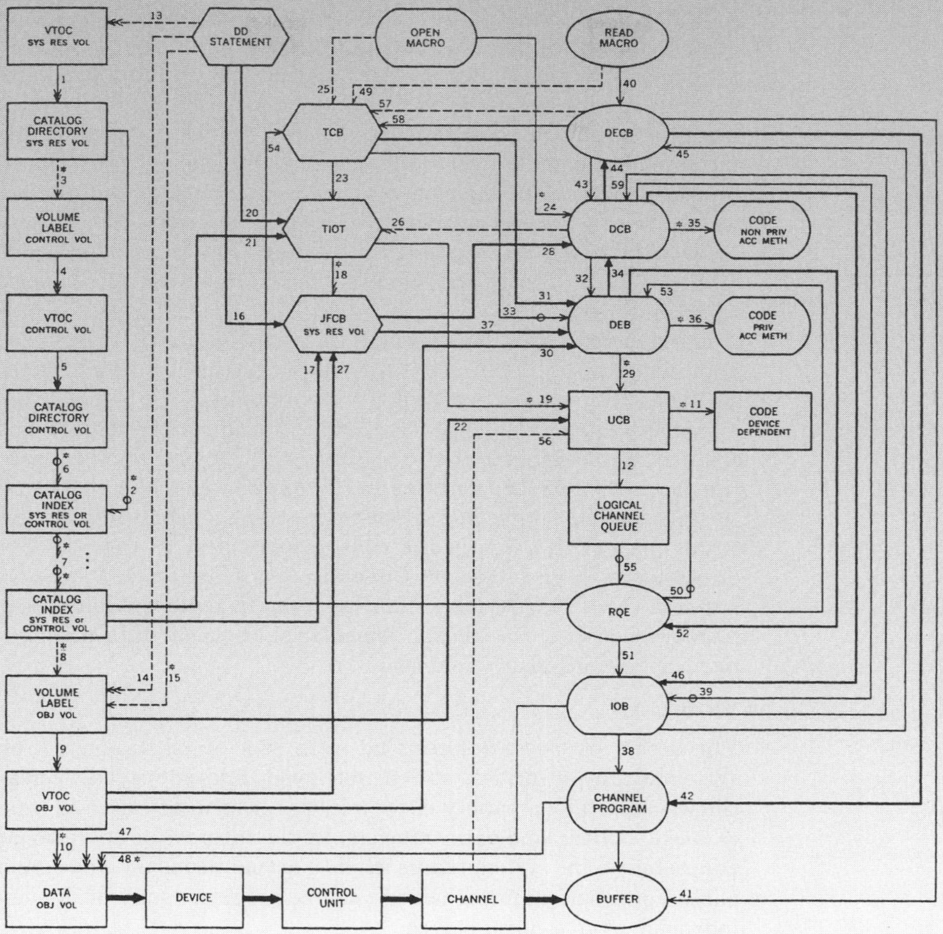

LEGEND

- MAIN STORAGE ADDRESS
- RELATIVE MAIN STORAGE ADDRESS
- DASD TRACK OR BLOCK ADDRESS
- RELATIVE DASD TRACK OR BLOCK ADDRESS
- VOLUME IDENTIFICATION NUMBER
- INTERRUPTION
- DATA SET NAME
- DATA DEFINITION STATEMENT NAME
- OBJECT DATA FLOW
- CONTROL DATA FLOW
- o INDICATES THAT ITEM POINTED TO IS ONE OF A CHAIN OF SIMILAR ITEMS
- * INDICATES THAT THE SOURCE OF A POINTER IS A TABLE WHICH IDENTIFIES SIMILAR ITEMS

device-dependent routines addressed by the UCB(11) are executed by the I/O supervisor.

When the IOB reaches the top of the logical channel queue and a related channel is free, the channel program associated with the IOB is logically prefixed with a short supervisory channel program and the result executed. The control unit is initialized by the supervisory channel program to inhibit the channel program from executing commands that might destroy information outside of the extent of the data set, leave the channel and control unit unused for significant periods, or attempt to write in a data set that is to be used in a read-only manner.[5] When the channel program finishes,(56) its completion is posted in the event control block within the DECB.(45)

At any time after issuing a READ macroinstruction, the program may issue a WAIT or CHECK macroinstruction which refers to the same DECB as the READ macroinstruction. Either of these macroinstructions suspends the task(57,58) until the READ operation has been completed, i.e., until the I/O supervisor posts the completion of the operation in the DECB.

Although the foregoing discussion applies specifically to the READ macroinstruction in the BSAM context and to the use of a direct-access device, the first three displays (Figures 1, 2, and 3) are applicable to other operations as well. In fact, the discussion introduces most of the control elements that bear on data-management operations in any context.

Summary

The design of OS/360 assures that data sets of all kinds can be systematically identified, stored, retrieved, and surveyed. Versatility is served by a variety of techniques for structuring data sets, catalogs, buffers, and data transfers. In the interest of operational adaptability, the system tailors itself to actual needs on a dynamic basis. For programming efficiency, source programs may be device-independent to a novel degree.

CITED REFERENCE AND FOOTNOTES

1. A. S. Noble, Jr., "Design of an integrated programming and operating system, Part I, system considerations and the monitor," *IBM Systems Journal* **2**, 153–161 (June 1963).
2. Although the CHECK macroinstruction includes the effect of the WAIT macroinstruction, the latter may also be used prior to CHECK.
3. Ordinarily, the results of a catalog search include the device type, the identification number of the desired volume, and label verification information. If the data set is a generation of a generation group (a case not considered in the main discussion), the results are the location of an index of generations and an archetype data-set label.
4. Generally, "logical channel" and physical channel are indistinguishable. The logical channel is taken to be the set of physical channels by which a device is accessible. All devices (independent of their type) that share exactly the same set of physical channels are associated with the same logical channel queue. For example, a set of tape drives attached to physical channels 1 and 2 would share a logical channel distinct from that of a printer attached only to physical channel 1.

5. In general, the control unit is initialized to inhibit seek operations that move the access mechanism. More stringent restrictions are placed on channel programs that actually refer to cylinders shared by two or more data sets. This is not to say that inter-cylinder seek operations are disallowed; rather, the I/O supervisor verifies that these operations refer to areas within the extent of the data set. During inter-cylinder seek operations, the channel and control unit are freed for other uses.

Mice and Windows

Alan Kay
Vice President of Research and Development
The Walt Disney Company, Glendale
Alan.Kay@squekland.org

Ph.D., University of Utah

Xerox PARC: Co-founder

ACM Software Systems Award,
J.D. Warnier Prix d'Informatique

Fellow of American Academy of Arts and Sciences, National Academy of Engineering and Royal Society of Arts

ARPANet: Designer

Major contributions: personal computing, graphical user interface

Current interests: Smalltalk, overlapping window interface, Desktop Publishing

Alan Kay

Graphical User Interfaces

Alan Kay presented the history of man-machine interfaces, in particular graphical user interfaces. His lecture gained its vividness mainly from demonstrations and several historical videos. This is hardly reproducible in a paper. Therefore, we refer the reader to the DVD recording of Alan Kay's talk (DVD No. 2).

B-Trees and Relational Database Systems

Rudolf Bayer
Professor of Informatics,
Technische Universität München
www3.in.tum.de
bayer@in.tum.de

Ph.D. in Mathematics,
University of Illinois

Research Scientist,
Boeing Scientific Research Laboratories

Associate Professor,
Purdue University

ACM SIGMOD Innovations Award 2001

Major contribution:
B-tree data structure

Current interests:
multimedia databases, parallel and distributed database systems, information retrieval, expert systems

Rudolf Bayer

B-Trees and Databases, Past and Future

B-Trees are a data structure for organizing and managing very large data volumes on peripheral computer stores. B-Trees were developed in 1969 by R. Bayer and E. McCreight; they have superb properties with respect to performance and scalability in multiuser environments.

The relational database model was developed by E.F. Codd also in 1969. Its revolutionary design is based on the mathematical theories of relations, relational algebra, and predicate logic.

During the 1970s it turned out (project System R at IBM) that B-Trees and relational database systems form a perfect symbiosis; therefore, all relational database systems are based on B-Trees today. Since the first commercial introduction of relational database systems in the 1980s they have grown into a multibillion dollar commercial success. The talk will also discuss UB-Trees, a new multidimensional variant of B-Trees for the next generation of databases.

Storage Technology

This paper describes the development of B-Trees and its relationship to databases, in particular to relational databases. Both technologies were developed in the fall of 1969. At that time computing technology was in its infancy as the following comparison of technological characteristics shows:

	1969	2001	Factor
Main memory	200 KB	200 MB	10^3
Cache	20 KB	20 MB	10^3
Cache pages	20	5000	$< 10^3$
Disk size	7.5 MB	20 GB	$3*10^3$
Disk/memory size	40	100	- 2.5
Transfer rate	150 KB/s	15 MB/s	10^2
Random access	50 ms	5 ms	10
Scanning full disk	130 s	1300 s	- 10

In 1969 backup store was mainly magnetic tape; magnetic disk was an immature, emerging technology. Observe that disk capacity grows slightly faster than main memory capacity. However, accessibility, i.e. the time it takes to completely scan the contents of a full disk, has been decreasing steadily. Therefore, good indexing techniques are more and more important, whereas main storage databases are quite elusive.

In 1969 the challenge of applications came from a few fields like the space industry (engineering computations for the Supersonic Transport SST, C5A, Boeing 747) and manufacturing (parts explosion, parts management). All business applications (bank cheque management, credit card management) were back office batch processing, so there was not much push from the commercial side, in particular there were no ATMs, no online booking of flights, no EC card payments. Since all data were managed on magnetic tapes, algorithms (like for parts explosion) had to be specifically designed under the constraint of reading the data sequentially.

B-Tree Basics

But disk technology and other storage media like magnetic bubbles were developing fast. Random access stores emerged and brought with them a vision of interactive, online databases. This raised the fundamental question of how to manage data on peripheral stores. Let me quote from the summary of the original B-Tree paper:

Organization and maintenance of an index for a dynamic random access file are considered. It is assumed that the index must be kept on some pseudo random access backup store like a disc or a drum… Experiments have been performed with indexes of up to 100,000 keys.

To solve this problem, we, i.e. myself and Ed McCreight, developed B-Trees. Figures 1 and 2 show the insertion process, in particular those pages of a B-Tree that are affected by a leaf split and the corresponding path split all the way back to the root if key 9 is inserted into the B-Tree of Fig. 1.

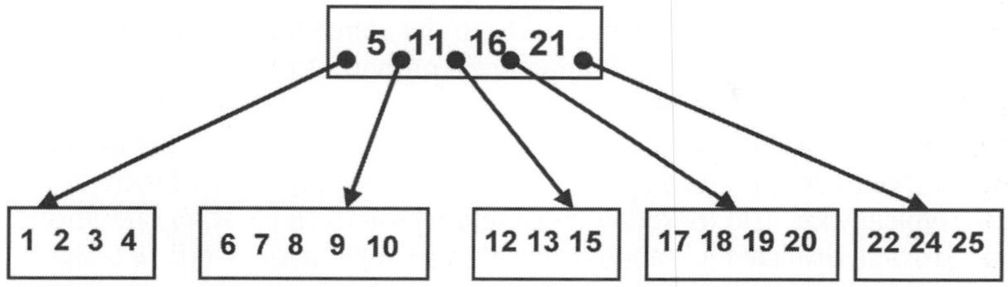

Fig. 1. A two-level B-Tree

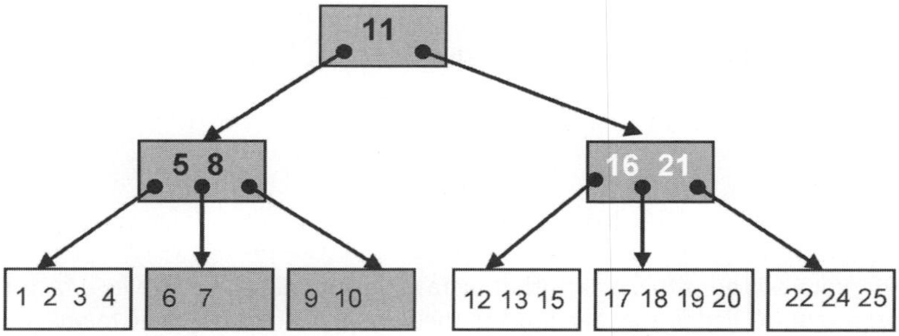

Fig. 2. Leaf split and B-Tree transformation upon insertion of key 9 into the B-Tree of Fig. 1, transformed pages are shaded

I remember vividly that this splitting process with its recursive propagation towards the root including a split of the root itself was the key algorithmic and structural insight for developing B-Trees. This guaranteed dynamic height balancing without the periodic reorganization which was required at that time for other data organizations. Today the idea of B-Trees may seem simple, but in 1969 there were several mental hurdles to take:

- Until then, all trees used in Informatics grew like natural trees, i.e. growing new leaves at the end of a branch instead of splitting leaves into two at the same level.
- Splitting branches all the way back to the root; this usually kills a tree instead of supporting growth.
- The danger of wasting up to 50% of the available extremely scarce storage seemed like a sacrilege in 1969. This could be corrected quickly and easily, but the spectre was there at the moment of invention, and there was the temptation of dumping the idea without further reflection. Fortunately, I resisted.

This idea of growing the tree by splitting, as simple as it might look today, seemed to be difficult to digest. Even in 1984, long after the paper of Doug Comer in the *Computing Surveys* [5], another description of B-Trees in *Scientific American* got the idea completely wrong and grew new leaves out of branches instead of splitting leaves.

The tree transformations associated with the elementary operations *insert* and *delete* yield some of the fundamental properties of B-Trees:

- **Time and I/O complexity** both are of $O(\log_k n)$ where n is the number of tree nodes or data pages and k is the branching factor of the tree, (typically $k > 400$ for today's technology) for all elementary operations.
- **Storage utilization** is about 83% with a simple page overflow technique between brother pages.

The following table shows the growth behaviour of a B-Tree, i.e. its height, the number of nodes and the corresponding file size for pages of 8 KB.

Height	Nodes	Size in bytes
1	1	8 KB
2	400	3.2 MB
3	$16*10^4$	1.3 GB
4	$64*10^6$	512 GB

In practical terms this means that increasing your database size by a factor of about 400 adds only one level to the B-Tree. At the same time main memory probably increases enough to cache one additional layer of the intermediate nodes, and therefore there is no slowdown in accessing this vastly increased database. *Remember: almost all B-Trees require less than 4 logical I/O per elementary operation.*

The top part of a B-Tree index excluding the two lowest levels requires less than 0.01% of the total file size, and this can easily be cached. *Therefore, all elementary operations require at most 2 physical I/O today!!*

Databases

In 1969 there were three competing database models:

IMS: is still today the commercially very successful product of IBM. Its hierarchical structure favours depth first left to right processing by linearization of a hierarchy in order to map it to sequential magnetic tape. The data manipulation language is DL/I embedded into host languages like COBOL, FORTRAN or PL/I.

CODASYL: allows an arbitrary network to get rid of the asymmetry in the hierarchical model. The DB-language is based on very low level procedural pointer chasing, which leads to terrible spaghetti code, usually embedded into COBOL.

Relational model: This very high level model is based on the theory of mathematical relations. It uses nonprocedural searching and matching of values instead of chasing pointers (addresses). Neither the relational nor the CODASYL models are suitable for processing from tape, and disks were still up in the sky!

The relational DB-model [6] was developed by Ted Codd in 1969 and published in 1970 in the CACM. It is based on the Relational Algebra, consisting of tables and a small set of generic operators. Today the usual set of operators are *projection, product, selection and set operators*. This algebra allows algebraic laws to improve expressions (algebraic query optimization). Surprisingly, however, Codd does not mention this aspect. In his original paper Codd only describes a quite restricted set of relational algebra:

- **Projection** p.
- Lossless **natural join** i.
- **Tie:** matching the last column of a relation with the first column, obviously in order to mirror the so-called *set construct* of the CODASYL model, which was so popular at the time and the main competition for Codd's model. Bachmann basically received the 1974 Turing Award for promoting the CODASYL model.
- **Restriction** by table $R_{L/M}S$, not by a general predicate with tuple variables. This was probably motivated by the puristic algebraic approach and forces one to think in terms of a restricting relation, which seems somewhat unnatural.

It seems that Codd was more motivated by theoretical aspects than the practical relevance of the operations of his algebra:

- He spends much time on cyclic joins.
- He defines composition, which he discards later as dispensable.
- He spends very little effort on restriction and defines it as an operation between two relations allowing only equality for comparison with an existential quantifier, but not e.g. a predicate like *divisible by 3* or *greater than 3* since this would require an infinite relation to specify the expression.

In proposing his relational model Ted Codd showed tremendous courage and foresight: he stood up as a single researcher against two powerful establishments, the commercially very successful IMS group within his own company (IBM) and the politically very powerful CODASYL lobby, which was already standardizing its own network database model.

At the time of writing Codd does not seem to recognize the relevance of logic (predicate calculus) for relational systems as a query language.
He discusses the notion of *derivability* of a relation from a given set of relations together with a set of operations (expressions). Codd only discusses 1NF with the goal to map typical hierarchical constructs and seems extremely conscious that he had to be able to map IMS hierarchies

to relations. 3NF is mentioned under the concept of *strong redundancy*, but not treated in detail. Codd makes the questionable statement that *strong redundancy increases storage space and CPU load*. This lead to a normalization euphoria in the 1980s and 1990s which sold a lot of disks and CPU power for IBM, but which is a bad idea in many cases.

Codd discusses the idea of consistency and consistency constraints . As options he proposes to

1. check for consistency at every insertion, deletion and update;
2. tolerate inconsistency for some reasonable time interval, then either the user or someone responsible for the security and integrity of the data is notified; or
3. conduct consistency checking as a batch operation once a day or less frequently.

Obviously, despite his revolutionary approach Codd did not have the vision of relational OLTP systems supporting thousands of transactions per second for many simultaneous users, or he would have realized that there is a serious problem with his proposals to deal with consistency. The key notion of *transaction* is missing here, and all proposals above are quite inadequate for OLTP. This problem was later solved by the System R team, in particular by Jim Gray, and it won him the Turing Award.

Codd quoted Church: *An Introduction to Mathematical Logic*, but did not see predicate calculus as a query language yet, or maybe intentionally did not mention it. This very important idea was taken up in his later DSL/Alpha paper [7].

Thus, there were two alternatives for query languages:

- An imperative, procedural language using algebraic expressions
- A declarative, nonprocedural language based on applied predicate calculus, DSL/Alpha [7]

But there was no implementation of acceptable efficiency in sight for either of those two approaches!

System R

From 1969 until 1974 there were several hard questions:

- Which database model?
- Which query language?
- Which implementation?

At that time Leonard Liu was heading a failing OS project at IBM Watson Research Lab. So he redirected work into a promising new field: *databases*. A member of his group, Don Chamberlin, was a specialist and a fan of the CODASYL approach at that time. However, Codd converted Chamberlin with DSL/Alpha and relational algebra to his approach.

Liu and Frank King moved to San Jose, Chamberlin and others followed. They pursued both CODASYL and relational work. Then there was this showdown panel at the 1974 SIGMOD conference in Ann Arbor: relational versus CODASYL or Codd versus Bachmann respectively, which developed into an almost religious schism in the database community.

It seems that there was also considerable infighting within IBM, between Mike Senko, Ted Codd and other groups (IMS). Codd finally moved to the Systems Department. Of course, management did not know what to do at all. So they came up with a very wise decision, namely, to defer the hard decisions and to build a low level storage system that could hopefully support all models and languages. This system was originally named *Research Storage System* and later became the *Relational Storage System (RSS)*.

In the meantime there was very controversial but serious work being done on a relational system under King. The reaction of the people in the Systems Department to Codd's paper was: "We do not understand, there's too much math."

Then there was a workshop in the tiny and obscure village Cargese on the mediterranean island of Corsica: the IFIP-TC2 conference on database management (April 1-5, 1974). Codd presented his paper "Seven Steps to Rendezvous with the Casual User" [8]. There I met King, who knew my B-Tree work and understood that it was vital for a successful implementation of relational systems. Frank invited me to San Jose to join his System R group to help them with the design of B-Trees for the RSS system.

At the time I was a rather young full professor of informatics at Technische Universität in Munich, frustrated by the bureaucracy at a German university. I took a leave of absence from the University, and went back to the US for one year, certainly one of the happiest and most productive years of my professional career.

Let me return to the hard questions: which model, which language, which implementation?

There were the following eight languages from the indicated authors in the running:

DL/I:	IMS group in IBM
CODASYL:	COBOL + pointer chasing and currency indicators, Senko, Bachmann
Relational Algebra:	Codd
DSL/Alpha:	Codd
SQUARE:	Chamberlin, et al.
SEQUEL:	Chamberlin, Boyce, Reisner
QBE:	Moshe Zloof
Rendezvous:	Codd

There are three survivors today: DL/I, SQL, and Relational Algebra.

DL/I was and still is the language of IMS to traverse and to manipulate hierarchies. Everybody was convinced that DL/I had to be supported even by a relational system. SQL is well known. Relational Algebra is the international level for algebraic optimization and the target language for SQL compilers.

For me the language issue was not critical, because all of them would need B-Trees to reach acceptable efficiency. There was never any competition for B-Trees, although there were a lot of open research issues about them, like concurrency. Note: none of these languages was computationally complete: there were no conditional statements, no iteration or recursion; this would be left to the field of *deductive databases* in the 1980s.

The question of the language level needed by the *non-programmer* was hotly disputed and investigated. Here is a quote from a System R meeting [9].

> Glenn Bacon, who had the System Department, used to wonder how Ted could justify that everybody would be able to write this language that was based on mathematical predicate calculus, with universal quantifiers and existential quantifiers and variables and really, really hairy stuff.

Well, Chamberlin and Boyce got rid of the "hairy stuff" of predicate calculus and transformed it into SEQUEL or SQL. But this is a different story, and as far as I know Codd had no hand in this.

A reaction of the IMS people to Codd's revolutionary and visionary proposals was: "It will take 10 years before we see anything." They were exactly right.

In contrast to several other scientifically ambitious approaches in Informatics, relational database systems became a spectacular success. Let me look at what I judge to be the essential factors of this success story.

- A simple, highly formalized model as proposed by Codd, which allows mathematical analysis and formal treatment of central questions
- Declarative SQL instead of an algorithmic programming interface. Yet SQL never became the simple end-user interface it was intended to be
- Relational algebra with laws for optimization
- B-Trees, which guarantee excellent worst-case performance of crucial algorithms [1], [4]
- Multiuser support: transactions and concurrency on B-Trees [3]
- Robustness: transactions plus recovery and self-organization of B-Trees
- Scalability: B-Trees with logarithmic growth and high potential for parallelism

Certainly most admirable was the courage of Ted Codd to propose a very simple and clean model for databases, which was based on sound mathematical foundations, even without being sure that it could be implemented reasonably well.

Most of these success factors are understood and well known by now. So let me just elaborate on performance, scalability and concurrency, in that order.

Performance: Performance can be substantially improved by so-called Prefix B-Trees. Again the basic idea is not difficult: when splitting a leaf page e.g. between the keys (Smith, Bernie) and (Smith, Henry), it suffices to store a shortest separator – here (Smith, C) – in the father page to guide the future search process correctly. Storing shortest separators in intermediate nodes leads to the concept of simple Prefix B-Trees. An additional improvement is to trim common prefixes from all entries in a subtree; this results in genuine Prefix B-Trees [4].

The overall effect is a higher degree of branching of interior nodes, thus, a lower tree and less I/O. This of course leads to higher overall performance with respect to both I/O and CPU load. B-Trees, particularly Prefix B-Trees, reduce I/O and improve overall performance to a level where a DBMS typically becomes CPU bound. A showdown comparison between QBE and System R gave System R the advantage primarily because of overall performance in query processing.

Scalability: The scalability of relational systems is ideal thanks to B-Trees. Increasing a database by a factor of about 400 adds only one level to a B-Tree and as argued before, does not degrade performance at all.

Concurrency on B-Trees: But much more difficult and important than performance and scalability was the question of concurrent usage of B-Trees. We had to cope somehow with the phenomenon that every access to a B-Tree passes through the root, therefore the root is a hotspot and restrictive root locking was not acceptable. On the other hand the root seldom changes, but still it sometimes changes, and algorithms had to take care of this correctly.

The solution we came up with is described in the well-known joint paper with Mario Schkolnick: *Concurrency of Operations on B-Trees* [3]. The interesting point about this paper is not so much the final result, but the story of how it was developed:

- There are many straightforward, correct, but hopelessly inefficient protocols.
- We needed a protocol that works efficiently for thousands of parallel transactions.
- I think we developed more than ten algorithms, showed them to colleagues like Jim Gray, who worked on transactions and on concurrency, Mike Blasgen, who actually wrote the B-Tree code in System R, and several other colleagues. We received a lot of recognition, encouragement, praise and moral support from our colleagues. Every time it

turned out – reflecting the protocol in the wee hours of the morning – that it was wrong.

Finally, Schkolnick and I and everybody else were convinced that we had the solution. It was clear there was absolutely no point in implementing and testing, we would never be able to generate the critical cases in a concurrent environment; we could not not even build up a tree of height 5, the minimal reasonable general test bed. Remember that a B-Tree of height 4 might have about 500 GB or $5 * 10^{11}$ Bytes. An extremely fast loader might load 1 MB/s into a B-Tree, so loading 500 GB would take at least 6 days, building a tree of height 5 about 2400 days.

We wrote down the algorithm and wrote up a paper to submit it for publication. And we decided, we must have a correctness proof. We tried, tried harder and harder, we had to consider more and more cases, and the proof approached the length of the paper itself. To understand the proof approach better, we tried to construct counter examples, and we found one which showed that our protocol was wrong! But by then, trying proofs and counter examples, we had such a good grasp of the problem that the right solution became easy and the correctness proof fell out as an aside.

This paper triggered a whole branch of research, and my latest count in the bibliography of Michael Ley [11] produced 28 papers published in high quality journals and conferences between 1977 and 1996, most of them between 1980 and 1994.

The issue here really is "shared access to a database by many users" or "concurrent multiuser support". Codd talked of "shared data banks" in the title of his paper, but he really meant a simple model and a simple language for non-programmers, primarily in a read only mode for casual users. To support this he used the notion of *consistency constraints* and proposed that *consistency be checked once a day* or even less frequently by a batch program, which should notify the DB-administrator if it finds consistency violations.

It was essentially the Turing Award winning work of Jim Gray to invent a suitable transaction model which allows fully automatic concurrency control at very high rates of more than 1000 transactions/s or roughly 100 million transactions/day.

Several critical success factors – namely performance, concurrency, robustness and scalability – could be met by B-Trees and so it turned out that relational DB-systems and B-Trees are a near perfect match, which created a multibillion dollar commercial market.

Relational systems would probably not have made it in the market without B-Trees, and vice versa, B-Trees would not have gained that much popularity without relational DBMS.

The following list shows some commercial products that were all built on the basis of B-Trees:

- IBM System R, 1976/77 (prototype)
- Oracle 1979
- IBM SQL/DS
- Ingres, later versions
- IBM DB2
- Informix
- Sybase
- Microsoft, SQL-Server
- MS Windows file system

UB-Trees: Multidimensional Variants of B-Trees

Now, was that it? What about the future? First let me make some very general remarks:

- Present trends are likely to continue for some time.
- The ratio of peripheral storage to main memory capacity will remain invariant or even increase. Just think of networked computers in your LAN or even in the Internet. Their total storage can be considered as peripheral to your computer.

The bottom line of this is that good indexing techniques will become not less important, but more important in the future.

So much about plain storage of data, but what about future applications? Let me stress a simple observation, of which only a few seem to be aware:

All databases are multidimensional!

This is obvious for geographic databases and datawarehousing, even for XML databases. Even more, all relational databases are multidimensional.

To explain this claim, just take E/R modelling and two entities, connected by an n:m relationship. This n:m relationship is turned into a 2-dimensional relational table, with the 2 dimensions corresponding to the 2 foreign keys.

In general, every tuple in a relation with n attributes can be considered as a point in an n-dimensional space, for which the n attribute values correspond to the n coordinates.

To index such data one needs multidimensional indexing techniques. Classical B-Trees in combination with space filling curves are a truly superb multidimensional index. I call them **UB-Trees** and we have investigated this approach in the **MISTRAL** project in my research group in Munich. The issue here is not the absolute optimal performance of the index, but the engineering problem of integrating such an index into an existing DBMS.

This is the main goal of the MISTRAL project, and we reported our first commercial success at VLDB 2000 in Cairo [10]. For details, look up my home page *http://www3.in.tum.de* or *http://mistral.in.tum.de.*

Trees in nature have a long life. Our performance results with UB-Trees are so superb that I am convinced that after 30 years of B-Trees the multi-dimensional variant UB-Trees will enter commercial usage quickly and I hope, of course, that they will have a long life and a great future ahead of them, too.

References:

[1] R. Bayer, E. McCreight: Organization and maintenance of large ordered indexes. Acta Informatica 1, 173-189 (1972).

[2] R. Bayer: Symmetric binary B-Trees: data structure and maintenance algorithms. Acta Informatica 1, 290-306 (1972).

[3] R. Bayer, M. Schkolnick: Concurrency of operations on B-Trees. IBM Res. Report R5 1791 (May 1976), and Acta Informatica 9, 1-21 (1977).

[4] R. Bayer, K. Unterauer: Prefix B-Trees. TODS 2, 11-26 (1977), ACM.

[5] D. Comer: The ubiquitous B-Tree: Computing Surveys 11, 121-137 (1979).

[6] E. F. Codd: A relational model of data for large shared data banks. Communications of the ACM, 13, 6 (June 1970).

[7] E. F. Codd: A data base sublanguage founded on the relational calculus. In: Proceedings, 1971 ACM SIGFIDET workshop on Data Description, Access and Control, ACM.

[8] E. F. Codd: Seven steps to rendezvous with the casual user. In: Proceedings, IFIP-TC2 Conference on Database Management (April 1-5, 1974), 179-200, Cargese, Corsica.

[9] System R Reunion: http://ftp.digital.com/pub/DEC/SRC/technical-notes/SRC-1997-018-html/sqlr95.html

[10] F. Ramsak, V. Markl, R. Fenk, M. Zirkel, K. Elhardt, R. Bayer: Integrating the UB-Tree into a database system kernel. In: Proceedings, VLDB 2000 (Sept. 2000), 263-272, Cairo, Egypt.

[11] M. Ley: http://www.informatik.uni-trier.de/~ley/db/index.html

Rudolf Bayer

R. Bayer, E. McCreight
Organization and Maintenance of Large Ordered Indexes

Acta Informatica, Vol. 1, Fasc. 3, 1972
pp. 173-189

Acta Informatica 1, 173–189 (1972)

Organization and Maintenance of Large Ordered Indexes

R. Bayer and E. McCreight

Received September 29, 1971

Summary. Organization and maintenance of an index for a dynamic random access file is considered. It is assumed that the index must be kept on some pseudo random access backup store like a disc or a drum. The index organization described allows retrieval, insertion, and deletion of keys in time proportional to $\log_k I$ where I is the size of the index and k is a device dependent natural number such that the performance of the scheme becomes near optimal. Storage utilization is at least 50% but generally much higher. The pages of the index are organized in a special data-structure, so-called B-trees. The scheme is analyzed, performance bounds are obtained, and a near optimal k is computed. Experiments have been performed with indexes up to 100000 keys. An index of size 15000 (100000) can be maintained with an average of 9 (at least 4) transactions per second on an IBM 360/44 with a 2311 disc.

1. Introduction

In this paper we consider the problem of organizing and maintaining an index for a dynamically changing random access file. By an *index* we mean a collection of index elements which are pairs (x, α) of fixed size physically adjacent data items, namely a key x and some associated information α. The key x identifies a unique element in the index, the associated information is typically a pointer to a record or a collection of records in a random access file. For this paper the associated information is of no further interest.

We assume that the index itself is so voluminous that only rather small parts of it can be kept in main store at one time. Thus the bulk of the index must be kept on some backup store. The class of backup stores considered are *pseudo random access devices* which have a rather long access or wait time—as opposed to a true random access device like core store—and a rather high data rate once the transmission of physically sequential data has been initiated. Typical pseudo random access devices are: fixed and moving head discs, drums, and data cells.

Since the data file itself changes, it must be possible not only to search the index and to retrieve elements, but also to delete and to insert keys—more accurately index elements—economically. The index organization described in this paper always allows retrieval, insertion, and deletion of keys in time proportional to $\log_k I$ or better, where I is the size of the index, and k is a device dependent natural number which describes the page size such that the performance of the maintenance and retrieval scheme becomes near optimal.

In more illustrative terms theoretical analysis and actual experiments show that it is possible to maintain an index of size 15000 with an average of 9 retrievals, insertions, and deletions per second in real time on an IBM 360/44 with a 2311 disc as backup store. According to our theoretical analysis, it should be possible to maintain an index of size 1500000 with at least two transactions per second on such a configuration in real time.

The index is organized in pages of fixed size capable of holding up to $2k$ keys, but pages need only be partially filled. Pages are the blocks of information transferred between main store and backup store.

The pages themselves are the nodes of a rather specialized tree, a so-called B-tree, described in the next section. In this paper these trees grow and contract in only one way, namely nodes split off a brother, or two brothers are merged or "catenated" into a single node. The splitting and catenation processes are initiated at the leaves only and propagate toward the root. If the root node splits, a new root must be introduced, and this is the only way in which the height of the tree can increase. The opposite process occurs if the tree contracts.

There are, of course, many competitive schemes, e.g., hash-coding, for organizing an index. For a large class of applications the scheme presented in this paper offers significant advantages over others:

i) Storage utilization is at least 50% at any time and should be considerably better in the average.

ii) Storage is requested and released as the file grows and contracts. There is no congestion problem or degradation of performance if the storage occupancy is very high.

iii) The natural order of the keys is maintained and allows processing based on that order like: find predecessors and successors; search the file sequentially to answer queries; skip, delete, retrieve a number of records starting from a given key.

iv) If retrievals, insertions, and deletions come in batches, very efficient essentially sequential processing of the index is possible by presorting the transactions on their keys and by using a simple prepaging algorithm.

Several other schemes try to solve the same or very similar problems. AVL-trees described in [1] and [2] guarantee performance in time $\log_2 I$, but they are suitable only for a one-level store. The schemes described in [3] and [4] do not have logarithmic performance. The solution presented in this paper is new and is related to those described in [1–4] only in the sense that the problem to be solved is similar and that it uses a data organization involving tree structures.

2. *B*-Trees

Def. 2.1. Let $h \geqq 0$ be an integer, k a natural number. A directed tree T is in the class $\tau(k, h)$ of *B-trees* if T is either empty ($h = 0$) or has the following properties:

i) Each path from the root to any leaf has the same length h, also called the *height* of T, i.e., $h =$ number of nodes in path.

ii) Each node except the root and the leaves has at least $k+1$ sons. The root is a leaf or has at least two sons.

iii) Each node has at most $2k+1$ sons.

Number of Nodes in B-Trees. Let $N_{\min}$ and $N_{\max}$ be the minimal and maximal number of nodes in a B-tree $T \in \tau(k, h)$. Then

$$N_{\min} = 1 + 2\left((k+1)^0 + (k+1)^1 + \cdots + (k+1)^{h-2}\right) = 1 + \frac{2}{k}\left((k+1)^{h-1} - 1\right)$$

for $h \geqq 2$. This also holds for $h=1$. Similarly one obtains

$$N_{\max} = \sum_{i=0}^{h-1} (2k+1)^i = \frac{1}{2k}\left((2k+1)^h - 1\right); \quad h \geqq 1.$$

Upper and lower bounds for the number $N(T)$ of nodes of $T \in \tau(k, h)$ are given by:

$$N(T) = 0 \quad \text{if } T \in \tau(k, 0); \tag{2.1}$$

$$1 + \frac{2}{k}\left((k+1)^{h-1} - 1\right) \leqq N(T) \leqq \frac{1}{2k}\left((2k+1)^h - 1\right) \quad \text{otherwise.}$$

Note that the classes $\tau(k, h)$ need not be disjoint.

3. The Data Structure and Retrieval Algorithm

To repeat, the pages on which the index is stored are the nodes of a B-tree $T \in \tau(k, h)$ and can hold up to $2k$ keys. In addition the data structure for the index has the following properties:

i) Each page holds between k and $2k$ keys (index elements) except the root page which may hold between 1 and $2k$ keys.

ii) Let the number of keys on a page P, which is not a leaf, be l. Then P has $l+1$ sons.

iii) Within each page P the keys are sequential in increasing order: $x_1, x_2, \ldots, x_l$; $k \leqq l \leqq 2k$ except for the root page for which $1 \leqq l \leqq 2k$. Furthermore, P contains $l+1$ pointers $p_0, p_1, \ldots, p_l$ to the sons of P. On leaf pages these pointers are undefined. Logically a page is then organized as shown in Fig. 1.

p_0	x_1	α_1	p_1	x_2	α_2	p_2	...	x_l	α_l	p_l	unused space

Fig. 1. Organization of a page

The α_i are the associated information in the index element (x_i, α_i). The triple (x_i, α_i, p_i) or—omitting α_i—the pair (x_i, p_i) is also called an *entry*.

iv) Let $P(p_i)$ be the page to which p_i points, let $K(p_i)$ be the set of keys on the pages of that maximal subtree of which $P(p_i)$ is the root. Then for the B-trees considered here the following conditions shall always hold:

$$(\forall y \in K(p_0))(y < x_1), \tag{3.1}$$

$$(\forall y \in K(p_i))(x_i < y < x_{i+1}); \quad i = 1, 2, \ldots, l-1, \tag{3.2}$$

$$(\forall y \in K(p_l))(x_l < y). \tag{3.3}$$

Fig. 2 is an example of a B-tree in $\tau(2, 3)$ satisfying all the above conditions. In the figure the α_i are not shown and the page pointers are represented graphically. The boxes represent pages and the numbers outside are page numbers to be used later.

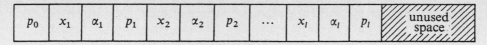

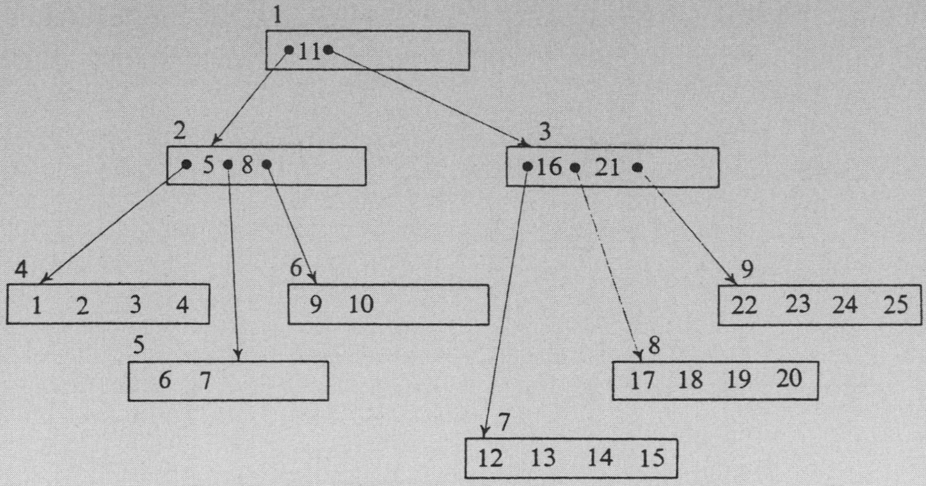

Fig. 2. A data structure in $\tau(2, 3)$ for an index

Retrieval Algorithm. The flowchart in Fig. 3 is an algorithm for retrieving a key y. Let p, r, s be pointer variables which can also assume the value "undefined" denoted as u. r points to the root and is u if the tree is empty, s does not serve any purpose for retrieval, but will be used in the insertion algorithm. Let $P(p)$ be the page to which p is pointing, then $x_1, \ldots, x_l$ are the keys in $P(p)$ and $p_0, \ldots, p_l$ the page pointers in $P(p)$.

The retrieval algorithm is simple logically, but to program it for a computer one would use an efficient technique, e.g., a binary search, to scan a page.

Cost of Retrieval. Let h be the height of the page tree. Then at most h pages must be scanned and therefore fetched from backup store to retrieve a key y. We will now derive bounds for h for a given index of size I. The minimum and maximum number I_{min} and I_{max} of keys in a B-tree of pages in $\tau(k, h)$ are:

$$I_{min} = 1 + k\left(2\,\frac{(k+1)^{h-1}-1}{k}\right) = 2(k+1)^{h-1} - 1$$

$$I_{max} = 2k\left(\frac{(2k+1)^h - 1}{2k}\right) = (2k+1)^h - 1 .$$

This is immediate from (2.1) for $h \geqq 1$. Thus we have as sharp bounds for the height h:

$$\log_{2k+1}(I+1) \leqq h \leqq 1 + \log_{k+1}\left(\frac{I+1}{2}\right) \quad \text{for} \quad I \geqq 1, \qquad h = 0 \quad \text{for} \quad I = 0. \tag{3.1}$$

4. Key Insertion

The algorithm in Fig. 4 inserts a single key y into an index described in Section 3. The variable s is a page pointer set by the retrieval algorithm pointing to the last page that was scanned or having the value u if the page tree is empty.

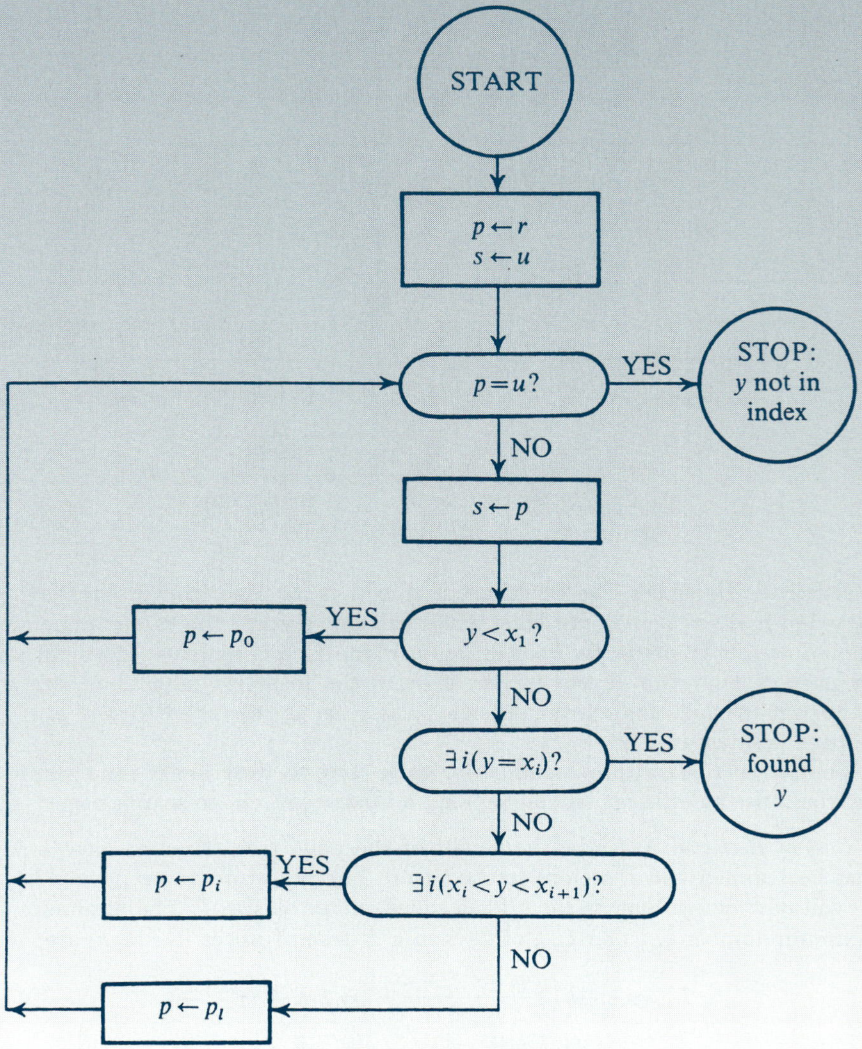

Fig. 3. Retrieval algorithm

Splitting a Page. If a page P in which an entry should be inserted is already full, it will be split into two pages. Logically first insert the entry into the sequence of entries in P—which is assumed to be in main store—resulting in a sequence

$$p_0, (x_1, p_1), (x_2, p_2), \ldots, (x_{2k+1}, p_{2k+1}).$$

Now put the subsequence $p_0, (x_1, p_1), \ldots, (x_k, p_k)$ into P and introduce a new page P' to contain the subsequence

$$p_{k+1}, (x_{k+2}, p_{k+2}), (x_{k+3}, p_{k+3}), \ldots, (x_{2k+1}, p_{2k+1}).$$

Let Q be the father page of P. Insert the entry (x_{k+1}, p'), where p' points to P', into Q. Thus P' becomes a brother of P.

Inserting (x_{k+1}, p') into Q may, of course, cause Q to split too, and so on, possibly up to the root. If the splitting page P is the root, then we introduce a new root page Q containing p, (x_{k+1}, p') where p points to P and p' to P'.

Note that this insertion process maps B-trees with parameter k into B-trees with parameter k, and preserves properties (3.1), (3.2), and (3.3).

To illustrate the insertion process, insertion of key 9 into the tree in Fig. 5 with parameter $k=2$ results in the tree in Fig. 2.

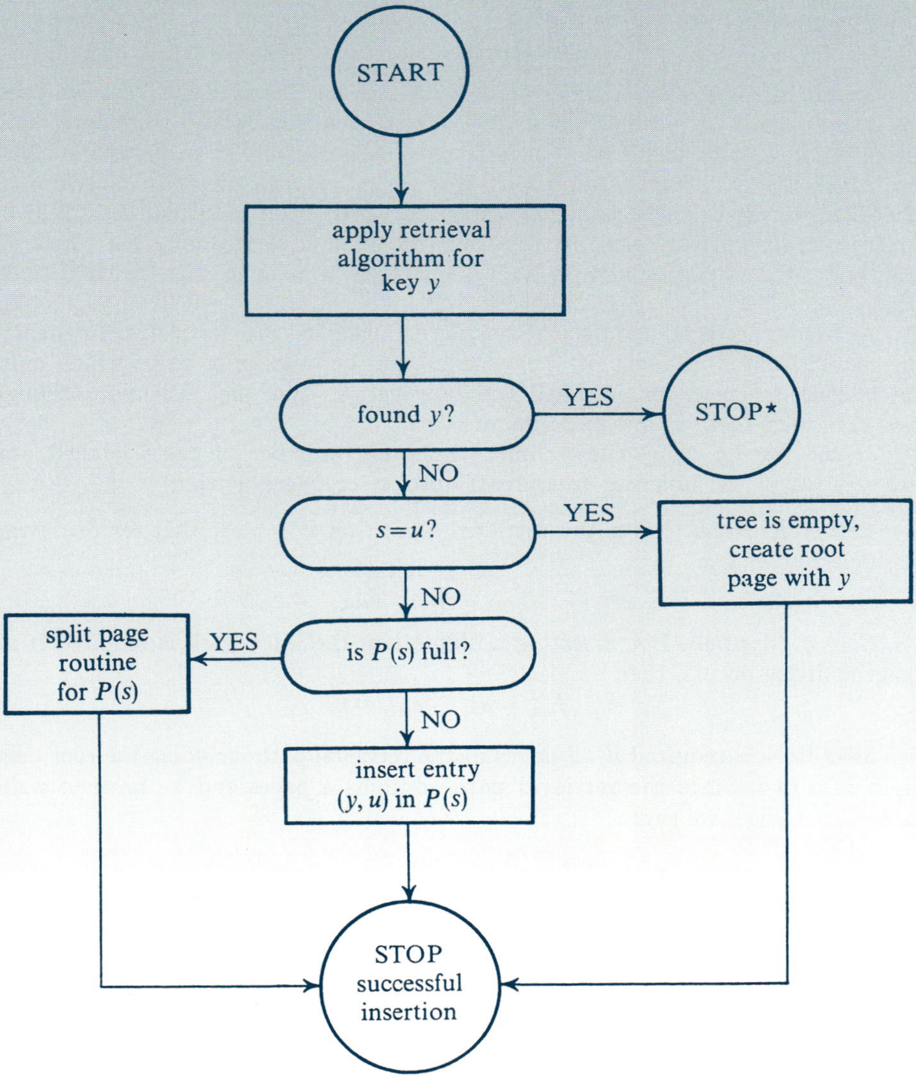

* Key y is already in index, take appropriate action.

Fig. 4. Insertion algorithm

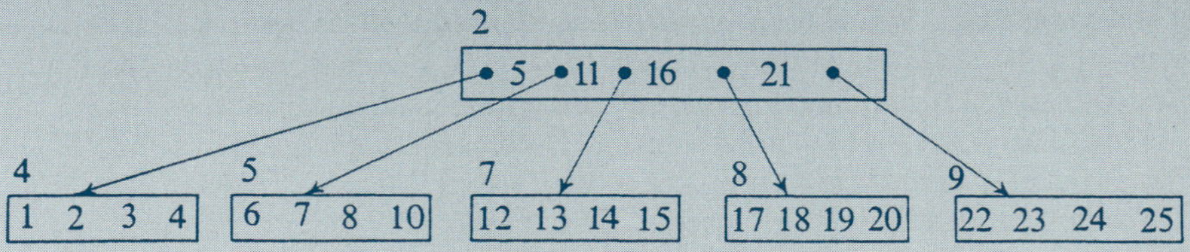

Fig. 5. Index structure in $\tau(2, 2)$

5. Cost of Retrievals and Insertions

To analyze the cost of maintaining an index and retrieving keys we need to know how many pages must be fetched from the backup store into main store and how many pages must be written onto the backup store. For our analysis we make the following assumption: Any page, whose content is examined or modified during a single retrieval, insertion, or deletion of a key, is fetched or paged out respectively exactly once. It will become clear during the course of this paper that a paging area to hold $h+1$ pages in main store is sufficient to do this.

Any more powerful paging scheme, like e.g., keeping the root page permanently locked in main store, will, of course, decrease the number of pages which must be fetched or paged out. We will not, however, analyze such schemes, although we have used them in our experiments.

Denote by $f_{\min}$ ($f_{\max}$) the minimal (maximal) number of pages fetched, and by $w_{\min}$ ($w_{\max}$) the minimal (maximal) number of pages written.

Cost of Retrieval. From the retrieval algorithm it is clear that for retrieving a single key we get

$$f_{\min}=1; \quad f_{\max}=h; \quad w_{\min}=w_{\max}=0.$$

Cost of Insertion. For inserting a single key the least work is required if no page splitting occurs, then

$$f_{\min}=h; \quad w_{\min}=1.$$

Most work is required if all pages in the retrieval path including the root page split into two. Since the retrieval path contains h pages and we have to write a new root page, we get:

$$f_{\max}=h; \quad w_{\max}=2h+1.$$

Note that h always denotes the height of the old tree. Although this worst bound is sharp, it is not a good measure for the amount of work which must generally be done for inserting one key.

If we consider an index in which keys are only retrieved or inserted, but no keys are deleted, then we can derive a bound for the average amount of work to be done for building an index of I keys as follows:

Each page split causes one (or two if the root page splits) new pages to be created. Thus the number of page splits occurring in building an index of I items is bounded by $n(I)-1$, where $n(I)$ is the number of pages in the tree. Since each page has at least k keys, except the root page which may have only 1, we get: $n(I) \leqq \frac{I-1}{k}+1$. Each single page split causes at most 2 additional pages to be written. Thus the average number of pages written per single key insertion due to page splitting is bounded by

$$(n(I)-1)\cdot\frac{2}{I}<\frac{2}{k}.$$

A page split does not require any additional page retrievals. Thus in the average for an index without deletions we get for a single insertion:

$$f_a = h; \quad w_a < 1+\frac{2}{k}.$$

6. Deletion Process

In a dynamically changing index it must be necessary to delete keys. The algorithm of Fig. 6 deletes one key y from an index and maintains our data structure properly. It first locates the key, say y_i. To maintain the data structure properly, y_i is deleted if it is on a leaf, otherwise it must be replaced by the smallest key in the subtree whose root is $P(p_i)$. This smallest key is found by going from $P(p_i)$ along the p_0 pointers to the leaf page, say L, and taking the first key in L. Then this key, say x_1, is deleted from L. As a consequence L may contain fewer than k keys and a catenation or underflow between L and an adjacent brother is performed.

Catenation. Two pages P and P' are called *adjacent brothers* if they have the same father Q and are pointed to by adjacent pointers in Q. P and P' can be catenated, if together they have fewer than $2k$ keys, as follows: The three pages of the form

$$Q: \quad \boxed{\ldots, (y_{j-1}, p), (y_j, p'), (y_{j+1}, p_{j-1}) \ldots,}$$

$$P: \quad \boxed{p_0, (x_1, p_1), \ldots, (x_l, p_l)} \qquad P': \quad \boxed{p'_0, (x_{l+1}, p_{l+1}), \ldots}$$

can be replaced by two pages of the form:

$$Q: \quad \boxed{\ldots, (y_{j-1}, p), (y_{j+1}, p_{j+1}), \ldots}$$

$$\downarrow P$$

$$\boxed{p_0, (x_1, p_1), \ldots, (x_l, p_l), (y_j, p'_0), (x_{l+1}, p_{l+1}), \ldots}$$

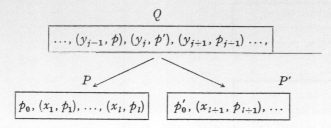

As a consequence of deleting the entry (y_j, p') from Q it is now possible that Q contains fewer than k keys and special action must be taken for Q. This process may propagate up to the root of the tree.

Underflow. If the sum of the number of keys in P and P' is greater than $2k$, then the keys in P and P' can be equally distributed, the process being called an underflow, as follows:

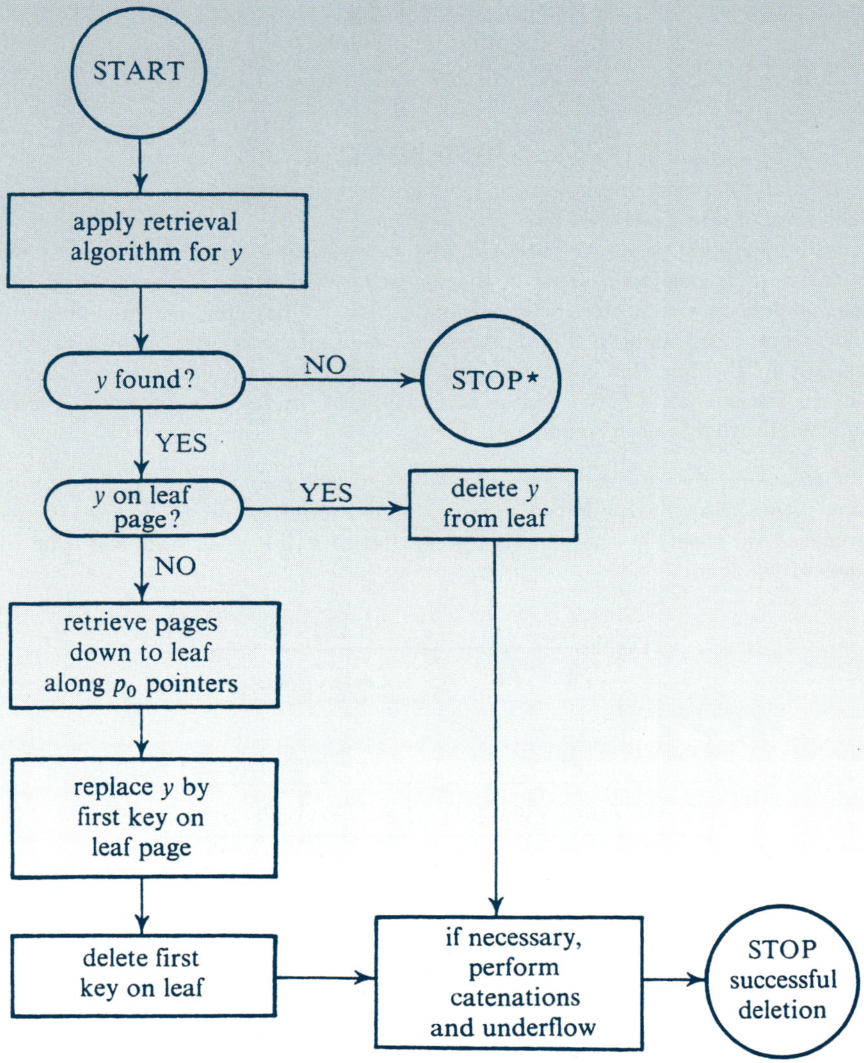

* The key to be deleted is not in index, take appropriate action.

Fig. 6. Deletion algorithm

Perform the catenation between P and P' resulting in too large a P. This is possible since P is in main store. Now split P "in the middle" as described in Section 4 with some obvious minor modifications.

Note that underflows do not propagate. Q is modified, but the number of keys in it is not changed.

To illustrate the deletion process consider the index in Fig. 2. Deleting key 9 results in the index in Fig. 5.

7. Cost of Deletions

For a successful deletion, i.e., if the key y to be deleted is in the index, the least amount of work is required if no catenations or underflows are performed and y is in a leaf. This requires:

$$f_{\min} = h; \qquad w_{\min} = 1.$$

If y is not in a leaf and no catenations or underflows occur, then

$$f = h; \qquad w = 2.$$

A maximal amount of work must be done if all but the first two pages in the retrieval path are catenated, the son of the root in the retrieval path has an underflow, and the root is modified. This requires:

$$f_{\max} = 2h - 1; \qquad w_{\max} = h + 1.$$

As in the case of the insertion process the bounds obtained are sharp, but very far apart and assumed rarely except in pathological examples. To obtain a more useful measure for the average amount of work necessary to delete a key, let us consider a "pure deletion process" during which all keys in an index I are deleted, but no keys are inserted.

Disregarding for the moment catenations and underflows we may get $f_1 = h$ and $w_1 = 2$ for each deletion at worst. But this is the best bound obtainable if one considers an example in which keys are always deleted from the root page.

Each deletion causes at most one underflow, requiring $f_2 = 1$ additional fetches and $w_2 = 2$ additional writes.

The total number of possible catenations is bounded by $n(I) - 1$, which is at most $\frac{I-1}{k}$. Each catenation causes 1 additional fetch and 2 additional writes, which results in an average

$$f_3 = \frac{1}{I}\left(\frac{I-1}{k}\right) < \frac{1}{k}$$

$$w_3 = \frac{2}{I}\left(\frac{I-1}{k}\right) < \frac{2}{k}.$$

Thus in the average we get:

$$f_a \leqq f_1 + f_2 + f_3 < h + 1 + \frac{1}{k}$$

$$w_a \leqq w_1 + w_2 + w_3 < 2 + 2 + \frac{2}{k} = 4 + \frac{2}{k}.$$

8. Page Overflow and Storage Utilization

In the scheme described so far utilization of back-up store may be as low as 50% in extreme cases—disregarding the root page—if all pages contain only k keys. This could be improved by avoiding certain page splits.

An *overflow* between two adjacent brother pages P and P' can be performed as follows: Assume that a key must be inserted in P and P is already full, but P' is not full. Then the key is inserted into the key-sequence in P and an underflow as described in Section 6 between the resulting sequence and P' is performed. This avoids the need to split P into two pages. Thus a page will be split only if both adjacent brothers are full, otherwise an overflow occurs.

In an index without deletions overflows will increase the storage utilization in the worst cases to about 66%. If both insertions and deletions occur, then the storage utilization may of course again be as low as 50%. For most practical applications, however, storage utilization should be improved appreciably with overflows.

One could, of course, consider a larger neighborhood of pages than just the adjacent brothers as candidates for overflows, underflows, and catenations and increase the minimal storage occupancy accordingly.

Bounds for the cost of insertions for a scheme with overflows are easily derived as:

$$f_{\min} = h; \qquad w_{\min} = 1;$$

$$f_{\max} = 3h - 2; \quad w_{\max} = 2h + 1.$$

For a pure insertion process one obtains as bounds for the average cost:

$$f_a < h + 2 + \frac{2}{k}; \qquad w_a < 3 + \frac{2}{k}.$$

It is easy to construct examples in which each insertion causes an overflow, thus these bounds cannot be improved very much without special assumptions about the insertion process.

9. Maintenance Cost for Index with Insertions and Deletions

The main purpose of this paper is to develop a data structure which allows economical maintenance of an index in which retrievals, insertions, and deletions must be done in any order. We will now derive bounds on the processing cost in such an environment.

The derivation of bounds for retrieval cost did not make any assumptions about the order of insertions or deletions, so they are still valid. Also, the minimal and maximal bounds for the cost of insertions and deletions were derived without any such assumptions and are still valid. The bounds derived for the average cost, however, are no longer valid if insertions and deletions are mixed.

The following example shows that the upper bounds for the average cost cannot be improved appreciably over the upper bounds of the cost derived for a single retrieval or deletion.

Example. Consider the trees T_2 in Fig. 2 and T_5 in Fig. 5. Deleting key 9 from T_2 leads to T_5, and inserting key 9 in T_5 leads back to T_2. Consider a sequence of alternating deletions and insertions of key 9 being applied starting with T_2.

Case 1. No page overflows, but only page splits occur:

i) Each deletion of key 9 from T_2 requires:
3 retrievals to locate key 9, namely pages 1, 2, 6.
1 retrieval of brother 5 of page 6 to find out that pages 5 and 6 can be catenated.
2 pages, namely 5 and 2 are modified and must be written. Pages 6 and 3 are deleted from the tree T_2.
Thus $f=5$ and $w=2$. But $f=5=2h-1=f_{max}$ and $w=2=h-1=w_{max}-2$.

ii) Each insertion of key 9 into T_5 requires:
2 retrievals to locate slot for 9 in page 5.
5 pages must be written, namely 1, 2, 3, 5, 6.
Thus

$$f=2=h=f_{max}$$

$$w=5=2h+1=w_{max}.$$

Case 2. Consider a scheme with page overflows.

i) Deletion of key 9 leads to the same results as in Case 1.

ii) Insertion of key 9 requires:
2 retrievals to locate slot for 9 on page 5.
2 retrievals of brothers 4 and 7 of 5 to find out that 5 must be split.
5 pages must be written as in Case 1.
Thus:

$$f=4=3h-2=f_{max}$$

$$w=5=2h+1=w_{max}.$$

Analogous examples can be constructed for arbitrary h and k.

From the analysis it is clear that the performance of our scheme depends on the actual sequence of insertions and deletions. The interference between insertions and deletions may degrade the performance of the scheme as opposed to doing insertions or deletions only. But even in the worst cases this interference degrades the performance at most by a factor of 3.

It is an open question how important this interference is in any actual applications and how relevant our worst case analysis is. Although the derivable cost bounds are worse, the scheme with overflows performed better in our experiments than the scheme without overflows.

10. Choice of k

The performance of our scheme depends on the parameter k. Thus care should be taken in choosing k to make the performance as good as possible.

To obtain a very rough approximation to the performance of the scheme we make the following assumptions:

	Retrieval	Insertion in index without deletions and without overflows	Deletion in index without insertions, with or without overflows	Insertion in index without deletions, but with overflow	Insertion in index with deletions, without overflow	Deletion in index with insertions, with or without overflows	Insertion in index with deletion, with overflow
min	$f=1$ $w=0$	$f=h$ $w=1$	$f=h$ $w=1$	$f=h$ $w=1$	$f=h$ $w=1$	$f=h$ $w=1$	$f=h$ $w=1$
Average as derived in paper	$f\leqq h$ $w=0$	$f=h$ $w<1+\frac{2}{k}$	$f<h+1+\frac{1}{k}$ $w<4+\frac{2}{k}$	$f\leqq h+2+\frac{2}{k}$ $w\leqq 3+\frac{2}{k}$	$f=h$ $w\leqq 2h+1$	$f\leqq 2h-1$ $h-1\leqq u \leqq h+1$	$f\leqq 3h-2$ $w\leqq 2h+1$
max	$f=h$ $w=0$	$f=h$ $w=2h+1$	$f=2h-1$ $w=h+1$	$f=3h-2$ $w=2h+1$	$f=h$ $w=2h+1$	$f=2h-1$ $w=h+1$	$f=3h-2$ $w=2h+1$

f = number of pages fetched h = height of B-tree
w = number of pages written k = parameter of B-tree of pages
I = size of index set u = best upper bound obtainable for w

Fig. 7. Table of costs for a single retrieval, insertion, or deletion of a key

i) The time spent for each page which is written or fetched can be expressed in the form:

$$\alpha+\beta(2k+1)+\gamma\ln(\nu k+1)$$

α fixed time spent per page, e.g., average disc seek time plus fixed CPU overhead, etc.

β transfer time per page entry.

γ constant for the logarithmic part of the time, e.g., for a binary search.

ν factor for average page occupancy, $1\leqq\nu\leqq 2$.

We assume that modifying a page does not require moving keys within a page, but that the necessary channel subcommands are generated to write a page by concatenating several pieces of information in main store. This is the reason for our assumption that fetching and writing a page takes the same time.

i) The average number of pages fetched and written per single transaction in an environment of mixed retrievals, insertions, and deletions is approximately proportional—see Fig. 7—to h, say δh. The total time T spent per transaction can then be approximated by:

$$T\approx\delta h\left(\alpha+\beta(2k+1)+\gamma\ln(\nu k+1)\right).$$

Approximating h itself by: $h\approx\log_{\nu k+1}(I+1)$ where I is the size of the index, we get: $T\approx T_a=\delta\log_{\nu k+1}(I+1)\left(\alpha+\beta(2k+1)+\gamma\ln(\nu k+1)\right)$.

Now one easily obtains the minimum of T_a if k is chosen such that:

$$\frac{\alpha}{\beta}=2\left(\frac{(\nu k+1)}{\nu}\ln(\nu k+1)\right)-(2k+1)=f(k,\nu).$$

Neglecting CPU time, k is a number which is characteristic for the device used as backup store. To obtain a near optimal page size for our test examples we assumed $\alpha = 50$ ms and $\beta = 90$ µs. According to the table in Fig. 8 an acceptable choice should be $64 < k < 128$. For reasons of programming convenience we chose $k = 60$ resulting in a page size of 120 entries.

k	$f(k, 1)$	$f(k, 1.5)$	$f(k, 2)$
2.00000E+00	1.59167E+00	2.39356E+00	3.04718E+00
4.00000E+00	7.09437E+00	9.16182E+00	1.07750E+01
8.00000E+00	2.25500E+01	2.74591E+01	3.11646E+01
1.60000E+01	6.33292E+01	7.42958E+01	8.23847E+01
3.20000E+01	1.65769E+02	1.89265E+02	2.06334E+02
6.40000E+01	4.13670E+02	4.62662E+02	4.97915E+02
1.28000E+02	9.96831E+02	1.09726E+03	1.16911E+03
2.56000E+02	2.33922E+03	2.54299E+03	2.68826E+03
5.12000E+02	5.37752E+03	5.78842E+03	6.08075E+03
1.02400E+03	1.21625E+04	1.29881E+04	1.35748E+04
2.04800E+03	2.71506E+04	2.88062E+04	2.99818E+04
4.09600E+03	5.99647E+04	6.32806E+04	6.56343E+04
8.19200E+03	1.31269E+05	1.37906E+05	1.42617E+05
1.63840E+04	2.85235E+05	2.98514E+05	3.07938E+05
3.27680E+04	6.15877E+05	6.42442E+05	6.61292E+05
6.55360E+04	1.32258E+06	1.37572E+06	1.41342E+06

Fig. 8. The function $f(k, v)$ for optimal choice of k

The size of the index which can be stored for $k = 60$ in a page tree of a certain height can be seen from Fig. 9.

Height of page tree	Minimum index size	Maximum index size
1	1	120
2	121	14640
3	7441	1771560
4	453961	214358880

Fig. 9. Height of page tree and index size

11. Experimental Results

The algorithms presented here were programmed and their performance measured during various experiments. The programs were run on an IBM 360/44 computer with a 2311 disc unit as a backup store. For the index element size chosen (14 8-bit characters) and index size generally used (about 10000 index elements), the average access mechanism delay for this unit is about 50 ms, after which information transfer takes place at the rate of about 90 µs per index element. From these two parameters, our analysis predicts an optimal page size ($2k$) on the order of 120 index elements.

The programming included a simple demand paging scheme to take advantage of available core storage (about 1250 index elements' worth) and thus to attempt to reduce the number of physical disc operations. In the following section by *virtual disc read* we mean a request to the paging scheme that a certain disc page be available in core; a virtual disc read will result in a physical disc read only of there is no copy of the requested disc page already in the paging area of core storage. A *virtual disc write* is defined analogously.

At the time of this writing ten experiments had been performed. These experiments were intended to give us an idea of what kind of performance to expect, what kind of storage utilization to expect, and so forth. For us the specification of an experiment consists of choosing

1) whether or not to permit overflows on insertion,
2) a number of index elements per page, and
3) a sequence of transactions to be made against an initially empty index.

At several points during the performance of an experiment certain performance variables are recorded. From these the performance of the algorithms according to various performance measures can be deduced; to wit

1) % storage utilization
2) average number of virtual disc reads/transaction
3) average number of physical disc reads/transaction
4) average number of virtual disc writes/insertion or deletion
5) average number of physical disc writes/insertion or deletion
6) average number of transactions/second.

We now summarize the experiments. Each experiment was divided into several phases, and at the end of each of these the performance variables were measured. Phases are denoted by numbers within parentheses.

$E1$: 25 elements/page, overflow permitted.
- (1) 10000 insertions sequential by key,
- (2) 50 insertions, 50 retrievals, and 100 deletions uniformly random in the key space.

$E2$: 120 elements/page; otherwise identical to $E1$.

$E3$: 250 elements/page; otherwise identical to $E1$.

$E4$: 120 elements/page, overflow permitted.
- (1) 10000 insertions sequential by key,
- (2) 1000 retrievals uniformly random in key space,
- (3) 10000 sequential deletions.

$E5$: 120 elements/page, overflow *not* permitted.
- (1) 5000 insertions uniformly random in key space,
- (2) 1000 retrievals uniformly random in key space,
- (3) 5000 deletions uniformly random in key space.

$E6$: Overflow permitted; otherwise identical to $E5$.

$E7$: 120 elements/page, overflow permitted.
- (1) 5000 insertions sequential by key,
- (2) 6000 each insertions, retrievals, and deletions uniformly random in key space.

*E*8: 120 elements/page, overflow permitted.

(1) 15000 insertions uniformly random in key space,

(2) 100 each insertions, deletions, and retrievals uniformly random in key space.

*E*9: 250 elements/page; otherwise identical to *E*8.

*E*10: 120 elements/page, overflow permitted.

(1) 100000 insertions sequential by key,

(2) 1000 each insertions, deletions, and retrievals uniformly random in key space,

(3) 100 group retrievals uniformly random in key space, where a group is a sequence of 100 consecutive keys (statistics on the basis of 10000 transactions),

(4) 10000 insertions sequential by key, to merge uniformly with the elements inserted in phase (1).

	% Storage used	VR/T*	PR/T	VW/I or D	PW/I or D	T/sec
*E*1 (1)	99.8	2.2	0	2.3	0.04	66.1
*E*1 (2)	91.5	4.4	1.62	2.7	1.5	6.6
*E*2 (1)	99.2	1.0	0	1.0	0.008	94.5
*E*2 (2)	87.3	2.5	1.15	1.3	1.1	6.7
*E*3 (1)	97.6	1.0	0	1.0	0.004	100.0
*E*3 (2)	84.7	2.4	1.08	1.3	1.1	5.2
*E*4 (1)	99.2	1.0	0	1.0	0.008	94.5
*E*4 (2)	99.2	2.0	—	—	—	19.5
*E*4 (3)	—	2.0	0.01	2.0	0	74.1
*E*5 (1)	67.1	1.0	0.55	1.0	0.56	17.0
*E*5 (2)	67.1	2.0	0.83	—	—	18.2
*E*5 (3)	—	4.0	0.68	2.2	0.65	12.4
*E*6 (1)	86.7	1.1	0.55	1.1	0.54	17.1
*E*6 (2)	86.7	2.0	0.79	—	—	24.3
*E*6 (3)	—	4.0	0.65	2.2	0.62	13.4
*E*7 (1)	96.9	1.0	0	1.0	0.008	111.9
*E*7 (2)	76.8	2.3	0.83	1.3	0.88	13.1
*E*8 (1)	84.5	1.3	0.87	1.3	0.85	10.1
*E*8 (2)	83.9	3.7	1.00	3.0	1.00	9.5
*E*9 (1)	86.4	1.1	0.84	1.0	0.82	8.5
*E*9 (2)	85.2	2.3	0.94	1.1	0.96	8.2
*E*10 (1)	99.8	1.9	0	1.9	0.008	91.7
*E*10 (2)	82.1	4.1	1.94	1.8	1.54	4.2
*E*10 (3)	82.1	4.0	0.03	—	—	75.7
*E*10 (4)	83.8	2.2	0.10	2.2	0.11	38.0

* This statistic is unnecessarily large for deletions, due to the way deletions were programmed. To find the *necessary* number of virtual reads, for sequential deletions subtract one from the number shown, and for random deletions subtract one and multiply the result by about 0.5.

References

1. Adelson-Velskii, G. M., Landis, E. M.: An information organization algorithm. DANSSSR, **146**, 263–266 (1962).
2. Foster, C. C.: Information storage and retrieval using AVL trees. Proc. ACM 20th Nat'l. Conf. 192–205 (1965).
3. Landauer, W. I.: The balanced tree and its utilization in information retrieval. IEEE Trans. on Electronic Computers, Vol. EC-12, No. 6, December 1963.
4. Sussenguth, E. H., Jr.: The use of tree structures for processing files. Comm. ACM, **6**, No. 5, May 1963.

Prof. Dr. R. Bayer
Dept. of Computer Science
Purdue University
Lafayette, Ind. 47907
U.S.A.

Dr. E M. McCreight
Palo Alto Research Center
3180 Porter Drive
Palo Alto, Calif. 94304
U.S.A.

Rudolf Bayer

E.F. Codd
A Relational Model of Data for Large Shared Data Banks

Communications of the ACM, Vol. 13 (6), 1970
pp. 377-387

A Relational Model of Data for Large Shared Data Banks

E. F. Codd
IBM Research Laboratory, San Jose, California

Future users of large data banks must be protected from having to know how the data is organized in the machine (the internal representation). A prompting service which supplies such information is not a satisfactory solution. Activities of users at terminals and most application programs should remain unaffected when the internal representation of data is changed and even when some aspects of the external representation are changed. Changes in data representation will often be needed as a result of changes in query, update, and report traffic and natural growth in the types of stored information.

Existing noninferential, formatted data systems provide users with tree-structured files or slightly more general network models of the data. In Section 1, inadequacies of these models are discussed. A model based on *n*-ary relations, a normal form for data base relations, and the concept of a universal data sublanguage are introduced. In Section 2, certain operations on relations (other than logical inference) are discussed and applied to the problems of redundancy and consistency in the user's model.

KEY WORDS AND PHRASES: data bank, data base, data structure, data organization, hierarchies of data, networks of data, relations, derivability, redundancy, consistency, composition, join, retrieval language, predicate calculus, security, data integrity

CR CATEGORIES: 3.70, 3.73, 3.75, 4.20, 4.22, 4.29

1. Relational Model and Normal Form

1.1. Introduction

This paper is concerned with the application of elementary relation theory to systems which provide shared access to large banks of formatted data. Except for a paper by Childs [1], the principal application of relations to data systems has been to deductive question-answering systems. Levein and Maron [2] provide numerous references to work in this area.

In contrast, the problems treated here are those of *data independence*—the independence of application programs and terminal activities from growth in data types and changes in data representation—and certain kinds of *data inconsistency* which are expected to become troublesome even in nondeductive systems.

The relational view (or model) of data described in Section 1 appears to be superior in several respects to the graph or network model [3, 4] presently in vogue for non-inferential systems. It provides a means of describing data with its natural structure only—that is, without superimposing any additional structure for machine representation purposes. Accordingly, it provides a basis for a high level data language which will yield maximal independence between programs on the one hand and machine representation and organization of data on the other.

A further advantage of the relational view is that it forms a sound basis for treating derivability, redundancy, and consistency of relations—these are discussed in Section 2. The network model, on the other hand, has spawned a number of confusions, not the least of which is mistaking the derivation of connections for the derivation of relations (see remarks in Section 2 on the "connection trap").

Finally, the relational view permits a clearer evaluation of the scope and logical limitations of present formatted data systems, and also the relative merits (from a logical standpoint) of competing representations of data within a single system. Examples of this clearer perspective are cited in various parts of this paper. Implementations of systems to support the relational model are not discussed.

1.2. Data Dependencies in Present Systems

The provision of data description tables in recently developed information systems represents a major advance toward the goal of data independence [5, 6, 7]. Such tables facilitate changing certain characteristics of the data representation stored in a data bank. However, the variety of data representation characteristics which can be changed *without logically impairing some application programs* is still quite limited. Further, the model of data with which users interact is still cluttered with representational properties, particularly in regard to the representation of collections of data (as opposed to individual items). Three of the principal kinds of data dependencies which still need to be removed are: ordering dependence, indexing dependence, and access path dependence. In some systems these dependencies are not clearly separable from one another.

1.2.1. *Ordering Dependence.* Elements of data in a data bank may be stored in a variety of ways, some involving no concern for ordering, some permitting each element to participate in one ordering only, others permitting each element to participate in several orderings. Let us consider those existing systems which either require or permit data elements to be stored in at least one total ordering which is closely associated with the hardware-determined ordering of addresses. For example, the records of a file concerning parts might be stored in ascending order by part serial number. Such systems normally permit application programs to assume that the order of presentation of records from such a file is identical to (or is a subordering of) the stored ordering. Those application programs which take advantage of the stored ordering of a file are likely to fail to operate correctly if for some reason it becomes necessary to replace that ordering by a different one. Similar remarks hold for a stored ordering implemented by means of pointers.

It is unnecessary to single out any system as an example, because all the well-known information systems that are marketed today fail to make a clear distinction between order of presentation on the one hand and stored ordering on the other. Significant implementation problems must be solved to provide this kind of independence.

1.2.2. *Indexing Dependence.* In the context of formatted data, an index is usually thought of as a purely performance-oriented component of the data representation. It tends to improve response to queries and updates and, at the same time, slow down response to insertions and deletions. From an informational standpoint, an index is a redundant component of the data representation. If a system uses indices at all and if it is to perform well in an environment with changing patterns of activity on the data bank, an ability to create and destroy indices from time to time will probably be necessary. The question then arises: Can application programs and terminal activities remain invariant as indices come and go?

Present formatted data systems take widely different approaches to indexing. TDMS [7] unconditionally provides indexing on all attributes. The presently released version of IMS [5] provides the user with a choice for each file: a choice between no indexing at all (the hierarchic sequential organization) or indexing on the primary key only (the hierarchic indexed sequential organization). In neither case is the user's application logic dependent on the existence of the unconditionally provided indices. IDS [8], however, permits the file designers to select attributes to be indexed and to incorporate indices into the file structure by means of additional chains. Application programs taking advantage of the performance benefit of these indexing chains must refer to those chains by name. Such programs do not operate correctly if these chains are later removed.

1.2.3. *Access Path Dependence.* Many of the existing formatted data systems provide users with tree-structured files or slightly more general network models of the data. Application programs developed to work with these systems tend to be logically impaired if the trees or networks are changed in structure. A simple example follows.

Suppose the data bank contains information about parts and projects. For each part, the part number, part name, part description, quantity-on-hand, and quantity-on-order are recorded. For each project, the project number, project name, project description are recorded. Whenever a project makes use of a certain part, the quantity of that part committed to the given project is also recorded. Suppose that the system requires the user or file designer to declare or define the data in terms of tree structures. Then, any one of the hierarchical structures may be adopted for the information mentioned above (see Structures 1–5).

Structure 1. Projects Subordinate to Parts

File	*Segment*	*Fields*
F	PART	part #
		part name
		part description
		quantity-on-hand
		quantity-on-order
	PROJECT	project #
		project name
		project description
		quantity committed

Structure 2. Parts Subordinate to Projects

File	*Segment*	*Fields*
F	PROJECT	project #
		project name
		project description
	PART	part #
		part name
		part description
		quantity-on-hand
		quantity-on-order
		quantity committed

Structure 3. Parts and Projects as Peers
Commitment Relationship Subordinate to Projects

File	*Segment*	*Fields*
F	PART	part #
		part name
		part description
		quantity-on-hand
		quantity-on-order
G	PROJECT	project #
		project name
		project description
	PART	part #
		quantity committed

Structure 4. Parts and Projects as Peers
Commitment Relationship Subordinate to Parts

File	*Segment*	*Fields*
F	PART	part #
		part description
		quantity-on-hand
		quantity-on-order
	PROJECT	project #
		quantity committed
G	PROJECT	project #
		project name
		project description

Structure 5. Parts, Projects, and
Commitment Relationship as Peers

File	*Segment*	*Fields*
F	PART	part #
		part name
		part description
		quantity-on-hand
		quantity-on-order
G	PROJECT	project #
		project name
		project description
H	COMMIT	part #
		project #
		quantity committed

Now, consider the problem of printing out the part number, part name, and quantity committed for every part used in the project whose project name is "alpha." The following observations may be made regardless of which available tree-oriented information system is selected to tackle this problem. If a program P is developed for this problem assuming one of the five structures above—that is, P makes no test to determine which structure is in effect—then P will fail on at least three of the remaining structures. More specifically, if P succeeds with structure 5, it will fail with all the others; if P succeeds with structure 3 or 4, it will fail with at least 1, 2, and 5; if P succeeds with 1 or 2, it will fail with at least 3, 4, and 5. The reason is simple in each case. In the absence of a test to determine which structure is in effect, P fails because an attempt is made to exceute a reference to a nonexistent file (available systems treat this as an error) or no attempt is made to execute a reference to a file containing needed information. The reader who is not convinced should develop sample programs for this simple problem.

Since, in general, it is not practical to develop application programs which test for all tree structurings permitted by the system, these programs fail when a change in structure becomes necessary.

Systems which provide users with a network model of the data run into similar difficulties. In both the tree and network cases, the user (or his program) is required to exploit a collection of user access paths to the data. It does not matter whether these paths are in close correspondence with pointer-defined paths in the stored representation—in IDS the correspondence is extremely simple, in TDMS it is just the opposite. The consequence, regardless of the stored representation, is that terminal activities and programs become dependent on the continued existence of the user access paths.

One solution to this is to adopt the policy that once a user access path is defined it will not be made obsolete until all application programs using that path have become obsolete. Such a policy is not practical, because the number of access paths in the total model for the community of users of a data bank would eventually become excessively large.

1.3. A Relational View of Data

The term *relation* is used here in its accepted mathematical sense. Given sets $S_1, S_2, \cdots, S_n$ (not necessarily distinct), R is a relation on these n sets if it is a set of n-tuples each of which has its first element from S_1, its second element from S_2, and so on.[1] We shall refer to S_j as the jth *domain* of R. As defined above, R is said to have *degree* n. Relations of degree 1 are often called *unary*, degree 2 *binary*, degree 3 *ternary*, and degree n *n-ary*.

[1] More concisely, R is a subset of the Cartesian product $S_1 \times S_2 \times \cdots \times S_n$.

For expository reasons, we shall frequently make use of an array representation of relations, but it must be remembered that this particular representation is not an essential part of the relational view being expounded. An array which represents an n-ary relation R has the following properties:

(1) Each row represents an n-tuple of R.
(2) The ordering of rows is immaterial.
(3) All rows are distinct.
(4) The ordering of columns is significant—it corresponds to the ordering $S_1, S_2, \cdots, S_n$ of the domains on which R is defined (see, however, remarks below on domain-ordered and domain-unordered relations).
(5) The significance of each column is partially conveyed by labeling it with the name of the corresponding domain.

The example in Figure 1 illustrates a relation of degree 4, called *supply,* which reflects the shipments-in-progress of parts from specified suppliers to specified projects in specified quantities.

supply	(*supplier*	*part*	*project*	*quantity*)
	1	2	5	17
	1	3	5	23
	2	3	7	9
	2	7	5	4
	4	1	1	12

FIG. 1. A relation of degree 4

One might ask: If the columns are labeled by the name of corresponding domains, why should the ordering of columns matter? As the example in Figure 2 shows, two columns may have identical headings (indicating identical domains) but possess distinct meanings with respect to the relation. The relation depicted is called *component.* It is a ternary relation, whose first two domains are called *part* and third domain is called *quantity.* The meaning of *component* (x, y, z) is that part x is an immediate component (or subassembly) of part y, and z units of part x are needed to assemble one unit of part y. It is a relation which plays a critical role in the parts explosion problem.

component	*(part*	*part*	*quantity)*
	1	5	9
	2	5	7
	3	5	2
	2	6	12
	3	6	3
	4	7	1
	6	7	1

FIG. 2. A relation with two identical domains

It is a remarkable fact that several existing information systems (chiefly those based on tree-structured files) fail to provide data representations for relations which have two or more identical domains. The present version of IMS/360 [5] is an example of such a system.

The totality of data in a data bank may be viewed as a collection of time-varying relations. These relations are of assorted degrees. As time progresses, each n-ary relation may be subject to insertion of additional n-tuples, deletion of existing ones, and alteration of components of any of its existing n-tuples.

In many commercial, governmental, and scientific data banks, however, some of the relations are of quite high degree (a degree of 30 is not at all uncommon). Users should not normally be burdened with remembering the domain ordering of any relation (for example, the ordering *supplier*, then *part*, then *project*, then *quantity* in the relation *supply*). Accordingly, we propose that users deal, not with relations which are domain-ordered, but with *relationships* which are their domain-unordered counterparts.[2] To accomplish this, domains must be uniquely identifiable at least within any given relation, without using position. Thus, where there are two or more identical domains, we require in each case that the domain name be qualified by a distinctive *role name*, which serves to identify the role played by that domain in the given relation. For example, in the relation *component* of Figure 2, the first domain *part* might be

[2] In mathematical terms, a relationship is an equivalence class of those relations that are equivalent under permutation of domains (see Section 2.1.1).

qualified by the role name *sub*, and the second by *super*, so that users could deal with the relationship *component* and its domains—*sub.part super.part, quantity*—without regard to any ordering between these domains.

To sum up, it is proposed that most users should interact with a relational model of the data consisting of a collection of time-varying relationships (rather than relations). Each user need not know more about any relationship than its name together with the names of its domains (role qualified whenever necessary).[3] Even this information might be offered in menu style by the system (subject to security and privacy constraints) upon request by the user.

There are usually many alternative ways in which a relational model may be established for a data bank. In order to discuss a preferred way (or normal form), we must first introduce a few additional concepts (active domain, primary key, foreign key, nonsimple domain) and establish some links with terminology currently in use in information systems programming. In the remainder of this paper, we shall not bother to distinguish between relations and relationships except where it appears advantageous to be explicit.

Consider an example of a data bank which includes relations concerning parts, projects, and suppliers. One relation called *part* is defined on the following domains:

(1) part number
(2) part name
(3) part color
(4) part weight
(5) quantity on hand
(6) quantity on order

and possibly other domains as well. Each of these domains is, in effect, a pool of values, some or all of which may be represented in the data bank at any instant. While it is conceivable that, at some instant, all part colors are present, it is unlikely that all possible part weights, part names, and part numbers are. We shall call the set of

[3] Naturally, as with any data put into and retrieved from a computer system, the user will normally make far more effective use of the data if he is aware of its meaning.

values represented at some instant the *active domain* at that instant.

Normally, one domain (or combination of domains) of a given relation has values which uniquely identify each element (n-tuple) of that relation. Such a domain (or combination) is called a *primary key*. In the example above, part number would be a primary key, while part color would not be. A primary key is *nonredundant* if it is either a simple domain (not a combination) or a combination such that none of the participating simple domains is superfluous in uniquely identifying each element. A relation may possess more than one nonredundant primary key. This would be the case in the example if different parts were always given distinct names. Whenever a relation has two or more nonredundant primary keys, one of them is arbitrarily selected and called *the* primary key of that relation.

A common requirement is for elements of a relation to cross-reference other elements of the same relation or elements of a different relation. Keys provide a user-oriented means (but not the only means) of expressing such cross-references. We shall call a domain (or domain combination) of relation R a *foreign key* if it is not the primary key of R but its elements are values of the primary key of some relation S (the possibility that S and R are identical is not excluded). In the relation *supply* of Figure 1, the combination of *supplier, part, project* is the primary key, while each of these three domains taken separately is a foreign key.

In previous work there has been a strong tendency to treat the data in a data bank as consisting of two parts, one part consisting of entity descriptions (for example, descriptions of suppliers) and the other part consisting of relations between the various entities or types of entities (for example, the *supply* relation). This distinction is difficult to maintain when one may have foreign keys in any relation whatsoever. In the user's relational model there appears to be no advantage to making such a distinction (there may be some advantage, however, when one applies relational concepts to machine representations of the user's set of relationships).

So far, we have discussed examples of relations which are defined on simple domains—domains whose elements are atomic (nondecomposable) values. Nonatomic values can be discussed within the relational framework. Thus, some domains may have relations as elements. These relations may, in turn, be defined on nonsimple domains, and so on. For example, one of the domains on which the relation *employee* is defined might be *salary history*. An element of the salary history domain is a binary relation defined on the domain *date* and the domain *salary*. The *salary history* domain is the set of all such binary relations. At any instant of time there are as many instances of the *salary history* relation in the data bank as there are employees. In contrast, there is only one instance of the *employee* relation.

The terms attribute and repeating group in present data base terminology are roughly analogous to simple domain and nonsimple domain, respectively. Much of the confusion in present terminology is due to failure to distinguish between type and instance (as in "record") and between components of a user model of the data on the one hand and their machine representation counterparts on the other hand (again, we cite "record" as an example).

1.4. Normal Form

A relation whose domains are all simple can be represented in storage by a two-dimensional column-homogeneous array of the kind discussed above. Some more complicated data structure is necessary for a relation with one or more nonsimple domains. For this reason (and others to be cited below) the possibility of eliminating nonsimple domains appears worth investigating.[4] There is, in fact, a very simple elimination procedure, which we shall call normalization.

Consider, for example, the collection of relations exhibited in Figure 3(a). *Job history* and *children* are nonsimple domains of the relation *employee*. *Salary history* is a nonsimple domain of the relation *job history*. The tree in Figure 3(a) shows just these interrelationships of the nonsimple domains.

[4] M. E. Sanko of IBM, San Jose, independently recognized the desirability of eliminating nonsimple domains.

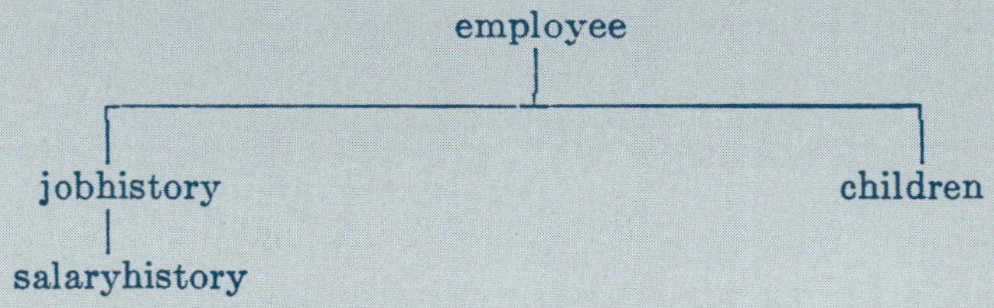

employee (*man#*, name, birthdate, jobhistory, children)
jobhistory (*jobdate*, title, salaryhistory)
salaryhistory (*salarydate*, salary)
children (*childname*, birthyear)

FIG. 3(a). Unnormalized set

employee′ (*man#*, name, birthdate)
jobhistory′ (*man#*, *jobdate*, title)
salaryhistory′ (*man#*, *jobdate*, *salarydate*, salary)
children′ (*man#*, *childname*, birthyear)

FIG. 3(b). Normalized set

Normalization proceeds as follows. Starting with the relation at the top of the tree, take its primary key and expand each of the immediately subordinate relations by inserting this primary key domain or domain combination. The primary key of each expanded relation consists of the primary key before expansion augmented by the primary key copied down from the parent relation. Now, strike out from the parent relation all nonsimple domains, remove the top node of the tree, and repeat the same sequence of operations on each remaining subtree.

The result of normalizing the collection of relations in Figure 3(a) is the collection in Figure 3(b). The primary key of each relation is italicized to show how such keys are expanded by the normalization.

If normalization as described above is to be applicable, the unnormalized collection of relations must satisfy the following conditions:

(1) The graph of interrelationships of the nonsimple domains is a collection of trees.

(2) No primary key has a component domain which is nonsimple.

The writer knows of no application which would require any relaxation of these conditions. Further operations of a normalizing kind are possible. These are not discussed in this paper.

The simplicity of the array representation which becomes feasible when all relations are cast in normal form is not only an advantage for storage purposes but also for communication of bulk data between systems which use widely different representations of the data. The communication form would be a suitably compressed version of the array representation and would have the following advantages:

(1) It would be devoid of pointers (address-valued or displacement-valued).

(2) It would avoid all dependence on hash addressing schemes.

(3) It would contain no indices or ordering lists.

If the user's relational model is set up in normal form, names of items of data in the data bank can take a simpler form than would otherwise be the case. A general name would take a form such as

$$R(g).r.d$$

where R is a relational name; g is a generation identifier (optional); r is a role name (optional); d is a domain name. Since g is needed only when several generations of a given relation exist, or are anticipated to exist, and r is needed only when the relation R has two or more domains named d, the simple form $R.d$ will often be adequate.

1.5. Some Linguistic Aspects

The adoption of a relational model of data, as described above, permits the development of a universal data sublanguage based on an applied predicate calculus. A first-order predicate calculus suffices if the collection of relations is in normal form. Such a language would provide a yardstick of linguistic power for all other proposed data languages, and would itself be a strong candidate for embedding (with appropriate syntactic modification) in a variety of host languages (programming, command- or problem-oriented). While it is not the purpose of this paper to describe such a language in detail, its salient features would be as follows.

Let us denote the data sublanguage by R and the host language by H. R permits the declaration of relations and their domains. Each declaration of a relation identifies the primary key for that relation. Declared relations are added to the system catalog for use by any members of the user community who have appropriate authorization. H permits supporting declarations which indicate, perhaps less permanently, how these relations are represented in storage. R permits the specification for retrieval of any subset of data from the data bank. Action on such a retrieval request is subject to security constraints.

The universality of the data sublanguage lies in its descriptive ability (not its computing ability). In a large data bank each subset of the data has a very large number of possible (and sensible) descriptions, even when we assume (as we do) that there is only a finite set of function subroutines to which the system has access for use in qualifying data for retrieval. Thus, the class of qualification expressions which can be used in a set specification must have the descriptive power of the class of well-formed formulas of an applied predicate calculus. It is well known that to preserve this descriptive power it is unnecessary to express (in whatever syntax is chosen) every formula of the selected predicate calculus. For example, just those in prenex normal form are adequate [9].

Arithmetic functions may be needed in the qualification or other parts of retrieval statements. Such functions can be defined in H and invoked in R.

A set so specified may be fetched for query purposes only, or it may be held for possible changes. Insertions take the form of adding new elements to declared relations without regard to any ordering that may be present in their machine representation. Deletions which are effective for the community (as opposed to the individual user or subcommunities) take the form of removing elements from declared relations. Some deletions and updates may be triggered by others, if deletion and update dependencies between specified relations are declared in R.

One important effect that the view adopted toward data has on the language used to retrieve it is in the naming of data elements and sets. Some aspects of this have been discussed in the previous section. With the usual network view, users will often be burdened with coining and using more relation names than are absolutely necessary, since names are associated with paths (or path types) rather than with relations.

Once a user is aware that a certain relation is stored, he will expect to be able to exploit[5] it using any combination of its arguments as "knowns" and the remaining arguments as "unknowns," because the information (like Everest) is there. This is a system feature (missing from many current information systems) which we shall call (logically) *symmetric exploitation* of relations. Naturally, symmetry in performance is not to be expected.

To support symmetric exploitation of a single binary relation, two directed paths are needed. For a relation of degree n, the number of paths to be named and controlled is n factorial.

Again, if a relational view is adopted in which every n-ary relation ($n > 2$) has to be expressed by the user as a nested expression involving only binary relations (see Feldman's LEAP System [10], for example) then $2n - 1$ names have to be coined instead of only $n + 1$ with direct n-ary notation as described in Section 1.2. For example, the 4-ary relation *supply* of Figure 1, which entails 5 names in n-ary notation, would be represented in the form

$$P\ (supplier,\ Q\ (part,\ R\ (project,\ quantity)))$$

in nested binary notation and, thus, employ 7 names.

A further disadvantage of this kind of expression is its asymmetry. Although this asymmetry does not prohibit symmetric exploitation, it certainly makes some bases of interrogation very awkward for the user to express (consider, for example, a query for those parts and quantities related to certain given projects via Q and R).

1.6. Expressible, Named, and Stored Relations

Associated with a data bank are two collections of relations: the *named set* and the *expressible set*. The named set

[5] Exploiting a relation includes query, update, and delete.

is the collection of all those relations that the community of users can identify by means of a simple name (or identifier). A relation R acquires membership in the named set when a suitably authorized user declares R; it loses membership when a suitably authorized user cancels the declaration of R.

The expressible set is the total collection of relations that can be designated by expressions in the data language. Such expressions are constructed from simple names of relations in the named set; names of generations, roles and domains; logical connectives; the quantifiers of the predicate calculus;[6] and certain constant relation symbols such as $=$, $>$. The named set is a subset of the expressible set—usually a very small subset.

Since some relations in the named set may be time-independent combinations of others in that set, it is useful to consider associating with the named set a collection of statements that define these time-independent constraints. We shall postpone further discussion of this until we have introduced several operations on relations (see Section 2).

One of the major problems confronting the designer of a data system which is to support a relational model for its users is that of determining the class of stored representations to be supported. Ideally, the variety of permitted data representations should be just adequate to cover the spectrum of performance requirements of the total collection of installations. Too great a variety leads to unnecessary overhead in storage and continual reinterpretation of descriptions for the structures currently in effect.

For any selected class of stored representations the data system must provide a means of translating user requests expressed in the data language of the relational model into corresponding—and efficient—actions on the current stored representation. For a high level data language this presents a challenging design problem. Nevertheless, it is a problem which must be solved—as more users obtain con-

[6] Because each relation in a practical data bank is a finite set at every instant of time, the existential and universal quantifiers can be expressed in terms of a function that counts the number of elements in any finite set.

current access to a large data bank, responsibility for providing efficient response and throughput shifts from the individual user to the data system.

2. Redundancy and Consistency

2.1. Operations on Relations

Since relations are sets, all of the usual set operations are applicable to them. Nevertheless, the result may not be a relation; for example, the union of a binary relation and a ternary relation is not a relation.

The operations discussed below are specifically for relations. These operations are introduced because of their key role in deriving relations from other relations. Their principal application is in noninferential information systems—systems which do not provide logical inference services—although their applicability is not necessarily destroyed when such services are added.

Most users would not be directly concerned with these operations. Information systems designers and people concerned with data bank control should, however, be thoroughly familiar with them.

2.1.1. *Permutation.* A binary relation has an array representation with two columns. Interchanging these columns yields the converse relation. More generally, if a permutation is applied to the columns of an n-ary relation, the resulting relation is said to be a *permutation* of the given relation. There are, for example, $4! = 24$ permutations of the relation *supply* in Figure 1, if we include the identity permutation which leaves the ordering of columns unchanged.

Since the user's relational model consists of a collection of relationships (domain-unordered relations), permutation is not relevant to such a model considered in isolation. It is, however, relevant to the consideration of stored representations of the model. In a system which provides symmetric exploitation of relations, the set of queries answerable by a stored relation is identical to the set answerable by any permutation of that relation. Although it is logically unnecessary to store both a relation and some permutation of it, performance considerations could make it advisable.

2.1.2. *Projection.* Suppose now we select certain columns of a relation (striking out the others) and then remove from the resulting array any duplication in the rows. The final array represents a relation which is said to be a *projection* of the given relation.

A selection operator π is used to obtain any desired permutation, projection, or combination of the two operations. Thus, if L is a list of k indices[7] $L = i_1, i_2, \cdots, i_k$ and R is an n-ary relation ($n \geq k$), then $\pi_L(R)$ is the k-ary relation whose jth column is column i_j of R ($j = 1, 2, \cdots, k$) except that duplication in resulting rows is removed. Consider the relation *supply* of Figure 1. A permuted projection of this relation is exhibited in Figure 4. Note that, in this particular case, the projection has fewer n-tuples than the relation from which it is derived.

2.1.3. *Join.* Suppose we are given two binary relations, which have some domain in common. Under what circumstances can we combine these relations to form a ternary relation which preserves all of the information in the given relations?

The example in Figure 5 shows two relations R, S, which are joinable without loss of information, while Figure 6 shows a join of R with S. A binary relation R is *joinable* with a binary relation S if there exists a ternary relation U such that $\pi_{12}(U) = R$ and $\pi_{23}(U) = S$. Any such ternary relation is called a *join* of R with S. If R, S are binary relations such that $\pi_2(R) = \pi_1(S)$, then R is joinable with S. One join that always exists in such a case is the *natural join* of R with S defined by

$$R*S = \{(a, b, c): R(a, b) \wedge S(b, c)\}$$

where $R(a, b)$ has the value *true* if (a, b) is a member of R and similarly for $S(b, c)$. It is immediate that

$$\pi_{12}(R*S) = R$$

and

$$\pi_{23}(R*S) = S.$$

[7] When dealing with relationships, we use domain names (role-qualified whenever necessary) instead of domain positions.

Π_{31}(*supply*)	(*project*	*supplier*)
	5	1
	5	2
	1	4
	7	2

FIG. 4. A permuted projection of the relation in Figure 1

R	(*supplier*	*part*)
	1	1
	2	1
	2	2

S	(*part*	*project*)
	1	1
	1	2
	2	1

FIG. 5. Two joinable relations

R*S	(*supplier*	*part*	*project*)
	1	1	1
	1	1	2
	2	1	1
	2	1	2
	2	2	1

FIG. 6. The natural join of R with S (from Figure 5)

U	(*supplier*	*part*	*project*)
	1	1	2
	2	1	1
	2	2	1

FIG. 7. Another join of R with S (from Figure 5)

Note that the join shown in Figure 6 is the natural join of R with S from Figure 5. Another join is shown in Figure 7.

Inspection of these relations reveals an element (element 1) of the domain *part* (the domain on which the join is to be made) with the property that it possesses more than one relative under R and also under S. It is this element which gives rise to the plurality of joins. Such an element in the joining domain is called a *point of ambiguity* with respect to the joining of R with S.

If either $\pi_{21}(R)$ or S is a function,[8] no point of ambiguity can occur in joining R with S. In such a case, the natural join of R with S is the only join of R with S. Note that the reiterated qualification "of R with S" is necessary, because S might be joinable with R (as well as R with S), and this join would be an entirely separate consideration. In Figure 5, none of the relations R, $\pi_{21}(R)$, S, $\pi_{21}(S)$ is a function.

Ambiguity in the joining of R with S can sometimes be resolved by means of other relations. Suppose we are given, or can derive from sources independent of R and S, a relation T on the domains *project* and *supplier* with the following properties:

$$(1)\ \pi_1(T) = \pi_2(S),$$

$$(2)\ \pi_2(T) = \pi_1(R),$$

$$(3)\ T(j, s) \rightarrow \exists p(R(S, p) \wedge S(p, j)),$$

$$(4)\ R(s, p) \rightarrow \exists j(S(p, j) \wedge T(j, s)),$$

$$(5)\ S(p, j) \rightarrow \exists s(T(j, s) \wedge R(s, p)),$$

then we may form a three-way join of R, S, T; that is, a ternary relation such that

$$\pi_{12}(U) = R, \qquad \pi_{23}(U) = S, \qquad \pi_{31}(U) = T.$$

Such a join will be called a *cyclic 3-join* to distinguish it from a *linear 3-join* which would be a quaternary relation V such that

$$\pi_{12}(V) = R, \qquad \pi_{23}(V) = S, \qquad \pi_{34}(V) = T.$$

While it is possible for more than one cyclic 3-join to exist (see Figures 8, 9, for an example), the circumstances under which this can occur entail much more severe constraints than those for a plurality of 2-joins. To be specific, the relations R, S, T must possess points of ambiguity with respect to joining R with S (say point x), S with T (say y), and T with R (say z), and, furthermore, y must be a relative of x under S, z a relative of y under T, and x a relative of z under R. Note that in Figure 8 the points $x = a$; $y = d$; $z = 2$ have this property.

[8] A function is a binary relation, which is one-one or many-one, but not one-many.

R	(s	p)		S	(p	j)		T	(j	s)
	1	a			a	d			d	1
	2	a			a	e			d	2
	2	b			b	d			e	2
					b	e			e	2

FIG. 8. Binary relations with a plurality of cyclic 3-joins

U	(s	p	j)		U'	(s	p	j)
	1	a	d			1	a	d
	2	a	e			2	a	d
	2	b	d			2	a	e
	2	b	e			2	b	d
						2	b	e

FIG. 9. Two cyclic 3-joins of the relations in Figure 8

The natural linear 3-join of three binary relations R, S, T is given by

$$R*S*T = \{(a, b, c, d):R(a, b) \wedge S(b, c) \wedge T(c, d)\}$$

where parentheses are not needed on the left-hand side because the natural 2-join $(*)$ is associative. To obtain the cyclic counterpart, we introduce the operator γ which produces a relation of degree $n - 1$ from a relation of degree n by tying its ends together. Thus, if R is an n-ary relation $(n \geq 2)$, the *tie* of R is defined by the equation

$$\gamma(R) = \{(a_1, a_2, \cdots, a_{n-1}):R(a_1, a_2, \cdots, a_{n-1}, a_n) \wedge a_1 = a_n\}.$$

We may now represent the natural cyclic 3-join of R, S, T by the expression

$$\gamma(R*S*T).$$

Extension of the notions of linear and cyclic 3-join and their natural counterparts to the joining of n binary relations (where $n \geq 3$) is obvious. A few words may be appropriate, however, regarding the joining of relations which are not necessarily binary. Consider the case of two relations R (degree r), S (degree s) which are to be joined on

p of their domains ($p < r$, $p < s$). For simplicity, suppose these p domains are the last p of the r domains of R, and the first p of the s domains of S. If this were not so, we could always apply appropriate permutations to make it so. Now, take the Cartesian product of the first r-p domains of R, and call this new domain A. Take the Cartesian product of the last p domains of R, and call this B. Take the Cartesian product of the last s-p domains of S and call this C.

We can treat R as if it were a binary relation on the domains A, B. Similarly, we can treat S as if it were a binary relation on the domains B, C. The notions of linear and cyclic 3-join are now directly applicable. A similar approach can be taken with the linear and cyclic n-joins of n relations of assorted degrees.

2.1.4. *Composition.* The reader is probably familiar with the notion of composition applied to functions. We shall discuss a generalization of that concept and apply it first to binary relations. Our definitions of composition and composability are based very directly on the definitions of join and joinability given above.

Suppose we are given two relations R, S. T is a *composition* of R with S if there exists a join U of R with S such that $T = \pi_{13}(U)$. Thus, two relations are composable if and only if they are joinable. However, the existence of more than one join of R with S does not imply the existence of more than one composition of R with S.

Corresponding to the natural join of R with S is the *natural composition*[9] of R with S defined by

$$R \cdot S = \pi_{13}(R * S).$$

Taking the relations R, S from Figure 5, their natural composition is exhibited in Figure 10 and another composition is exhibited in Figure 11 (derived from the join exhibited in Figure 7).

[9] Other writers tend to ignore compositions other than the natural one, and accordingly refer to this particular composition as *the* composition—see, for example, Kelley's "General Topology."

R·S	(*project*	*supplier*)
	1	1
	1	2
	2	1
	2	2

FIG. 10. The natural composition of R with S (from Figure 5)

T	(*project*	*supplier*)
	1	2
	2	1

FIG. 11. Another composition of R with S (from Figure 5)

R	(*supplier*	*part*)	S	(*part*	*project*)
	1	a		a	g
	1	b		b	f
	1	c		c	f
	2	c		c	g
	2	d		d	g
	2	e		e	f

FIG. 12. Many joins, only one composition

When two or more joins exist, the number of distinct compositions may be as few as one or as many as the number of distinct joins. Figure 12 shows an example of two relations which have several joins but only one composition. Note that the ambiguity of point *c* is lost in composing *R* with *S*, because of unambiguous associations made via the points *a*, *b*, *d*, *e*.

Extension of composition to pairs of relations which are not necessarily binary (and which may be of different degrees) follows the same pattern as extension of pairwise joining to such relations.

A lack of understanding of relational composition has led several systems designers into what may be called the *connection trap*. This trap may be described in terms of the following example. Suppose each supplier description is linked by pointers to the descriptions of each part supplied by that supplier, and each part description is similarly linked to the descriptions of each project which uses that part. A conclusion is now drawn which is, in general, erroneous: namely that, if all possible paths are followed from a given supplier via the parts he supplies to the projects using those parts, one will obtain a valid set of all projects supplied by that supplier. Such a conclusion is correct only in the very special case that the target relation between projects and suppliers is, in fact, the natural composition of the other two relations—and we must normally add the phrase "for all time," because this is usually implied in claims concerning path-following techniques.

2.1.5. *Restriction*. A subset of a relation is a relation. One way in which a relation S may act on a relation R to generate a subset of R is through the operation *restriction* of R by S. This operation is a generalization of the restriction of a function to a subset of its domain, and is defined as follows.

Let L, M be equal-length lists of indices such that $L = i_1, i_2, \cdots, i_k$, $M = j_1, j_2, \cdots, j_k$ where $k \leqq$ degree of R and $k \leqq$ degree of S. Then the L, M restriction of R by S denoted $R_L|_M S$ is the maximal subset R' of R such that

$$\pi_L(R') = \pi_M(S).$$

The operation is defined only if equality is applicable between elements of $\pi_{i_h}(R)$ on the one hand and $\pi_{j_h}(S)$ on the other for all $h = 1, 2, \cdots, k$.

The three relations R, S, R' of Figure 13 satisfy the equation $R' = R_{(2,3)}|_{(1,2)}S$.

R (s	p	j)
1	a	A
2	a	A
2	a	B
2	b	A
2	b	B

S (p	j)
a	A
c	B
b	B

R′ (s	p	j)
1	a	A
2	a	A
2	b	B

FIG. 13. Example of restriction

We are now in a position to consider various applications of these operations on relations.

2.2. REDUNDANCY

Redundancy in the named set of relations must be distinguished from redundancy in the stored set of representations. We are primarily concerned here with the former.

To begin with, we need a precise notion of derivability for relations.

Suppose θ is a collection of operations on relations and each operation has the property that from its operands it yields a unique relation (thus natural join is eligible, but join is not). A relation R is *θ-derivable* from a set S of relations if there exists a sequence of operations from the collection θ which, for all time, yields R from members of S. The phrase "for all time" is present, because we are dealing with time-varying relations, and our interest is in derivability which holds over a significant period of time. For the named set of relationships in noninferential systems, it appears that an adequate collection θ_1 contains the following operations: projection, natural join, tie, and restriction. Permutation is irrelevant and natural composition need not be included, because it is obtainable by taking a natural join and then a projection. For the stored set of representations, an adequate collection θ_2 of operations would include permutation and additional operations concerned with subsetting and merging relations, and ordering and connecting their elements.

2.2.1. *Strong Redundancy.* A set of relations is *strongly redundant* if it contains at least one relation that possesses a projection which is derivable from other projections of relations in the set. The following two examples are intended to explain why strong redundancy is defined this way, and to demonstrate its practical use. In the first example the collection of relations consists of just the following relation:

employee (*serial #*, *name*, *manager#*, *managername*)

with *serial#* as the primary key and *manager#* as a foreign key. Let us denote the active domain by Δ_t, and suppose that

$$\Delta_t(manager\#) \subset \Delta_t(serial\#)$$

and

$$\Delta_t(managername) \subset \Delta_t(name)$$

for all time t. In this case the redundancy is obvious: the domain *managername* is unnecessary. To see that it is a strong redundancy as defined above, we observe that

$$\pi_{34}(employee) = \pi_{12}(employee)_1|_1\pi_3(employee).$$

In the second example the collection of relations includes a relation S describing suppliers with primary key $s\#$, a relation D describing departments with primary key $d\#$, a relation J describing projects with primary key $j\#$, and the following relations:

$$P(s\#, d\#, \cdots), \qquad Q(s\#, j\#, \cdots), \qquad R(d\#, j\#, \cdots),$$

where in each case $\cdots$ denotes domains other than $s\#$, $d\#$, $j\#$. Let us suppose the following condition C is known to hold independent of time: supplier s supplies department d (relation P) if and only if supplier s supplies some project j (relation Q) to which d is assigned (relation R). Then, we can write the equation

$$\pi_{12}(P) = \pi_{12}(Q) \cdot \pi_{21}(R)$$

and thereby exhibit a strong redundancy.

An important reason for the existence of strong redundancies in the named set of relationships is user convenience. A particular case of this is the retention of semi-obsolete relationships in the named set so that old programs that refer to them by name can continue to run correctly. Knowledge of the existence of strong redundancies in the named set enables a system or data base administrator greater freedom in the selection of stored representations to cope more efficiently with current traffic. If the strong redundancies in the named set are directly reflected in strong redundancies in the stored set (or if other strong redundancies are introduced into the stored set), then, generally speaking, extra storage space and update time are consumed with a potential drop in query time for some queries and in load on the central processing units.

2.2.2. *Weak Redundancy*. A second type of redundancy may exist. In contrast to strong redundancy it is not characterized by an equation. A collection of relations is *weakly redundant* if it contains a relation that has a projection which is not derivable from other members but is at all times a projection of *some* join of other projections of relations in the collection.

We can exhibit a weak redundancy by taking the second example (cited above) for a strong redundancy, and assuming now that condition C does not hold at all times. The relations $\pi_{12}(P)$, $\pi_{12}(Q)$, $\pi_{12}(R)$ are complex[10] relations with the possibility of points of ambiguity occurring from time to time in the potential joining of any two. Under these circumstances, none of them is derivable from the other two. However, constraints do exist between them, since each is a projection of some cyclic join of the three of them. One of the weak redundancies can be characterized by the statement: for all time, $\pi_{12}(P)$ is *some* composition of $\pi_{12}(Q)$ with $\pi_{21}(R)$. The composition in question might be the natural one at some instant and a nonnatural one at another instant.

Generally speaking, weak redundancies are inherent in the logical needs of the community of users. They are not removable by the system or data base administrator. If they appear at all, they appear in both the named set and the stored set of representations.

2.3. Consistency

Whenever the named set of relations is redundant in either sense, we shall associate with that set a collection of statements which define all of the redundancies which hold independent of time between the member relations. If the information system lacks—and it most probably will—detailed semantic information about each named relation, it cannot deduce the redundancies applicable to the named set. It might, over a period of time, make attempts to *in*duce the redundancies, but such attempts would be fallible.

Given a collection C of time-varying relations, an associated set Z of constraint statements and an instantaneous value V for C, we shall call the state (C, Z, V) *consistent* or *inconsistent* according as V does or does not satisfy Z. For example, given stored relations R, S, T together with the constraint statement "$\pi_{12}(T)$ is a composition of $\pi_{12}(R)$ with $\pi_{12}(S)$", we may check from time to time that

[10] A binary relation is complex if neither it nor its converse is a function.

the values stored for R, S, T satisfy this constraint. An algorithm for making this check would examine the first two columns of each of R, S, T (in whatever way they are represented in the system) and determine whether

(1) $\pi_1(T) = \pi_1(R)$,

(2) $\pi_2(T) = \pi_2(S)$,

(3) for every element pair (a, c) in the relation $\pi_{12}(T)$ there is an element b such that (a, b) is in $\pi_{12}(R)$ and (b, c) is in $\pi_{12}(S)$.

There are practical problems (which we shall not discuss here) in taking an instantaneous snapshot of a collection of relations, some of which may be very large and highly variable.

It is important to note that consistency as defined above is a property of the instantaneous state of a data bank, and is independent of how that state came about. Thus, in particular, there is no distinction made on the basis of whether a user generated an inconsistency due to an act of omission or an act of commission. Examination of a simple example will show the reasonableness of this (possibly unconventional) approach to consistency.

Suppose the named set C includes the relations S, J, D, P, Q, R of the example in Section 2.2 and that P, Q, R possess either the strong or weak redundancies described therein (in the particular case now under consideration, it does not matter which kind of redundancy occurs). Further, suppose that at some time t the data bank state is consistent and contains no project j such that supplier 2 supplies project j and j is assigned to department 5. Accordingly, there is no element $(2, 5)$ in $\pi_{12}(P)$. Now, a user introduces the element $(2, 5)$ into $\pi_{12}(P)$ by inserting some appropriate element into P. The data bank state is now inconsistent. The inconsistency could have arisen from an act of omission, if the input $(2, 5)$ is correct, and there does exist a project j such that supplier 2 supplies j and j is assigned to department 5. In this case, it is very likely that the user intends in the near future to insert elements into Q and R which will have the effect of introducing $(2, j)$ into $\pi_{12}(Q)$ and $(5, j)$ in $\pi_{12}(R)$. On the other hand, the input $(2, 5)$

might have been faulty. It could be the case that the user intended to insert some other element into P—an element whose insertion would transform a consistent state into a consistent state. The point is that the system will normally have no way of resolving this question without interrogating its environment (perhaps the user who created the inconsistency).

There are, of course, several possible ways in which a system can detect inconsistencies and respond to them. In one approach the system checks for possible inconsistency whenever an insertion, deletion, or key update occurs. Naturally, such checking will slow these operations down. If an inconsistency has been generated, details are logged internally, and if it is not remedied within some reasonable time interval, either the user or someone responsible for the security and integrity of the data is notified. Another approach is to conduct consistency checking as a batch operation once a day or less frequently. Inputs causing the inconsistencies which remain in the data bank state at checking time can be tracked down if the system maintains a journal of all state-changing transactions. This latter approach would certainly be superior if few nontransitory inconsistencies occurred.

2.4. Summary

In Section 1 a relational model of data is proposed as a basis for protecting users of formatted data systems from the potentially disruptive changes in data representation caused by growth in the data bank and changes in traffic. A normal form for the time-varying collection of relationships is introduced.

In Section 2 operations on relations and two types of redundancy are defined and applied to the problem of maintaining the data in a consistent state. This is bound to become a serious practical problem as more and more different types of data are integrated together into common data banks.

Many questions are raised and left unanswered. For example, only a few of the more important properties of the data sublanguage in Section 1.4 are mentioned. Neither the purely linguistic details of such a language nor the implementation problems are discussed. Nevertheless, the material presented should be adequate for experienced systems programmers to visualize several approaches. It is also hoped that this paper can contribute to greater precision in work on formatted data systems.

Acknowledgment. It was C. T. Davies of IBM Poughkeepsie who convinced the author of the need for data independence in future information systems. The author wishes to thank him and also F. P. Palermo, C. P. Wang, E. B. Altman, and M. E. Senko of the IBM San Jose Research Laboratory for helpful discussions.

RECEIVED SEPTEMBER, 1969; REVISED FEBRUARY, 1970

REFERENCES

1. CHILDS, D. L. Feasibility of a set-theoretical data structure —a general structure based on a reconstituted definition of relation. Proc. IFIP Cong., 1968, North Holland Pub. Co., Amsterdam, p. 162–172.
2. LEVEIN, R. E., AND MARON, M. E. A computer system for inference execution and data retrieval. *Comm. ACM 10*, 11 (Nov. 1967), 715–721.
3. BACHMAN, C. W. Software for random access processing. *Datamation* (Apr. 1965), 36–41.
4. McGEE, W. C. Generalized file processing. In *Annual Review in Automatic Programming 5*, 13, Pergamon Press, New York, 1969, pp. 77–149.
5. Information Management System/360, Application Description Manual H20-0524-1. IBM Corp., White Plains, N. Y., July 1968.
6. GIS (Generalized Information System), Application Description Manual H20-0574. IBM Corp., White Plains, N. Y., 1965.
7. BLEIER, R. E. Treating hierarchical data structures in the SDC time-shared data management system (TDMS). Proc. ACM 22nd Nat. Conf., 1967, MDI Publications, Wayne, Pa., pp. 41–49.
8. IDS Reference Manual GE 625/635, GE Inform. Sys. Div., Pheonix, Ariz., CPB 1093B, Feb. 1968.
9. CHURCH, A. *An Introduction to Mathematical Logic I*. Princeton U. Press, Princeton, N.J., 1956.
10. FELDMAN, J. A., AND ROVNER, P. D. An Algol-based associative language. Stanford Artificial Intelligence Rep. AI-66, Aug. 1, 1968.

Entity-Relationship Modeling

Peter Chen
Foster Professor of Computer Science
Louisiana State University
chen@bit.csc.lsu.edu

Ph.D. in Computer Science, Harvard

Individual Achievement Award
from DAMA International,
Data Resource Management
Technology Award, Stevens Award 2001

Fellow of ACM, IEEE, AAAS

Major contribution:
entity-relationship model

Current interests: databases, software engineering,
Internet security, XML

Peter Chen

Entity-Relationship Modeling: Historical Events, Future Trends, and Lessons Learned

This paper describes the historical developments of the Entity-Relationship (ER) model from the 1970s to recent years. It starts with a discussion of the motivations and environmental factors in the early days. Then, the paper points out the role of the ER model in the Computer-Aided Software Engineering (CASE) movement in the late 1980s and early 1990s. It also describes the possible role of the author's Chinese cultural heritage in the development of the ER model. In that context, the relationships between natural languages (including Ancient Egyptian hieroglyphics) and ER concepts are explored. Finally, the lessons learned and future directions are presented.

1. Introduction

Entity-Relationship (ER) modeling is an important step in information system design and software engineering. In this paper, we will describe not only the history of the development of the ER approach but also the reactions and new developments since then. In this perspective, this paper may be a little different from other papers in this volume, for we are not just talking about historical events that happened 20 or 30 years ago, but also about the consequences and relevant developments in the past 25 years. At the end, we will talk about lessons learned during this time period. In particular, we intend to show that it is possible that one concept such as the ER model can be applied to many different things across a long time horizon (for more than 25 years), even in this quickly changing information technology area.

This paper is divided into eight sections. Section 1 is this introduction. In Section 2, the historical background and events that happened around 25 years ago will be explained. For example, descriptions of what happened at that time, what the competing forces were, and what triggered researchers like myself to work on this topic. Section 3 describes the initial reactions in the first five years from 1976 to 1981. For example, how did the academic world and industry initially view the ER model? Section 4 states the developments in the next 20 years from 1981 to 2001. In particular, the role of the ER model in Computer-Aided Software Engineering (CASE) will be discussed. Section 5 describes a possible factor that led me to come up with the ER modeling idea, that is, my Chinese cultural heritage. I had not considered this particular possibility until about 15 years ago. Section 6 presents my view of the future of ER modeling. Section 7 states the lessons learned. For those of you who have had similar experiences in the past 25 years, you probably will recognize similar principles and lessons in this section. For those who started their professional careers recently, I hope the lessons I learned will also be helpful. Section 8 is the conclusion.

2. Historical Background

In this section, we will look at the competing forces, the needs of the computer industry at that time, how the ER model was developed, and the main differences between the ER model and the relational model.

2.1 Competing Forces

First, let us look at the competing forces in the computer software field at that time. What were the competing forces then? What triggered people like myself to research this area (data models) and this particular topic (ER modeling)? In the following, we will discuss the competing forces in the industry and in the academic world in the early 1970s.

Competing Forces in Industry. There were several competing data models that had been implemented as commercial products in the early 1970s: the file system model, the hierarchical model (such as IBM's IMS database system), and the network model (such as Honeywell's IDS database system). The network model (also known as the CODASYL model) was developed by Charles Bachman, who received the ACM Turing Award in 1973. Most organizations at that time used file systems, and not too many used database systems. Some people were working on developing better data or index structures for storing and retrieving data such as the B+-tree by Bayer and McCreight [1].

Competing Forces in the Academic World. In 1970, the relational model was proposed, and it generated considerable interest in the academic community. It is correct to say that in the early 1970s, most people in the academic world performed research on the relational model instead of other models. One of the main reasons is that many professors had a difficult time understanding the long and dry manuals of commercial database management systems, and Codd's relational model paper [2] was written in a much more concise and scientific style. For his contributions in the development of the relational model, Codd received the ACM Turing Award in 1981.

Most people were working on DBMS Prototypes. Many people at that time in the academic world or in industry worked on the implementation of database management system prototypes. Most of them were based on the relational model.

Most people in academia were investigating the definitions and algorithms for the normal forms of relations. Many academic people worked on normalization of relations because only mathematical skills were needed to work on this subject. They could work on the improvement of existing algorithms for well-defined normal forms, or they could work on new normal forms. The speed of research moved very fast in the development of normal forms and can be illustrated by the following scenario. Let us say that several people were ready to publish their results on normal forms. Assuming that one person published a paper on 4th normal form and another person had written a paper on 4th normal form but had not published it yet, the 2nd person would have changed the title of the paper from 4th normal form to 5th normal form. Then, the rest would work on the 6th normal form. This became an endless game till one day somebody wrote a paper claiming that he had an infinity-th normal form and arguing that it did not make any sense to continue this game. Most practitioners also said loudly that any relational normal form higher than 3rd or 4th won't have practical significance. As a result, the game of pursuing the next normal form finally ran out of steam.

2.2 Needs of the System Software in the Early 1970s

The Needs of the Hardware/Software Vendors. In terms of software vendors at that time, there were urgent needs for (1) integration of various file and database formats and (2) more incorporation of "data semantics" into the data models.

The Needs of the User Organizations. For user organizations such as General Motors and Citibank, there was urgent need for (1) a unified methodology for file and database design for various file and database systems available in the commercial market and (2) incorporation of more data semantics, including business rules, into the requirements and design specifications.

2.3 How the ERM Was Developed

Here, I will give a personal history of the development of the ER model: where I was and what I did in the early 1970s, particularly how I developed the ER model.

Harvard (September 1969 to June 1973). After I received a B.S. in electrical engineering from National Taiwan University in 1968, I received a fellowship to study computer science (at that time, it was a part of applied mathematics) at Harvard graduate school. I received the Ph.D. degree in 1973. The thesis was very mathematically oriented – focusing on the file allocation problems in a storage hierarchy using the queuing theory and mathematical programming techniques. The knowledge I learned in electrical engineering, computer science and applied math was crucial in the development of the ER model in subsequent years.

Honeywell and Digital (June 1973 to August 1974). I joined Honeywell Information Systems in Waltham, MA in June 1973. I participated in the "next-generation computer system" project to develop a computer system based on distributed system architecture. There were about ten people in the team, and most of them were at least 20 years older than myself. The team consisted of several well-known computer experts including Charles Bachman. One of the requirements of such a "distributed system" was to make the files and databases in different nodes of the network compatible with each other. The ER model was motivated by this requirement. Even though I started to crystallize the concepts in my mind when I worked for Honeywell, I did not write or speak to anyone about this concept then. Around June of 1994, Honeywell abandoned the "next-generation computer system" project, and all the project team members went different ways. I then spent three months at Digital Equipment Corporation in Maynard, MA to develop a computer performance model for the PDP-10 system.

MIT Sloan School of Management (1974 – 1978). In September 1974, I joined MIT Sloan School of Management as an assistant professor. This was the place where I first stated my ER ideas in an article. Being a professor in a business/management school provided me with many opportunities to interact with the user organizations. I was particularly impressed by a common need of many organizations to have a unified methodology for file structure and database design. This observation certainly influenced the development of the ER model. As a result, the first ER paper was presented at the First International Conference on Very Large Databases in 1975 and subsequently published in the first issue of *ACM Transactions on Database Systems* [3] in March of 1976.

2.4 Fulfilling the Needs

How did the ER model fulfill the needs of the vendor and user organizations at that time? I will start with the graphical representation and theoretical foundations of the ER model. Then, I will explain the significant differences between the ER model and the relational model.

The Concepts of Entity, Relationship, Types, and Roles. In Fig. 1, there are two entities; both of them are of the "Person" type. There is a relationship called, "is-married-to," between these two persons. In this relationship, each of these two Person entities has a role. One person plays the role of "husband," and another person plays the role of "wife."

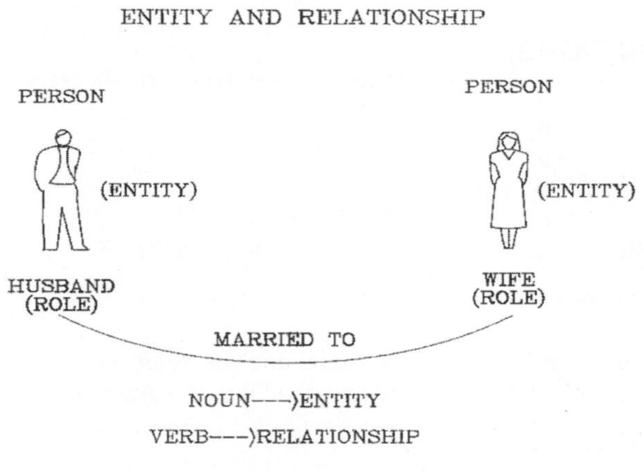

Fig. 1. The concept of entity and relationship

The ER Diagram. One of the key techniques in ER modeling is to document the entity and relationship types in a graphical form, called an ER diagram. Figure 2 is a typical ER diagram. The entity types such as EMP and PROJ are depicted as rectangular boxes, and the relationship types such as WORK-FOR are depicted as a diamond-shaped box. The value sets (domains) such as EMP#, NAME, and PHONE are depicted as circles, while attributes are the "mappings" from entity and relationship types to the value sets. The cardinality information of relationship is also expressed. For example, the "1" or "N" on the lines between the entity types and relationship types indicates the upper limit of the entities of that entity type participating in that relationship.

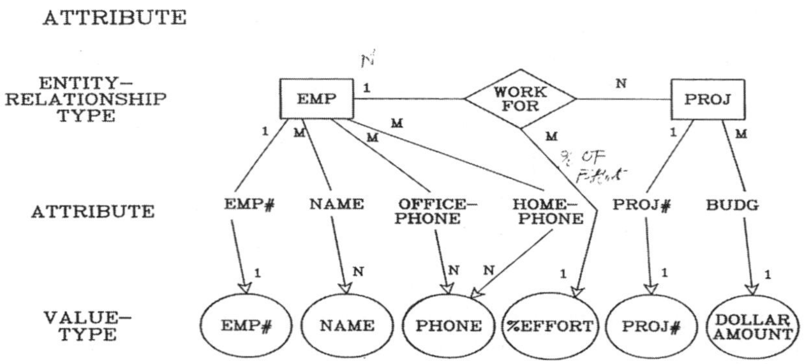

Fig. 2. An ER diagram

ER Model is Based on Strong Mathematical Foundations. The ER model is based on (1) set theory, (2) mathematical relations, (3) modern algebra, (4) logic, and (5) lattice theory. A formal definition of the entity and relationship concepts can be found in Fig. 3.

SET THEORY (DEFINITIONS)

A RELATIONSHIP SET DEFINED AS A "MATHEMATICAL RELATION" ON ENTITY SETS

ENTITY   e
ENTITY SET   E; e∈E
VALUE   v
VALUE SET   V; v∈V
RELATIONSHIP   r
RELATIONSHIP SET   R; r∈R

$R = \{r_1, r_2, \ldots, r_n\}$

$r_i = [e_{i_1}, e_{i_2}, \ldots, e_{i_n}] | e_{i_1} \in E_1, \ldots, e_{i_n} \in E_n$

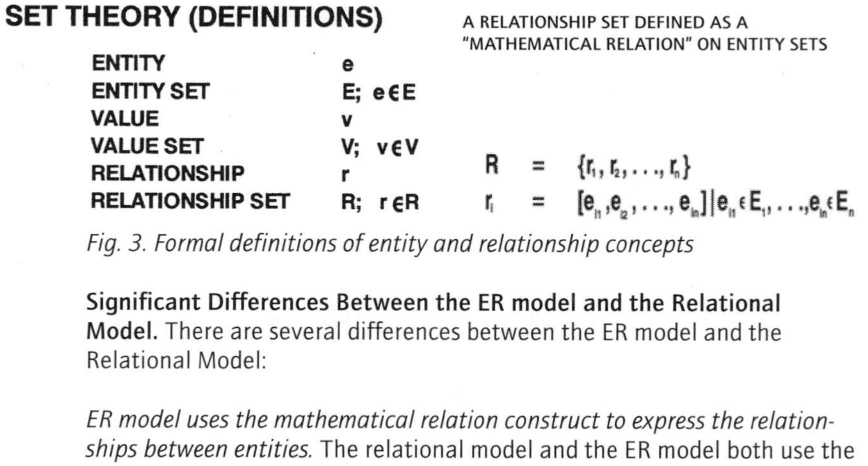

Fig. 3. Formal definitions of entity and relationship concepts

Significant Differences Between the ER model and the Relational Model. There are several differences between the ER model and the Relational Model:

ER model uses the mathematical relation construct to express the relationships between entities. The relational model and the ER model both use the mathematical structure called Cartesian product. In some way, both models look the same – both use the mathematical structure that utilizes the Cartesian product of something. As can be seen in Fig. 3, a relationship in the ER model is defined as an ordered tuple of "entities." In the relational model, a Cartesian product of data "domains" is a "relation," while in the ER model a Cartesian product of "entities" is a "relationship." In other words, in the relational model the mathematical relation construct is used to express the "structure of data values," while in the ER model the same construct is used to express the "structure of entities."

ER model contains more semantic information than the relational model. By the original definition of relation by Codd, any table is a relation. There is very little in the semantics of what a relation is or should be. The ER model adds the semantics of data to a data structure. Several years later, Codd developed a data model called RM/T, which incorporated some of the concepts of the ER model.

ER model has explicit linkage between entities. As can be seen in Fig. 2 and 4, the linkage between entities is explicit in the ER model while in the relational model it is implicit. In addition, the cardinality information is explicit in the ER model, and some of the cardinality information is not captured in the relational model.

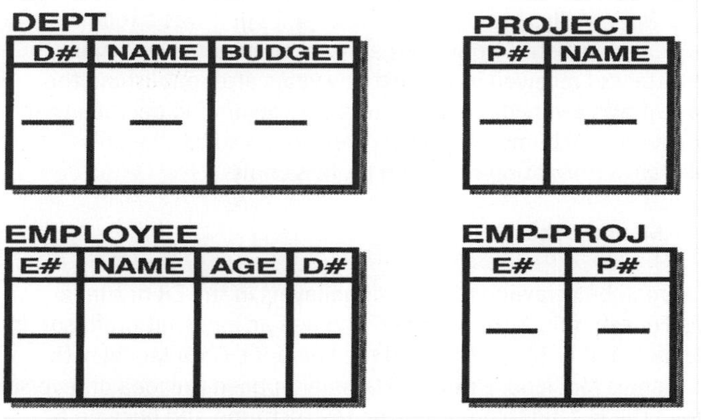

Fig. 4. Relational model of data

3. Initial Reactions and Reactions in the First Five Years (1976 – 1981)

3.1 First Paper Published and Codd's Reactions

As stated before, the first ER model paper was published in 1976. Codd wrote a long letter to the editor of *ACM Transactions on Database Systems* criticizing my paper. I was not privileged to see the letter. The editor of the journal told me that the letter was very long and single-spaced. In any case, Dr. Codd was not pleased with the ER model paper. Ironically, several years later, Codd proposed a new version of the relational data model called RM/T, which incorporated some concepts of the ER model. Perhaps, the first paper on the ER model was not as bad as Codd initially thought. Furthermore, in the 1990s, the Codd and Date consulting group invited me to serve as a keynote speaker (together with Codd) several times in their database symposia in London. This indicates that the acceptance of the ER model was so widespread so that initial unbelievers either became convinced or found it difficult to ignore.

3.2 Other Initial Reactions and Advice

During that time, a "religious war" emerged between different camps of data models. In particular, there was a big debate between the supporters of the relational model and that of the network model. Suddenly, this young assistant professor had written a paper talking about a "unified data model." In some sense, I was a "new kid on the block" thrown into the midst of a battle between two giants. An example of the type of advice I received at that time was: "Why don't you do research on the n-th normal form like most other researchers do? It would be much easier to get your normal form papers published." Even though this advice was based on good intentions and wisdom, I did not follow that type of advice because I believed that I could make a more significant contribution to the field by continuing to work on this topic (for example, [4-13]). It was a tough choice for a person just starting his career. You can imagine how many problems or attacks I received in the first few years after publishing the first ER paper. It was a very dangerous but a very rewarding decision that had a significant impact not only on my career, but also on the daily practices of many information-modeling professionals.

3.3 IDEF, ICAM, and Other Believers

There was a small but growing number of believers in the ER or similar data models. For example, Mike Hammer, who was an assistant professor at the EECS department of MIT, developed the Semantic Data Model with his student, Dennis McCleod. Later on, Hammer applied the idea of reverse engineering in the IT field to organization restructuring and became a management guru. Outside of the academic world, the industry and government agencies began to see the potential benefits of ER modeling. In the late 1970s, I served as a consultant in a team that developed the data modeling methodology for the ICAM (Integrated Computer-Aided Manufacturing) project sponsored by the U.S. Air Force. One of the objectives was to develop at least two modeling methodologies for modeling the aircraft manufacturing processes and data: one methodology for process modeling and one for data modeling. The data modeling methodology was called IDEF1 methodology and has been used widely in U.S. military projects.

3.4 Starting a Series of ER Conferences

The first ER conference was held at UCLA in 1979. We were expecting 50 people, but 250–300 people showed up. That was a big surprise. Initially, the ER conference was a bi-annual event, but now it is an annual event held in different parts of the world [14]. In November of this year (Year 2001), it will be held in Japan [15], and next year (Year 2002) it will be held in Finland. This series of conferences has become a major annual forum for exchanging ideas between researchers and practitioners in conceptual modeling.

4. The Next 20 Years (1981 –2001)

4.1 ER Model Adopted as a Standard for Repository Systems and ANSI IRDS.

In the 1980s, many vendors and user organizations recognized the need for a repository system to keep track of information resources in an organization and to serve as the focal point for planning, tracking, and monitoring the changes of hardware and software in various information systems in an organization. It turned out that the ER model was a good data model for repository systems. Around 1987, ANSI adopted the ER model as the data model for Information Resource Directory Systems (IRDS) standards. Several repository systems were implemented based on the ER model including IBM's Repository Manager for DB2 and DEC's CDD+ system.

4.2 ER Model as a Driving Force for Computer-Aided Software Engineering (CASE) Tools and Industry

Software development has been a nightmare for many years since the 1950s. In the late 1980s, IBM and others recognized the need for methodologies and tools for Computer-Aided Software Engineering (CASE). IBM proposed a software development framework and repository system called AD Cycle and the Repository Manager that used the ER model as the data model. I was one of the leaders who actively preached the technical approach and practical applications of CASE. In 1987, Digital Consulting Inc. (DCI) in Andover, Mass., founded by Dr. George Schussel, organized the First Symposium on CASE in Atlanta and invited the author to be one of the two keynote speakers. To everybody's surprise, the symposium was a huge commercial success, and DCI grew from a small company to a major force in the symposium and trade show business.

4.3 Object-Oriented (OO) Analysis Techniques are Partly Based on the ER Concepts

It is commonly acknowledged that one major component of the object-oriented (OO) analysis techniques is based on ER concepts. However, the "relationship" concept in the OO analysis techniques is still hierarchy-oriented and not yet equal to the general relationship concept advocated in the ER model. It is noticeable in the past few years that OO analysis techniques are moving toward the direction of adopting a more general relationship concept.

4.4 Data Mining is a Way to Discover Hidden Relationships

Many of you have heard about data mining. If you think deeply about what data mining actually does, you will see the linkage between data mining and the ER model. What is data mining? What does data mining really do? In our view, it is a discovery of "hidden relationships" between data entities. The relationships exist already, and we need only to discover them and then take advantage of them. This is different from conventional database design in which the database designers identify the relationships. In data mining, algorithms instead of humans are used to discover the hidden relationships.

5. In Retrospect: Another Important Factor – Chinese Cultural Heritage

5.1 Chinese Cultural Heritage

Many people asked me how I derived the idea of the ER model. After receiving many questions of this nature, I hypothesized that it might be related to something that many people in western culture may not have. After some soul searching, I thought it could be related to my Chinese cultural heritage. There are some concepts in Chinese character development and evolution that are closely related to modeling of things in the real world.

Here is an example. Figure 5 shows the Chinese characters of "sun", "moon," and "person". As you can see, these characters closely resemble the real world entities. Initially, many of the lines in the characters were made of curves. Because it was easier to cut straight lines on oracle bones, the curves became straight lines. Therefore, the current forms of the Chinese characters are of different shapes.

Fig. 5. Chinese characters that represent the real-world entities

Chinese characters also have several principles for "composition". For example, Fig. 6 shows how two characters, sun and moon, are composed into a new character. How do we know the meaning of the new character? Let us first think: what do sun and moon have in common? If your answer is: both reflect lights, it is not difficult to guess that the meaning of the new character is "brightness". There are other principles of composing Chinese characters [10].

日 (sun) + 月 (moon) = 明 (Bright/ Brightness by light)

Fig. 6. Composition of two chinese characters into a new chinese character

What does the Chinese character construction principles have to do with ER modeling? The answer is: both Chinese characters and the ER model are trying to model the world – trying to use graphics to represent the entities in the real world. Therefore, there should be some similarities in their constructs.

5.2 Ancient Egyptian Hieroglyphics

Besides Chinese characters, there are other languages having graphic characters. Ancient Egyptian is one of them. It turns out that there are several ancient Egyptian characters that are virtually the same as the Chinese characters. One is "sun", another is "mouth", and the third one is "water." It is amazing that both the Egyptian people and the Chinese people developed very similar characters even though they were thousands of miles away and had virtually no communication at that time. Ancient Egyptian hieroglyphics also have the concept of composition. Interested readers should refer to [11].

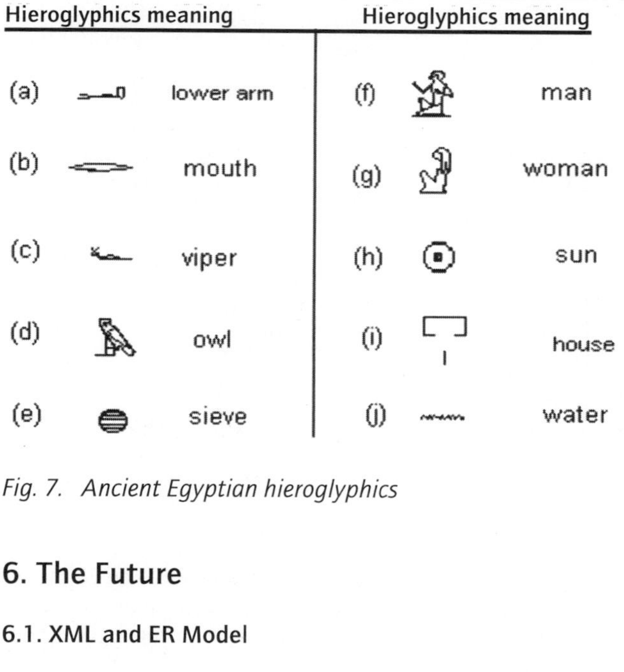
Hieroglyphics meaning
Hieroglyphics meaning
(a) lower arm
(b) mouth
(c) viper
(d) owl
(e) sieve
(f) man
(g) woman
(h) sun
(i) house
(j) water

Fig. 7. Ancient Egyptian hieroglyphics

6. The Future

6.1. XML and ER Model

In the past few years, I have been involved in developing the "standards" for XML. I have participated in two XML Working Groups of the World Wide Web Consortium (W3C) as an invited expert. During this involvement, some similarities between XML and the ER model were discovered including the following:

RDF and the ER Model. There are several components in the XML family. One of them is RDF, which stands for Resource Definition Framework. This is a technology that Tim Berners-Lee, the Director of W3C, pushes very hard as a tool for describing the meta-data of the web. There are some similarities and differences between RDF and the ER model, and Mr. Berners-Lee has written several articles discussing this issue. In a joint meeting of the RDF and Schema Working Groups in 1999, they issued the Cambridge Communiqué [16] that states: "RDF can be viewed as a member of the Entity-Relationship model family…"

XLink and the ER model. Most of us are familiar with the hyperlink in HTML. The XLink Working Group of W3C has been trying to develop a new kind of hyperlink for XML. In HTML, the hyperlink is basically a "physical pointer" because it specifies the exact URL of the target. In XLink, the new link is one step closer to a "logical pointer." In the evolution of operating systems, we have been moving from physical pointers to logical pointers. The XLink Working Group proposed a new structure called, "extended link." For example, Fig. 8 is an extended link for five remote resources. The extended link concept in XML is very similar to the n-ary relationship concept in the ER model. Figure 8 can be viewed as a relationship type defined on five entity types.

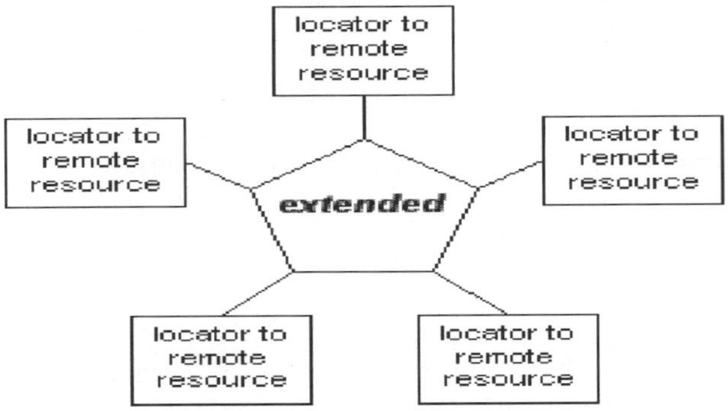

Fig. 8. "Extended link" in XML is similar to the n-ary relationship concept in the ER model

6.2. Theory of the Web

One thing that is still missing today is the theory of the Web. The ER model could be one of the foundations for the theory of the Web. I plan to work on that topic and would encourage the readers to work on the subject, too.

7. Lessons Learned

7.1 Reflections on Career Choices

In the past 25 years, I made some tough career choices as some of the other authors in this volume did. It is my hope that our experience will be useful to some other people who just started their professional careers and are making their career choices. Here are some reflections based on my own experience:

Right idea, right place, right time, and believe in yourself. In order to have your idea be accepted by other people, you need not only to have the right idea but also to present it at the right place and right time. You also need "persistence." In other words, you need to believe in yourself. This is probably the most difficult part because you have to endure some unnecessary pressures and criticisms when you are persistent about your idea and try to push it forward. Hopefully, some day in the future, you will be proven right. At that time, you will be happy that you persisted.

Getting Fresh Ideas from Unconventional Places. After working on a particular area for a while, you may run out of "big" ideas. You may still have some "good" ideas to get you going, but those ideas are not "earth-shattering." At that time, you need to look for ideas in different subject areas and to talk to new people. For example, most of us are immersed in western culture, and learning another culture may trigger new ways of thinking. Similarly, you may look into some fields outside of information technology such as physics, chemistry, biology, or architecture to find fresh ideas. By looking at the theories, techniques, and approaches used in other fields, you may get very innovative ideas to make a breakthrough in the IT field.

7.2 Implications of the Similarities and Differences Between the Chinese Characters and Ancient Egyptian Hieroglyphics on Software Engineering and Systems Development Methodologies

As we pointed out earlier, there are several Chinese characters that are almost the same as their counterparts in ancient Egyptian hieroglyphics. What does this mean? One possible answer is that human beings think alike even though there was virtually no communication between ancient Chinese people and ancient Egyptian people. It is very likely that the way to conceptualize basic things in the real world is common across races and cultures. As was discussed earlier, the construction and development of other characters are different in Chinese and in Ancient Egyptian hieroglyphics. It is valid to say that the language developments were dependent on the local environment and culture. What is the implication of the similarities and differences in character developments on the development of software engineering and information system development methodologies? The answer could be: some basic concepts and guidelines of software engineering and system development methodologies can be uniformly applied to all people in the world while some other parts of the methodologies may need to be adapted to local cultures and customs.

8. Conclusions

I was very fortunate to have the opportunity to meet the right people and to be given the opportunity to develop the Entity-Relationship (ER) model at the time and environment such a model was needed. I am very grateful to many other researchers who have continued to advance the theory of the ER approach and to many software professionals who have practiced ER modeling in their daily jobs in the past 25 years. I believe that the concepts of entity and relationship are very fundamental concepts in software engineering and information system development. In the future, we will see new applications of these concepts in the Web and other frontiers of the software world.

References

[1] Bayer, R. and McCreight, E., "Organization and Maintenance of Large Ordered Indexes," *Acta Informatica, Vol. 1*, Fasc. 3, 1972, pp. 173-189.

[2] Codd, E. F. "The Relational Model of Data for Large Shared Data Banks," *Communications of the ACM, Vol. 13, No. 6*, 1970, pp. 377-387.

[3] Chen, P.P., "The Entity-Relationship Model: Toward a Unified View of Data," *ACM Trans. on Database Systems, Vol.1, No.1*, March 1976, pp. 1-36.

[4] Chen, P. P., "An Algebra for a Directional Binary Entity-Relationship Model," *IEEE First International Conference on Data Engineering*, Los Angeles, April 1984, pp. 37-40.

[5] Chen, P. P., "Database Design Using Entities and Relationships," in: S. B. Yao (ed.), *Principles of Data Base Design*, Prentice-Hall, NJ, 1985, pp. 174-210.

[6] Chen, P. P., "The Time-Dimension in the Entity-Relationship Model," in: H. -J. Kugler (ed.), *Information Processing '86*, North-Holland, Amsterdam, 1986, pp. 387-390.

[7] Chen, P. P. and Zvieli, A., "Entity-Relationship Modeling of Fuzzy Data," *Proceedings of 2nd International Conference on Data Engineering*, Los Angeles, February 1986, pp. 320-327.

[8] Chen, P. P. and Li, M., "The Lattice Concept in Entity Set," in: S. Spaccapietra (ed.), *Entity-Relationship Approach*, North-Holland, Amsterdam, 1987, pp. 311-326.

[9] Chandrasekaran, N., Iyengar, S.S., and Chen, P. P., "The Denotational Semantics of the Entity-Relationship Model," *International Journal of Computer Mathematics*, 1988, pp. 1-15.

[10] Chen, P. P., "English, Chinese and ER Diagrams," *Data and Knowledge Engineering, Vol. 23, No. 1*, June 1997, pp. 5-16.

[11] Chen, P. P., "From Ancient Egyptian Language to Future Conceptual Modeling," in: Chen, P.P., et al. (eds), *Conceptual Modeling: Current Issues and Future Directions*, Lecture Notes in Computer Science, No. 1565, Springer, Berlin, 1998, pp. 57-66.

[12] Yang, A. and Chen, P. P., "Efficient Data Retrieval and Manipulation using Boolean Entity Lattice," *Data and Knowledge Engineering, Vol. 20*, 1996, pp. 211-226.

[13] http://www.csc.lsu.edu/~chen/

[14] http://www.er2000.byu.edu/

[15] *http://www.arislab.dnj.ynu.ac.jp/ER2001*

[16] http://www.w3.org/TR/schema-arch

Peter Chen

The Entity Relationship Model – Toward a Unified View of Data

ACM Transactions on Database Systems
Vol. 1 (1), 1976

The Entity-Relationship Model—Toward a Unified View of Data

PETER PIN-SHAN CHEN

Massachusetts Institute of Technology

A data model, called the entity-relationship model, is proposed. This model incorporates some of the important semantic information about the real world. A special diagrammatic technique is introduced as a tool for database design. An example of database design and description using the model and the diagrammatic technique is given. Some implications for data integrity, information retrieval, and data manipulation are discussed.

The entity-relationship model can be used as a basis for unification of different views of data: the network model, the relational model, and the entity set model. Semantic ambiguities in these models are analyzed. Possible ways to derive their views of data from the entity-relationship model are presented.

Key Words and Phrases: database design, logical view of data, semantics of data, data models, entity-relationship model, relational model, Data Base Task Group, network model, entity set model, data definition and manipulation, data integrity and consistency
CR Categories: 3.50, 3.70, 4.33, 4.34

1. INTRODUCTION

The logical view of data has been an important issue in recent years. Three major data models have been proposed: the network model [2, 3, 7], the relational model [8], and the entity set model [25]. These models have their own strengths and weaknesses. The network model provides a more natural view of data by separating entities and relationships (to a certain extent), but its capability to achieve data independence has been challenged [8]. The relational model is based on relational theory and can achieve a high degree of data independence, but it may lose some important semantic information about the real world [12, 15, 23]. The entity set model, which is based on set theory, also achieves a high degree of data independence, but its viewing of values such as "3" or "red" may not be natural to some people [25].

This paper presents the entity-relationship model, which has most of the advantages of the above three models. The entity-relationship model adopts the more natural view that the real world consists of entities and relationships. It

A version of this paper was presented at the International Conference on Very Large Data Bases, Framingham, Mass., Sept. 22–24, 1975.
Author's address: Center for Information System Research, Alfred P. Sloan School of Management, Massachusetts Institute of Technology, Cambridge, MA 02139.

incorporates some of the important semantic information about the real world (other work in database semantics can be found in [1, 12, 15, 21, 23, and 29]). The model can achieve a high degree of data independence and is based on set theory and relation theory.

The entity-relationship model can be used as a basis for a unified view of data. Most work in the past has emphasized the difference between the network model and the relational model [22]. Recently, several attempts have been made to reduce the differences of the three data models [4, 19, 26, 30, 31]. This paper uses the entity-relationship model as a framework from which the three existing data models may be derived. The reader may view the entity-relationship model as a generalization or extension of existing models.

This paper is organized into three parts (Sections 2–4). Section 2 introduces the entity-relationship model using a framework of multilevel views of data. Section 3 describes the semantic information in the model and its implications for data description and data manipulation. A special diagrammatric technique, the entity-relationship diagram, is introduced as a tool for database design. Section 4 analyzes the network model, the relational model, and the entity set model, and describes how they may be derived from the entity-relationship model.

2. THE ENTITY-RELATIONSHIP MODEL

2.1 Multilevel Views of Data

In the study of a data model, we should identify the levels of logical views of data with which the model is concerned. Extending the framework developed in [18, 25], we can identify four levels of views of data (Figure 1):

(1) Information concerning entities and relationships which exist in our minds.

(2) Information structure—organization of information in which entities and relationships are represented by data.

(3) Access-path-independent data structure—the data structures which are not involved with search schemes, indexing schemes, etc.

(4) Access-path-dependent data structure.

In the following sections, we shall develop the entity-relationship model step by step for the first two levels. As we shall see later in the paper, the network model, as currently implemented, is mainly concerned with level 4; the relational model is mainly concerned with levels 3 and 2; the entity set model is mainly concerned with levels 1 and 2.

2.2 Information Concerning Entities and Relationships (Level 1)

At this level we consider entities and relationships. An *entity* is a "thing" which can be distinctly identified. A specific person, company, or event is an example of an entity. A *relationship* is an association among entities. For instance, "father-son" is a relationship between two "person" entities.[1]

---

[1] It is possible that some people may view something (e.g. marriage) as an entity while other people may view it as a relationship. We think that this is a decision which has to be made by the enterprise administrator [27]. He should define what are entities and what are relationships so that the distinction is suitable for his environment.

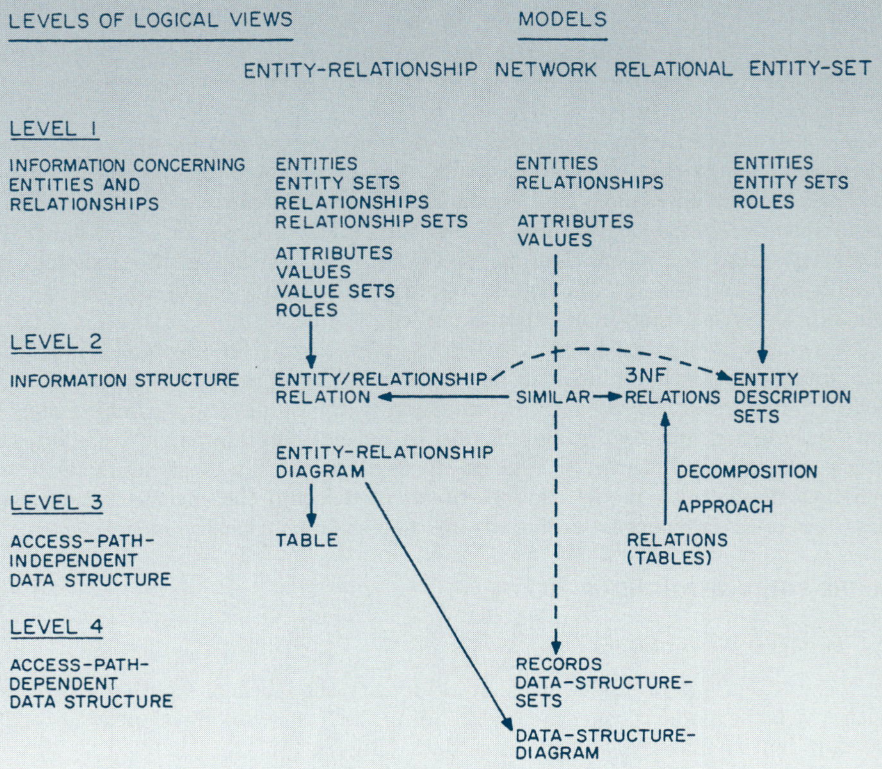

Fig. 1. Analysis of data models using multiple levels of logical views

The database of an enterprise contains relevant information concerning entities and relationships in which the enterprise is interested. A complete description of an entity or relationship may not be recorded in the database of an enterprise. It is impossible (and, perhaps, unnecessary) to record every potentially available piece of information about entities and relationships. From now on, we shall consider only the entities and relationships (and the information concerning them) which are to enter into the design of a database.

2.2.1 Entity and Entity Set. Let e denote an entity which exists in our minds. Entities are classified into different *entity sets* such as EMPLOYEE, PROJECT, and DEPARTMENT. There is a predicate associated with each entity set to test whether an entity belongs to it. For example, if we know an entity is in the entity set EMPLOYEE, then we know that it has the properties common to the other entities in the entity set EMPLOYEE. Among these properties is the aforementioned test predicate. Let E_i denote entity sets. Note that entity sets may not be mutually disjoint. For example, an entity which belongs to the entity set MALE–PERSON also belongs to the entity set PERSON. In this case, MALE–PERSON is a subset of PERSON.

2.2.2 Relationship, Role, and Relationship Set. Consider associations among entities. A *relationship set*, R_i, is a mathematical relation [5] among n entities, each taken from an entity set:

$$\{[e_1, e_2, \ldots, e_n] \mid e_1 \in E_1, e_2 \in E_2, \ldots, e_n \in E_n\},$$

and each tuple of entities, $[e_1, e_2, \ldots, e_n]$, is a *relationship*. Note that the E_i in the above definition may not be distinct. For example, a "marriage" is a relationship between two entities in the entity set PERSON.

The *role* of an entity in a relationship is the function that it performs in the relationship. "Husband" and "wife" are roles. The ordering of entities in the definition of relationship (note that square brackets were used) can be dropped if roles of entities in the relationship are explicitly stated as follows: $(r_1/e_1, r_2/e_2, \ldots, r_n/e_n)$, where r_i is the role of e_i in the relationship.

2.2.3 Attribute, Value, and Value Set. The information about an entity or a relationship is obtained by observation or measurement, and is expressed by a set of attribute-value pairs. "3", "red", "Peter", and "Johnson" are values. Values are classified into different *value sets*, such as FEET, COLOR, FIRST-NAME, and LAST-NAME. There is a predicate associated with each value set to test whether a value belongs to it. A value in a value set may be equivalent to another value in a different value set. For example, "12" in value set INCH is equivalent to "1" in value set FEET.

An *attribute* can be formally defined as a function which maps from an entity set or a relationship set into a value set or a Cartesian product of value sets:

$$f: E_i \text{ or } R_i \rightarrow V_i \text{ or } V_{i_1} \times V_{i_2} \times \cdots \times V_{i_n}.$$

Figure 2 illustrates some attributes defined on entity set PERSON. The attribute AGE maps into value set NO-OF-YEARS. An attribute can map into a Cartesian product of value sets. For example, the attribute NAME maps into value sets FIRST-NAME, and LAST-NAME. Note that more than one attribute may map from the same entity set into the same value set (or same group of value sets). For example, NAME and ALTERNATIVE-NAME map from the entity set EMPLOYEE into value sets FIRST-NAME and LAST-NAME. Therefore, attribute and value set are different concepts although they may have the same name in some cases (for example, EMPLOYEE-NO maps from EMPLOYEE to value set EMPLOYEE-NO). This distinction is not clear in the network model and in many existing data management systems. Also note that an attribute is defined as a function. Therefore, it maps a given entity to a single value (or a single tuple of values in the case of a Cartesian product of value sets).

Note that relationships also have attributes. Consider the relationship set PROJECT-WORKER (Figure 3). The attribute PERCENTAGE-OF-TIME, which is the portion of time a particular employee is committed to a particular project, is an attribute defined on the relationship set PROJECT-WORKER. It is neither an attribute of EMPLOYEE nor an attribute of PROJECT, since its meaning depends on both the employee and project involved. The concept of attribute of relationship is important in understanding the semantics of data and in determining the functional dependencies among data.

2.2.4 Conceptual Information Structure. We are now concerned with how to organize the information associated with entities and relationships. The method proposed in this paper is to separate the information about entities from the infor-

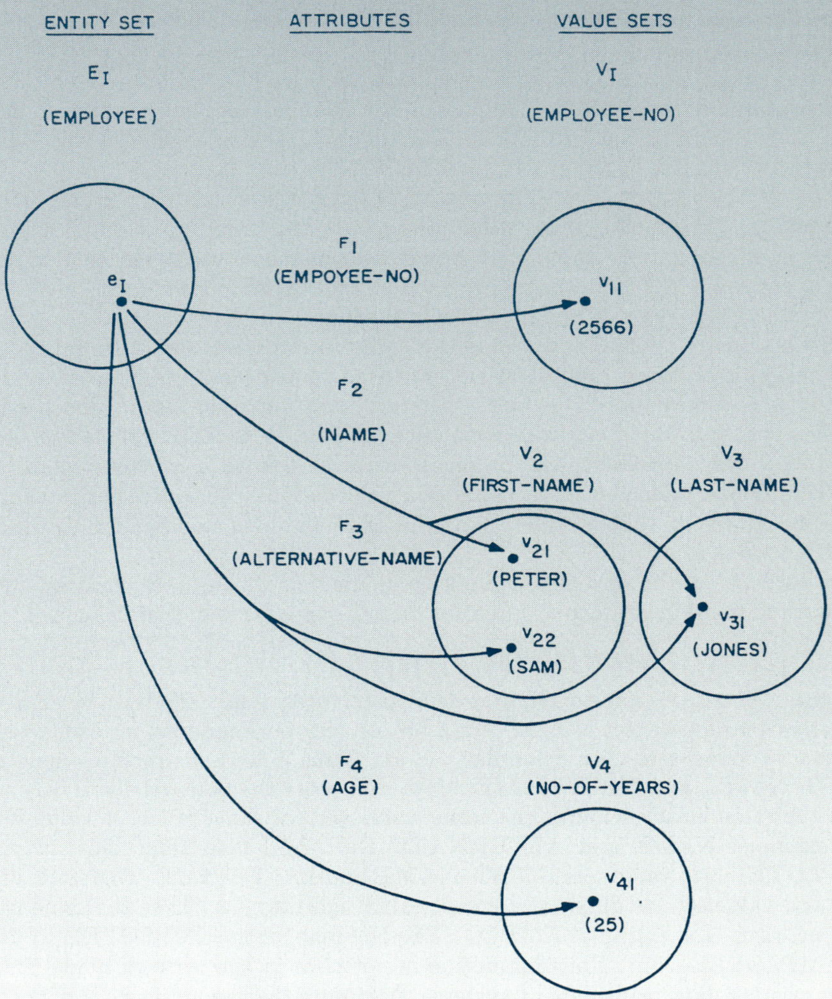

Fig. 2. Attributes defined on the entity set PERSON

mation about relationships. We shall see that this separation is useful in identifying functional dependencies among data.

Figure 4 illustrates in table form the information about entities in an entity set. Each row of values is related to the same entity, and each column is related to a value set which, in turn, is related to an attribute. The ordering of rows and columns is insignificant.

Figure 5 illustrates information about relationships in a relationship set. Note that each row of values is related to a relationship which is indicated by a group of entities, each having a specific role and belonging to a specific entity set.

Note that Figures 4 and 2 (and also Figures 5 and 3) are different forms of the same information. The table form is used for easily relating to the relational model.

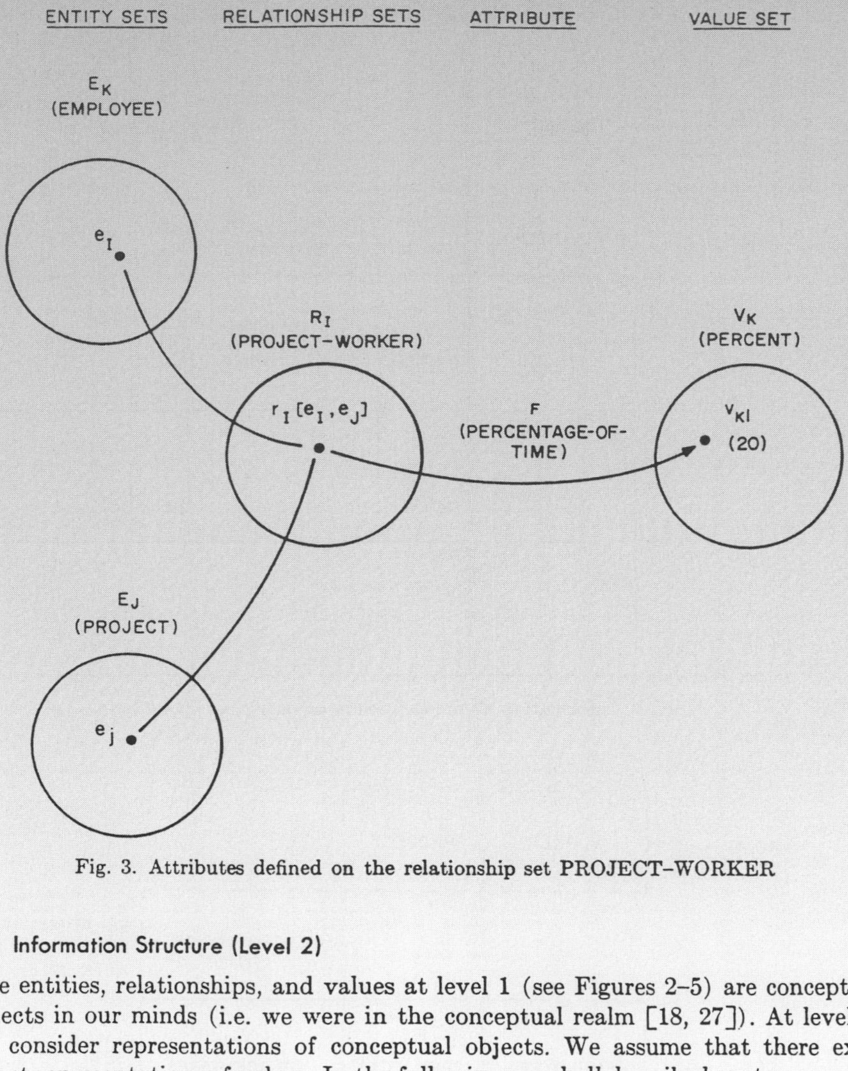

Fig. 3. Attributes defined on the relationship set PROJECT–WORKER

2.3 Information Structure (Level 2)

The entities, relationships, and values at level 1 (see Figures 2–5) are conceptual objects in our minds (i.e. we were in the conceptual realm [18, 27]). At level 2, we consider representations of conceptual objects. We assume that there exist direct representations of values. In the following, we shall describe how to represent entities and relationships.

2.3.1 Primary Key. In Figure 2 the values of attribute EMPLOYEE–NO can be used to identify entities in entity set EMPLOYEE if each employee has a different employee number. It is possible that more than one attribute is needed to identify the entities in an entity set. It is also possible that several groups of attributes may be used to identify entities. Basically, an *entity key* is a group of attributes such that the mapping from the entity set to the corresponding group of value sets is one-to-one. If we cannot find such one-to-one mapping on available data, or if simplicity in identifying entities is desired, we may define an artificial attribute and a value set so that such mapping is possible. In the case where

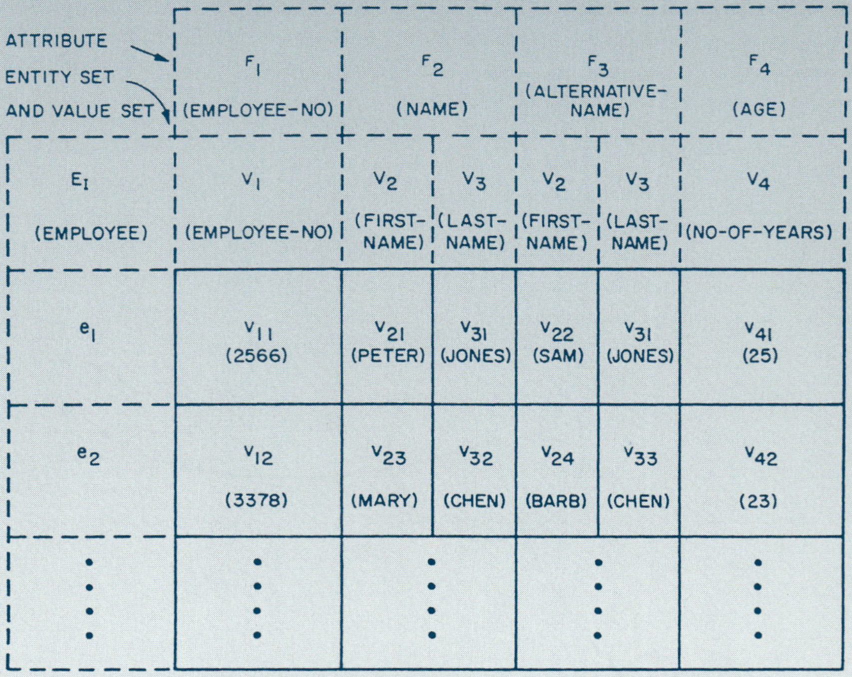

Fig. 4. Information about entities in an entity set (table form)

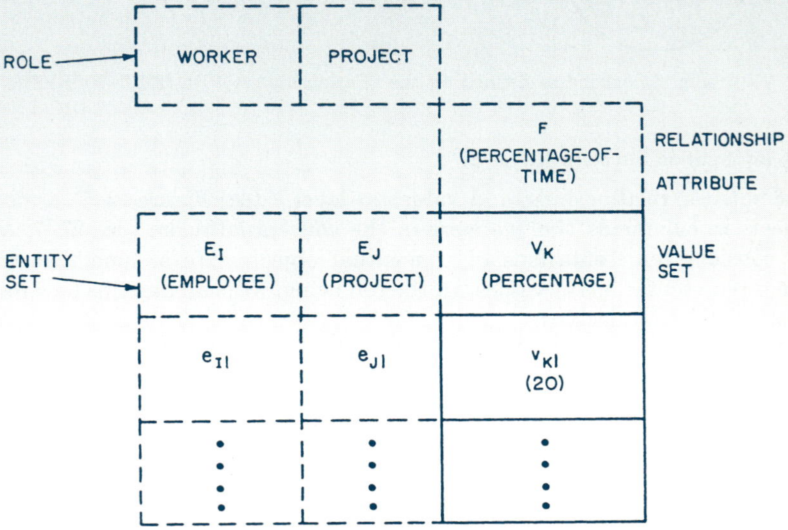

Fig. 5. Information about relationships in a relationship set (table form)

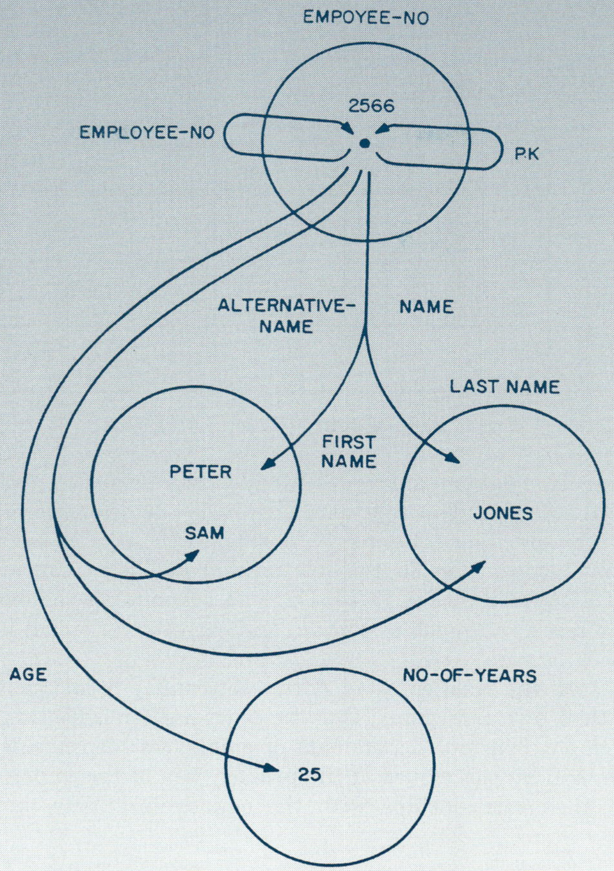

Fig. 6. Representing entities by values (employee numbers)

several keys exist, we usually choose a semantically meaningful key as the *entity primary key* (PK).

Figure 6 is obtained by merging the entity set EMPLOYEE with value set EMPLOYEE-NO in Figure 2. We should notice some semantic implications of Figure 6. Each value in the value set EMPLOYEE-NO represents an entity (employee). Attributes map from the value set EMPLOYEE-NO to other value sets. Also note that the attribute EMPLOYEE-NO maps from the value set EMPLOYEE-NO to itself.

2.3.2 Entity/Relationship Relations. Information about entities in an entity set can now be organized in a form shown in Figure 7. Note that Figure 7 is similar to Figure 4 except that entities are represented by the values of their primary keys. The whole table in Figure 7 is an *entity relation*, and each row is an *entity tuple*.

Since a relationship is identified by the involved entities, the *primary key of a relationship* can be represented by the primary keys of the involved entities. In

PRIMARY KEY: EMPLOYEE-NO; RELATIONAL VIEW: VALUE SET (DOMAIN) and ENTITY (TUPLE) rows

	EMPLOYEE-NO	NAME		ALTERNATIVE-NAME		AGE
ATTRIBUTE	EMPLOYEE-NO	NAME		ALTERNATIVE-NAME		AGE
VALUE SET (DOMAIN)	EMPLOYEE-NO	FIRST-NAME	LAST-NAME	FIRST-NAME	LAST-NAME	NO-OF-YEARS
ENTITY (TUPLE)	2566	PETER	JONES	SAM	JONES	25
	3378	MARY	CHEN	BARB	CHEN	23
	⋮	⋮	⋮	⋮	⋮	⋮

Fig. 7. Regular entity relation EMPLOYEE

Figure 8, the involved entities are represented by their primary keys EMPLOYEE–NO and PROJECT–NO. The role names provide the semantic meaning for the values in the corresponding columns. Note that EMPLOYEE–NO is the primary key for the involved entities in the relationship and is not an attribute of the relationship. PERCENTAGE–OF–TIME is an attribute of the relationship. The table in Figure 8 is a *relationship relation*, and each row of values is a *relationship tuple*.

In certain cases, the entities in an entity set cannot be uniquely identified by the values of their own attributes; thus we must use a relationship(s) to identify them. For example, consider dependents of employees: dependents are identified by their names and by the values of the primary key of the employees supporting them (i.e. by their relationships with the employees). Note that in Figure 9,

PRIMARY KEY: EMPLOYEE-NO, PROJECT-NO; RELATIONAL VIEW: ENTITY ATTRIBUTE through RELATIONSHIP TUPLE rows

ENTITY RELATION NAME	EMPLOYEE	PROJECT		
ROLE	WORKER	PROJECT		
ENTITY ATTRIBUTE	EMPLOYEE-NO	PROJECT-NO	PERCENTAGE-OF-TIME	RELATIONSHIP ATTRIBUTE
VALUE SET (DOMAIN)	EMPLOYEE-NO	PROJECT-NO	PERCENTAGE	
RELATIONSHIP TUPLE	2566	31	20	
	2173	25	100	
	⋮	⋮	⋮	

Fig. 8. Regular relationship relation PROJECT–WORKER

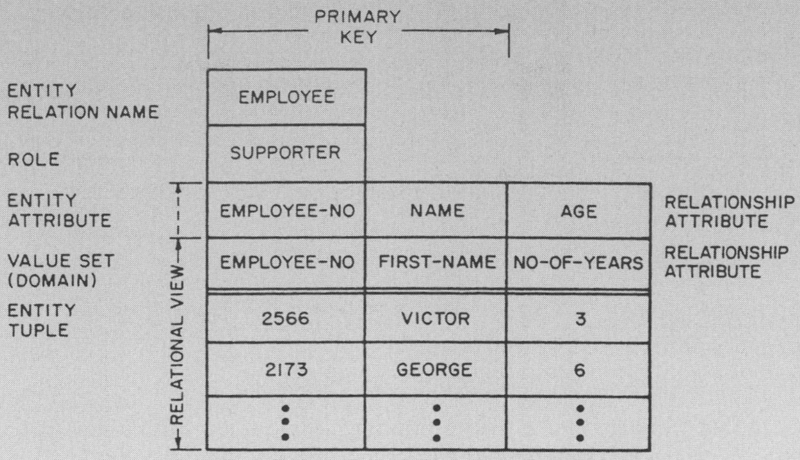

Fig. 9. A weak entity relation DEPENDENT

EMPLOYEE–NO is not an attribute of an entity in the set DEPENDENT but is the primary key of the employees who support dependents. Each row of values in Figure 9 is an entity tuple with EMPLOYEE–NO and NAME as its primary key. The whole table is an entity relation.

Theoretically, any kind of relationship may be used to identify entities. For simplicity, we shall restrict ourselves to the use of only one kind of relationship: the binary relationships with $1{:}n$ mapping in which the existence of the n entities on one side of the relationship depends on the existence of one entity on the other side of the relationship. For example, one employee may have n $(= 0, 1, 2, \ldots)$ dependents, and the existence of the dependents depends on the existence of the corresponding employee.

This method of identification of entities by relationships with other entities can be applied recursively until the entities which can be identified by their own attribute values are reached. For example, the primary key of a department in a company may consist of the department number and the primary key of the division, which in turn consists of the division number and the name of the company.

Therefore, we have two forms of entity relations. If relationships are used for identifying the entities, we shall call it a *weak entity relation* (Figure 9). If relationships are not used for identifying the entities, we shall call it a *regular entity relation* (Figure 7). Similarly, we also have two forms of relationship relations. If all entities in the relationship are identified by their own attribute values, we shall call it a *regular relationship relation* (Figure 8). If some entities in the relationship are identified by other relationships, we shall call it a *weak relationship relation*. For example, any relationships between DEPENDENT entities and other entities will result in weak relationship relations, since a DEPENDENT entity is identified by its name and its relationship with an EMPLOYEE entity. The distinction between regular (entity/relationship) relations and weak (entity/relationship) relations will be useful in maintaining data integrity.

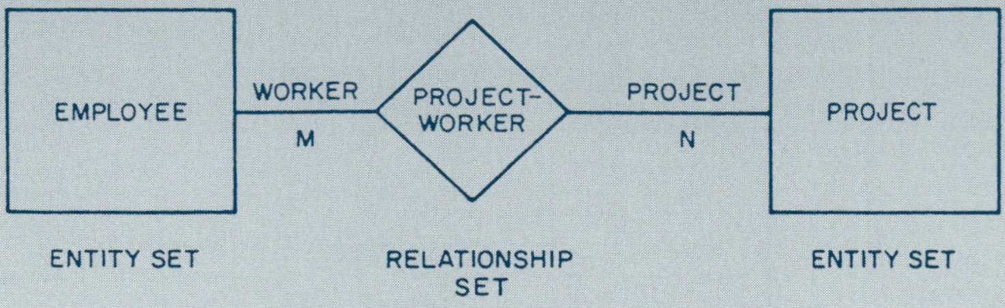

Fig. 10. A simple entity-relationship diagram

3. ENTITY-RELATIONSHIP DIAGRAM AND INCLUSION OF SEMANTICS IN DATA DESCRIPTION AND MANIPULATION

3.1 System Analysis Using the Entity-Relationship Diagram

In this section we introduce a diagrammatic technique for exhibiting entities and relationships: the entity-relationship diagram.

Figure 10 illustrates the relationship set PROJECT–WORKER and the entity sets EMPLOYEE and PROJECT using this diagrammatic technique. Each entity set is represented by a rectangular box, and each relationship set is represented by a diamond-shaped box. The fact that the relationship set PROJECT–WORKER is defined on the entity sets EMPLOYEE and PROJECT is represented by the lines connecting the rectangular boxes. The roles of the entities in the relationship are stated.

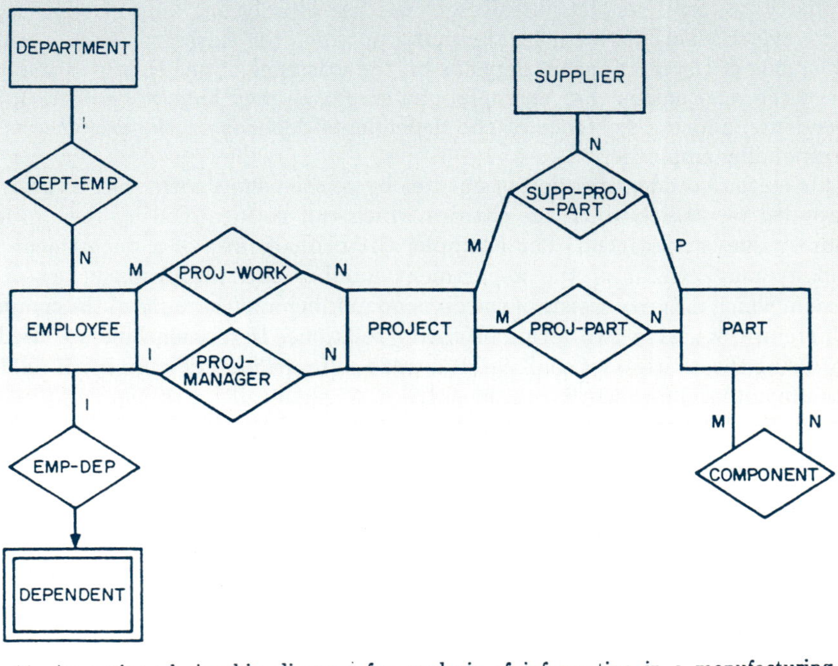

Fig. 11. An entity-relationship diagram for analysis of information in a manufacturing firm

Figure 11 illustrates a more complete diagram of some entity sets and relationship sets which might be of interest to a manufacturing company. DEPARTMENT, EMPLOYEE, DEPENDENT, PROJECT, SUPPLIER, and PART are entity sets. DEPARTMENT–EMPLOYEE, EMPLOYEE–DEPENDENT, PROJECT–WORKER, PROJECT–MANAGER, SUPPLIER–PROJECT–PART, PROJECT–PART, and COMPONENT are relationship sets. The COMPONENT relationship describes what subparts (and quantities) are needed in making superparts. The meaning of the other relationship sets need not be explained.

Several important characteristics about relationships in general can be found in Figure 11:

(1) A relationship set may be defined on more than two entity sets. For example, the SUPPLIER–PROJECT–PART relationship set is defined on three entity sets: SUPPLIER, PROJECT, and PART.

(2) A relationship set may be defined on only one entity set. For example, the relationship set COMPONENT is defined on one entity set, PART.

(3) There may be more than one relationship set defined on given entity sets. For example, the relationship sets PROJECT–WORKER and PROJECT–MANAGER are defined on the entity sets PROJECT and EMPLOYEE.

(4) The diagram can distinguish between $1{:}n$, $m{:}n$, and $1{:}1$ mappings. The relationship set DEPARTMENT–EMPLOYEE is a $1{:}n$ mapping, that is, one department may have n ($n = 0, 1, 2, \ldots$) employees and each employee works for only one department. The relationship set PROJECT–WORKER is an $m{:}n$ mapping, that is, each project may have zero, one, or more employees assigned to it and each employee may be assigned to zero, one, or more projects. It is also possible to express 1:1 mappings such as the relationship set MARRIAGE. Information about the number of entities in each entity set which is allowed in a relationship set is indicated by specifying "1", "m", "n" in the diagram. The relational model and the entity set model[2] do not include this type of information; the network model cannot express a 1:1 mapping easily.

(5) The diagram can express the *existence dependency* of one entity type on another. For example, the arrow in the relationship set EMPLOYEE–DEPENDENT indicates that existence of an entity in the entity set DEPENDENT depends on the corresponding entity in the entity set EMPLOYEE. That is, if an employee leaves the company, his dependents may no longer be of interest.

Note that the entity set DEPENDENT is shown as a special rectangular box. This indicates that at level 2 the information about entities in this set is organized as a weak entity relation (using the primary key of EMPLOYEE as a part of its primary key).

3.2 An Example of a Database Design and Description

There are four steps in designing a database using the entity-relationship model: (1) identify the entity sets and the relationship sets of interest; (2) identify semantic information in the relationship sets such as whether a certain relationship

---

[2] This mapping information is included in DIAM II [24].

set is an $1{:}n$ mapping; (3) define the value sets and attributes; (4) organize data into entity/relationship relations and decide primary keys.

Let us use the manufacturing company discussed in Section 3.1 as an example. The results of the first two steps of database design are expressed in an entity-relationship diagram as shown in Figure 11. The third step is to define value sets and attributes (see Figures 2 and 3). The fourth step is to decide the primary keys for the entities and the relationships and to organize data as entity/relationship relations. Note that each entity/relationship set in Figure 11 has a corresponding entity/relationship relation. We shall use the names of the entity sets (at level 1) as the names of the corresponding entity/relationship relations (at level 2) as long as no confusion will result.

At the end of the section, we illustrate a schema (data definition) for a small part of the database in the above manufacturing company example (the syntax of the data definition is not important). Note that value sets are defined with specifications of representations and allowable values. For example, values in EMPLOYEE-NO are represented as 4-digit integers and range from 0 to 2000. We then declare three entity relations: EMPLOYEE, PROJECT, and DEPENDENT. The attributes and value sets defined on the entity sets as well as the primary keys are stated. DEPENDENT is a weak entity relation since it uses EMPLOYEE.PK as part of its primary key. We also declare two relationship relations: PROJECT-WORKER and EMPLOYEE-DEPENDENT. The roles and involved entities in the relationships are specified. We use EMPLOYEE.PK to indicate the name of the entity relation (EMPLOYEE) and whatever attribute-value-set pairs are used as the primary keys in that entity relation. The maximum number of entities from an entity set in a relation is stated. For example, PROJECT-WORKER is an $m{:}n$ mapping. We may specify the values of m and n. We may also specify the minimum number of entities in addition to the maximum number. EMPLOYEE-DEPENDENT is a weak relationship relation since one of the related entity relations, DEPENDENT, is a weak entity relation. Note that the existence dependence of the dependents on the supporter is also stated.

```
DECLARE    VALUE-SETS     REPRESENTATION    ALLOWABLE-VALUES
           EMPLOYEE-NO    INTEGER (4)       (0,2000)
           FIRST-NAME     CHARACTER (8)     ALL
           LAST-NAME      CHARACTER (10)    ALL
           NO-OF-YEARS    INTEGER (3)       (0,100)
           PROJECT-NO     INTEGER (3)       (1,500)
           PERCENTAGE     FIXED (5.2)       (0,100.00)

DECLARE    REGULAR ENTITY RELATION EMPLOYEE
           ATTRIBUTE/VALUE-SET:
               EMPLOYEE-NO/EMPLOYEE-NO
               NAME/(FIRST-NAME, LAST-NAME)
               ALTERNATIVE-NAME/(FIRST-NAME,LAST-NAME)
               AGE/NO-OF-YEARS
           PRIMARY KEY:
               EMPLOYEE-NO
```

```
DECLARE    REGULAR ENTITY RELATION PROJECT
           ATTRIBUTE/VALUE-SET:
                PROJECT-NO/PROJECT-NO
           PRIMARY KEY:
                PROJECT-NO

DECLARE    REGULAR RELATIONSHIP RELATION PROJECT-WORKER
           ROLE/ENTITY-RELATION.PK/MAX-NO-OF-ENTITIES
                WORKER/EMPLOYEE.PK/m
                PROJECT/PROJECT.PK/n        (m:n mapping)
           ATTRIBUTE/VALUE-SET:
                PERCENTAGE-OF-TIME/PERCENTAGE

DECLARE    WEAK RELATIONSHIP RELATION EMPLOYEE-DEPENDENT
           ROLE/ENTITY-RELATION.PK/MAX-NO-OF-ENTITIES
                SUPPORTER/EMPLOYEE.PK/1
                DEPENDENT/DEPENDENT.PK/n
           EXISTENCE OF DEPENDENT DEPENDS ON
           EXISTENCE OF SUPPORTER

DECLARE    WEAK ENTITY RELATION DEPENDENT
           ATTRIBUTE/VALUE-SET:
                NAME/FIRST-NAME
                AGE/NO-OF-YEARS
           PRIMARY KEY:
                NAME
                EMPLOYEE.PK THROUGH EMPLOYEE-DEPENDENT
```

3.3 Implications on Data Integrity

Some work has been done on data integrity for other models [8, 14, 16, 28]. With explicit concepts of entity and relationship, the entity-relationship model will be useful in understanding and specifying constraints for maintaining data integrity. For example, there are three major kinds of constraints on values:

(1) Constraints on *allowable values* for a value set. This point was discussed in defining the schema in Section 3.2.

(2) Constraints on *permitted* values for a certain attribute. In some cases, not all allowable values in a value set are permitted for some attributes. For example, we may have a restriction of ages of employees to between 20 and 65. That is,

$$\text{AGE}(e) \in (20,65), \text{ where } e \in \text{EMPLOYEE}.$$

Note that we use the level 1 notations to clarify the semantics. Since each entity/relationship set has a corresponding entity/relationship relation, the above expression can be easily translated into level 2 notations.

(3) Constraints on *existing values* in the database. There are two types of constraints:

(i) Constraints between sets of existing values. For example,

$$\{\text{NAME}(e) \mid e \in \text{MALE-PERSON}\} \subseteq \{\text{NAME}(e) \mid e \in \text{PERSON}\}.$$

(ii) Constraints between particular values. For example,

$$\text{TAX}(e) \leq \text{SALARY}(e),\ e \in \text{EMPLOYEE}$$

or

$$\text{BUDGET}(e_i) = \sum \text{BUDGET}(e_j), \text{ where } e_i \in \text{COMPANY}$$
$$e_j \in \text{DEPARTMENT}$$
$$\text{and } [e_i, e_j] \in \text{COMPANY–DEPARTMENT}.$$

3.4 Semantics and Set Operations of Information Retrieval Requests

The semantics of information retrieval requests become very clear if the requests are based on the entity-relationship model of data. For clarity, we first discuss the situation at level 1. Conceptually, the information elements are organized as in Figures 4 and 5 (on Figures 2 and 3). Many information retrieval requests can be considered as a combination of the following basic types of operations:

(1) Selection of a subset of values from a value set.

(2) Selection of a subset of entities from an entity set (i.e. selection of certain rows in Figure 4). Entities are selected by stating the values of certain attributes (i.e. subsets of value sets) and/or their relationships with other entities.

(3) Selection of a subset of relationships from a relationship set (i.e. selection of certain rows in Figure 5). Relationships are selected by stating the values of certain attribute(s) and/or by identifying certain entities in the relationship.

(4) Selection of a subset of attributes (i.e. selection of columns in Figures 4 and 5).

An information retrieval request like "What are the ages of the employees whose weights are greater than 170 and who are assigned to the project with PROJECT-NO 254?" can be expressed as:

$$\{\text{AGE}(e) \mid e \in \text{EMPLOYEE},\ \text{WEIGHT}(e) > 170,$$
$$[e, e_j] \in \text{PROJECT–WORKER},\ e_j \in \text{PROJECT},$$
$$\text{PROJECT–NO}(e_j) = 254\};$$

or

{AGE(EMPLOYEE) | WEIGHT(EMPLOYEE) > 170,
[EMPLOYEE,PROJECT] ∈ PROJECT–WORKER,
PROJECT–NO(EMPLOYEE) = 254}.

To retrieve information as organized in Figure 6 at level 2, "entities" and "relationships" in (2) and (3) should be replaced by "entity PK" and "relationship PK." The above information retrieval request can be expressed as:

{AGE(EMPLOYEE.PK) | WEIGHT(EMPLOYEE.PK) > 170
(WORKER/EMPLOYEE.PK,PROJECT/PROJECT.PK) ∈ {PROJECT–WORKER.PK},
PROJECT–NO (PROJECT.PK) = 254}.

To retrieve information as organized in entity/relationship relations (Figures 7, 8, and 9), we can express it in a SEQUEL-like language [6]:

```
SELECT    AGE
FROM      EMPLOYEE
WHERE     WEIGHT > 170
```

Table I. Insertion

level 1	level 2
operation: insert an entity to an entity set	*operation:* create an entity tuple with a certain entity-PK *check:* whether PK already exists or is acceptable
operation: insert a relationship in a relationship set *check:* whether the entities exist	*operation:* create a relationship tuple with given entity PKs *check:* whether the entity PKs exist
operation: insert properties of an entity or a relationship *check:* whether the value is acceptable	*operation:* insert values in an entity tuple or a relationship tuple *check:* whether the values are acceptable

```
AND         EMPLOYEE.PK =
            SELECT    WORKER/EMPLOYEE.PK
            FROM      PROJECT-WORKER
            WHERE     PROJECT-NO = 254.
```

It is possible to retrieve information about entities in two different entity sets without specifying a relationship between them. For example, an information retrieval request like "List the names of employees and ships which have the same

Table II. Updating

level 1	level 2
operation: • change the value of an entity attribute	*operation:* • update a value *consequence:* • if it is not part of an entity PK, no consequence • if it is part of an entity PK, •• change the entity PKs in all related relationship relations •• change PKs of other entities which use this value as part of their PKs (for example, DEPENDENTS' PKs use EMPLOYEE'S PK)
operation: • change the value of a relationship attribute	*operation:* • update a value (note that a relationship attribute will not be a relationship PK)

Table III. Deletion

level 1	level 2
operation: • delete an entity *consequences:* • delete any entity whose existence depends on this entity • delete relationships involving this entity • delete all related properties	*operation:* • delete an entity tuple *consequences* (applied recursively): • delete any entity tuple whose existence depends on this entity tuple • delete relationship tuples associated with this entity
operation: • delete a relationship *consequences:* • delete all related properties	*operation:* • delete a relationship tuple

age" can be expressed in the level 1 notation as:

$$\{(\text{NAME}(e_i),\text{NAME}(e_j)) \mid e_i \in \text{EMPLOYEE}, e_j \in \text{SHIP}, \text{AGE}(e_i) = \text{AGE}(e_j)\}.$$

We do not further discuss the language syntax here. What we wish to stress is that information requests may be expressed using set notions and set operations [17], and the request semantics are very clear in adopting this point of view.

3.5 Semantics and Rules for Insertion, Deletion, and Updating

It is always a difficult problem to maintain data consistency following insertion, deletion, and updating of data in the database. One of the major reasons is that the semantics and consequences of insertion, deletion, and updating operations usually are not clearly defined; thus it is difficult to find a set of rules which can enforce data consistency. We shall see that this data consistency problem becomes simpler using the entity-relationship model.

In Tables I–III, we discuss the semantics and rules[3] for insertion, deletion, and updating at both level 1 and level 2. Level 1 is used to clarify the semantics.

4. ANALYSIS OF OTHER DATA MODELS AND THEIR DERIVATION FROM THE ENTITY-RELATIONSHIP MODEL

4.1 The Relational Model

4.1.1 The Relational View of Data and Ambiguity in Semantics. In the relational model, *relation*, R, is a mathematical relation defined on sets $X_1, X_2, \ldots, X_n$:

$$R = \{(x_1, x_2, \ldots, x_n) \mid x_1 \in X_1, x_2 \in X_2, \ldots, x_n \in X_n\}.$$

The sets $X_1, X_2, \ldots, X_n$ are called *domains*, and $(x_1, x_2, \ldots, x_n)$ is called a *tuple*. Figure 12 illustrates a relation called EMPLOYEE. The domains in the relation

[3] Our main purpose is to illustrate the semantics of data manipulation operations. Therefore, these rules may not be complete. Note that the consequence of operations stated in the tables can be performed by the system instead of by the users.

ROLE		LEGAL	LEGAL	ALTERNATIVE	ALTERNATIVE	
DOMAIN	EMPLOYEE-NO	FIRST-NAME	LAST-NAME	FIRST-NAME	LAST-NAME	NO-OF-YEARS
TUPLE	2566	PETER	JONES	SAM	JONES	25
	3378	MARY	CHEN	BARB	CHEN	23

Fig. 12. Relation EMPLOYEE

are EMPLOYEE-NO, FIRST-NAME, LAST-NAME, FIRST-NAME, LAST-NAME, NO-OF-YEAR. The ordering of rows and columns in the relation has no significance. To avoid ambiguity of columns with the same domain in a relation, domain names are qualified by *roles* (to distinguish the role of the domain in the relation). For example, in relation EMPLOYEE, domains FIRST-NAME and LAST-NAME may be qualified by roles LEGAL or ALTERNATIVE. An *attribute name* in the relational model is a domain name concatenated with a role name [10]. Comparing Figure 12 with Figure 7, we can see that "domains" are basically equivalent to value sets. Although "role" or "attribute" in the relational model seems to serve the same purpose as "attribute" in the entity-relationship model, the semantics of these terms are different. The "role" or "attribute" in the relational model is mainly used to distinguish domains with the same name in the same relation, while "attribute" in the entity-relationship model is a function which maps from an entity (or relationship) set into value set(s).

Using relational operators in the relational model may cause semantic ambiguities. For example, the join of the relation EMPLOYEE with the relation EMPLOYEE-PROJECT (Figure 13) on domain EMPLOYEE-NO produces the

PROJECT-NO	EMPLOYEE-NO
7	2566
3	2566
7	3378

Fig. 13. Relation EMPLOYEE-PROJECT

		LEGAL	LEGAL	ALTERNATIVE	ALTERNATIVE	
PROJECT-NO	EMPLOYEE-NO	FIRST-NAME	LAST-NAME	FIRST-NAME	LAST-NAME	NO-OF-YEARS
7	2566	PETER	JONES	SAM	JONES	25
3	2566	PETER	JONES	SAM	JONES	25
7	3378	MARY	CHEN	BARB	CHEN	23

Fig. 14. Relation EMPLOYEE–PROJECT′ as a "join" of relations EMPLOYEE and EMPLOYEE–PROJECT

relation EMPLOYEE–PROJECT′ (Figure 14). But what is the meaning of a join between the relation EMPLOYEE with the relation SHIP on the domain NO–OF–YEARS (Figure 15)? The problem is that the same domain name may have different semantics in different relations (note that a role is intended to distinguish domains in a given relation, not in all relations). If the domain NO–OF–YEAR of the relation EMPLOYEE is not allowed to be compared with the domain NO–OF–YEAR of the relation SHIP, different domain names have to be declared. But if such a comparison is acceptable, can the database system warn the user?

In the entity-relationship model, the semantics of data are much more apparent. For example, one column in the example stated above contains the values of AGE of EMPLOYEE and the other column contains the values of AGE of SHIP. If this semantic information is exposed to the user, he may operate more cautiously (refer to the sample information retrieval requests stated in Section 3.4). Since the database system contains the semantic information, it should be able to warn the user of the potential problems for a proposed "join-like" operation.

4.1.2 Semantics of Functional Dependencies Among Data. In the relational model, "attribute" B of a relation is *functionally dependent* on "attribute" A of the same relation if each value of A has no more than one value of B associated with it in the relation. Semantics of functional dependencies among data become clear

SHIP-NO	NAME	NO-OF-YEARS
037	MISSOURI	25
056	VIRGINIA	10

Fig. 15. Relation SHIP

in the entity-relationship model. Basically, there are two major types of functional dependencies:

(1) Functional dependencies related to description of entities or relationships. Since an attribute is defined as a function, it maps an entity in an entity set to a single value in a value set (see Figure 2). At level 2, the values of the primary key are used to represent entities. Therefore, nonkey value sets (domains) are functionally dependent on primary-key value sets (for example, in Figures 6 and 7, NO-OF-YEARS is functionally dependent on EMPLOYEE-NO). Since a relation may have several keys, the nonkey value sets will functionally depend on any key value set. The key value sets will be mutually functionally dependent on each other. Similarly, in a relationship relation the nonkey value sets will be functionally dependent on the prime-key value sets (for example, in Figure 8, PERCENTAGE is functionally dependent on EMPLOYEE-NO and PROJECT-NO).

(2) Functional dependencies related to entities in a relationship. Note that in Figure 11 we identify the types of mappings ($1{:}n$, $m{:}n$, etc.) for relationship sets. For example, PROJECT-MANAGER is a $1{:}n$ mapping. Let us assume that PROJECT-NO is the primary key in the entity relation PROJECT. In the relationship relation PROJECT-MANAGER, the value set EMPLOYEE-NO will be functionally dependent on the value set PROJECT-NO (i.e. each project has only one manager).

The distinction between level 1 (Figure 2) and level 2 (Figures 6 and 7) and the separation of entity relation (Figure 7) from relationship relation (Figure 8) clarifies the semantics of functional dependencies among data.

4.1.3 3NF Relations Versus Entity-Relationship Relations. From the definition of "relation," any grouping of domains can be considered to be a relation. To avoid undesirable properties in maintaining relations, a normalization process is proposed to transform arbitrary relations into the first normal form, then into the second normal form, and finally into the third normal form (3NF) [9, 11]. We shall show that the entity and relationship relations in the entity-relationship model are similar to 3NF relations but with clearer semantics and without using the transformation operation.

Let us use a simplified version of an example of normalization described in [9]. The following three relations are in first normal form (that is, there is no domain whose elements are themselves relations):

EMPLOYEE (<u>EMPLOYEE-NO</u>)
PART (<u>PART-NO</u>, PART-DESCRIPTION, QUANTITY-ON-HAND)
PART-PROJECT (<u>PART-NO</u>, <u>PROJECT-NO</u>, PROJECT-DESCRIPTION,
PROJECT-MANAGER-NO, QUANTITY-COMMITTED).

Note that the domain PROJECT-MANAGER-NO actually contains the EMPLOYEE-NO of the project manager. In the relations above, primary keys are underlined.

Certain rules are applied to transform the relations above into third normal form:

EMPLOYEE' (<u>EMPLOYEE-NO</u>)
PART' (<u>PART-NO</u>, PART-DESCRIPTION, QUANTITY-ON-HAND)

PROJECT' (PROJECT-NO, PROJECT-DESCRIPTION, PROJECT-MANAGER-NO)
PART-PROJECT' (PART-NO, PROJECT-NO, QUANTITY-COMMITTED).

Using the entity-relationship diagram in Figure 11, the following entity and relationship relations can be easily derived:

entity relations	PART''(PART-NO, PART-DESCRIPTION, QUANTITY-ON-HAND) PROJECT''(PROJECT-NO, PROJECT-DESCRIPTION) EMPLOYEE''(EMPLOYEE-NO)
relationship relations	PART-PROJECT''(PART/PART-NO, PROJECT/PROJECT-NO, QUANTITY-COMMITTED) PROJECT-MANAGER''(PROJECT/PROJECT-NO, MANAGER/EMPLOYEE-NO).

The role names of the entities in relationships (such as MANAGER) are indicated. The entity relation names associated with the PKs of entities in relationships and the value set names have been omitted.

Note that in the example above, entity/relationship relations are similar to the 3NF relations. In the 3NF approach, PROJECT–MANAGER–NO is included in the relation PROJECT' since PROJECT–MANAGER–NO is assumed to be functionally dependent on PROJECT–NO. In the entity-relationship model, PROJECT–MANAGER–NO (i.e. EMPLOYEE–NO of a project manager) is included in a relationship relation PROJECT–MANAGER since EMPLOYEE–NO is considered as an entity PK in this case.

Also note that in the 3NF approach, changes in functional dependencies of data may cause some relations not to be in 3NF. For example, if we make a new assumption that one project may have more than one manager, the relation PROJECT' is no longer a 3NF relation and has to be split into two relations as PROJECT'' and PROJECT–MANAGER''. Using the entity-relationship model, no such change is necessary. Therefore, we may say that by using the entity-relationship model we can arrange data in a form similar to 3NF relations but with clear semantic meaning.

It is interesting to note that the decomposition (or transformation) approach described above for normalization of relations may be viewed as a bottom-up approach in database design.[4] It starts with arbitrary relations (level 3 in Figure 1) and then uses some semantic information (functional dependencies of data) to transform them into 3NF relations (level 2 in Figure 1). The entity-relationship model adopts a top-down approach, utilizing the semantic information to organize data in entity/relationship relations.

4.2 The Network Model

4.2.1 Semantics of the Data-Structure Diagram. One of the best ways to explain the network model is by use of the *data-structure diagram* [3]. Figure 16(a) illustrates a data-structure diagram. Each rectangular box represents a record type.

[4] Although the decomposition approach was emphasized in the relational model literature, it is a procedure to obtain 3NF and may not be an intrinsic property of 3NF.

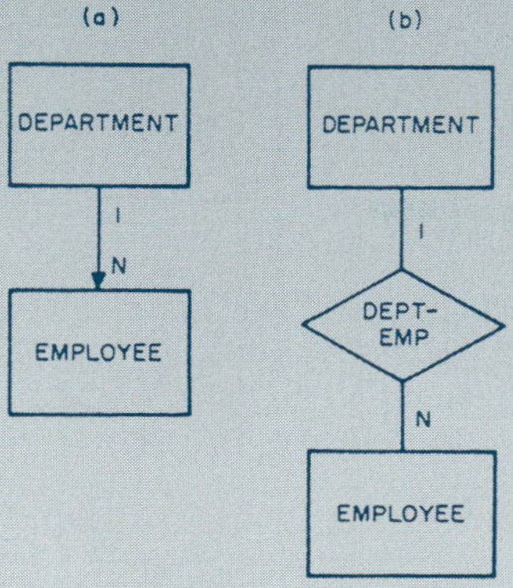

Fig. 16. Relationship DEPARTMENT–EMPLOYEE
(a) data structure diagram
(b) entity-relationship diagram

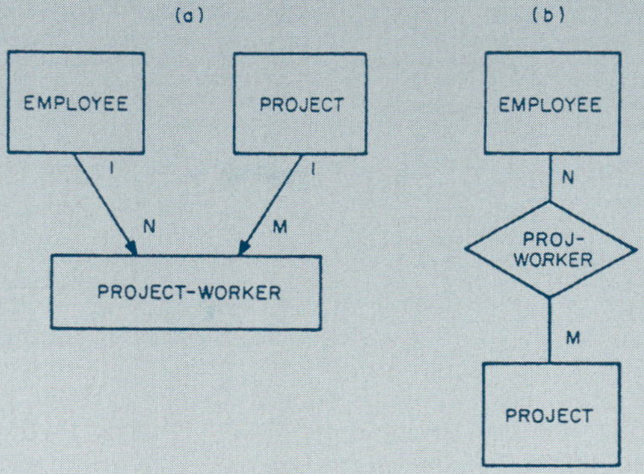

Fig. 17. Relationship PROJECT–WORKER
(a) data structure diagram
(b) entity–relationship diagram

The arrow represents a data-structure-set in which the DEPARTMENT record is the *owner-record*, and one owner-record may own n ($n = 0, 1, 2, \ldots$) *member-records*. Figure 16(b) illustrates the corresponding entity-relationship diagram. One might conclude that the arrow in the data-structure diagram represents a relationship between entities in two entity sets. This is not always true. Figures 17(a) and 17(b) are the data-structure diagram and the entity-relationship diagram expressing the relationship PROJECT–WORKER between two entity types EMPLOYEE and PROJECT. We can see in Figure 17(a) that the relationship PROJECT–WORKER becomes another record type and that the arrows no longer represent relationships between entities. What are the real meanings of tbe arrows in data-structure diagrams? The answer is that an arrow represents an $1{:}n$ relationship between two *record* (not entity) types and also implies the existence of an access path from the owner record to the member records. The data-structure diagram is a representation of the organization of records (level 4 in Figure 1) and is not an exact representation of entities and relationships.

4.2.2 Deriving the Data-Structure Diagram. Under what conditions does an arrow in a data-structure diagram correspond to a relationship of entities? A close comparison of the data-structure diagrams with the corresponding entity-relationship diagrams reveals the following rules:

1. For $1{:}n$ binary relationships an arrow is used to represent the relationship (see Figure 16(a)).

2. For $m{:}n$ binary relationships a "relationship record" type is created to represent the relationship and arrows are drawn from the "entity record" type to the "relationship record" type (see Figure 17(a)).

3. For k-ary ($k \geq 3$) relationships, the same rule as (2) applies (i.e. creating a "relationship record" type).

Since DBTG [7] does not allow a data-structure-set to be defined on a single record type (i.e. Figure 18 is not allowed although it has been implemented in [13]), a "relationship record" is needed to implement such relationships (see

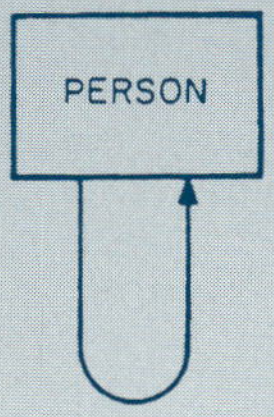

Fig. 18. Data-structure-set defined on the same record type

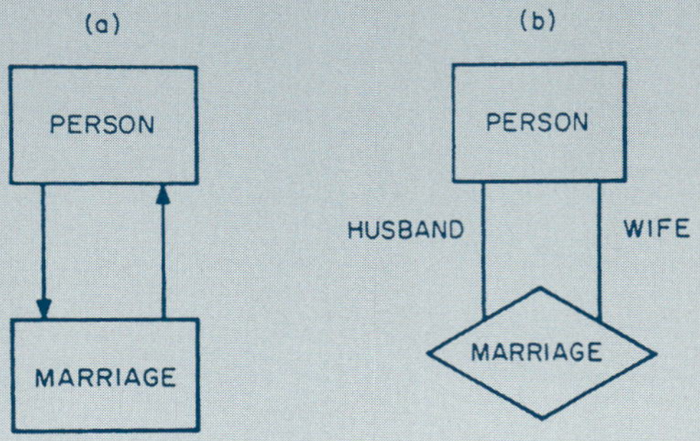

Fig. 19. Relationship MARRIAGE (a) data structure diagram (b) entity–relationship diagram

Figure 19(a)) [20]. The corresponding entity-relationship diagram is shown in Figure 19(b).

It is clear now that arrows in a data-structure diagram do not always represent relationships of entities. Even in the case that an arrow represents a $1{:}n$ relationship, the arrow only represents a unidirectional relationship [20] (although it is possible to find the owner-record from a member-record). In the entity-relationship model, both directions of the relationship are represented (the roles of both entities are specified). Besides the semantic ambiguity in its arrows, the network model is awkward in handling changes in semantics. For example, if the relationship between DEPARTMENT and EMPLOYEE changes from a $1{:}n$ mapping to an $m{:}n$ mapping (i.e. one employee may belong to several departments), we must create a relationship record DEPARTMENT–EMPLOYEE in the network model.

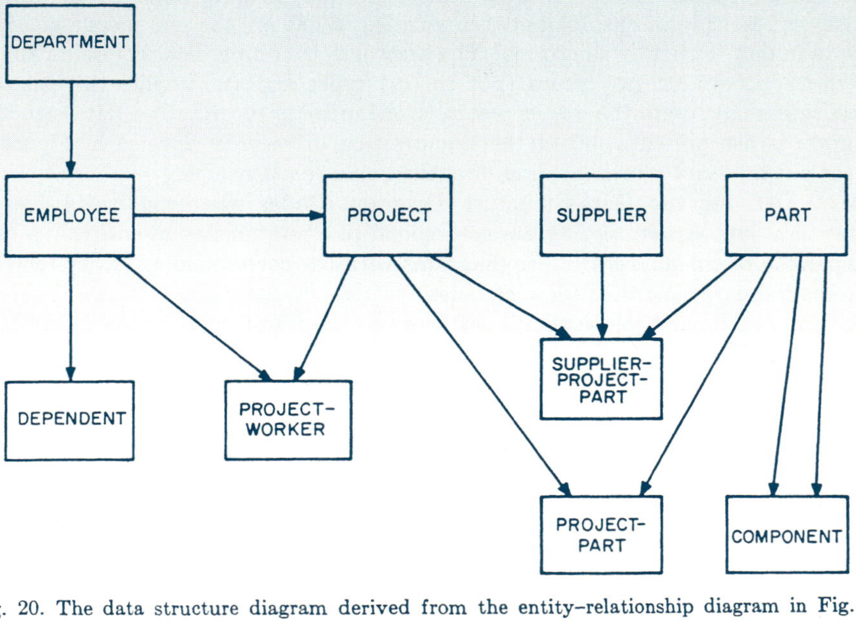

Fig. 20. The data structure diagram derived from the entity–relationship diagram in Fig. 11

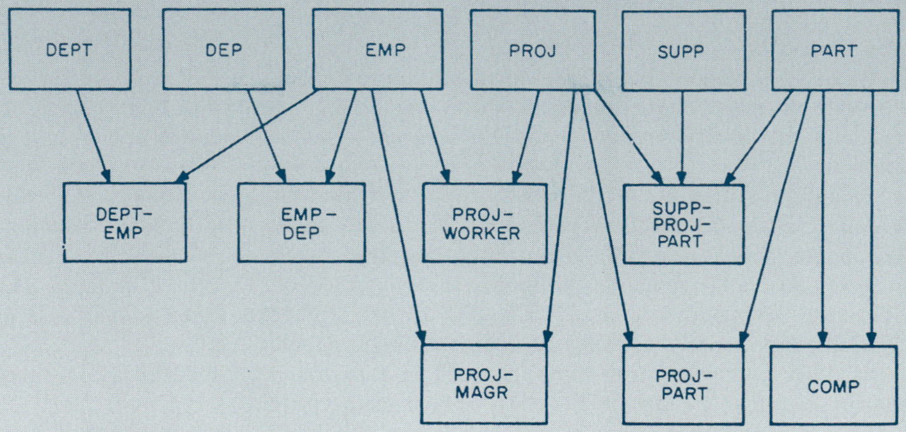

Fig. 21. The "disciplined" data structure diagram derived from the entity–relationship diagram in Fig. 11

In the entity-relationship model, all kinds of mappings in relationships are handled uniformly.

The entity-relationship model can be used as a tool in the structured design of databases using the network model. The user first draws an entity-relationship diagram (Figure 11). He may simply translate it into a data-structure diagram (Figure 20) using the rules specified above. He may also follow a discipline that every entity or relationship must be mapped onto a record (that is, "relationship records" are created for all types of relationships no matter that they are $1{:}n$ or $m{:}n$ mappings). Thus, in Figure 11, all one needs to do is to change the diamonds to boxes and to add arrowheads on the appropriate lines. Using this approach three more boxes—DEPARTMENT-EMPLOYEE, EMPLOYEE-DEPENDENT, and PROJECT-MANAGER—will be added to Figure 20 (see Figure 21). The validity constraints discussed in Sections 3.3–3.5 will also be useful.

4.3 The Entity Set Model

4.3.1 The Entity Set View. The basic element of the entity set model is the entity. Entities have names (*entity names*) such as "Peter Jones", "blue", or "22". Entity names having some properties in common are collected into an *entity-name-set*, which is referenced by the *entity-name-set-name* such as "NAME", "COLOR", and "QUANTITY".

An entity is represented by the entity-name-set-name/entity-name pair such as NAME/Peter Jones, EMPLOYEE–NO/2566, and NO–OF–YEARS/20. An entity is described by its association with other entities. Figure 22 illustrates the entity set view of data. The "DEPARTMENT" of entity EMPLOYEE–NO/2566 is the entity DEPARTMENT–NO/405. In other words, "DEPARTMENT" is the role that the entity DEPARTMENT–NO/405 plays to describe the entity EMPLOYEE–NO/2566. Similarly, the "NAME", "ALTERNATIVE–NAME", or "AGE" of EMPLOYEE–NO/2566 is "NAME/Peter Jones", "NAME/Sam Jones", or "NO–OF–YEARS/20", respectively. The description of the entity EMPLOYEE–

NO/2566 is a collection of the related entities and their roles (the entities and roles circled by the dotted line). An example of the *entity description* of "EMPLOYEE–NO/2566" (in its full-blown, unfactored form) is illustrated by the set of role-name/entity-name-set-name/entity-name triplets shown in Figure 23. Conceptually, the entity set model differs from the entity-relationship model in the following ways:

(1) In the entity set model, everything is treated as an entity. For example, "COLOR/BLACK" and "NO–OF–YEARS/45" are entities. In the entity-relationship model, "blue" and "36" are usually treated as values. Note treating values as entities may cause semantic problems. For example, in Figure 22, what is the difference between "EMPLOYEE–NO/2566", "NAME/Peter Jones", and "NAME/Sam Jones"? Do they represent different entities?

(2) Only binary relationships are used in the entity set model,[5] while n-ary relationships may be used in the entity-relationship model.

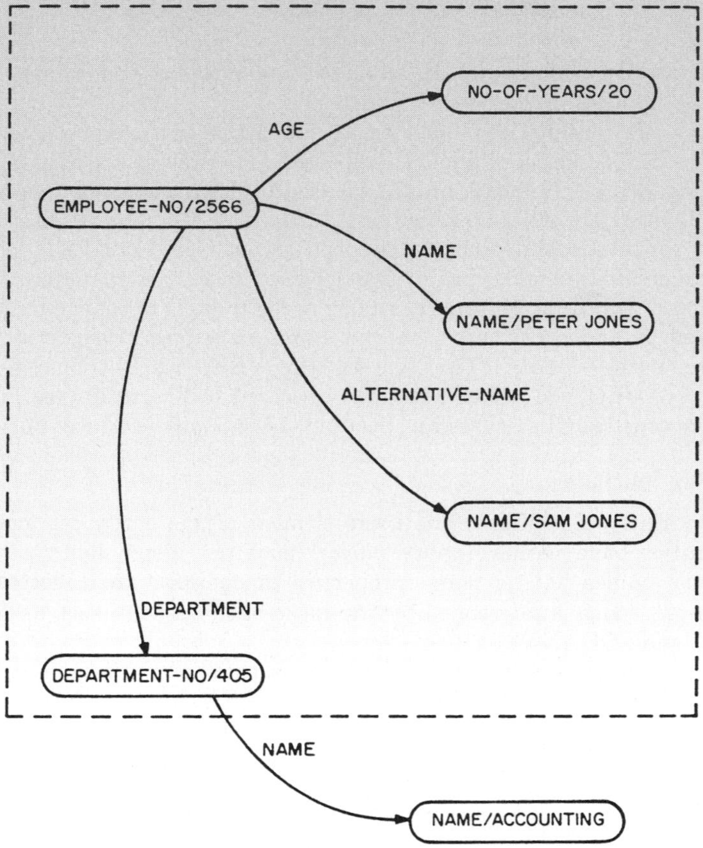

Fig. 22. The entity-set view

[5] In DIAM II [24], n-ary relationships may be treated as special cases of identifiers.

THE ENTITY-RELATIONSHIP MODEL TERMINOLOGY	ATTRIBUTE OR ROLE	VALUE SET	VALUE
THE ENTITY SET MODEL TERMINOLOGY	"ROLE-NAME"	"ENTITY-NAME-SET-NAME"	"ENTITY-NAME"
	IDENTIFIER	EMPLOYEE-NO	2566
	NAME	NAME	PETER JONES
	NAME	NAME	SAM JONES
	AGE	NO-OF-YEARS	25
	DEPARTMENT	DEPARTMENT-NO	405

Fig. 23. An "entity description" in the entity-set model

4.3.2 Deriving the Entity Set View. One of the main difficulties in understanding the entity set model is due to its world view (i.e. identifying values with entities). The entity-relationship model proposed in this paper is useful in understanding and deriving the entity set view of data. Consider Figures 2 and 6. In Figure 2, entities are represented by e_i's (which exist in our minds or are pointed at with fingers). In Figure 6, entities are represented by values. The entity set model works both at level 1 and level 2, but we shall explain its view at level 2 (Figure 6). The entity set model treats all value sets such as NO-OF-YEARS as "entity-name-sets" and all values as "entity-names." The attributes become role names in the entity set model. For binary relationships, the translation is simple: the role of an entity in a relationship (for example, the role of "DEPARTMENT" in the relationship DEPARTMENT-EMPLOYEE) becomes the role name of the entity in describing the other entity in the relationship (see Figure 22). For n-ary ($n > 2$) relationships, we must create artificial entities for relationships in order to handle them in a binary relationship world.

ACKNOWLEDGMENTS

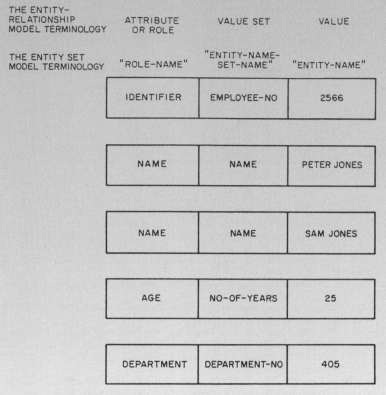

The author wishes to express his thanks to George Mealy, Stuart Madnick, Murray Edelberg, Susan Brewer, Stephen Todd, and the referees for their valuable suggestions (Figure 21 was suggested by one of the referees). This paper was motivated by a series of discussions with Charles Bachman. The author is also indebted to E.F. Codd and M.E. Senko for their valuable comments and discussions in revising this paper.

REFERENCES

1. Abrial, J.R. Data semantics. In *Data Base Management*, J.W. Klimbie and K.L. Koffeman, Eds., North-Holland Pub. Co., Amsterdam, 1974, pp. 1–60.
2. Bachman, C.W. Software for random access processing. *Datamation 11* (April 1965), 36–41.
3. Bachman, C.W. Data structure diagrams. *Data Base 1*, 2 (Summer 1969), 4–10.
4. Bachman, C.W. Trends in database management—1975. Proc., AFIPS 1975 NCC, Vol. 44, AFIPS Press, Montvale, N.J., pp. 569–576.
5. Birkhoff, G., and Bartee, T.C. *Modern Applied Algebra*. McGraw-Hill, New York, 1970.
6. Chamberlin, D.D., and Raymond, F.B. SEQUEL: A structured English query language. Proc. ACM-SIGMOD 1974, Workshop, Ann Arbor, Michigan, May, 1974.
7. CODASYL. Data base task group report. ACM, New York, 1971.
8. Codd, E.F. A relational model of data for large shared data banks. *Comm. ACM 13*, 6 (June 1970), 377–387.
9. Codd, E.F. Normalized data base structure: A brief tutorial. Proc. ACM-SIGFIDET 1971, Workshop, San Diego, Calif., Nov. 1971, pp. 1–18.
10. Codd, E.F. A data base sublanguage founded on the relational calculus. Proc. ACM-SIGFIDET 1971, Workshop, San Diego, Calif., Nov. 1971, pp. 35–68.
11. Codd, E.F. Recent investigations in relational data base systems. Proc. IFIP Congress 1974, North-Holland Pub. Co., Amsterdam, pp. 1017–1021.
12. Deheneffe, C., Hennebert, H., and Paulus, W. Relational model for data base. Proc. IFIP Congress 1974, North-Holland Pub. Co., Amsterdam, pp. 1022–1025.
13. Dodd, G.G. APL—a language for associate data handling in PL/I. Proc. AFIPS 1966 FJCC, Vol. 29, Spartan Books, New York, pp. 677–684.
14. Eswaran, K.P., and Chamberlin, D.D. Functional specifications of a subsystem for data base integrity. Proc. Very Large Data Base Conf., Framingham, Mass., Sept. 1975, pp. 48–68.
15. Hainaut, J.L., and Lecharlier, B. An extensible semantic model of data base and its data language. Proc. IFIP Congress 1974, North-Holland Pub. Co., Amsterdam, pp. 1026–1030.
16. Hammer, M.M., and McLeod, D.J. Semantic integrity in a relation data base system. Proc. Very Large Data Base Conf., Framingham, Mass., Sept. 1975, pp. 25–47.
17. Lindgreen, P. Basic operations on information as a basis for data base design. Proc. IFIP Congress 1974, North-Holland Pub. Co., Amsterdam, pp. 993–997.
18. Mealy, G.H. Another look at data base. Proc. AFIPS 1967 FJCC, Vol. 31, AFIPS Press, Montvale, N.J., pp. 525–534.
19. Nijssen, G.M. Data structuring in the DDL and the relational model. In *Data Base Management*, J.W. Klimbie and K.L. Koffeman, Eds., North-Holland Pub. Co., Amsterdam, 1974, pp. 363–379.
20. Olle, T.W. Current and future trends in data base management systems. Proc. IFIP Congress 1974, North-Holland Pub. Co., Amsterdam, pp. 998–1006.
21. Roussopoulos, N., and Mylopoulos, J. Using semantic networks for data base management. Proc. Very Large Data Base Conf., Framingham, Mass., Sept. 1975, pp. 144–172.
22. Rustin, R. (Ed.). Proc. ACM-SOGMOD 1974—debate on data models. Ann Arbor, Mich., May 1974.
23. Schmid, H.A., and Swenson, J.R. On the semantics of the relational model. Proc. ACM-SIGMOD 1975, Conference, San Jose, Calif., May 1975, pp. 211–233.
24. Senko, M.E. Data description language in the concept of multilevel structured description: DIAM II with FORAL. In *Data Base Description*, B.C.M. Dougue, and G.M. Nijssen, Eds., North-Holland Pub. Co., Amsterdam, pp. 239–258.

25. Senko, M.E., Altman, E.B., Astrahan, M.M., and Fehder, P.L. Data structures and accessing in data-base systems. *IBM Syst. J. 12*, 1 (1973), 30–93.
26. Sibley, E.H. On the equivalence of data base systems. Proc. ACM-SIGMOD 1974 debate on data models, Ann Arbor, Mich., May 1974, pp. 43–76.
27. Steel, T.B. Data base standardization—a status report. Proc. ACM-SIGMOD 1975, Conference, San Jose, Calif., May 1975, pp. 65–78.
28. Stonebraker, M. Implementation of integrity constraints and views by query modification. Proc. ACM-SIGMOD 1975, Conference, San Jose, Calif., May 1975, pp. 65–78.
29. Sundgren, B. Conceptual foundation of the infological approach to data bases. In *Data Base Management*, J.W. Klimbie and K.L. Koffeman, Eds., North-Holland Pub. Co., Amsterdam, 1974, pp. 61–95.
30. Taylor, R.W. Observations on the attributes of database sets. In *Data Base Description*, B.C.M. Dougue and G.M. Nijssen, Eds., North-Holland Pub. Co., Amsterdam, pp. 73–84.
31. Tsichritzis, D. A network framework for relation implementation. In *Data Base Description*, B.C.M. Douge and G.M. Nijssen, Eds., North-Holland Pub. Co., Amsterdam, pp. 269–282.

From "Goto Considered Harmful" to Structured Programming

Edsger W. Dijkstra
Department of Computer Sciences
The University of Texas at Austin
Austin, TX 78712-1188, USA
dijkstra@cs.utexas.edu

Ph.D. in Computing Science, University of Amsterdam

Professor of Mathematics, Eindhoven University of Technology:

ACM Turing Award, AFIPS Harry Good Memorial Award

Member of the Royal Netherlands Academy of Arts and Sciences

Distinguished Fellow of the British Computer Society

Major contributions: implementation of ALGOL 60, THE operating system, structured programming

Current interests: formal derivation of proofs and programs, streamlining of the mathematical argument

Edsger W. Dijkstra

EWD 1308: What Led to "Notes on Structured Programming"

The purpose of this historical note is to describe the experiences which with hindsight seem to have influenced me when I wrote EWD 249 "Notes on Structured Programming" in 1969. The note is based on what I remember; I am sure my memory has been selective and hence don't claim the objectivity of a professional historian.

I was introduced to programming at the 1951 Summer School, given in Cambridge by Wilkes, Wheeler and Gill, and in 1952 became on a part-time basis – initially two days per week – the programmer of the Mathematical Centre in Amsterdam; the rest of the week I studied theoretical physics in Leyden.

My only model was the program organization for the EDSAC in Cambridge; I followed it closely when designing program notation, input, output and library organisation for the ARRA in Amsterdam. For the next machines, the FERTA, the ARMAC and the X1, program notation and input would very much follow the same pattern: I clearly was a conservative programmer. Add to this the ARMAC's instruction buffer with a capacity of one track of the drum having destroyed the store's homogeneity, then you will understand that I did not embark on adventurers like "autocoders".

In 1955 I made the decision not to become a theoretical physicist, but to become a programmer instead. I made that decision because I had concluded that of theoretical physics and programming, programming embodied the greater intellectual challenge. You see, in those days I did not suffer from intellectual modesty. It was a difficult decision, for I had been groomed as a first-class scientist and becoming a programmer looked like a farewell from science. When I explained my dilemma to A. van Wyngaarden, then my boss, he told me that computers were here to stay and that in the world of programming I could very well be the one to create the science that was still lacking. Getting my physics degree in Leyden became a formality to be done as quickly as possible. (As a matter of fact I no longer felt welcome in Leyden: the physicists considered me a deserter and the mathematicians, one of whom openly prided himself on "of course knowing nothing about computers", were just contemptuous.)

In the meantime, a pattern emerged for the co-operation between me and my hardware colleagues Bram J. Loopstra and Carel S. Scholten. After the functional specification of the next machine had been written down (usually by me), that document served as a kind of contract between us: it told them what machine to design and construct, while I knew what I could count on while writing all the basic software for the machine. The aim of this division of labour was that my programs would be ready by the time the construction of the machine had been completed.

Looking back I now observe that the above arrangement has had a profound influence on how I grew up as a programmer: I found it perfectly normal to program for not yet existing machines. As a byproduct it became firmly ingrained in my mind that I programmed for the abstract machine as specified in the original document, and not for the actual piece of hardware: the original document was not a description but a prescription, and in the case of a discrepancy, not the text but the actual hardware would be at fault.

At the time I regarded this division of labour and the resulting practice of programming for non-existing machines as perfectly normal. Later I read an American article on why software was always late; I remember being very amazed when I read that limited availability of the hardware was a main cause, and I concluded that the circumstances under which I had learned programming had been less common than I had assumed.

Of course I could not exclude from my designs typographical errors and similar blemishes, but such shortcomings did not matter as long as the machine was not ready yet, and after the completion of the machine they could be readily identified as soon as they manifested themselves, but this last comforting thought was denied to me in 1957 with the introduction of the real-time interrupt. When Loopstra and Scholten suggested this feature for the X1, our next machine, I had visions of my program causing irreproducible errors and I panicked.

Eventually, Loopstra and Scholten flattered me out of my resistance and I studied their proposal. The first thing I investigated was whether I could demonstrate that the machine state could be saved and restored in such a way that interrupted programs could be continued as if nothing had happened. I demonstrated instead that it could not be done and my friends had to change their proposal. Admittedly, the scenarios under which the original proposal would fail were very unlikely, but this could have only strengthened my conviction that I had to rely on arguments rather than on experiments. At the time that conviction was apparently not so widespread, for up to seven years later I would find flaws in the interrupt hardware of new commercial machines.

I had a very illuminating experience in 1959, when I had posed the following problem to my colleagues at the Mathematical Centre. Consider two programs that can communicate via atomic reads and writes in a shared store. Can they be programmed in such a way that the executions of their critical sections exclude each other in time? Solutions came pouring in, but all wrong, so people tried more complicated "solutions". As these required more and more elaborate counter examples for their refutation, I had to change the rules: besides the programs they had to hand in an argument why the solution was correct. Within a few hours T. J. Dekker handed in a true solution with its correctness argument. Dekker had first analysed the proof obligations, then chosen the shape of an argument that would meet them, and then constructed the program to which this argument was applicable. It was a very clear example of how much one loses when the role of mathematics is confined to a posteriori verification; in 1968 I would publish a paper titled "A constructive approach to the problem of program correctness".

And then there was ALGOL 60. We saw it coming in 1959 and implemented it in the first half of 1960. I was terribly afraid, this implementation was then by far my most ambitious project: ALGOL 60 was so far ahead of its time that even its designers did not know how to implement it. I had never written a compiler and had to achieve my goal with a machine that had only 4096 words of storage. (The latter constraint was of course a blessing in disguise, but I don't remember seeing that when we started.)

By today's standards we did not know what we were doing; we did not dream of giving any guarantee that our implementation would be correct because we knew full well that we lacked the theoretical knowledge that would be needed for that. We did the only thing we could do under the circumstances, namely, to keep our design as simple and systematic as we could and to check that we had avoided all the mistakes we could think of. Eventually we learned that we had made mistakes we had not thought of, and after all the repairs the compiler was no longer as clean as we had originally hoped (F.E.J. Kruseman Aretz still found and repaired a number of errors after I had left the Mathematical Centre in 1962).

Right from the start we expected two very different types of errors, writing errors, whose repair is trivial, and thinking errors that would send us back to the drawing board, and the distinction has helped us because one combats them with different techniques. J.A. Zonneveld and I combatted the writing errors by coding together, each with his own text in front of him. When we were done, both our texts were punched independently, the two tapes were compared mechanically and about two dozen discrepancies – if I remember correctly – showed up. The thinking errors we had tried to prevent by convincing each other why the proposed section would work. With this reasoning we would mainly discuss the workings of the compiler while the program compiled was treated as data, and this experience was helpful for later, as it made us accustomed to non-operational considerations of program texts.

The whole experience made me receptive to what later would be called modularization or divide-and-rule or abstraction. It also sensitized me to the care with which interfaces have to be chosen and to the potential scope of the programming challenge in general. It heavily contributed to my subsequent opinion that creating confidence in the correctness of the design was the most important but hardest aspect of the programmer's task. In a world obsessed with speed, this was not a universally popular notion.

I remember from those days two design principles that have served me well ever since:

(i) before really embarking on a sizeable project, in particular before starting the large investment of coding, try to kill the project first, and
(ii) start with the most difficult, most risky parts first.

My first test program was almost the empty block, say

```
begin real x end,
```

not the most difficult example, but my 4th test was a double summation in which the nested calls of the summation routine were introduced via the parameter mechanism, while the summation routine itself had been defined recursively. In passing we had demonstrated the validity of what became known as Jensen's Device.

After this implementation interlude I returned in fairly general terms to the still open problem of the proper coordination of, in principle, asynchronous components. Without being directly involved I had witnessed a few years earlier the efforts of coupling all sorts of punched card equipment to the X1 and had been horrified by the degree of complication introduced by the inclusion of real-time commitments. For the sake of simplicity I therefore insisted on "timeless" designs, the correctness of which could be established by discrete reasoning only.

Almost unavoidably the model of Cooperating Sequential Processes emerged: sequential processes with (by definition!) undefined relative speeds and hence, for the sake of their cooperation, equipped with some primitives for synchronization.

Another opportunity for simplification was presented when we recognized that the timing aspects between a piece of communication equipment and the program that used it were completely symmetrical and independent of whether we had an input or output device. Needless to say, this unification helped a lot.

When we got involved in the design of the THE Multiprogramming System, scaling up slowly became a more and more explicit concern. It had to. Within IBM, and possibly elsewhere as well, circulated the concept as a supposed law of nature that "system complexity" in some informal sense would grow as the square of the number of components; the reasoning behind it was simple – each component could interfere with every other component – but if it were true it would de facto rule out systems beyond a certain size. This evoked my interest in systems structured in such a way that "system complexity" in the same informal sense would not grow more than linearly with the size. In 1967, the expression "layers of abstraction" entered the computer lingo.

Let me close the discussion of this episode by quoting the last two sentences of EWD 123 "Cooperating Sequential Processing" (September 1965):

> If this monograph gives any reader a clearer indication of what kind of hierarchical ordering can be expected to be relevant, I have reached one of my clearest goals. And may we not hope that a confrontation with the intricacies of Multiprogramming gives us a clearer understanding of what Uniprogramming is all about?

In 1968 I suffered from a deep depression, partly caused by the Department, which did not accept Informatics as relevant to its calling and disbanded the group I had built up, and partly caused by my own hesitation about what to do next. I knew that in retrospect, the ALGOL implementation and the THE Multiprogramming System had only been agility exercises and that now I had to tackle the real problem of How to Do Difficult Things. In my depressed state it took me months to gather the courage to write (for therapeutic reasons) EWD 249 "Notes on Structured Programming" (August 1969); it marked the beginning of my recovery.

EWD 249 tries to synthesize the above mentioned ingredients of the preceding decade. It mentions on an early page "the program structure in connection with a convincing demonstration of the correctness of the program", mentions as our mental aids "(1) Enumeration, 82) Mathematical Induction, (3) Abstraction", and about the first and the last I quote (from EWD 249-14):

> Enumerative reasoning is all right as far as it goes, but as we are rather slow-witted it does not go very far. Enumerative reasoning is only an adequate mental tool under the severe boundary condition that we only use it very moderately. We should appreciate abstraction as our main mental technique to reduce the demands made upon enumerative reasoning.

I had had two external stimuli: the 1968 NATO Conference on "Software Engineering" in Garmisch-Partenkirchen and the founding of the IFIP Working Group on „Programming Methodology". Thanks to the ubiquitous Xerox machine, my typewritten text could spread like wildfire, and it did so, probably because people found it refreshing in the prevailing culture characterized by the 1968 IBM advertisement in Datamation, which presented in full colour a beaming Susie Mayer who had solved all her programming problems by switching to PL/I. Apparently, IBM did not like the popularity of my text; it stole the term "Structured Programming" and under its auspices Harlan D. Mills trivialized the original concept to the abolishment of the goto statement.

Looking back I cannot fail to observe my fear of formal mathematics at the time. In 1970, I had spent more than a decade hoping and then arguing that programming would and should become a mathematical activity; I had (re)arranged the programming task so as to make it better amenable to mathematical treatment, but carefully avoided creating the required mathematics myself. I had to wait for Bob Floyd, who laid the foundation, for Jim King, who showed me the first example that convinced me, and for Tony Hoare, who showed how semantics could be defined in terms of the axioms needed for the proofs of properties of programs, and even then I did not see the significance of their work immediately. I was really slow.

Finally a short story for the record. In 1968, the *Communications of the ACM* published a text of mine under the title "The goto statement considered harmful", which in later years would be most frequently referenced, regrettably, however, often by authors who had seen no more of it than its title. This text became a cornerstone of my fame by becoming a template: we would see all sorts of articles under the title "X considered harmful" for almost any X, including one titled "Dijkstra considered harmful". But what had happened? I had submitted a paper under the title "A case against the goto statement", which, in order to speed up its publication, the editor had changed into a "Letter to the Editor", and in the process he had given it a new title of his own invention! The editor was Niklaus Wirth.

Edsger W. Dijkstra

Solution of a Problem in Concurrent Programming Control

Communications of the ACM,
Vol. 8 (9), 1965
pp. 569

Solution of a Problem in Concurrent Programming Control

E. W. DIJKSTRA
Technological University, Eindhoven, The Netherlands

A number of mainly independent sequential-cyclic processes with restricted means of communication with each other can be made in such a way that at any moment one and only one of them is engaged in the "critical section" of its cycle.

Introduction

Given in this paper is a solution to a problem for which, to the knowledge of the author, has been an open question since at least 1962, irrespective of the solvability. The paper consists of three parts: the problem, the solution, and the proof. Although the setting of the problem might seem somewhat academic at first, the author trusts that anyone familiar with the logical problems that arise in computer coupling will appreciate the significance of the fact that this problem indeed can be solved.

The Problem

To begin, consider N computers, each engaged in a process which, for our aims, can be regarded as cyclic. In each of the cycles a so-called "critical section" occurs and the computers have to be programmed in such a way that at any moment only one of these N cyclic processes is in its critical section. In order to effectuate this mutual exclusion of critical-section execution the computers can communicate with each other via a common store. Writing a word into or nondestructively reading a word from this store are undividable operations; i.e., when two or more computers try to communicate (either for reading or for writing) simultaneously with the same common location, these communications will take place one after the other, but in an unknown order.

The solution must satisfy the following requirements.

(a) The solution must be symmetrical between the N computers; as a result we are not allowed to introduce a static priority.

(b) Nothing may be assumed about the relative speeds of the N computers; we may not even assume their speeds to be constant in time.

(c) If any of the computers is stopped well outside its critical section, this is not allowed to lead to potential blocking of the others.

(d) If more than one computer is about to enter its critical section, it must be impossible to devise for them such finite speeds, that the decision to determine which one of them will enter its critical section first is postponed until eternity. In other words, constructions in which "After you"-"After you"-blocking is still possible, although improbable, are not to be regarded as valid solutions.

We beg the challenged reader to stop here for a while and have a try himself, for this seems the only way to get a feeling for the tricky consequences of the fact that each computer can only request one one-way message at a time. And only this will make the reader realize to what extent this problem is far from trivial.

The Solution

The common store consists of:

"**Boolean array** b, $c[1{:}N]$; **integer** k"

The integer k will satisfy $1 \leq k \leq N$, $b[i]$ and $c[i]$ will only be set by the ith computer; they will be inspected by the others. It is assumed that all computers are started well outside their critical sections with all Boolean arrays mentioned set to **true**; the starting value of k is immaterial.

The program for the ith computer ($1 \leq i \leq N$) is:

```
"integer j;
Li0: b[i] := false;
Li1: if k ≠ i then
Li2: begin c[i] := true;
Li3: if b[k] then k := i;
     go to Li1
     end
       else
```

```
Li4: begin c[i] := false;
       for j := 1 step 1 until N do
         if j ≠ i and not c[j] then go to Li1
     end;
     critical section;
     c[i] := true; b[i] := true;
     remainder of the cycle in which stopping is allowed;
     go to Li0"
```

The Proof

We start by observing that the solution is safe in the sense that no two computers can be in their critical section simultaneously. For the only way to enter its critical section is the performance of the compound statement *Li*4 without jumping back to *Li*1, i.e., finding all other *c*'s **true** after having set its own *c* to **false**.

The second part of the proof must show that no infinite "After you"-"After you"-blocking can occur; i.e., when none of the computers is in its critical section, of the computers looping (i.e., jumping back to *Li*1) at least one—and therefore exactly one—will be allowed to enter its critical section in due time.

If the *k*th computer is not among the looping ones, $b[k]$ will be **true** and the looping ones will all find $k \neq i$. As a result one or more of them will find in *Li*3 the Boolean $b[k]$ **true** and therefore one or more will decide to assign "$k := i$". After the first assignment "$k := i$", $b[k]$ becomes **false** and no new computers can decide again to assign a new value to k. When all decided assignments to k have been performed, k will point to one of the looping computers and will not change its value for the time being, i.e., until $b[k]$ becomes **true**, viz., until the *k*th computer has completed its critical section. As soon as the value of k does not change any more, the *k*th computer will wait (via the compound statement *Li*4) until all other *c*'s are **true**, but this situation will certainly arise, if not already present, because all other looping ones are forced to set their *c* **true**, as they will find $k \neq i$. And this, the author believes, completes the proof.

Letters to the Editor

Edsger W. Dijkstra

Go To Statement Considered Harmful

Communications of the ACM,
Vol. 11 (2), 1968
pp. 147-148

Go To Statement Considered Harmful

Key Words and Phrases: go to statement, jump instruction, branch instruction, conditional clause, alternative clause, repetitive clause, program intelligibility, program sequencing
CR Categories: 4.22, 5.23, 5.24

EDITOR:

For a number of years I have been familiar with the observation that the quality of programmers is a decreasing function of the density of **go to** statements in the programs they produce. More recently I discovered why the use of the **go to** statement has such disastrous effects, and I became convinced that the **go to** statement should be abolished from all "higher level" programming languages (i.e. everything except, perhaps, plain machine code). At that time I did not attach too much importance to this discovery; I now submit my considerations for publication because in very recent discussions in which the subject turned up, I have been urged to do so.

My first remark is that, although the programmer's activity ends when he has constructed a correct program, the process taking place under control of his program is the true subject matter of his activity, for it is this process that has to accomplish the desired effect; it is this process that in its dynamic behavior has to satisfy the desired specifications. Yet, once the program has been made, the "making" of the corresponding process is delegated to the machine.

My second remark is that our intellectual powers are rather geared to master static relations and that our powers to visualize processes evolving in time are relatively poorly developed. For that reason we should do (as wise programmers aware of our limitations) our utmost to shorten the conceptual gap between the static program and the dynamic process, to make the correspondence between the program (spread out in text space) and the process (spread out in time) as trivial as possible.

Let us now consider how we can characterize the progress of a process. (You may think about this question in a very concrete manner: suppose that a process, considered as a time succession of actions, is stopped after an arbitrary action, what data do we have to fix in order that we can redo the process until the very same point?) If the program text is a pure concatenation of, say, assignment statements (for the purpose of this discussion regarded as the descriptions of single actions) it is sufficient to point in the program text to a point between two successive action descriptions. (In the absence of **go to** statements I can permit myself the syntactic ambiguity in the last three words of the previous sentence: if we parse them as "successive (action descriptions)" we mean successive in text space; if we parse as "(successive action) descriptions" we mean successive in time.) Let us call such a pointer to a suitable place in the text a "textual index."

When we include conditional clauses (**if** B **then** A), alternative clauses (**if** B **then** $A1$ **else** $A2$), choice clauses as introduced by C. A. R. Hoare (case[i] of($A1$, $A2$, $\cdots$, An)), or conditional expressions as introduced by J. McCarthy ($B1 \rightarrow E1$, $B2 \rightarrow E2$, $\cdots$, $Bn \rightarrow En$), the fact remains that the progress of the process remains characterized by a single textual index.

As soon as we include in our language procedures we must admit that a single textual index is no longer sufficient. In the case that a textual index points to the interior of a procedure body the dynamic progress is only characterized when we also give to which call of the procedure we refer. With the inclusion of procedures we can characterize the progress of the process via a sequence of textual indices, the length of this sequence being equal to the dynamic depth of procedure calling.

Let us now consider repetition clauses (like, **while** B **repeat** A or **repeat** A **until** B). Logically speaking, such clauses are now superfluous, because we can express repetition with the aid of recursive procedures. For reasons of realism I don't wish to exclude them: on the one hand, repetition clauses can be implemented quite comfortably with present day finite equipment; on the other hand, the reasoning pattern known as "induction" makes us well equipped to retain our intellectual grasp on the processes generated by repetition clauses. With the inclusion of the repetition clauses textual indices are no longer sufficient to describe the dynamic progress of the process. With each entry into a repetition clause, however, we can associate a so-called "dynamic index," inexorably counting the ordinal number of the corresponding current repetition. As repetition clauses (just as procedure calls) may be applied nestedly, we find that now the progress of the process can always be uniquely characterized by a (mixed) sequence of textual and/or dynamic indices.

The main point is that the values of these indices are outside programmer's control; they are generated (either by the write-up of his program or by the dynamic evolution of the process) whether he wishes or not. They provide independent coordinates in which to describe the progress of the process.

Why do we need such independent coordinates? The reason is—and this seems to be inherent to sequential processes—that we can interpret the value of a variable only with respect to the progress of the process. If we wish to count the number, n say, of people in an initially empty room, we can achieve this by increasing n by one whenever we see someone entering the room. In the in-between moment that we have observed someone entering the room but have not yet performed the subsequent increase of n, its value equals the number of people in the room minus one!

The unbridled use of the **go to** statement has an immediate consequence that it becomes terribly hard to find a meaningful set of coordinates in which to describe the process progress. Usually, people take into account as well the values of some well chosen variables, but this is out of the question because it is relative to the progress that the meaning of these values is to be understood! With the **go to** statement one can, of course, still describe the progress uniquely by a counter counting the number of actions performed since program start (viz. a kind of normalized clock). The difficulty is that such a coordinate, although unique, is utterly unhelpful. In such a coordinate system it becomes an extremely complicated affair to define all those points of progress where, say, n equals the number of persons in the room minus one!

The **go to** statement as it stands is just too primitive; it is too much an invitation to make a mess of one's program. One can regard and appreciate the clauses considered as bridling its use. I do not claim that the clauses mentioned are exhaustive in the sense that they will satisfy all needs, but whatever clauses are suggested (e.g. abortion clauses) they should satisfy the requirement that a programmer independent coordinate system can be maintained to describe the process in a helpful and manageable way.

It is hard to end this with a fair acknowledgment. Am I to judge by whom my thinking has been influenced? It is fairly obvious that I am not uninfluenced by Peter Landin and Christopher Strachey. Finally I should like to record (as I remember it quite distinctly) how Heinz Zemanek at the pre-ALGOL meeting in early 1959 in Copenhagen quite explicitly expressed his doubts whether the **go to** statement should be treated on equal syntactic footing with the assignment statement. To a modest extent I blame myself for not having then drawn the consequences of his remark.

The remark about the undesirability of the **go to** statement is far from new. I remember having read the explicit recommendation to restrict the use of the **go to** statement to alarm exits, but I have not been able to trace it; presumably, it has been made by C. A. R. Hoare. In [1, Sec. 3.2.1.] Wirth and Hoare together make a remark in the same direction in motivating the case construction: "Like the conditional, it mirrors the dynamic structure of a program more clearly than **go to** statements and switches, and it eliminates the need for introducing a large number of labels in the program."

In [2] Guiseppe Jacopini seems to have proved the (logical) superfluousness of the **go to** statement. The exercise to translate an arbitrary flow diagram more or less mechanically into a jump-less one, however, is not to be recommended. Then the resulting flow diagram cannot be expected to be more transparent than the original one.

REFERENCES:

1. WIRTH, NIKLAUS, AND HOARE, C. A. R. A contribution to the development of ALGOL. *Comm. ACM 9* (June 1966), 413–432.
2. BÖHM, CORRADO, AND JACOPINI, GUISEPPE. Flow diagrams, Turing machines and languages with only two formation rules. *Comm. ACM 9* (May 1966), 366–371.

EDSGER W. DIJKSTRA
Technological University
Eindhoven, The Netherlands

Software Fundamentals

C.A.R. Hoare
Senior Researcher, Microsoft Research Ltd.
Cambridge, England
thoare@microsoft.com

Certificate in Statistics, Oxford

Professor of Computing Science,
Queen's University Belfast

Professor of Computation,
Oxford

ACM Turing Award (1980)

Major contributions:
Quicksort, provable correctness,
Z specification language,
CSP programming model

Current interests:
programming theory and concurrency,
exploitation of assertions by programming tools,
languages and compilers

C.A.R. Hoare

Assertions: A Personal Perspective

It was my early experience in industry that triggered my interest in assertions and their in-program proofs; and my subsequent research at university extended the idea into a methodology for the specification and design of programs. Now that I have returned to work in industry, I have looked into the current role of assertions in industrial program development. My personal perspective illustrates the complementary roles of pure research, aimed at academic ideals of excellence, and the unexpected ways in which the results of such research contribute to the gradual improvement of engineering practice.

1. Early Days in Industry, 1960-1968

It is a pleasure and a privilege to talk to such a large audience of successful programmers in the modern software industry. My first job was in industry, as a programmer for a small British computer manufacturer, Elliott Brothers of London at Borehamwood. My task was to write library programs in decimal machine code [1] for the company's new 803 computer.

After a preliminary exercise which gave my boss confidence in my skill, I was entrusted with the task of implementing a new sorting method recently invented and published by Shell [2]. I really enjoyed optimizing the inner loops of my program to take advantage of the most ingenious instructions of the machine code. I also enjoyed documenting the code according to the standards laid down for programs to be delivered to customers as part of our library. Even testing the program was fun; tracing the errors was like solving mathematical puzzles. How wonderful that programmers get paid for that too! In fairness, surely the programmers should pay back to their employers the cost for removal of their own mistakes.

But not such fun was the kind of error that caused my test programs to run wild (crash); quite often, they even overwrote the data needed to diagnose the cause of the error. Was the crash due perhaps to a jump into the data space, or to an instruction overwritten by a number? The only way to find out was to add extra output instructions to the program, tracing its behaviour up to the moment of the crash. But the sheer volume of the output only added to the confusion. Remember, in those days the lucky programmer was one who had access to the computer just once a day. Even 40 years later, the problem of crashing programs is not altogether solved.

When I had been in my job for six months, an even more important task was given to me: that of designing a new high-level programming language for the projected new and faster members of the company's range of computers. By great good fortune, there came into my hands a copy of Peter Naur's 'report on the algorithmic language ALGOL 60' [3], which had recently been designed by an international committee of experts; we decided to implement a subset of that language, which I selected with the goal of efficient implementation on the Elliott computers. In the end, I thought of an efficient way of implementing nearly the whole language.

An outstanding merit of Peter Naur's report was that it was only 21 pages long. Yet it gave enough accurate information for an implementer to compile the language without any communication with the language designers. Furthermore, a user could program in the language without any communication either with the implementers or with the designers. Even so the program worked the very first time it was submitted to the compiler. Apart from a small error in the character codes, this is what actually happened one day at an exhibition of an Elliott 803 computer in eastern Europe. Few languages designed since then have matched such an achievement.

Part of the credit for this success was the very compact yet precise notation for defining the grammar or syntax of the language, the class of texts that are worthy of consideration as meaningful programs. This syntactic notation was due originally to the great linguist, psychologist and philosopher Noam Chomsky [4]. It was first applied to programming languages by John Backus [5], in a famous article on 'The syntax and the semantics of the proposed international algorithmic language of the Zurich ACM-GAMM conference', Paris, 1959. After dealing with the syntax, the author looked forward to a continuation article on the semantics. It never appeared: in fact it laid down a challenge of finding a precise and elegant formal definition of the meaning of programs, which inspired good research in computer science right up to the present day.

The syntactic definition of the language served as a pattern for the structure of the whole of our ALGOL compiler, which used a method now known as recursive descent. As a result, it was logically impossible (almost) for any error in the syntax of a submitted program to escape detection by the compiler. If a successfully compiled program went wrong, the programmer had complete confidence that this was not the result of a misprint that made the program meaningless. Chomsky's syntactic definition method was soon more widely applied to earlier and to later programming languages, with results that were rarely as attractive as for ALGOL 60. I thought that this failure reflected the intrinsic irregularity and ugliness of the syntax of these other languages. One purpose of a good formal definition method is to guide the designer to improve the quality of the language it is used to define.

In designing the machine code to be output by the Elliott ALGOL compiler [6], I took it as an over-riding principle that no program compiled from the high-level language could ever run wild. Our customers had to accept a significant performance penalty, because every subscripted array access had to be checked at run time against both upper and lower array bounds; they knew how often such a check fails in a production run, and they told me later that they did not even want the option to remove the check. As a result, programs written in ALGOL would never run wild, and debugging was relatively simple, because the effect of every program could be inferred from the source text of the program itself, without knowing anything about the compiler, and knowing only a little or nothing about the machine on which it was running. If only we had a formal semantics to complement the formal syntax of the language, perhaps the compiler would be able to help in detecting and averting other kinds of programming errors as well.

Interest in semantics was widespread. In 1964, a conference took place in Vienna on Formal Language Description Languages for Computer Programming [7]. It was attended by 51 scientists from 12 nations. One of the papers was entitled 'The definition of programming languages by their compilers' [8], by Jan Garwick, pioneer of computing science in Norway. The title horrified me, because it suggested that the meaning of any program is determined by selecting a standard implementation of that language on a particular machine. So if you wanted to know the meaning of a Fortran program, for example, you would run it on an IBM 709, and see what happened. Such a proposal seemed to me grossly unfair to all computer manufacturers other than IBM, at that time the world-dominant computing company. It would be impossibly expensive and counter-productive on an Elliott 803, with a word length of 39 bits, to give the same numerical answers as the IBM machine, which had only 36 bits in a word – we could more efficiently give greater accuracy and range. Even more unfair was the consequence that the IBM compiler was by definition correct; but any other manufacturer would be compelled to reproduce all of its errors – they would have to be called just anomalies, because errors would be logically impossible. Since then, I have always avoided operational approaches to programming language semantics. The principle that 'a program is what a program does' is not a good basis for exploration of the concept of program correctness.

I did not make a presentation at the Vienna conference, but I did make one comment: I thought that the most important attribute of a formal definition of semantics should be to leave certain aspects of the language carefully undefined. As a result, each implementation would have carefully circumscribed freedom to make efficient choices in the interests of its users and in the light of the characteristics of a particular machine architecture. I was very encouraged that this comment was applauded, and even Garwick expressed his agreement. In fact, I had misinterpreted his title: his paper called for an abstract compiler for an abstract machine, rather than selection of an actual commercial product as standard.

The inspiration of my remark in Vienna dates back to 1952, when I went to Oxford University as an undergraduate student. Some of my neighbours in college were mathematicians, and I joined them in a small unofficial evening reading party to study mathematical logic from the text book by Quine [9]. Later, a course in the philosophy of mathematics pursued more deeply this interest in axioms and proofs, as an explanation of the unreasonable degree of certainty which accompanies the contemplation of mathematical truth. It was this background that led me to propose the axiomatic method for defining the semantics of a programming language, while preserving a carefully controlled vagueness in certain aspects. I drew the analogy with the foundations of the various branches of mathematics, like projective geometry or group theory; each branch is in effect defined by the set of axioms that are used without further justification in all proofs of the theorems of that branch. The axioms are written in the common notations of mathematics, but they also contain a number of undefined terms, like lines and points in projective geometry, or units and products in group theory; these constitute the conceptual framework of that branch. I was convinced that an axiomatic presentation of the basic concepts of programming could be much simpler than any runnable compiler of any language for any computer, however abstract.

My first proposal for such an axiom set took the form of equations, as encountered in school texts on algebra, but with fragments of program on the left and right hand sides of the equation instead of numbers and numeric expressions. The same idea was explored earlier and more thoroughly in a doctoral dissertation by Shigeru Igarashi at the University of Tokyo [10]. I showed my first pencilled draft of a paper on the axiomatic approach to Peter Lucas; he was leading a project at the IBM Research Laboratory in Vienna to give a formal definition to IBM's new programming language, known as PL/I [11]. He was attracted by the proposal, but he rapidly abandoned the attempt to apply it to PL/I. The designers of PL/I had a very operational view of what each construct of the language would do, and they had no inclination to support a level of abstraction necessary for an attractive or helpful axiomatic presentation of the semantics. I was not disappointed: in the arrogance of idealism, I was confirmed in my view that a good formal definition method would be one that clearly reveals the quality of a programming language, whether bad or good; and the axiomatic method had at least shown its capability revealing badness. Other evidence for the badness of PL/I was its lack of protection against crashing programs.

2. Research in Belfast, 1968-1977

By 1968, it was evident that research into programming language semantics was going to take a long time before it found application in industry; and in those days it was accepted that long-term research should take place in universities. I therefore welcomed the opportunity to take up a post as Professor of Computer Science at the Queen's University in Belfast. By a happy coincidence, as I was moving house, I came across a preprint of Robert Floyd's paper on 'Assigning Meanings to Programs' [12]. Floyd adopted the same philosophy as I had: that the meaning of a programming language is defined by the rules that can be used for reasoning about programs in the language. These could include not only equations, but also rules of inference. By this means, he presented an effective method of proving the total correctness of programs, not just their equality to other programs. I saw this as the achievement of the ultimate goal of a good formal semantics for a good programming language, namely, the complete avoidance of programming error. Furthermore, the quality of the language was now the subject of objective scientific assessment, based on simplicity of the axioms and the guidance they give for program construction. The axiomatic method is a way to avoid the dogmatism and controversy that so often accompanies programming language design, particularly by committees.

For a general-purpose programming language, correctness can be defined only relative to the intention of a particular program. In many cases, the intention can be expressed as a post-condition of the program, that is an assertion about the values of the variables of the program that is intended to be true when the program terminates. The proof of this fact usually depends on annotating the program with additional assertions in the middle of the program text; these are expected to be true whenever execution of the program reaches the point where the assertion is written. At least one assertion, called an invariant, is needed in each loop: it is intended to be true before and after every execution of the body of the loop. Often, the correct working of a program depends on the assumption of some precondition, which must be true before the program starts. Floyd gave the proof rules whose application could guarantee the validity of all the assertions except the precondition, which had to be assumed. He even looked forward to the day when a verifying compiler could actually check the validity of all the assertions automatically before allowing the program to be run. This would be the ultimate solution to the problem of programming error, making it logically impossible in a running program; though I correctly predicted its achievement would be some time after I had retired from academic life, which would be in 30 year's time.

So I started my life-long project by first extending the set of axioms and rules to cover all the familiar constructions of a conventional high-level programming language. These included iterations, procedures and parameters, recursion, functions, and even jumps [13-18]. Eventually, there were enough proof rules to cover almost all of a reasonable programming language, like Pascal, for which I developed a proof calculus in collaboration with Niklaus Wirth [19]. Since then, the axiomatic method has been explicitly used to guide the design of languages like Euclid and Eiffel [20, 21]. These languages were prepared to accept the restrictions on the generality of expression that are necessary to make the axioms consistent with efficient program execution. For example, the body of an iteration (for statement) should not assign a new value to the controlled variable; the parameters of a procedure should all be distinct from each other (no aliases); and all jumps should be forward rather than backward. I recommended that these restrictions should be incorporated in the design of any future programming language; they were all of a kind that could be enforced by a compiler, so as to avert the risk of programming error. Restrictions that contribute to provability, I claimed, are what make a programming language good.

I was even worried that my axiomatic method was too powerful, because it could deal with jumps, which Dijkstra had pointed out to be a bad feature of the conventional programming of the day [22]. My consolation was that the proof rule for jumps relies on a subsidiary hypothesis, and is inherently more complicated than the rules for structured programming constructs. Subsequent wide adoption of structured programming confirmed my view that simplicity of the relevant proof rule is an objective measure of quality in a programming language feature. Further confirmation is now provided by program analysis tools like Lint [23] and PREfix [24], applied to less disciplined languages such as C; they identify just those constructions that would invalidate the simple and obvious proof methods, and warn against their use.

A common objection to Floyd's method of program proving was the need to supply additional assertions at intermediate points in the program. It is very difficult to look at an existing program and guess what these assertions should be. I thought this was an entirely mistaken objection. It was not sensible to try to prove the correctness of existing programs, partly because they were mostly going to be incorrect anyway. I followed Dijkstra's constructive approach [25] to the task of programming: the obligation of ultimate correctness should be the driving force in designing programs that were going to be correct by construction. In this top-down approach, the starting point for a software project should always be the specification, and the proof of the program should be developed along with the program itself. Thus the most effective proofs are those constructed before the program is written. This philosophy has been beautifully illustrated in Dijkstra's own book on 'A Discipline of Programming' [26], and in many subsequent textbooks on formal approaches to software engineering [27].

In all my work on the formalisation of proof methods for sequential programming languages, I knew that I was only preparing the way for a much more serious challenge, which was to extend the proof technology into the realm of concurrent program execution. I took as my first model of concurrency a kind of quasi-parallel programming (co-routines), which was introduced by Ole-Johan Dahl and Kristen Nygaard into Simula (and later Simula 67) for purposes of discrete event simulation [28, 29]. But you have already heard the story of Simula from one of its originators, so I can end my account of early history at this point.

3. Back in Industry, 1999-

In 1999 I reached the expected age of retirement for an academic in Britain, and I was fortunate to receive and accept an offer of employment back in industry, this time for Microsoft Research Ltd., in Cambridge, England. So now I have a chance to test in person the accuracy of my prediction at the start of my academic career: that my research might begin to be relevant for industry at the end of my career.

In one way, my predictions have come true. Assertions figure very strongly in Microsoft code. A recent count discovered over one quarter of a million of them in the code for Office. The primary role of an assertion today is as a test oracle, defining the circumstances under which a program under test is considered to fail. A collection of aptly placed assertions is what permits a massive suite of test cases to be run overnight, in the absence of human intervention. Failure of an assertion triggers a dump of the program state, to be analysed by the programmer on the following morning. Apart from merely indicating the fact of failure, the place where the first assertion fails is likely to give a good indication of where and why the program is going wrong. And this indication is given in advance of any crash, thus avoiding the risk that the necessary diagnostic information is overwritten. So assertions have already found their major application, as a partial solution to the problems of program crashes, which I first encountered as a new programmer in 1960. The other solution is the ubiquitous personal workstation, which reduces the turn-around interval for program correction from days to minutes.

Assertions are usually compiled differently for test runs and for code that is shipped to the customer. In ship code, the assertions are often omitted, to avoid the run-time penalty and the confusion that would follow from an error diagnostic or a checkpoint dump in view of the customer. Ideally, the only assertions to be omitted are those that have been subjected to proof. But more practically, many teams leave the assertions in ship code to generate an exception when false; to continue execution in such an unexpected and untested circumstance would run a grave risk of crash. So instead, the handler for the exception makes a recovery that is sensible to the customer in the environment of use.

Assertions are also used to advantage by program analysis tools like PREfix [23]; this is being developed within Microsoft, for application to the maintenance of legacy code. The value of such tools is limited if they give so many warning messages that the programmer cannot afford the time to examine them. Ideally, each warning should be accompanied by an automatically generated test case that would reveal the bug; but that will depend on further advances in model checking and theorem proving. Assertions and assumptions provide a means for the programmer to explain that a certain error cannot occur, or is irrelevant, and the tool will suppress the corresponding sheaf of error reports. This is another motivating factor for programmers to include more and stronger assertions in their code. Another acknowledged motive is to inform programmers engaged in subsequent program modification that certain properties of the program must be maintained.

However, I must freely admit that there is one purpose for which assertions are never used, the purpose that actually inspired my whole life's research – the design of a better programming language. That remains an inspiration and a challenge for those who come after.

4. Challenges

Assertional techniques for writing correct programs still need to be extended from the simple sequential languages like PASCAL to cover the complexities of object orientation, including inheritance, over-riding, and pointer-swinging manipulations. Disciplined patterns of usage need to be formalised to simplify correctness reasoning; and the rules need to be policed by compile-time checks. The problems of concurrent programming, race conditions, deadlock and livelock need to be analysed and solved in the highly unpredictable environment of the World Wide Web. Dynamic configuration and reconfiguration, mobile code, transactions and even exceptions are essential to modern systems software and applications; the errors to which they are most liable must be brought under control. To incorporate the results of this research into practical tools, further advances are required in automatic theorem proving and in automatic test case generation.

I still nourish the vision that pure academic research will eventually be applied to the design of a better programming language. The ideas that crystallise from theoretical research into correctness of programs are often first subjected to practical evaluation in the context of an experimental programming language like Euclid. For purposes of experiment, such a language must have a pure and simple semantics, achieved by exclusion of all extraneous features and complicating factors. To progress to the next stage towards industrial use, a proposed language feature can be promulgated in the form of a design pattern, so that it can be exploited by users of existing programming languages. The pattern will include advice on the specification of interfaces, systematic coding techniques, and on the construction of test harnesses, including assertions to guard against errors. The advice needs support from program analysis tools, which will check observance of the disciplines on which correctness depends.

Finally, at intervals measured in decades rather than years, there arises an opportunity to introduce a new programming language onto the marketplace. This will never be done by taking a recent experimental language from the laboratory. The challenge of language design is to combine a multitude of features that have proved sufficiently successful that they have been tested and incorporated in all generally accepted research languages. The name of any language that aims at commercial success will be devised to reflect current buzzwords, and its syntax will be crafted to reflect the latest fashionable craze.

But there are now strong signs that the actual quality of the language is beginning to matter. A similar transformation occurred some time ago in the market place for cars, where reliability and safety once took second place to styling, chrome plating, and acceleration. In the progression from C to C++ and then to Java, each language designer has given more explicit attention to removing the traps and insecurities of its predecessor. As with all evolutionary processes, progress has been exceedingly slow. One day, I expect the new programming language designer will learn how to use assertional methods as a design tool, to evaluate and refine the objectively evaluated quality of familiar language features, as well as averting the risks involved in the introduction of new features. To bring that day forward is surely still a worthy goal for research, both academic and industrial, both theoretical and experimental. I hope you and your company will be both a contributor and a beneficiary of the results of this research.

References

[1] *Elliott 803 Programming Manual*, Elliott Brothers (London) Ltd., Borehamwood, Herts (1960).

[2] D. Shell. A high-speed sorting procedure. *Comm. ACM 2*, 30-32 (1959).

[3] P. Naur (ed). Report on the algorithmic language ALGOL 60. *Comm. ACM* 3(5) 299-314 (1960).

[4] N. Chomsky. *Syntactic structures*, Mouton & Co, The Hague (1957).

[5] J.W. Backus. The syntax and the semantics of the proposed international algebraic language of the Zurich ACM-GAMM Conference. *ICIP Proceedings*, Paris, 125-132 (1959).

[6] T. Hoare, Report on the Elliott ALGOL translator. *Comput. J.* 5(4) 345-348 (1963).

[7] T.B. Steel Jr. (ed.), *Formal language description languages for computer programming*. North-Holland (1966).

[8] J.V. Garwick. The definition of programming languages by their compilers.

[9] W.V.O. Quine, *Mathematical Logic*. Revised edition, Harvard University Press (1955).

[10] S. Igarashi. An axiomatic approach to equivalence problems of algorithms with applications. PhD. Thesis, Tokyo University (1964).

[11] P. Lucas et al. Informal introduction to the abstract syntax and interpretation of PL/I, ULD version II, IBM TR 25.03 (1968).

[12] R.W. Floyd. Assigning meanings to programs. *Proc. Am. Soc. Symp. Appl. Math.* 19, 19-31 (1967).

[13] T. Hoare. An axiomatic basis for computer programming. *Comm. ACM* 12(10) 576-580, 583 (1969).

[14] T. Hoare. *Procedures and parameters: an axiomatic approach*. LNM 188. Springer (1971).

[15] T. Hoare and M. Foley. Proof of a recursive program: QUICKSORT. *Comput. J.* 14, 391-395 (1971).

[16] T. Hoare. Towards a theory of parallel programming. In *Operating Systems Techniques*, Academic Press (1972).

[17] T. Hoare and M. Clint. Program proving: jumps and functions. *Acta Inf. 1*, 214-224 (1972).

[18] T. Hoare. A note on the for statement. *BIT 12*(3) 334-341 (1972).

[19] T. Hoare, and N. Wirth. An axiomatic definition of the programming language PASCAL. *Acta Inf. 2*(4) 335-355 (1973).

[20] R.L. London et al. Proof rules for the programming language EUCLID. *Acta Inf. 10*, 1-26 (1978).

[21] B. Meyer. *Object-oriented software construction* (2nd ed.). Prentice Hall PTR (1997).

[22] E.W. Dijkstra. Go to statement considered harmful. *Comm. ACM* 11 147-148 (1968).

[23] W.R. Bush, J.D. Pincus, D.J. Sielaff. A static analyser for finding dynamic programming errors. *Software Prac. Exper.* 30, 775-802 (2000).

[24] S.C. Johnson. Lint: a C program checker. *UNIX Prog. Man. 4.2.* UC Berkeley (1984).

[25] E.W. Dijkstra. A constructive approach to the problem of program correctness. *BIT* 8, 174-186 (1968).

[26] E.W. Dijkstra. *A discipline of programming*. Prentice Hall (1976).

[27] C. Morgan. *Programming from Specifications*. Prentice Hall (1990).

C.A.R. Hoare

An Axiomatic Basis for Computer Programming

Communications of the ACM, Vol. 12 (10), 1969 pp. 576-580, 583

An Axiomatic Basis for Computer Programming

C. A. R. Hoare
The Queen's University of Belfast, Northern Ireland*

In this paper an attempt is made to explore the logical foundations of computer programming by use of techniques which were first applied in the study of geometry and have later been extended to other branches of mathematics. This involves the elucidation of sets of axioms and rules of inference which can be used in proofs of the properties of computer programs. Examples are given of such axioms and rules, and a formal proof of a simple theorem is displayed. Finally, it is argued that important advantages, both theoretical and practical, may follow from a pursuance of these topics.

KEY WORDS AND PHRASES: axiomatic method, theory of programming' proofs of programs, formal !anguage definition, programming language design, machine-independent programming, program documentation
CR CATEGORY: 4.0, 4.21, 4.22, 5.20, 5.21, 5.23, 5.24

1. Introduction

Computer programming is an exact science in that all the properties of a program and all the consequences of executing it in any given environment can, in principle, be found out from the text of the program itself by means of purely deductive reasoning. Deductive reasoning involves the application of valid rules of inference to sets of valid axioms. It is therefore desirable and interesting to elucidate the axioms and rules of inference which underlie our reasoning about computer programs. The exact choice of axioms will to some extent depend on the choice of programming language. For illustrative purposes, this paper is confined to a very simple language, which is effectively a subset of all current procedure-oriented languages.

* Department of Computer Science

2. Computer Arithmetic

The first requirement in valid reasoning about a program is to know the properties of the elementary operations which it invokes, for example, addition and multiplication of integers. Unfortunately, in several respects computer arithmetic is not the same as the arithmetic familiar to mathematicians, and it is necessary to exercise some care in selecting an appropriate set of axioms. For example, the axioms displayed in Table I are rather a small selection of axioms relevant to integers. From this incomplete set of axioms it is possible to deduce such simple theorems as:

$$x = x + y \times 0$$

$$y \leqslant r \supset r + y \times q = (r - y) + y \times (1 + q)$$

The proof of the second of these is:

A5 $(r - y) + y \times (1 + q)$

$= (r - y) + (y \times 1 + y \times q)$

A9 $= (r - y) + (y + y \times q)$

A3 $= ((r - y) + y) + y \times q$

A6 $= r + y \times q$ provided $y \leqslant r$

The axioms A1 to A9 are, of course, true of the traditional infinite set of integers in mathematics. However, they are also true of the finite sets of "integers" which are manipulated by computers provided that they are confined to *nonnegative* numbers. Their truth is independent of the size of the set; furthermore, it is largely independent of the choice of technique applied in the event of "overflow"; for example:

(1) Strict interpretation: the result of an overflowing operation does not exist; when overflow occurs, the offending program never completes its operation. Note that in this case, the equalities of A1 to A9 are strict, in the sense that both sides exist or fail to exist together.

(2) Firm boundary: the result of an overflowing operation is taken as the maximum value represented.

TABLE I

A1	$x + y = y + x$	addition is commutative
A2	$x \times y = y \times x$	multiplication is commutative
A3	$(x + y) + z = x + (y + z)$	addition is associative
A4	$(x \times y) \times z = x \times (y \times z)$	multiplication is associative
A5	$x \times (y + z) = x \times y + x \times z$	multiplication distributes through addition
A6	$y \leqslant x \supset (x - y) + y = x$	addition cancels subtraction
A7	$x + 0 = x$	
A8	$x \times 0 = 0$	
A9	$x \times 1 = x$	

TABLE II

1. Strict Interpretation

+	0	1	2	3
0	0	1	2	3
1	1	2	3	*
2	2	3	*	*
3	3	*	*	*

×	0	1	2	3
0	0	0	0	0
1	0	1	2	3
2	0	2	*	*
3	0	3	*	*

* nonexistent

2. Firm Boundary

+	0	1	2	3
0	0	1	2	3
1	1	2	3	3
2	2	3	3	3
3	3	3	3	3

×	0	1	2	3
0	0	0	0	0
1	0	1	2	3
2	0	2	3	3
3	0	3	3	3

3. Modulo Arithmetic

+	0	1	2	3
0	0	1	2	3
1	1	2	3	0
2	2	3	0	1
3	3	0	1	2

×	0	1	2	3
0	0	0	0	0
1	0	1	2	3
2	0	2	0	2
3	0	3	2	1

(3) Modulo arithmetic: the result of an overflowing operation is computed modulo the size of the set of integers represented.

These three techniques are illustrated in Table II by addition and multiplication tables for a trivially small model in which 0, 1, 2, and 3 are the only integers represented.

It is interesting to note that the different systems satisfying axioms A1 to A9 may be rigorously distinguished from each other by choosing a particular one of a set of mutually exclusive supplementary axioms. For example, infinite arithmetic satisfies the axiom:

$A10_I \quad \neg \exists x \forall y \qquad (y \leqslant x),$

where all finite arithmetics satisfy:

$A10_F \quad \forall x \qquad (x \leqslant \max)$

where "max" denotes the largest integer represented.

Similarly, the three treatments of overflow may be distinguished by a choice of one of the following axioms relating to the value of max + 1:

$A11_S \quad \neg \exists x \quad (x = \max + 1)$ (strict interpretation)

$A11_B \quad \max + 1 = \max$ (firm boundary)

$A11_M \quad \max + 1 = 0$ (modulo arithmetic)

Having selected one of these axioms, it is possible to use it in deducing the properties of programs; however, these properties will not necessarily obtain, unless the program is executed on an implementation which satisfies the chosen axiom.

3. Program Execution

As mentioned above, the purpose of this study is to provide a logical basis for proofs of the properties of a program. One of the most important properties of a program is whether or not it carries out its intended function. The intended function of a program, or part of a program, can be specified by making general assertions about the values which the relevant variables will take *after* execution of the program. These assertions will usually not ascribe particular values to each variable, but will rather specify certain general properties of the values and the relationships holding between them. We use the normal notations of mathematical logic to express these assertions, and the familiar rules of operator precedence have been used wherever possible to improve legibility.

In many cases, the validity of the results of a program (or part of a program) will depend on the values taken by the variables before that program is initiated. These initial preconditions of successful use can be specified by the same type of general assertion as is used to describe the results obtained on termination. To state the required connection between a precondition (P), a program (Q) and a description of the result of its execution (R), we introduce a new notation:

$$P\,\{Q\}\,R.$$

This may be interpreted "If the assertion P is true before initiation of a program Q, then the assertion R will be true on its completion." If there are no preconditions imposed, we write **true** $\{Q\}R$.[1]

The treatment given below is essentially due to Floyd [8] but is applied to texts rather than flowcharts.

3.1. Axiom of Assignment

Assignment is undoubtedly the most characteristic feature of programming a digital computer, and one that most clearly distinguishes it from other branches of mathematics. It is surprising therefore that the axiom governing our reasoning about assignment is quite as simple as any to be found in elementary logic.

Consider the assignment statement:

$$x := f$$

where

x is an identifier for a simple variable;
f is an expression of a programming language without side effects, but possibly containing x.

Now any assertion $P(x)$ which is to be true of (the value of) x *after* the assignment is made must also have been true of (the value of) the expression f, taken *before* the assignment is made, i.e. with the old value of x. Thus if $P(x)$ is to be true after the assignment, then $P(f)$ must be true before the assignment. This fact may be expressed more formally:

D0 Axiom of Assignment

$\vdash P_0\{x := f\}\, P$

where

x is a variable identifier;

f is an expression;

P_0 is obtained from P by substituting f for all occurrences of x.

[1] If this can be proved in our formal system, we use the familiar logical symbol for theoremhood: $\vdash P\ \{Q\}\ R$

It may be noticed that D0 is not really an axiom at all, but rather an axiom schema, describing an infinite set of axioms which share a common pattern. This pattern is described in purely syntactic terms, and it is easy to check whether any finite text conforms to the pattern, thereby qualifying as an axiom, which may validly appear in any line of a proof.

3.2. Rules of Consequence

In addition to axioms, a deductive science requires at least one rule of inference, which permits the deduction of new theorems from one or more axioms or theorems already proved. A rule of inference takes the form "If $\vdash X$ and $\vdash Y$ then $\vdash Z$", i.e. if assertions of the form X and Y have been proved as theorems, then Z also is thereby proved as a theorem. The simplest example of an inference rule states that if the execution of a program Q ensures the truth of the assertion R, then it also ensures the truth of every assertion logically implied by R. Also, if P is known to be a precondition for a program Q to produce result R, then so is any other assertion which logically implies P. These rules may be expressed more formally:

D1 Rules of Consequence

If $\vdash P\{Q\}R$ and $\vdash R \supset S$ then $\vdash P\{Q\}S$

If $\vdash P\{Q\}R$ and $\vdash S \supset P$ then $\vdash S\{Q\}R$

3.3. Rule of Composition

A program generally consists of a sequence of statements which are executed one after another. The statements may be separated by a semicolon or equivalent symbol denoting procedural composition: $(Q_1 ; Q_2 ; \cdots ; Q_n)$. In order to avoid the awkwardness of dots, it is possible to deal initially with only two statements $(Q_1 ; Q_2)$, since longer sequences can be reconstructed by nesting, thus $(Q_1 ; (Q_2 ; (\cdots (Q_{n-1} ; Q_n) \cdots)))$. The removal of the brackets of this nest may be regarded as convention based on the associativity of the ";-operator", in the same way as brackets are removed from an arithmetic expression $(t_1 + (t_2 + (\cdots (t_{n-1} + t_n) \cdots)))$.

The inference rule associated with composition states that if the proven result of the first part of a program is identical with the precondition under which the second part of the program produces its intended result, then the whole program will produce the intended result, provided that the precondition of the first part is satisfied.

In more formal terms:

D2 Rule of Composition

If $\vdash P\{Q_1\}R_1$ and $\vdash R_1\{Q_2\}R$ then $\vdash P\{(Q_1 ; Q_2)\}R$

3.4. Rule of Iteration

The essential feature of a stored program computer is the ability to execute some portion of program (S) repeatedly until a condition (B) goes false. A simple way of expressing such an iteration is to adapt the Algol 60 **while** notation:

while B **do** S

In executing this statement, a computer first tests the condition B. If this is false, S is omitted, and execution of the loop is complete. Otherwise, S is executed and B is tested again. This action is repeated until B is found to be false. The reasoning which leads to a formulation of an inference rule for iteration is as follows. Suppose P to be an assertion which is always true on completion of S, provided that it is also true on initiation. Then obviously P will still be true after any number of iterations of the statement S (even no iterations). Furthermore, it is known that the controlling condition B is false when the iteration finally terminates. A slightly more powerful formulation is possible in light of the fact that B may be assumed to be true on initiation of S:

D3 Rule of Iteration

If $\vdash P \wedge B\{S\}P$ then $\vdash P\{\textbf{while } B \textbf{ do } S\}\neg B \wedge P$

3.5. Example

The axioms quoted above are sufficient to construct the proof of properties of simple programs, for example, a routine intended to find the quotient q and remainder r obtained on dividing x by y. All variables are assumed to range over a set of nonnegative integers conforming to the axioms listed in Table I. For simplicity we use the trivial but inefficient method of successive subtraction. The proposed program is:

$$((r := x;\quad q := 0);\quad \textbf{while}\ y \leqslant r\ \textbf{do}\ (r := r - y;\quad q := 1 + q))$$

An important property of this program is that when it terminates, we can recover the numerator x by adding to the remainder r the product of the divisor y and the quotient q (i.e. $x = r + y \times q$). Furthermore, the remainder is less than the divisor. These properties may be expressed formally:

$$\textbf{true}\ \{Q\}\ \neg y \leqslant r \wedge x = r + y \times q$$

where Q stands for the program displayed above. This expresses a necessary (but not sufficient) condition for the "correctness" of the program.

A formal proof of this theorem is given in Table III. Like all formal proofs, it is excessively tedious, and it would be fairly easy to introduce notational conventions which would significantly shorten it. An even more powerful method of reducing the tedium of formal proofs is to derive general rules for proof construction out of the simple rules accepted as postulates. These general rules would be shown to be valid by demonstrating how every theorem proved with their assistance could equally well (if more tediously) have been proved without. Once a powerful set

TABLE III

Line number	Formal proof	Justification
1	$\mathbf{true} \supset x = x + y \times 0$	Lemma 1
2	$x = x + y \times 0 \{r := x\} x = r + y \times 0$	D0
3	$x = r + y \times 0 \{q := 0\} x = r + y \times q$	D0
4	$\mathbf{true} \{r := x\} x = r + y \times 0$	D1 (1, 2)
5	$\mathbf{true} \{r := x;\ q := 0\} x = r + y \times q$	D2 (4, 3)
6	$x = r + y \times q \wedge y \leqslant r \supset x = (r-y) + y \times (1+q)$	Lemma 2
7	$x = (r-y) + y \times (1+q)\{r := r-y\} x = r + y \times (1+q)$	D0
8	$x = r + y \times (1+q)\{q := 1+q\} x = r + y \times q$	D0
9	$x = (r-y) + y \times (1+q)\{r := r-y;\ q := 1+q\}\ x = r + y \times q$	D2 (7, 8)
10	$x = r + y \times q \wedge y \leqslant r\ \{r := r-y;\ q := 1+q\}\ x = r + y \times q$	D1 (6, 9)
11	$x = r + y \times q\ \{\mathbf{while}\ y \leqslant r\ \mathbf{do}\ (r := r-y;\ q := 1+q)\}\ \neg y \leqslant r \wedge x = r + y \times q$	D3 (10)
12	$\mathbf{true}\ \{((r := x;\ q := 0);\ \mathbf{while}\ y \leqslant r\ \mathbf{do}\ (r := r-y;\ q := 1+q))\}\ \neg y \leqslant r \wedge x = r + y \times q$	D2 (5, 11)

NOTES

1. The left hand column is used to number the lines, and the right hand column to justify each line, by appealing to an axiom, a lemma or a rule of inference applied to one or two previous lines, indicated in brackets. Neither of these columns is part of the formal proof. For example, line 2 is an instance of the axiom of assignment (D0); line 12 is obtained from lines 5 and 11 by application of the rule of composition (D2).

2. Lemma 1 may be proved from axioms A7 and A8.

3. Lemma 2 follows directly from the theorem proved in Sec. 2.

of supplementary rules has been developed, a "formal proof" reduces to little more than an informal indication of how a formal proof could be constructed.

4. General Reservations

The axioms and rules of inference quoted in this paper have implicitly assumed the absence of side effects of the evaluation of expressions and conditions. In proving properties of programs expressed in a language permitting side effects, it would be necessary to prove their absence in each case before applying the appropriate proof technique. If the main purpose of a high level programming language is to assist in the construction and verification of correct programs, it is doubtful whether the use of functional notation to call procedures with side effects is a genuine advantage.

Another deficiency in the axioms and rules quoted above is that they give no basis for a proof that a program successfully terminates. Failure to terminate may be due to an infinite loop; or it may be due to violation of an implementation-defined limit, for example, the range of numeric operands, the size of storage, or an operating system time limit. Thus the notation "$P\{Q\}R$" should be interpreted "provided that the program successfully terminates, the properties of its results are described by R." It is fairly easy to adapt the axioms so that they cannot be used to predict the "results" of nonterminating programs; but the actual use of the axioms would now depend on knowledge of many implementation-dependent features, for example, the size and speed of the computer, the range of numbers, and the choice of overflow technique. Apart from proofs of the avoidance of infinite loops, it is probably better to prove the "conditional" correctness of a program and rely on an implementation to give a warning if it has had to abandon execution of the program as a result of violation of an implementation limit.

Finally it is necessary to list some of the areas which have not been covered: for example, real arithmetic, bit and character manipulation, complex arithmetic, fractional arithmetic, arrays, records, overlay definition, files, input/

output, declarations, subroutines, parameters, recursion, and parallel execution. Even the characterization of integer arithmetic is far from complete. There does not appear to be any great difficulty in dealing with these points, provided that the programming language is kept simple. Areas which do present real difficulty are labels and jumps, pointers, and name parameters. Proofs of programs which made use of these features are likely to be elaborate, and it is not surprising that this should be reflected in the complexity of the underlying axioms.

5. Proofs of Program Correctness

The most important property of a program is whether it accomplishes the intentions of its user. If these intentions can be described rigorously by making assertions about the values of variables at the end (or at intermediate points) of the execution of the program, then the techniques described in this paper may be used to prove the correctness of the program, provided that the implementation of the programming language conforms to the axioms and rules which have been used in the proof. This fact itself might also be established by deductive reasoning, using an axiom set which describes the logical properties of the hardware circuits. When the correctness of a program, its compiler, and the hardware of the computer have all been established with mathematical certainty, it will be possible to place great reliance on the results of the program, and predict their properties with a confidence limited only by the reliability of the electronics.

The practice of supplying proofs for nontrivial programs will not become widespread until considerably more powerful proof techniques become available, and even then will not be easy. But the practical advantages of program proving will eventually outweigh the difficulties, in view of the increasing costs of programming error. At present, the method which a programmer uses to convince himself of the correctness of his program is to try it out in particular cases and to modify it if the results produced do not correspond to his intentions. After he has found a reasonably wide variety of example cases on which the program seems to work, he believes that it will always work. The time spent in this program testing is often more than half the time spent on the entire programming project; and with a realistic costing of machine time, two thirds (or more) of the cost of the project is involved in removing errors during this phase.

The cost of removing errors discovered after a program has gone into use is often greater, particularly in the case of items of computer manufacturer's software for which a large part of the expense is borne by the user. And finally, the cost of error in certain types of program may be almost incalculable—a lost spacecraft, a collapsed building, a crashed aeroplane, or a world war. Thus the practice of program proving is not only a theoretical pursuit, followed in the interests of academic respectability, but a serious recommendation for the reduction of the costs associated with programming error.

The practice of proving programs is likely to alleviate some of the other problems which afflict the computing world. For example, there is the problem of program documentation, which is essential, firstly, to inform a potential user of a subroutine how to use it and what it accomplishes, and secondly, to assist in further development when it becomes necessary to update a program to meet changing circumstances or to improve it in the light of increased knowledge. The most rigorous method of formulating the purpose of a subroutine, as well as the conditions of its proper use, is to make assertions about the values of variables before and after its execution. The proof of the correctness of these assertions can then be used as a lemma in the proof of any program which calls the subroutine. Thus, in a large program, the structure of the whole can be clearly mirrored in the structure of its proof. Furthermore, when it becomes necessary to modify a program, it will always be valid to replace any subroutine by another which satisfies the same criterion of correctness. Finally, when examining the detail of the algorithm, it seems probable that the proof will be helpful in explaining not only *what* is happening but *why*.

Another problem which can be solved, insofar as it is soluble, by the practice of program proofs is that of transferring programs from one design of computer to another. Even when written in a so-called machine-independent programming language, many large programs inadvertently take advantage of some machine-dependent property of a particular implementation, and unpleasant and expensive surprises can result when attempting to transfer it to another machine. However, presence of a machine-dependent feature will always be revealed in advance by the failure of an attempt to prove the program from machine-independent axioms. The programmer will then have the choice of formulating his algorithm in a machine-independent fashion, possibly with the help of environment enquiries; or if this involves too much effort or inefficiency, he can deliberately construct a machine-dependent program, and rely for his proof on some machine-dependent axiom, for example, one of the versions of **A11** (Section 2). In the latter case, the axiom must be explicitly quoted as one of the preconditions of successful use of the program. The program can still, with complete confidence, be transferred to any other machine which happens to satisfy the same machine-dependent axiom; but if it becomes necessary to transfer it to an implementation which does not, then all the places where changes are required will be clearly annotated by the fact that the proof at that point appeals to the truth of the offending machine-dependent axiom.

Thus the practice of proving programs would seem to lead to solution of three of the most pressing problems in software and programming, namely, reliability, documentation, and compatibility. However, program proving, certainly at present, will be difficult even for programmers of high caliber; and may be applicable only to quite simple program designs. As in other areas, reliability can be purchased only at the price of simplicity.

6. Formal Language Definition

A high level programming language, such as ALGOL, FORTRAN, or COBOL, is usually intended to be implemented on a variety of computers of differing size, configuration, and design. It has been found a serious problem to define these languages with sufficient rigour to ensure compatibility among all implementors. Since the purpose of compatibility is to facilitate interchange of programs expressed in the language, one way to achieve this would be to insist that all implementations of the language shall "satisfy" the axioms and rules of inference which underlie proofs of the properties of programs expressed in the language, so that all predictions based on these proofs will be fulfilled, except in the event of hardware failure. In effect, this is equivalent to accepting the axioms and rules of inference as the ultimately definitive specification of the meaning of the language.

Apart from giving an immediate and possibly even provable criterion for the correctness of an implementation, the axiomatic technique for the definition of programming language semantics appears to be like the formal syntax of the ALGOL 60 report, in that it is sufficiently simple to be understood both by the implementor and by the reasonably sophisticated user of the language. It is only by bridging this widening communication gap in a single document (perhaps even provably consistent) that the maximum advantage can be obtained from a formal language definition.

Another of the great advantages of using an axiomatic approach is that axioms offer a simple and flexible technique for leaving certain aspects of a language *undefined*, for example, range of integers, accuracy of floating point, and choice of overflow technique. This is absolutely essential for standardization purposes, since otherwise the language will be impossible to implement efficiently on differing hardware designs. Thus a programming language standard should consist of a set of axioms of universal applicability, together with a choice from a set of supplementary axioms describing the range of choices facing an implementor. An example of the use of axioms for this purpose was given in Section 2.

Another of the objectives of formal language definition is to assist in the design of better programming languages.

The regularity, clarity, and ease of implementation of the Algol 60 syntax may at least in part be due to the use of an elegant formal technique for its definition. The use of axioms may lead to similar advantages in the area of "semantics," since it seems likely that a language which can be described by a few "self-evident" axioms from which proofs will be relatively easy to construct will be preferable to a language with many obscure axioms which are difficult to apply in proofs. Furthermore, axioms enable the language designer to express his general *intentions* quite simply and directly, without the mass of detail which usually accompanies algorithmic descriptions. Finally, axioms can be formulated in a manner largely independent of each other, so that the designer can work freely on one axiom or group of axioms without fear of unexpected interaction effects with other parts of the language.

Acknowledgments. Many axiomatic treatments of computer programming [1, 2, 3] tackle the problem of proving the equivalence, rather than the correctness, of algorithms. Other approaches [4, 5] take recursive functions rather than programs as a starting point for the theory. The suggestion to use axioms for defining the primitive operations of a computer appears in [6, 7]. The importance of program proofs is clearly emphasized in [9], and an informal technique for providing them is described. The suggestion that the specification of proof techniques provides an adequate formal definition of a programming language first appears in [8]. The formal treatment of program execution presented in this paper is clearly derived from Floyd. The main contributions of the author appear to be: (1) a suggestion that axioms may provide a simple solution to the problem of leaving certain aspects of a language undefined; (2) a comprehensive evaluation of the possible benefits to be gained by adopting this approach both for program proving and for formal language definition.

However, the formal material presented here has only an expository status and represents only a minute proportion of what remains to be done. It is hoped that many of the fascinating problems involved will be taken up by others.

REFERENCES

1. YANOV, YU I. Logical operator schemes. *Kybernetika 1*, (1958).
2. IGARASHI, S. An axiomatic approach to equivalence problems of algorithms with applications. Ph.D. Thesis 1964. Rep. Compt. Centre, U. Tokyo, 1968, pp. 1–101.
3. DE BAKKER, J. W. Axiomatics of simple assignment statements. M.R. 94, Mathematisch Centrum, Amsterdam, June 1968.
4. MCCARTHY, J. Towards a mathematical theory of computation. Proc. IFIP Cong. 1962, North Holland Pub. Co., Amsterdam, 1963.
5. BURSTALL, R. Proving properties of programs by structural induction. Experimental Programming Reports: No. 17 DMIP, Edinburgh, Feb. 1968.
6. VAN WIJNGAARDEN, A. Numerical analysis as an independent science. *BIT 6* (1966), 66–81.
7. LASKI, J. Sets and other types. ALGOL Bull. 27, 1968.
8. FLOYD, R. W. Assigning meanings to programs. Proc. Amer. Math. Soc. Symposia in Applied Mathematics, Vol. 19, pp. 19–31.
9. NAUR, P. Proof of algorithms by general snapshots. *BIT 6* (1966), 310–316.

RECEIVED NOVEMBER, 1968; REVISED MAY, 1969

C.A.R. Hoare

Proof of Correctness of Data Representations

Acta Informatica, Vol. 1, Fasc. 4, 1972
pp. 271-281

Acta Informatica 1, 271—281 (1972)

Proof of Correctness of Data Representations

C. A. R. Hoare

Received February 16, 1972

Summary. A powerful method of simplifying the proofs of program correctness is suggested; and some new light is shed on the problem of functions with side-effects.

1. Introduction

In the development of programs by stepwise refinement [1–4], the programmer is encouraged to postpone the decision on the representation of his data until after he has designed his algorithm, and has expressed it as an "abstract" program operating on "abstract" data. He then chooses for the abstract data some convenient and efficient concrete representation in the store of a computer; and finally programs the primitive operations required by his abstract program in terms of this concrete representation. This paper suggests an automatic method of accomplishing the transition between an abstract and a concrete program, and also a method of proving its correctness; that is, of proving that the concrete representation exhibits all the properties expected of it by the "abstract" program. A similar suggestion was made more formally in algebraic terms in [5], which gives a general definition of simulation. However, a more restricted definition may prove to be more useful in practical program proofs.

If the data representation is proved correct, the correctness of the final concrete program depends only on the correctness of the original abstract program. Since abstract programs are usually very much shorter and easier to prove correct, the total task of proof has been considerably lightened by factorising it in this way. Furthermore, the two parts of the proof correspond to the successive stages in program development, thereby contributing to a constructive approach to the correctness of programs [6]. Finally, it must be recalled that in the case of larger and more complex programs the description given above in terms of two stages readily generalises to multiple stages.

2. Concepts and Notations

Suppose in an abstract program there is some abstract variable t which is regarded as being of type T (say a small set of integers). A concrete representation of t will usually consist of several variables $c_1, c_2, \ldots, c_n$ whose types are directly (or more directly) represented in the computer store. The primitive operations on the variable t are represented by procedures $p_1, p_2, \ldots, p_m$, whose bodies carry out on the variables $c_1, c_2, \ldots, c_n$ a series of operations directly (or more directly) performed by computer hardware, and which correspond to meaningful operations on the abstract variable t. The entire concrete representation of the type T can be expressed by declarations of these variables and procedures. For this we adopt the notation of the SIMULA 67 [7] class declaration, which specifies the association between an abstract type T and its concrete representation:

```
class T;
   begin ... declarations of c1, c2, ..., cn ...;
         procedure p1 <formal parameter part>; Q1;
         procedure p2 <formal parameter part>; Q2;
               .......                                          (1)
         procedure pm <formal parameter part>; Qm;
         Q
   end;
```

where Q is a piece of program which assigns initial values (if desired) to the variables $c_1, c_2, \ldots, c_n$. As in ALGOL 60, any of the p's may be functions; this is signified by preceding the procedure declaration by the type of the procedure.

Having declared a representation for a type T, it will be required to use this in the abstract program to declare all variables which are to be represented in that way. For this purpose we use the notation:

var $(T)\ t$;

or for multiple declarations:

var $(T)\ t_1, t_2, \ldots$;

The same notation may be used for specifying the types of arrays, functions, and parameters. Within the block in which these declarations are made, it will be required to operate upon the variables $t, t_1, \ldots$, in the manner defined by the bodies of the procedures $p_1, p_2, \ldots, p_m$. This is accomplished by introducing a compound notation for a procedure call:

$$t_i \cdot p_j \ \langle\text{actual parameter part}\rangle;$$

where t_i names the variable to be operated upon and p_j names the operation to be performed.

If p_j is a function, the notation displayed above is a function designator; otherwise it is a procedure statement. The form $t_i \cdot p_j$ is known as a *compound identifier*.

These concepts and notations have been closely modelled on those of SIMULA 67. The only difference is the use of **var** (T) instead of **ref** (T). This reflects the fact that in the current treatment, objects of declared classes are not expected to be addressed by reference; usually they will occupy storage space contiguously in the local workspace of the block in which they are declared, and will be addressed by offset in the same way as normal integer and real variables of the block.

3. Example

As an example of the use of these concepts, consider an abstract program which operates on several small sets of integers. It is known that none of these sets ever has more than a hundred members. Furthermore, the only operations actually used in the abstract program are the initial clearing of the set, and the insertion and removal of individual members of the set. These are denoted by procedure statements

$$s \cdot \text{insert}(i)$$

and

$$s \cdot \text{remove}(i).$$

There is also a function "$s \cdot \text{has}(i)$", which tests whether i is a member of s.

It is decided to represent each set as an array A of 100 integer elements, together with a pointer m to the last member of the set; m is zero when the set is empty. This representation can be declared:

```
class smallintset;
begin integer m; integer array A [1:100];

      procedure insert (i); integer i;
         begin integer j;
            for j:=1 step 1 until m do
               if A [j] =i then go to end insert;
               m:=m+1;
               A [m]:=i;

  end insert: end insert;

      procedure remove (i); integer i;
         begin integer j, k;
            for j:=1 step 1 until m do
               if A [j] =i then
                  begin for k:=j+1 step 1 until m do A [k-1]:=A [k];
                     comment close the gap over the removed member;
                     m:= m-1;
                     go to end remove
                  end;

end remove: end remove;

      Boolean procedure has (i); integer i;
         begin integer j;
               has:=false;
               for j:=1 step 1 until m do
                  if A [j] =i then
                     begin has:=true; go to end contains end;

end contains: end contains;

   m:=0; comment initialise set to empty;

end smallintset;
```

Note: as in SIMULA 67, simple variable parameters are presumed to be called by value.

4. Semantics and Implementation

The meaning of class declarations and calls on their constituent procedures may be readily explained by textual substitution; this also gives a useful clue to a practical and efficient method of implementation. A declaration:

$$\textbf{var}(T)t;$$

is regarded as equivalent to the unbracketed body of the class declaration with **begin** ... **end** brackets removed, after every occurrence of an identifier c_i or p_i declared in it has been prefixed by "$t\cdot$". If there are any initialising statements in the class declaration these are removed and inserted just in front of the compound tail of the block in which the declaration is made. Thus if T has the form displayed in (1), **var**$(T)t$ is equivalent to:

... declarations for $t\cdot c_1, t\cdot c_2, \ldots, t\cdot c_n \ldots$;
procedure $t\cdot p_1(\ldots); Q_1'$;
procedure $t\cdot p_2(\ldots); Q_2'$;
....
procedure $t\cdot p_m(\ldots); Q_m'$;

where $Q_1', Q_2', \ldots, Q_m', Q'$ are obtained from $Q_1, Q_2, \ldots, Q_m, Q$ by prefixing every occurrence of $c_1, c_2, \ldots, c_n, p_1, p_2, \ldots, p_m$ by "$t\cdot$". Furthermore, the initialising statement Q' will have been inserted just ahead of the statements of the block body.

If there are several variables of class T declared in the same block, the method described above can be applied to each of them. But in a practical implementation, only one copy of the procedure bodies will be translated. This would contain as an extra parameter an address to the block of $c_1, c_2, \ldots, c_n$ on which a particular call is to operate.

5. Criterion of Correctness

In an abstract program, an operation of the form

$$t_i\cdot p_j(a_1, a_2, \ldots, a_{n_j}) \tag{2}$$

will be expected to carry out some transformation on the variable t_i, in such a way that its resulting value is $f_j(t_i, a_1, a_2, \ldots, a_{n_j})$, where f_j is some primitive operation required by the abstract program. In other words the procedure statement is expected to be equivalent to the assignment

$$t_i := f_j(t_i, a_1, a_2, \ldots, a_{n_j});$$

When this equivalence holds, we say that p_j *models* f_j. A similar concept of modelling applies to functions. It is desired that the proof of the abstract program may be based on the equivalence, using the rule of assignment [8], so that for any propositional formula S, the abstract programmer may assume:

$$S^{t_i}_{f_j(t_i, a_1, a_2, \ldots, a_{n_j})}\{t_i\cdot p_j(a_1, a_2, \ldots, a_{n_j})\}S.$$[1]

1 S^x_y stands for the result of replacing all free occurrences of x in S by y: if any free variables of y would become bound in S by this substitution, this is avoided by preliminary systematic alteration of bound variables in S.

In addition, the abstract programmer will wish to assume that all declared variables are initialised to some designated value d_0 of the abstract space.

The criterion of correctness of a data representation is that every p_j models the intended f_j and that the initialisation statement "models" the desired initial value; and consequently, a program operating on abstract variables may validly be replaced by one carrying out equivalent operations on the concrete representation.

Thus in the case of smallintset, we require to prove that:

$$\begin{aligned}
&\textbf{var}\,(i)\,t \text{ initialises } t \text{ to } \{\ \} \text{ (the empty set)}\\
&t \cdot \text{insert } (i) \equiv t := t \cup \{i\}\\
&t \cdot \text{remove}\,(i) \equiv t := t \cap \neg\{i\}\\
&t \cdot \text{has}\,(i) \equiv i \in t.
\end{aligned} \qquad (3)$$

6. Proof Method

The first requirement for the proof is to define the relationship between the abstract space in which the abstract program is written, and the space of the concrete representation. This can be accomplished by giving a function $\mathscr{A}(c_1, c_2, \ldots, c_n)$ which maps the concrete variables into the abstract object which they represent. For example, in the case of smallintset, the representation function can be defined as

$$\mathscr{A}(m, A) = \{i\colon \text{integer} \mid \exists k\,(1 \leqq k \leqq m \;\&\; A\,[k] = i)\} \qquad (4)$$

or in words, "(m, A) represents the set of values of the first m elements of A". Note that in this and in many other cases $\mathscr{A}$ will be a many-one function. Thus there is no unique concrete value representing any abstract one.

Let t stand for the value of $\mathscr{A}(c_1, c_2, \ldots, c_m)$ before execution of the body Q_j of procedure p_j. Then what we must prove is that after execution of Q_j the following relation holds:

$$\mathscr{A}(c_1, c_2, \ldots, c_n) = f_j(t, v_1, v_2, \ldots, v_{nj})$$

where $v_1, v_2, \ldots, v_{nj}$ are the formal parameters of p_j.

Using the notations of [8], the requirement for proof may be expressed:

$$t = \mathscr{A}(c_1, c_2, \ldots, c_n)\ \{Q_j\}\ \mathscr{A}(c_1, c_2, \ldots, c_n) = f_j(t, v_1, v_2, \ldots, v_{nj})$$

where t is a variable which does not occur in Q_j. On the basis of this we may say: $t \cdot p_j(a_1, a_2, \ldots, a_n) \equiv t := f_j(t, a_1, a_2, \ldots, a_n)$ with respect to $\mathscr{A}$. This deduction depends on the fact that no Q_j alters or accesses any variables other than $c_1, c_2, \ldots, c_n$; we shall in future assume that this constraint has been observed.

In fact for practical proofs we need a slightly stronger rule, which enables the programmer to give an invariant condition $I(c_1, c_2, \ldots, c_n)$, defining some relationship between the constituent concrete variables, and thus placing a constraint on the possible combinations of values which they may take. Each operation (except initialisation) may assume that I is true when it is first entered; and each operation must in return ensure that it is true on completion.

In the case of smallintset, the correctness of all operations depends on the fact that m remains within the bounds of A, and the correctness of the remove operation is dependent on the fact that the values of $A[1], A[2], \ldots, A[m]$ are all different; a simple expression of this invariant is:

$$\text{size}\,(\mathscr{A}(m, A)) = m \leqq 100. \qquad (I)$$

One additional complexity will often be required; in general, a procedure body is not prepared to accept arbitrary combinations of values for its parameters, and its correctness therefore depends on satisfaction of some precondition $P(t, a_1, a_2, \ldots, a_n)$ before the procedure is entered. For example, the correctness of the insert procedure depends on the fact that the size of the resulting set is not greater than 100, that is

$$\text{size}\,(t \cup \{i\}) \leqq 100$$

This precondition (with t replaced by $\mathscr{A}$) may be assumed in the proof of the body of the procedure; but it must accordingly be proved to hold before every call of the procedure.

It is interesting to note that any of the p's that are functions may be permitted to change the values of the c's, on condition that it preserves the truth of the invariant, and also that it preserves unchanged the value of the abstract object $\mathscr{A}$. For example, the function "has" could reorder the elements of A; this might be an advantage if it is expected that membership of some of the members of the set will be tested much more frequently than others. The existence of such a concrete side-effect is wholly invisible to the abstract program. This seems to be a convincing explanation of the phenomenon of "benevolent side-effects", whose existence I was not prepared to admit in [8].

7. Proof of Smallintset

The proof may be split into four parts, corresponding to the four parts of the class declaration:

7.1. Initialisation

What we must prove is that after initialisation the abstract set is empty and that the invariant I is true:

$$\textbf{true}\,\{m := 0\}\,\{i \mid \exists k\,(1 \leqq k \leqq m \,\&\, A[k] = i)\} = \{\,\}$$
$$\&\ \text{size}\,(\mathscr{A}(m, a)) = m \leqq 100$$

Using the rule of assignment, this depends on the obvious truth of the lemma

$$\{i \mid \exists k\,(1 \leqq k \leqq 0 \,\&\, A[k] = i\} = \{\,\} \,\&\ \text{size}\,(\{\,\}) = 0 \leqq 100$$

7.2. Has

What we must prove is

$$\mathscr{A}(m, A) = k \,\&\, I\,\{Q_{\text{has}}\}\,\mathscr{A}(m, A) = k \,\&\, I \,\&\ \text{has} = i \in \mathscr{A}(m, A)$$

where Q_{has} is the body of has. Since Q_{has} does not change the value of m or A, the truth of the first two assertions on the right hand side follows directly from their truth beforehand. The invariant of the loop inside Q_{has} is:

$$j \leqq m \,\&\, \text{has} = i \in \mathscr{A}(j, A)$$

as may be verified by a proof of the lemma:

$$\begin{aligned} & j < m \,\&\, j \leqq m \,\&\, \text{has} = i \in \mathscr{A}(j, A) \\ & \supset \textbf{if } A[j+1] = i \textbf{ then } (\textbf{true} = i \in \mathscr{A}(m, A)) \\ & \qquad\qquad \textbf{else } \text{has} = i \in \mathscr{A}(j+1, A). \end{aligned}$$

Since the final value of j is m, the truth of the desired result follows directly from the invariant; and since the "initial" value of j is zero, we only need the obvious lemma

$$\textbf{false} = i \in \mathscr{A}(0, A)$$

7.3. *Insert*

What we must prove is:

$$P \,\&\, \mathscr{A}(m, A) = k \,\&\, I \{Q_{\text{insert}}\} \mathscr{A}(m, A) = (k \cup \{i\}) \,\&\, I,$$

where $P \equiv_{\text{df}} \text{size}\,(\mathscr{A}(m, A) \cup \{i\}) \leqq 100$.

The invariant of the loop is:

$$P \,\&\, \mathscr{A}(m, A) = k \,\&\, I \,\&\, i \,\bar{\in}\, \mathscr{A}(j, A) \,\&\, 0 \leqq j \leqq m \qquad (6)$$

as may be verified by the proof of the lemma

$$\begin{aligned} & \mathscr{A}(m, A) = k \,\&\, i \,\bar{\in}\, \mathscr{A}(j, A) \,\&\, 0 \leqq j \leqq m \,\&\, j < m \supset \\ & \quad \textbf{if } A[j+1] = i \textbf{ then } \mathscr{A}(m, A) = (k \cup \{i\}) \\ & \qquad\qquad \textbf{else } 0 \leqq j+1 \leqq m \,\&\, i \,\bar{\in}\, \mathscr{A}(j+1, A) \end{aligned}$$

(The invariance of $P \,\&\, \mathscr{A}(m, A) = k \,\&\, I$ follows from the fact that the loop does not change the values of m or A). That (6) is true before the loop follows from $i \,\bar{\in}\, \mathscr{A}(0, A)$.

We must now prove that the truth of (6), together with $j = m$ at the end of the loop, is adequate to ensure the required final condition. This depends on proof of the lemma

$$j = m \,\&\, (6) \cup \mathscr{A}(m+1, A') = (k \cup \{i\}) \,\&\, \text{size}\,(\mathscr{A}(m+1, A')) = m+1 \leqq 100$$

where $A' = (A, m+1 : i)$ is the new value of A after assignment of i to $A[m+1]$.

7.4. *Remove*

What we must prove is

$$\mathscr{A}(m, A) = k \,\&\, I \{Q_{\text{remove}}\} \mathscr{A}(m, A) = (k \cap \neg \{i\}) \,\&\, I.$$

The details of the proof are complex. Since they add nothing more to the purpose of this paper, they will be omitted.

8. Formalities

Let T be a class declared as shown in Section 2, and let $\mathscr{A}$, I, P_j, f_j be formulae as explained in Section 6 (free variable lists are omitted where convenient). Suppose also that the following $m+1$ theorems have been proved:

$$\textbf{true}\ \{Q\}\ I \& \mathscr{A} = d_0 \tag{7}$$

$$\mathscr{A} = t \& I \& P_j(t)\ \{Q_j\}\ I \& \mathscr{A} = f_j(t) \quad \text{for procedure bodies } Q_j \tag{8}$$

$$\mathscr{A} = t \& I \& P_j(t)\ \{Q_j\}\ I \& \mathscr{A} = t \& p_j = f_j(t) \quad \text{for function bodies } Q_j. \tag{9}$$

In this section we show that the proof of these theorems is a sufficient condition for the correctness of the data representation, in the sense explained in Section 5.

Let X be a program beginning with a declaration of a variable t of an abstract type, and initialising it to d_0. The subsequent operations on this variable are of the form

(1) $t := f_j(t, a_1, a_2, \ldots, a_{n_j})$ if Q_j is a procedure

(2) $f_j(t, a_1, a_2, \ldots, a_{n_j})$ if Q_j is a function.

Suppose also that $P_j(t, a_1, a_2, \ldots, a_{n_j})$ has been proved true before each such operation.

Let X' be a program formed from X by replacements described in Section 4, as well as the following (see Section 5):

(1) initialisation $t := d_0$ replaced by Q'

(2) $t := f_j(t, a_1, a_2, \ldots, a_{n_j})$ replaced by $t \cdot p_j\,(a_1, a_2, \ldots, a_{n_j})$

(4) $f_j(t, a_1, a_2, \ldots, a_{n_j})$ by $t \cdot p_j(a_1, a_2, \ldots, a_{n_j})$.

Theorem. Under conditions described above, if X and X' both terminate, the value of t on termination of X will be $\mathscr{A}(c_1, c_2, \ldots, c_n)$, where $c_1, c_2, \ldots, c_n$ are the values of these variables on termination of X'.

Corollary. If $R(t)$ has been proved true on termination of X, $R(\mathscr{A})$ will be true on termination of X'.

Proof. Consider the sequence S of operations on t executed during the computation of X, and let S' be the sequence of subcomputations of X' arising from execution of the procedure calls which have replaced the corresponding operations on t in X. We will prove that there is a close elementwise correspondence between the two sequences, and that

(a) each item of S' is the very procedure statement which replaced the corresponding operation in S.

(b) the values of all variables (and hence also the actual parameters) which are common to both "programs" are the same after each operation.

(c) the invariant I is true between successive items of S'.

(d) if the operations are function calls, their results in both sequences are the same.

(e) and if they are procedure calls (or the initialisation) the value of t immediately after the operation in S is given by $\mathscr{A}$, as applied to the values of c_1, $c_2, \ldots, c_n$ after the corresponding operation in S'.

It is this last fact, applied to the last item of the two sequences, that establishes the truth of the theorem.

The proof is by induction on the position of an item in S.

(1) *Basis*. Consider its first item of S, $t := d_0$. Since X and X' are identical up to this point, the first item of S' must be the subcomputation of the procedure Q which replaced it, proving (a). By (7), I is true after Q in S', and also $\mathscr{A} = d_0$, proving (c) and (e). (d) is not relevant. Q is not allowed to change any non-local variable, proving (b).

(2) *Induction step*. We may assume that conditions (a) to (e) hold immediately after the $(n-1)$-th item of S and S', and we establish that they are true after the n-th. Since the value of all other variables (and the result, if a function) were the same after the previous operation in both sequences, the subsequent course of the computation must also be the same until the very next point at which X' differs from X. This establishes (a) and (b). Since the only permitted changes to the values of $t \cdot c_1, t \cdot c_2, \ldots, t \cdot c_n$ occur in the subcomputations of S', and I contains no other variables, the truth of I after the previous subcomputation proves that it is true before the next. Since S contains *all* operations on t, the value of t is the same before the n-th as it was after the $(n-1)$-th operation, and it is still equal to $\mathscr{A}$. It is given as proved that the appropriate $P_j(t)$ is true before each call of f_j in S. Thus we have established that $\mathscr{A} = t \& I \& P_j(t)$ is true before the operation in S'. From (8) or (9) the truth of (c), (d), (e) follows immediately. (b) follows from the fact that the assignment in S changes the value of no other variable besides t; and similarly, Q_j is not permitted to change the value of any variable other than $t \cdot c_1, t \cdot c_2, \ldots, t \cdot c_n$.

This proof has been an informal demonstration of a fairly obvious theorem. Its main interest has been to show the necessity for certain restrictive conditions placed on class declarations. Fortunately these restrictions are formulated as scope rules, which can be rigorously checked at compile time.

9. Extensions

The exposition of the previous sections deals only with the simplest cases of the Simula 67 class concept; nevertheless, it would seem adequate to cover a wide range of practical data representations. In this section we consider the possibility of further extensions, roughly in order of sophistication.

9.1. Class Parameters

It is often useful to permit a class to have formal parameters which can be replaced by different actual parameters whenever the class is used in a declaration. These parameters may influence the method of representation, or the identity of the initial value, or both. In the case of smallintset, the usefulness of the definition could be enhanced if the maximum size of the set is a parameter, rather than being fixed at 100.

9.2. Dynamic Object Generation

In Simula 67, the value of a variable c of class C may be reinitialised by an assignment:

$c :=$ **new** C ⟨actual parameter part⟩;

This presents no extra difficulty for proofs.

9.3. Remote Identification

In many cases, a local concrete variable of a class has a meaningful interpretation in the abstract space. For example, the variable m of smallintset always stands for the size of the set. If the main program needs to test the size of the set, it would be possible to make this accessible by writing a function

integer procedure size; size$:= m$;

But it would be simpler and more convenient to make the variable more directly accessible by a compound identifier, perhaps by declaring it

public integer m;

The proof technique would specify that

$$m = \text{size}\left(\mathscr{A}(m, A)\right)$$

is part of the invariant of the class.

9.4. Class Concatenation

The basic mechanism for representing sets by arrays can be applied to sets with members of type or class other than just integers. It would therefore be useful to have a method of defining a class "smallset", which can then be used to construct other classes such as "smallrealset" or "smallcarset", where "car" is another class. In SIMULA 67, this effect can be achieved by the class/subclass and virtual mechanisms.

9.5. Recursive Class Declaration

In Simula 67, the parameters of a class, or of a local procedure of the class, and even the local variables of a class, may be declared as belonging to that very same class. This permits the construction of lists and trees, and their processing by recursive procedure activation. In proving the correctness of such a class, it will be necessary to assume the correctness of all "recursive" operations in the proofs of the bodies of the procedures. In the implementation of recursive classes, it will be necessary to represent variables by a null pointer (**none**) or by the *address* of their value, rather than by direct inclusion of space for their values in block workspace of the block to which they are local. The reason for this is that the amount of space occupied by a value of recursively defined type cannot be determined at compile time.

It is worthy of note that the proof-technique recommended above is valid only if the data structure is "well-grounded" in the sense that it is a pure tree, without cycles and without convergence of branches. The restrictions suggested in this paper make it impossible for local variables of a class to be updated except by the body of a procedure local to that very same activation of the class; and I believe that this will effectively prevent the construction of structures which are not well-grounded, provided that assignment is implemented by copying the complete value, not just the address.

I am deeply indebted to Doug Ross and to all authors of referenced works. Indeed, the material of this paper represents little more than my belated understanding and formalisation of their original work.

References

1. Wirth, N.: The development of programs by stepwise refinement. Comm. ACM. **14**, 221–227 (1971).
2. Dijkstra, E. W.: Notes on structured programming. In Structured Programming. Academic Press (1972).
3. Hoare, C. A. R.: Notes on data structuring. *Ibid.*
4. Dahl, O.-J.: Hierachical program structures. *Ibid.*
5. Milner, R.: An algebraic definition of simulation between programs. CS 205 Stanford University, February 1971.
6. Dijkstra, E. W.: A constructive approach to the problem of program correctness. BIT. **8**, 174–186 (1968).
7. Dahl, O.-J., Myhrhaug, B., Nygaard, K.: The SIMULA 67 common base language. Norwegian Computing Center, Oslo, Publication No. S-22, 1970.
8. Hoare, C. A. R.: An axiomatic approach to computer programming. Comm. ACM. **12**, 576–580, 583 (1969).

Prof. C. A. R. Hoare
Computer Science
The Queen's University of Belfast
Belfast BT 71 NN
Northern Ireland

On the Criteria to be Used in Decomposing Systems into Modules

David L. Parnas
Professor of Software Engineering
McMaster University
parnas@qusunt.cas.mcmaster.ca

Ph.D. in Electrical Engineering, Carnegie Mellon University

Advisor, Philips, IBM, United States Naval Research Laboratory, Atomic Energy Control Board of Canada

ACM SIGSOFT Outstanding Research Award, ACM Best Paper Award, ICSE Most Influential Paper Award

Honorary degree from the ETH Zurich and from the University of Louvain in Belgium

Fellow of ACM and Royal Society of Canada

Major contributions: information hiding principle, modularization, module specification

Current interests: safety-critical real-time software, computer system design

David L. Parnas

The Secret History of Information Hiding

The concept of "information-hiding" as a software design principle is widely accepted in academic circles. Many successful designs can be seen as successful applications of abstraction or information hiding. On the other hand, most industrial software developers do not apply the idea and many consider it unrealistic. This paper describes how the idea developed, discusses difficulties in its application, and speculates on why it has not been completely successful.

1. A Dutch Adventure

The question of how to divide systems into modules was not on my mind in 1969 when I took a one year leave-of-absence from my academic position to work with a start-up, Philips Computer Industry, in the Netherlands. I was interested in the use of simulation in software development and wanted to see if my ideas could be made to work in industry. I never found out! Instead, I came to understand that software development in industry was at a far more primitive level than my academic experience would have suggested. The ideas that I had gone to Holland to explore were obviously unrealistic.

I had heard and read many arguments in favour of modular programming and those arguments seemed *obviously* correct. The message was clear; software should be designed as a set of independently changeable components. With a background in electrical engineering, I could not even imagine anyone trying to build a complex product except as an assembly of smaller components or modules. Moreover, I was used to having modules with well-designed, relatively simple, interfaces. I was puzzled by the complexity and arbitrary nature of the interfaces that were being discussed.

At that time, in electrical engineering, the modules or components were presented to us; we learned about how to assemble them into larger circuits. We never had occasion to debate the nature of a component. It was exposure to actual industrial software development that made me realise that in software, the nature of a component, or module, is not obvious. In fact people are rarely able to define those terms unless the tools define them. In academia and papers I had seen a few well-organised systems (Dijkstra's THE system for example [5]), but as I became familiar with commercial software, I began to realise that it takes both deep understanding and a lot of effort to keep software simple and easy to understand.

2. Phrases Like "Clean Design" or "Beautiful" Are Not Meaningful

Philips gave me the opportunity to watch the design of a new product. As it developed, I became increasingly uneasy and started to confide my concerns to my supervisor. I repeatedly told him that the designs were not "clean" and that they should be "beautiful" in the way that Dijkstra's designs were beautiful. He did not find this to be a useful comment. He reminded me that we were engineers, not artists. In his mind I was simply expressing my taste and he wanted to know what criterion I was using in making my judgement. I was unable to tell him! Because I could not explain what principles were being violated, my comments were simply ignored and the work went on. Sensing that there must be a way to explain and provide motivation for my "taste", I began to try to find a principle that would distinguish good designs from bad ones. Engineers are very pragmatic; they may admire beauty, but they seek utility. I tried to find an explanation of why "beauty" was useful.

3. The Napkin of Doom

The first glimmer of understanding came during lunch. Two members of the software team began to discuss work. Jan, who was building a compiler, asked Johan, the database expert, "What does it mean to open a file anyway?" Johan and I looked at him strangely; surely he knew what "open" did. You could not do anything to a file until you had opened it. Johan said that, and Jan replied, "No, what does it *really* mean? I have to implement the open command and I need to know what code to generate." Johan grabbed a napkin from the dispenser on the table and began to draw a diagram of a control block. As he drew, he explained the meaning of each field and indicated where specific addresses and control bits were located.

I felt that drawing the picture was a mistake and tried to stop them. I told them that they would be sorry if they continued. When they asked why, I replied only that they should not have to exchange such information. They insisted that the compiler writers needed that information and tried to get rid of me because I was interfering with their progress. Because I could neither explain why I thought they were making a mistake, nor offer them a better way to solve the problem, I became uncharacteristically silent and watched as the discussion was completed. The napkin was taken upstairs, two holes were punched in it, and it was inserted in a binder. Later, other members of the compiler team made copies of the napkin and put them in their own files.

Months later, I had evidence that I had been right. Johan forgot about the napkin. When it was necessary to change the control block format, the napkin was not updated. Jan's team continued to work as if the control block had not changed. When the two products were integrated, they did not work properly and it took some time to figure out why and even longer to correct the compiler. Each change in the control block format triggered a change in both the compiler and the data base system; this undesired coupling often consumed valuable time.

At this point it was clear to me that I had been right but I still was bothered by what Jan and Johan had said. They too were correct. The compiler group did need some information that allowed the necessary code to be generated. I could not tell them what they could have done instead of exchanging the details on the napkin.

4. Information Distribution in Software

As my time in Europe came to an end, I had some of the concept but not all of it. My first paper on "information hiding" was presented at the IFIP conference in 1971 [8]. Although such conferences are intended to bring together people from all over the world, sessions are not widely attended nor are the proceedings well read. Nonetheless, the paper did explain that it was information distribution that made systems "dirty" by establishing almost invisible connections between supposedly independent modules.

5. Teaching an Early Software Engineering Course

When I returned to teaching, I was told that, since I was an engineer[1] and now had industrial software development experience, I should teach the new "software engineering" course. Nobody could tell me what the content of that course should be or how it should differ from an advanced programming course. I was told to teach whatever I thought was important.[2] After some thought, it became clear to me that information distribution, and

1 Actually, I was not an engineer but had engineering degrees. At that time I had not applied for a an engineering license as I was not practising engineering. Like most scientists, the department head did not understand the difference.

2 I fear that the same thing happens in many computer science departments today.

how to avoid it, had to be a big part of that course. I decided to do this by means of a project with limited information distribution and demonstrate the benefits of a "clean" design [11]. I also found a definition of "module" that was independent of any tools, a work assignment for an individual or a group.

To teach this first course, I decided to take the concept of modularity seriously. I designed a project that was divided into five modules and divided the class into five groups. Every individual in a group would have the same work assignment (module) but would use a different method to complete it. With four members per group, we hoped to get four implementations of each module. The implementations of a module were required to be interchangeable so that (if everything worked) we would end up with 1024 executable combinations. Twenty-five randomly selected combinations were tested and (eventually) ran.

The results of this experiment seemed significant to me because the construction of interchangeable software modules seemed unachievable by others at the time. In fact, even today I see how difficult it can be to install an upgrade in the software we all use. Exchanging one implementation of a module for another rarely goes smoothly. There are almost always complex interfaces and many tiny details that keep the "help lines" busy whenever new versions of minor modules are installed.

6. What Johan *Should* Have Said to Jan

In conducting the experiment as an educational exercise, I had finally found the answer to the issue raised a year earlier by Jan and Johan. Because each module had four really different implementations, my students could not exchange the kind of detailed information that Jan and Johan had discussed. Instead I designed an abstract interface, one that all versions of a module could have in common. We treated the description of the interface as a specification, i.e. all implementations were required to satisfy that description. Because we did not tell the students which combinations would be tested, they could not use any additional information. In effect, the implementation details were kept a secret.

The paper that described the experiment and its outcome was published at a major conference but has not been widely read or cited [12].

7. Two Journal Articles on Information Hiding

There were two journal articles that followed and these are widely cited though sometimes it appears that those who cite them have not read them and many who have read them refer to them without citing them. The better of the two articles [10] was first rejected by the journal. The reviewer stated, "Obviously Parnas does not know what he is talking about because nobody does it that way." I objected by pointing out that one should not reject an idea claiming to be new simply because it was new. The editor then published the article without major change.

7.1 Language - Often the Wrong Issue

Because it had taken so long to find a solution to the dilemma posed by Jan and Johan, my first thought was that it was the specification notation that was significant. Consequently, the first *Communications of the ACM* article [9] discussed the specification technique and illustrated the idea by publishing the specifications with many typographic errors introduced in the (manual) typesetting process.

It took several discussions with people who had read the article before I realised the significance of the (obvious) fact that the specification method would not work with an arbitrary set of modules. Most important, it would not have worked with conventionally designed modules. The secret of the experiment's success was not the specification notation but the decomposition.

Over the years I have proposed and studied more general approaches to module specification [1, 18]. Many others have also looked at this problem. Slowly I have come to the realisation that standard mathematical notation is better than most languages that computer scientists can invent. I have learned that it is more important to learn how to use existing mathematical concepts and notation than to invent new, ad hoc, notations.

7.2 What Really Matters Is Design

It was the fact that the modularisation limited the amount of information that had to be shared by module implementers that made the effort a success. After some reflection I realised that I had found the answer to the question I had been asked by my manager at Philips. The difference between the designs that I found "beautiful" and those that I found ugly was that the beautiful designs encapsulated changeable or arbitrary information. Because the interfaces contained only "solid" or lasting information, they appeared simple and could be called beautiful. To keep the interfaces "beautiful", one needs to structure the system so that all arbitrary or changeable or very detailed information is hidden.

The second article explained this although it never used the phrase "information hiding". It is possible to formalise the concept of information hiding using Shannon's theory of information. Interfaces are a set of assumptions that a user may make about the module. The less likely these assumptions are to change (i.e. the more arbitrary the assumptions are), the more "beautiful" they appear. Thus, interfaces that contain less information in Shannon's sense are more "beautiful". Unfortunately, this formalisation is not very useful. Applications of Shannon's theories require us to estimate the probability of the "signals" and we never have such data.

7.3 From Parnas-Modules to Information-Hiding Modules

The first large group to build on the information hiding idea was SRI International. They introduced the HDM or Hierarchical Development Method described in [19–22]. They referred to the modules as "Parnas-Modules". I requested that they switch to more descriptive terminology, "information-hiding" modules, and the term "information-hiding" is the one that has been accepted. This term is now often used in books and articles without either citation or explanation.

8. Information Hiding Is Harder Than It Looks

Information hiding, like dieting, is something that is more easily described than done. The principle is simple and is easily explained in terms of established mathematics and information theory. When it is applied properly it is usually successful. Abstractions such as Unix pipes, postscript or X-windows achieve the goal of allowing several implementations of the same interface. However, although the principle was published 30 years ago, today's software is full of places where information is not hidden, interfaces are complex, and changes are unreasonably difficult. In part, this can be attributed to a lack of educational standards for professional software developers. It is quite easy to get a job developing software with little or no exposure to the concept. Moreover, if one's manager and co-workers do not know the concept, they will not be supportive of the use of information hiding by someone who does understand it. They will argue that, "Nobody does it that way" and that they cannot afford to experiment with new ideas.

In addition to the fact that software engineering educational standards are much lower than those for other engineering fields, I have observed other reasons for failing to apply information hiding.

8.1 The Flow-Chart Instinct

My original experiments using the KWIC index revealed that most designers design software by thinking about the flow of control. Often they begin with a flowchart. The boxes in the flowcharts are then called modules. Unfortunately modules developed in this way exchange large amounts of information in complex data structures and have "ugly" interfaces. Flowchart-based decomposition almost always violates the information-hiding principle. The habit of beginning with a flowchart dies hard. I can teach the principle, warn against basing a modularisation on a flowchart, and watch my students do exactly that one hour later.

8.2 Planning for Change Seems Like Too Much Planning

Proper application of the information-hiding principle requires a great deal of analysis. One must study the class of applications and determine which aspects are unlikely to change while identifying changeable decisions.

Simple intuition is not adequate for this task. When designing software, I have identified certain facts as likely to change only to learn later that there were fundamental reasons why they would not change. Even worse, one might assume that certain facts will never change (e.g. the length of a local phone number or the length of a Postleitzahl). Differentiating what will not change from things that can be mentioned in interface descriptions takes a lot of time. Often a developer is much more concerned with reducing development cost, or time-to-market, than with reducing maintenance cost. This is especially true when the development group will not be responsible for maintenance. Why should a manager spend money from his/her own budget to save some other manager money in the future?

8.3 Bad Models

Engineers in any field are not always innovative. They often take an earlier product as a model making as few changes as possible. Most of our existing software products were designed without use of information hiding and serve as bad models when used in this way. Innovative information-hiding designs are often rejected because they are not conventional.

8.4 Extending Bad Software

Given the large base of installed software, most new products are developed as extensions to that base. If the old software has complex interfaces, the extended software will have the same problem. There are ways around this but they are not frequently applied. Usually, if information hiding has not been applied in the base software, future extensions will not apply it either.

8.5 Assumptions Often Go Unnoticed

My original example of information hiding was the simple KWIC index program described in [10]. This program is still used as an example of the principle. Only once has anyone noticed that it contains a flaw caused by a failure to hide an important assumption. Every module in the system was written with the knowledge that the data comprised strings of strings. This led to a very inefficient sorting algorithm because comparing two strings, an operation that would be repeated many times, is relatively slow. Considerable speed-up could be obtained if the words were sorted once and replaced by a set of integers with the property that if two words are alphabetically ordered, the integers representing them will have the same order. Sorting strings of integers can be done much more quickly than sorting strings of strings. The module interfaces described in [9] do not allow this simple improvement to be confined to one module.

My mistake illustrates how easy it is to distribute information unnecessarily. This is a very common error when people attempt to use the information-hiding principle. While the basic idea is to hide information that is likely to change, one can often profit by hiding other information as well because it allows the re-use of algorithms or the use of more efficient algorithms.

8.6 Reflecting the System Environment in the Software Structure

Several software design "experts" have suggested that one should reflect exciting business structures and file structures in the structure of the software. In my experience, this speeds up the software development process (by making decisions quickly) but leads to software that is a burden on its owners should they try to update their data structures or change their organisation. Reflecting changeable facts in software structure is a violation of the information-hiding principle.

In determining requirements it is very important to know about the environment but it is rarely the right "move" to reflect that environment in the program structure.

8.7 "Oh, Too Bad We Changed It"

Even when information hiding has been successfully applied, it often fails in the long run because of inadequate documentation. I have seen firms where the design criteria were not explained to the maintenance programmers. In these cases, the structure of the system quickly deteriorated because good design decisions, which made no sense to the maintainers, were reversed. In one case, I pointed out the advantages of a particular software structure only to see a glimmer of understanding in the eyes of my listener as he said, "Oh, too bad we changed it."

9. A Hierarchical Modular Structure for Complex Systems

My original articles used very small systems as examples. Systems with 5, 10 or 15 modules can be comprehended by most programmers. However, when applying the principle to a problem with hundreds of independently changeable design decisions, interface details and requirements, we recognised that it would be difficult for a maintenance programmer to have the overview needed to find relevant parts of the code when a change was needed. To overcome this, we organised the modules into a hierarchy. Each of the top-level modules had a clearly defined secret, and was divided into smaller modules using the same principle. This resulted in a tree structure. The programmer can find the relevant modules by selecting a module at each level and moving on to a more detailed decomposition. The structure and document are described in detail in [16].

10. There Are No "Levels of Abstraction"!

Some authors and teachers have used the phrase "levels of abstraction" when talking about systems in which abstraction or information hiding has been applied. Although the terminology seems innocent it is not clearly defined and can lead to serious design errors. The term "levels" can be used properly only if one has defined a relation that is loop free [13]. None of the authors who use this phrase have defined the relation, "more abstract than", which is needed to understand the levels. What would such a relation really mean? Is a module that hides the representation of angles more (or less) abstract than a program that hides the details of a peripheral device? If not, should both modules be on the same level? If so, what would that mean? Using the phrase "levels of abstraction" without a useful definition of the defining relation is a sign of confusion or carelessness.

Often the phrase "levels of abstraction" is used by those who do not clearly distinguish between modules (collections of programs) and the programs themselves. It has been shown useful to arrange programs in a hierarchical structure based on the relation "uses" where A uses B means that B must be present when A runs [5]. However, as shown in [15] it is not usually right to put all programs in a module at the same level in the "uses" hierarchy. Some parts of a module may use programs at a higher level than other programs in that module. An illustration of this can be found in [15].

11. Newer Ideas: Abstract Data Types, O-O, Components

The concept of information hiding is the basis of three more recent concepts. The following paragraphs briefly describe these.

11.1 Abstract Data Types

The notion of abstract data types corrected an important oversight in my original treatment of information hiding. In the examples that motivated my work, there was only one copy of each hidden data structure. Others (most notably Barbara Liskov and John Guttag) recognised that we might want many copies of such structures and that this would make it possible to implement new data types in which the implementation (representation plus algorithms) was hidden. Being able to use variables of these new, user-defined, abstract data types in exactly the way as we use variables of built-in data types is obviously a good idea. Unfortunately, I have never seen a language that achieved this.

11.2 Object-Oriented Languages

A side effect of the application of information hiding is the creation of new objects that store data. Objects can be found in Dijkstra's THE system [5] and many subsequent examples. Information-hiding objects can be created in any language; my first experiments used primitive versions of FORTRAN. Some subsequent languages such as Pascal actually made things a little harder by introducing restrictions on where variables could be declared making it necessary to make hidden data global. It was discovered that a family of languages initially designed for simulation purposes, SIMULA [3] and SIMULA 67 [4], proved to be convenient for this type of programming. The concept of process was introduced to model ongoing activities in the system being simulated. Each process could store data and there were ways for other processes to send "messages" that altered this data. Thus, the features of SIMULA languages were well suited to writing software that hid data and algorithms. Similar ideas were later found in Smalltalk [6]. In both of these languages, the concept of module (class) and process were linked, i.e. modules could operate in parallel. This introduces unnecessary overhead in many situations.

More recent languages have added new types of features (known as inheritance) designed to make it possible to share representations between objects. Often, these features are misused and result in a violation of information hiding and programs that are hard to change.

The most negative effect of the development of O-O languages has been to distract programmers from design principles. Many seem to believe that if they write their program in an O-O language, they will write O-O programs. Nothing could be further from the truth.

11.3 Component-Oriented Design

The old problems and dreams are still with us. Only the words are new.

12. Future Research: Standard Information – Hiding Designs

Thirty years after I was able to formulate the principle of information hiding, it is clear to me that this is the right principle. Further, I see few open questions about the principle. However, as outlined in Section 8, information hiding is not easy. As discussed in [2], [14] and [16], designing such modules requires a lot of research. I would hope that in the next decade we will see academic research turning in this (useful) direction and journals that present module structures and module interfaces for specific devices, data structures, requirements classes and algorithms. Examples of reusable module designs can be found in [17].

References

[1] Bartussek, W., Parnas, D.L., "Using Assertions About Traces to Write Abstract Specifications for Software Modules", UNC Report No. TR77-012, December 1977; Lecture Notes in Computer Science (65), *Information Systems Methodology, Proceedings ICS*, Venice, 1978, Springer Verlag, pp. 211-236.
Also in *Software Specification Techniques* edited by N. Gehani & A.D. McGettrick, AT&T Bell Telephone Laboratories, 1985, pp. 111-130 (QA 76.6 S6437).
Reprinted as Chapter 1 in item .

[2] Britton, K.H., Parker, R.A., Parnas, D.L., "A Procedure for Designing Abstract Interfaces for Device Interface Modules", *Proceedings of the 5th International Conference on Software Engineering*, 1981.

[3] Dahl, Ol.-J., Nygaard, K., "SIMULA - An Algol-based Simulation Language", *Communications of the ACM*, vol. 9, no. 9, pp. 671-678, 1966.

[4] Dahl, O-J., et al., "The Simula 67 Common Base Language", 1970, Norwegian Computing Centre, publication S22.

[5] Dijkstra, E.W., "The Structure of the THE Multiprogramming System", *Communications of the ACM*, vol. 11, no. 5, May 1968.

[6] Goldberg, A., "Smalltalk-80: The Interactive Programming Environment", Addison-Wesley, 1984.

[7] Hoffman, D.M., Weiss, D.M. (eds.), "*Software Fundamentals: Collected Papers by David L. Parnas*", Addison-Wesley, 2001, ISBN 0-201-70369-6.

[8] Parnas, D.L., "Information Distributions Aspects of Design Methodology", *Proceedings of IFIP Congress '71*, 1971, Booklet TA-3, pp. 26-30.

[9] Parnas, D.L., "A Technique for Software Module Specification with Examples", *Communications of the ACM*, vol. 15, no. 5, pp. 330-336, 1972;
Republished in Yourdon, E.N. (ed.), *Writings of the Revolution*, Yourdon Press, 1982, pp. 5-18:
in Gehani, N., McGettrick, A.D. (eds.), *Software Specification Techniques*, AT&T Bell Telephone Laboratories (QA 76.7 S6437), 1985, pp. 75-88 .

[10] Parnas, D.L., "On the Criteria to be Used in Decomposing Systems into Modules", *Communications of the ACM*, vol. 15, no. 12, pp. 1053-1058, 1972;.
Republished in Laplante P. (ed.), *Great Papers in Computer Science*, West Publishing Co, Minneapolis/St. Paul, 1996, pp. 433-441.
Translated into Japanese - BIT, vol. 14, no. 3, pp. 54-60, 1982.
Republished in Yourdon, E.N. (ed.), *Classics in Software Engineering*, Yourdon Press, 1979, pp. 141-150.
Reprinted as Chapter 7 in item [7].

[11] Parnas, D.L., "A Course on Software Engineering Techniques", in *Proceedings of the ACM SIGCSE*, Second Technical Symposium, 24-25 March 1972, pp. 1-6.

[12] Parnas, D.L., "Some Conclusions from an Experiment in Software Engineering Techniques", *Proceedings of the 1972 FJCC*, 41, Part I, pp. 325-330.

[13] Parnas, D.L., "On a 'Buzzword': Hierarchical Structure", *IFIP Congress '74*, North-Holland, 1974, pp. 336-339.
Reprinted as Chapter 8 in item [7].

[14] Parnas, D.L., "Use of Abstract Interfaces in the Development of Software for Embedded Computer Systems", NRL Report No. 8047, June 1977;
Reprinted in Infotech State of the Art Report, Structured System Development, Infotech International, 1979.

[15] Parnas D.L., "Designing Software for Ease of Extension and Contraction", *IEEE Transactions on Software Engineering*, Vol. SE-5 No. 2, pp. 128-138, March 1979;
In *Proceedings of the Third International Conference on Software Engineering*, May 1978, pp. 264-277.

[16] Parnas, D.L., Clements, P.C., Weiss, D.M., "The Modular Structure of Complex Systems", *IEEE Transactions on Software Engineering*, Vol. SE-11 No. 3, pp. 259-266, 1985;
In *Proceedings of 7th International Conference on Software Engineering*, March 1984, 408-417;
Reprinted in Peterson, G.E. (ed.), IEEE Tutorial: "Object-Oriented Computing", Vol. 2: Implementations, IEEE Computer Society Press, IEEE Catalog Number EH0264-2, ISBN 0-8186-4822-8, 1987, pp. 162-169;
Reprinted as Chapter 16 in item [7].

[17] Parker, R.A., Heninger, K.L., Parnas, D.L., Shore, J.E., "Abstract Interface Specifications for the A-7E Device Interface Modules", NRL Report 4385, November 1980.

[18] Parnas D.L., Wang, Y., "Simulating the Behaviour of Software Modules by Trace Rewriting Systems", *IEEE Transactions of Software Engineering*, Vol. 19, No. 10, pp. 750 - 759, 1994.

[19] Roubine, O., Robinson, L., "Special Reference Manual" Technical Report CSL-45, Computer Science Laboratory, Stanford Research Institute, August 1976.

[20] Robinson, L., Levitt K.N., Neumann P.G., Saxena, A.R., "A Formal Methodology for the Design of Operating System Software", in Yeh, R. (ed.), Current Trends in Programming Methodology I, Prentice-Hall, 1977, pp. 61-110.

[21] Robinson, L., Levitt K.N., "Proof Techniques for Hierarchically Structured Programs", *Communications of the ACM*, Vol. 20, No. 4, pp. 271-283, 1977.

[22] Robinson, L., Levitt K.N., Silverberg, B.A.,"The (HDM) Handbook", Computer Science Laboratory, SRI International, June 1979.

David L. Parnas

On the Criteria to Be Used in Decomposing Systems into Modules

Communications of the ACM, Vol. 15 (12), 1972 pp. 1053-1058

On the Criteria To Be Used in Decomposing Systems into Modules

D.L. Parnas
Carnegie-Mellon University

This paper discusses modularization as a mechanism for improving the flexibility and comprehensibility of a system while allowing the shortening of its development time. The effectiveness of a "modularization" is dependent upon the criteria used in dividing the system into modules. A system design problem is presented and both a conventional and unconventional decomposition are described. It is shown that the unconventional decompositions have distinct advantages for the goals outlined. The criteria used in arriving at the decompositions are discussed. The unconventional decomposition, if implemented with the conventional assumption that a module consists of one or more subroutines, will be less efficient in most cases. An alternative approach to implementation which does not have this effect is sketched.

Key Words and Phrases: software, modules, modularity, software engineering, KWIC index, software design

CR Categories: 4.0

Author's address: Department of Computer Science, Carnegie-Mellon University, Pittsburgh, PA 15213.

Introduction

A lucid statement of the philosophy of modular programming can be found in a 1970 textbook on the design of system programs by Gouthier and Pont [1, ¶10.23], which we quote below:[1]

> A well-defined segmentation of the project effort ensures system modularity. Each task forms a separate, distinct program module. At implementation time each module and its inputs and outputs are well-defined, there is no confusion in the intended interface with other system modules. At checkout time the integrity of the module is tested independently; there are few scheduling problems in synchronizing the completion of several tasks before checkout can begin. Finally, the system is maintained in modular fashion; system errors and deficiencies can be traced to specific system modules, thus limiting the scope of detailed error searching.

Usually nothing is said about the criteria to be used in dividing the system into modules. This paper will discuss that issue and, by means of examples, suggest some criteria which can be used in decomposing a system into modules.

A Brief Status Report

The major advancement in the area of modular programming has been the development of coding techniques and assemblers which (1) allow one module to be written with little knowledge of the code in another module, and (2) allow modules to be reassembled and replaced without reassembly of the whole system. This facility is extremely valuable for the production of large pieces of code, but the systems most often used as examples of problem systems are highly-modularized programs and make use of the techniques mentioned above.

[1] Reprinted by permission of Prentice-Hall, Englewood Cliffs, N.J.

Expected Benefits of Modular Programming

The benefits expected of modular programming are: (1) managerial—development time should be shortened because separate groups would work on each module with little need for communication: (2) product flexibility—it should be possible to make drastic changes to one module without a need to change others; (3) comprehensibility—it should be possible to study the system one module at a time. The whole system can therefore be better designed because it is better understood.

What Is Modularization?

Below are several partial system descriptions called *modularizations*. In this context "module" is considered to be a responsibility assignment rather than a subprogram. The *modularizations* include the design decisions which must be made *before* the work on independent modules can begin. Quite different decisions are included for each alternative, but in all cases the intention is to describe all "system level" decisions (i.e. decisions which affect more than one module).

Example System 1: A KWIC Index Production System

The following description of a KWIC index will suffice for this paper. The KWIC index system accepts an ordered set of lines, each line is an ordered set of words, and each word is an ordered set of characters. Any line may be "circularly shifted" by repeatedly removing the first word and appending it at the end of the line. The KWIC index system outputs a listing of all circular shifts of all lines in alphabetical order.

This is a small system. Except under extreme circumstances (huge data base, no supporting software), such a system could be produced by a good programmer within a week or two. Consequently, none of the difficulties motivating modular programming are important for this system. Because it is impractical to treat a large system thoroughly, we must go through the exercise of treating this problem as if it were a large project. We give one modularization which typifies current approaches, and another which has been used successfully in undergraduate class projects.

Modularization 1

We see the following modules:

Module 1: Input. This module reads the data lines from the input medium and stores them in core for processing by the remaining modules. The characters are packed four to a word, and an otherwise unused character is used to indicate the end of a word. An index is kept to show the starting address of each line.

Module 2: Circular Shift. This module is called after the input module has completed its work. It prepares an index which gives the address of the first character of each circular shift, and the original index of the line in the array made up by module 1. It leaves its output in core with words in pairs (original line number, starting address).

Module 3: Alphabetizing. This module takes as input the arrays produced by modules 1 and 2. It produces an array in the same format as that produced by module 2. In this case, however, the circular shifts are listed in another order (alphabetically).

Module 4: Output. Using the arrays produced by module 3 and module 1, this module produces a nicely formatted output listing all of the circular shifts. In a sophisticated system the actual start of each line will be marked, pointers to further information may be inserted, and the start of the circular shift may actually not be the first word in the line, etc.

Module 5: Master Control. This module does little more than control the sequencing among the other four modules. It may also handle error messages, space allocation, etc.

It should be clear that the above does not constitute a definitive document. Much more information would have to be supplied before work could start. The defining documents would include a number of pictures showing core formats, pointer conventions, calling conventions, etc. All of the interfaces between the four modules must be specified before work could begin.

This is a modularization in the sense meant by all proponents of modular programming. The system is divided into a number of modules with well-defined interfaces; each one is small enough and simple enough to be thoroughly understood and well programmed. Experiments on a small scale indicate that this is approximately the decomposition which would be proposed by most programmers for the task specified.

Modularization 2

We see the following modules:

Module 1: Line Storage. This module consists of a number of functions or subroutines which provide the means by which the user of the module may call on it. The function call $CHAR(r,w,c)$ will have as value an integer representing the cth character in the rth line, wth word. A call such as $SETCHAR(r,w,c,d)$ will cause the cth character in the wth word of the rth line to be the character represented by d (i.e. $CHAR(r,w,c) = d$). $WORDS(r)$ returns as value the number of words in line r. There are certain restrictions in the way that these routines may be called; if these restrictions are violated the routines "trap" to an error-handling subroutine which is to be provided by the users of the routine. Additional routines are available which reveal to the caller the number of words in any line, the number of lines currently stored, and the number of characters in any word. Functions *DELINE* and *DELWRD* are provided to delete portions of lines which have already been stored. A precise specification of a similar module has been given in [3] and [8] and we will not repeat it here.

Module 2: INPUT. This module reads the original lines from the input media and calls the line storage module to have them stored internally.

Module 3: Circular Shifter. The principal functions provided by this module are analogs of functions provided in module 1. The module creates the impression that we have created a line holder containing not all of the lines but all of the circular shifts of the lines. Thus the function call *CSCHAR(l,w,c)* provides the value representing the *c*th character in the *w*th word of the *l*th circular shift. It is specified that (1) if $i < j$ then the shifts of line i precede the shifts of line j, and (2) for each line the first shift is the original line, the second shift is obtained by making a one-word rotation to the first shift, etc. A function *CSSETUP* is provided which must be called before the other functions have their specified values. For a more precise specification of such a module see [8].

Module 4: Alphabetizer. This module consists principally of two functions. One, *ALPH*, must be called before the other will have a defined value. The second, *ITH*, will serve as an index. *ITH*(i) will give the index of the circular shift which comes ith in the alphabetical ordering. Formal definitions of these functions are given [8].

Module 5: Output. This module will give the desired printing of set of lines or circular shifts.

Module 6: Master Control. Similar in function to the modularization above.

Comparison of the Two Modularizations

General. Both schemes will work. The first is quite conventional; the second has been used successfully in a class project [7]. Both will reduce the programming to the relatively independent programming of a number of small, manageable, programs.

Note first that the two decompositions may share all data representations and access methods. Our discussion is about two different ways of cutting up what *may* be the same object. A system built according to decomposition 1 could conceivably be identical *after assembly* to one built according to decomposition 2. The differences between the two alternatives are in the way that they are divided into the work assignments, and the interfaces between modules. The algorithms used in both cases *might* be identical. The systems are substantially different even if identical in the runnable representation. This is possible because the runnable representation need only be used for running; other representations are used for changing, documenting, understanding, etc. The two systems will not be identical in those other representations.

Changeability. There are a number of design decisions which are questionable and likely to change under many circumstances. This is a partial list.

1. Input format.
2. The decision to have all lines stored in core. For large jobs it may prove inconvenient or impractical to keep all of the lines in core at any one time.
3. The decision to pack the characters four to a word. In cases where we are working with small amounts of data it may prove undesirable to pack the characters; time will be saved by a character per word layout. In other cases we may pack, but in different formats.
4. The decision to make an index for the circular shifts rather that actually store them as such. Again, for a small index or a large core, writing them out may be the preferable approach. Alternatively, we may choose to prepare nothing during *CSSETUP*. All computation could be done during the calls on the other functions such as *CSCHAR*.
5. The decision to alphabetize the list once, rather than either (a) search for each item when needed, or (b) partially alphabetize as is done in Hoare's FIND

[2]. In a number of circumstances it would be advantageous to distribute the computation involved in alphabetization over the time required to produce the index.

By looking at these changes we can see the differences between the two modularizations. The first change is confined to one module in both decompositions. For the first decomposition the second change would result in changes in every module! The same is true of the third change. In the first decomposition the format of the line storage in core must be used by all of the programs. In the second decomposition the story is entirely different. Knowledge of the exact way that the lines are stored is entirely hidden from all but module 1. Any change in the manner of storage can be confined to that module!

In some versions of this system there was an additional module in the decomposition. A symbol table module (as specified in [3]) was used within the line storage module. This fact was completely invisible to the rest of the system.

The fourth change is confined to the circular shift module in the second decomposition, but in the first decomposition the alphabetizer and the output routines will also know of the change.

The fifth change will also prove difficult in the first decomposition. The output module will expect the index to have been completed before it began. The alphabetizer module in the second decomposition was designed so that a user could not detect when the alphabetization was actually done. No other module need be changed.

Independent Development. In the first modularization the interfaces between the modules are the fairly complex formats and table organizations described above. These represent design decisions which cannot be taken lightly. The table structure and organization are essential to the efficiency of the various modules and must be designed carefully. The development of those formats will be a major part of the module development and that part must be a joint effort among the several development groups. In the second modularization the interfaces are more abstract; they consist primarily in the function names and the numbers and types of the parameters. These are relatively simple decisions and the independent development of modules should begin much earlier.

Comprehensibility. To understand the output module in the first modularization, it will be necessary to understand something of the alphabetizer, the circular shifter, and the input module. There will be aspects of the tables used by output which will only make sense because of the way that the other modules work. There will be constraints on the structure of the tables due to the algorithms used in the other modules. The system will only be comprehensible as a whole. It is my subjective judgment that this is not true in the second modularization.

The Criteria

Many readers will now see what criteria were used in each decomposition. In the first decomposition the criterion used was to make each major step in the processing a module. One might say that to get the first decomposition one makes a flowchart. This is the most common approach to decomposition or modularization. It is an outgrowth of all programmer training which teaches us that we should begin with a rough flowchart and move from there to a detailed implementation. The flowchart was a useful abstraction for systems with on the order of 5,000–10,000 instructions, but as we move beyond that it does not appear to be sufficient; something additional is needed.

The second decomposition was made using "information hiding" [4] as a criterion. The modules no longer correspond to steps in the processing. The line storage module, for example, is used in almost every action by the system. Alphabetization may or may not correspond to a phase in the processing according to the method used. Similarly, circular shift might, in some circumstances, not make any table at all but calculate each character as demanded. Every module in the second decomposition is characterized by its knowledge of a design decision which it hides from all others. Its interface or definition was chosen to reveal as little as possible about its inner workings.

Improvement in Circular Shift Module

To illustrate the impact of such a criterion let us take a closer look at the design of the circular shift module from the second decomposition. Hindsight now suggests that this definition reveals more information than necessary. While we carefully hid the method of storing or calculating the list of circular shifts, we specified an order to that list. Programs could be effectively written if we specified only (1) that the lines indicated in circular shift's current definition will all exist in the table, (2) that no one of them would be included twice, and (3) that an additional function existed which would allow us to identify the original line given the shift. By prescribing the order for the shifts we have given more information than necessary and so unnecessarily restricted the class of systems that we can build without changing the definitions. For example, we have not allowed for a system in which the circular shifts were produced in alphabetical order, *ALPH* is empty, and *ITH* simply returns its argument as a value. Our failure to do this in constructing the systems with the second decomposition must clearly be classified as a design error.

In addition to the general criteria that each module hides some design decision from the rest of the system, we can mention some specific examples of decompositions which seem advisable.

1. A *data structure*, its internal linkings, *accessing procedures and modifying procedures* are part of a single module. They are not shared by many modules as is conventionally done. This notion is perhaps just an elaboration of the assumptions behind the papers of Balzer [9] and Mealy [10]. Design with this in mind is clearly behind the design of BLISS [11].
2. *The sequence of instructions necessary to call a given routine and the routine itself are part of the same module.* This rule was not relevant in the Fortran systems used for experimentation but it becomes essential for systems constructed in an assembly language. There are no perfect general calling sequences for real machines and consequently they tend to vary as we continue our search for the ideal sequence. By assigning responsibility for generating the call to the person responsible for the routine we make such improvements easier and also make it more feasible to have several distinct sequences in the same software structure.
3. The *formats of control blocks* used in queues in operating systems and similar programs *must be hidden* within a "control block module." It is conventional to make such formats the interfaces between various modules. Because design evolution forces frequent changes on control block formats such a decision often proves extremely costly.
4. *Character codes, alphabetic orderings, and similar data should be hidden* in a module for greatest flexibility.
5. The sequence in which certain items will be processed should (as far as practical) be hidden within a single module. Various changes ranging from equipment additions to unavailability of certain resources in an operating system make sequencing extremely variable.

Efficiency and Implementation

If we are not careful the second decomposition will prove to be much less efficient than the first. If each of the functions is actually implemented as a procedure with an elaborate calling sequence there will be a great deal of such calling due to the repeated switching between modules. The first decomposition will not suffer from this problem because there is relatively infrequent transfer of control between modules.

To save the procedure call overhead, yet gain the advantages that we have seen above, we must implement these modules in an unusual way. In many cases the routines will be best inserted into the code by an assembler; in other cases, highly specialized and efficient transfers would be inserted. To successfully and efficiently make use of the second type of decomposition will require a tool by means of which programs may be written as if the functions were subroutines, but assembled by whatever implementation is appropriate. If such a technique is used, the separation between modules may not be clear in the final code. For that reason additional program modification features would also be useful. In other words, the several representations of the program (which were mentioned earlier) must be maintained in the machine together with a program performing mapping between them.

A Decomposition Common to a Compiler and Interpretor for the Same Language

In an earlier attempt to apply these decomposition rules to a design project we constructed a translator for a Markov algorithm expressed in the notation described in [6]. Although it was not our intention to investigate the relation between compiling and interpretive translators of a langugage, we discovered that our decomposition was valid for a pure compiler and several varieties of interpretors for the language. Although there would be deep and substantial differences in the final running representations of each type of compiler, we found that the decisions implicit in the early decomposition held for all.

This would not have been true if we had divided responsibilities along the classical lines for either a compiler or interpretor (e.g. syntax recognizer, code generator, run time routines for a compiler). Instead the decomposition was based upon the hiding of various decisions as in the example above. Thus register representation, search algorithm, rule interpretation etc. were modules and these problems existed in both compiling and interpretive translators. Not only was the decomposition valid in all cases, but many of the routines could be used with only slight changes in any sort of translator.

This example provides additional support for the statement that the order in time in which processing is expected to take place should not be used in making the decomposition into modules. It further provides evidence that a careful job of decomposition can result in considerable carryover of work from one project to another.

A more detailed discussion of this example was contained in [8].

Hierarchical Structure

We can find a program hierarchy in the sense illustrated by Dijkstra [5] in the system defined according to decomposition 2. If a symbol table exists, it functions without any of the other modules, hence it is on level 1. Line storage is on level 1 if no symbol table is used or it is on level 2 otherwise. Input and Circular Shifter require line storage for their functioning. Output and Alphabetizer will require Circular Shifter, but since Circular Shifter and line holder are in some sense compatible, it would be easy to build a parameterized version of those routines which could be used to alphabetize or print out either the original lines or the circular shifts. In the first usage they would not require

Circular Shifter; in the second they would. In other words, our design has allowed us to have a single representation for programs which may run at either of two levels in the hierarchy.

In discussions of system structure it is easy to confuse the benefits of a good decomposition with those of a hierarchical structure. We have a hierarchical structure if a certain relation may be defined between the modules or programs and that relation is a partial ordering. The relation we are concerned with is "uses" or "depends upon." It is better to use a relation between programs since in many cases one module depends upon only part of another module (e.g. Circular Shifter depends only on the output parts of the line holder and not on the correct working of *SETWORD*). It is conceivable that we could obtain the benefits that we have been discussing without such a partial ordering, e.g. if all the modules were on the same level. The partial ordering gives us two additional benefits. First, parts of the system are benefited (simplified) because they use the services of lower[2] levels. Second, we are able to cut off the upper levels and still have a usable and useful product. For example, the symbol table can be used in other applications; the line holder could be the basis of a question answering system. The existence of the hierarchical structure assures us that we can "prune" off the upper levels of the tree and start a new tree on the old trunk. If we had designed a system in which the "low level" modules made some use of the "high level" modules, we would not have the hierarchy, we would find it much harder to remove portions of the system, and "level" would not have much meaning in the system.

Since it is conceivable that we could have a system with the type of decomposition shown in version 1 (important design decisions in the interfaces) but retaining a hierarchical structure, we must conclude

[2] Here "lower" means "lower numbered."

that hierarchical structure and "clean" decomposition are two desirable but *independent* properties of a system structure.

Conclusion

We have tried to demonstrate by these examples that it is almost always incorrect to begin the decomposition of a system into modules on the basis of a flowchart. We propose instead that one begins with a list of difficult design decisions or design decisions which are likely to change. Each module is then designed to hide such a decision from the others. Since, in most cases, design decisions transcend time of execution, modules will not correspond to steps in the processing. To achieve an efficient implementation we must abandon the assumption that a module is one or more subroutines, and instead allow subroutines and programs to be assembled collections of code from various modules.

Received August 1971; revised November 1971

References

1. Gauthier, Richard, and Pont, Stephen. *Designing Systems Programs*, (*C*), Prentice-Hall, Englewood Cliffs, N.J., 1970.
2. Hoare, C. A. R. Proof of a program, FIND. *Comm. ACM 14*, 1 (Jan. 1971), 39–45.
3. Parnas, D. L. A technique for software module specification with examples. *Comm. ACM 15*, 5 (May, 1972), 330–336.
4. Parnas, D. L. Information distribution aspects of design methodology. Tech. Rept., Depart. Computer Science, Carnegie-Mellon U., Pittsburgh, Pa., 1971. Also presented at the IFIP Congress 1971, Ljubljana, Yugoslavia.
5. Dijkstra, E. W. The structure of "THE"-multiprogramming system. *Comm. ACM 11*, 5 (May 1968), 341–346.
6. Galler, B., and Perlis, A. J. *A View of Programming Languages*, Addison-Wesley, Reading, Mass., 1970.
7. Parnas, D. L. A course on software engineering. Proc. SIGCSE Technical Symposium, Mar. 1972.

8. Parnas, D. L. On the criteria to be used in decomposing systems into modules. Tech. Rept., Depart. Computer Science, Carnegie-Mellon U., Pittsburgh, Pa., 1971.
9. Balzer, R. M. Dataless programming. Proc. AFIPS 1967 FJCC, Vol. 31, AFIPS Press, Montvale, N.J., pp. 535–544.
10. Mealy, G. H. Another look at data. Proc. AFIPS 1967 FJCC, Vol. 31, AFIPS Press, Montvale, N.J., pp. 525–534.
11. Wulf, W. A., Russell, D. B., and Habermann, A. N. BLISS, A language for systems programming. *Comm. ACM 14*, 12 (Dec. 1971), 780–790.

David L. Parnas

On a "Buzzword": Hierarchical Structure

Informaion Processing 74, IFIP congress 74,
North-Holland, 1974
pp. 336-339

ON A 'BUZZWORD': HIERARCHICAL STRUCTURE

David PARNAS

Technische Hochschule Darmstadt, Fachbereich Informatik
Research Group on Operating Systems I
Steubenplatz 12, 61 Darmstadt, West Germany

This paper discusses the use of the term "hierarchically structured" to describe the design of operating systems. Although the various uses of this term are often considered to be closely related, close examination of the use of the term shows that it has a number of quite different meanings. For example, one can find two different senses of "hierarchy" in a single operating system [3] and [6]. An understanding of the different meanings of the term is essential, if a designer wishes to apply recent work in Software Engineering and Design Methodology. This paper attempts to provide such an understanding.

INTRODUCTION

The phrase "hierarchical structure" has become a buzzword in the computer field. For many it has acquired a connotation so positive that it is akin to the quality of being a good mother. Others have rejected it as being an unrealistic restriction on the system [1]. This paper attempts to give some meaning to the term by reviewing some of the ways that the term has been used in various operating systems (e.g. T.H.E. [3], MULTICS [12], and the RC4ooo [8]) and providing some better definitions. Uses of the term, which had been considered equivalent or closely related, are shown to be independant. Discussions of the advantages and disadvantages of the various hierarchical restrictions are included.

GENERAL PROPERTIES OF ALL USES OF THE PHRASE "HIERARCHICAL STRUCTURE"

As discussed earlier [2], the word "structure" refers to a partial description of a system showing it as a collection of parts and showing some relations between the parts. We can term such a structure hierarchical, if a relation or predicate on pairs of the parts ($R(\alpha,\beta)$) allows us to define levels by saying that

1. Level 0 is the set of parts α such that there does not exist a β such that $R(\alpha,\beta)$, and

2. Level i is the set of parts α such that
   a) there exists a β on level i-1 such that $R(\alpha,\beta)$ and
   b) if $R(\alpha,\gamma)$ then γ is on level i-1 <u>or lower</u>.

This is possible with a relation R only if the directed graph representing R has no loops.

The above definition is the most precise reasonably simple definition, which encompasses all uses of the word in the computer literature. This suggests that the statement "our Operating System has a hierarchical structure" carries no information at all. <u>Any</u> system can be represented as a hierarchical system with one level and one part; more importantly, it is possible to divide <u>any</u> system into parts and contrive a relation such that the system has a hierarchical structure. Before such a statement can carry any information at all, the way that the system is divided into parts and the nature of the relation must be specified.

The decision to produce a hierarchically structured system may restrict the class of possible systems, and may, therefore, introduce disadvantages as well as the desired advantages. In the remainder of this paper we shall introduce a variety of definitions for "hierarchical structure", and mention some advantages and disadvantages of the restriction imposed by these definitions.

1. <u>THE PROGRAM HIERARCHY</u>

Prof. E.W. Dijkstra in his paper on the T.H.E. system and in later papers on structured programming [3] and [4] has demonstrated the value of programming using layers of abstract machines. We venture the following definition for this program hierarchy. The parts of the system are subprograms, which may be called as if they were procedures.* We assume that each such program has a specified purpose (e.g. FNO :: = find next odd number in sequence or invoke DONE if there is none). The relation "uses" may be defined by USES(p_i,p_j)=<u>iff</u> p_i calls p_j <u>and</u> p_i will be considered incorrect <u>if</u> p_j does not function properly.

* They may be expanded as MACROS.

With the last clause we intend to imply that, our example, FNO does not "use" DONE in the sense defined here. The task of FNO is to invoke DONE; the purpose and "correctness" of DONE is irrelevant to FNO. Without excepting such calls, we could not consider a program to be higher in the hierarchy than the machine, which it uses. Most machines have "trap" facilities, and invoke software routines, when trap conditions occur.

A program divided into a set of subprograms may be said to be hierarchically structured, when the relation "uses" defines levels as described above. The term "abstract machine" is commonly used, because the relation between the lower level programs and the higher level programs is analoguous to the relation between hardware and software.

A few remarks are necessary here. First, we do not claim that the only good programs are hierarchically structured programs. Second, we point out that the way that the program is divided into subprograms can be rather arbitrary. For any program, some decompositions into subprograms may reveal a hierarchical structure, while other decompositions may show a graph with loops in it. As demonstrated in the simple example above, the specification of each program's purpose is critical!

The purpose of the restriction on program structure implied by this definition, is twofold. First, the calling program should be able to ignore the internal workings of the called program; the called program should make no assumptions about the internal structure of the calling program. Allowing the called program to call its user, might make this more difficult since each would have to be designed to work properly in the situations where it could be called by the other.

The second purpose might be termed "ease of subsetting". When a program has this "program hierarchy", the lower levels may always be used without the higher levels, when the higher levels are not ready or their services are not needed. An example of non-hierarchical systems would be one in which the "lower level" scheduling programs made use of the "high level" file system for storage of information about the tasks that it schedules. Assuming that nothing useful could be done without the scheduler, no subset of the system that did not include the file system could exist. The file system (usually a complex and "buggy" piece of software) could not be developped using the remainder of the system as a "virtual machine".

For those who argue that the hierarchical structuring proposed in this section prevents the use of recursive programming techniques, we remind them of the freedom available in choosing a decomposition into subprograms. If there exists a subset of the programs, which call each other recursively, we can view the group as a single program for this analysis and then consider the remaining structure to see, if it is hierarchical. In looking for possible subsets of a system, we must either include or exclude this group of programs as a single program.

One more remark: please, note that the division of the program into levels by the above discussed relation has no necessary connection with the division of the programs into modules as discussed in [5]. This is discussed further later (section 6).

2. THE "HABERMANN" HIERARCHY IN THE T.H.E. SYSTEM

The T.H.E. system was also hierarchical in another sense. In order to make the system relatively insensitive to the number of processors and their relative speeds, the system was designed as a set of "parallel sequential processes". The activities in the system were organized into "processes" such that the sequence of events within a process was relatively easy to predict, but the sequencing of events in different processes were considered unpredictable (the relative speeds of the processes were considered unknown). Resource allocation was done in terms of the processes and the processes exchanged work assignments and information. In carrying out a task, a process could assign part of the task to another process in the system.

One important relation between the processes in such a system is the relation "gives work to". In his thesis [6] Habermann assumed that "gives work to" defined a hierarchy to prove "harmonious cooperation". If we have an Operating System we want to show that a request of the system will generate only a finite (and reasonably small) number of requests to individual processes before the original request is satisfied. If the relation "gives work to" defines a hierarchy, we can prove our result by examining each process seperately to make sure that every request to it results in only a finite number of requests to other processes. If the relation is not hierarchical, a more difficult, "global", analysis would be required.

Restricting "gives work to" so that it defines a hierarchy helps in the establishment of the "well-behavedness", but it is certainly not a necessary condition for "harmonious cooperation".*

In the T.H.E. system the two hierarchies described above coincided. Every level of abstraction was achieved by the introduction of parallel processes and these processes only gave work to those written to implement lower levels in the program hierarchy. One should not draw general conclusions about system structure on the basis of this coincidence. For example, the remark that "building a system with more levels than were found in the T.H.E. system is undesirable, because it introduces more queues" is often heard because of this coincidence. The later work by Dijkstra on structured programming [21] shows that the levels of abstraction are useful when there is only one process. Further, the "Habermann hierarchy" is useful, when the processes are controlled by badly structured programs. Adding levels in the program hierarchy need not introduce new processes or queues. Adding processes can be done without writing new programs.

The "program hierarchy" is only significant at times when humans are working with the program (e.g. when the program is being constructed or changed). If the programs were all implemented as macros, there would be no trace of this hierarchy in the running system. The "Habermann hierarchy" is a restriction on the run time behavior of the system. The theorems proven

*This restriction is also valuable in human organizations. Where requests for administrative work flow only in one direction things go relatively smoothly, but in departments where the "leader" constantly refers requests "downward" to committees (which can themselves send requests to the "leader") we often find the system filling up with uncompleted tasks and a correspondingly large increase in overhead.

by Habermann would hold even if a process that is controlled by a program written at a low level in the program hierarchy "gave work to" a process which was controlled by a program originally written at a higher level in the program hierarchy. There are also no detrimental effects on the program hierarchy provided that the programs written at the lower level are not written in terms of programs at the higher level. Readers are referred to "Flatland" [7].

3. HIERARCHICAL STRUCTURES RELATING TO RESOURCE OWNERSHIP AND ALLOCATION

The RC4ooo system [8] and [9] enforced a hierarchical relation based upon the ownership of memory. A generalization of that hierarchical structure has been proposed by Varney [1o] and similar hierarchical relationships are to be found in various commercial operating systems, though they are not often formally described.

In the RC4ooo system the objects were processes and the relation was "allocated a memory region to". Varney proposes extending the relation so that the hierarchical structure controlled the allocation of other resources as well. (In the RC4ooo systems specific areas of memory were allocated, but that was primarily a result of the lack of virtual memory hardware; in most systems of interest now, we can allocate quantities of a resource without allocating the specific physical resources until they are actually used). In many commercial systems we also find that resources are not allocated directly to the processes which use them. They are allocated to administrative units, who, in turn, may allocate them to other processes. In these systems we do not find any loops in the graph of "allocates resources to", and the relation defines a hierarchy, which is closely related to the RC4ooo structure.

This relation was not a significant one in the T.H.E. system, where allocating was done by a central allocator called a BANKER. Again this sense of hierarchy is not strongly related to the others, and if it is present with one or more of the others, they need not coincide.

The disadvantage of a non-trivial hierarchy (the hierarchy is present in a trivial form even in the

T.H.E. system) of this sort are (1) poor resource utilization that may occur when some processes in the system are short of resources while other processes, under a different allocator in the hierarchy, have an excess; (2) high overhead that occurs when resources are tight. Requests for more resources must always go up all the levels of the hierarchy before being denied or granted. The central "banker" does not have these disadvantages. A central resource allocator, however, becomes complicated in situations where groups of related processes wish to dynamically share resources without influence by other such groups. Such situations can arise in systems that are used in real time by independant groups of users. The T.H.E. system did not have such problems and as a result, centralized resource allocation was quite natural.

It is this particular hierarchical relation which the Hydra group rejected. They did not mean to reject the general notion of hierarchical structure as suggested in the original report [1] and [11].

4. PROTECTION HIERARCHIES A LA MULTICS

Still another hierarchy can be found in the MULTICS system. The conventional two level approach to operating systems (low level called the supervisor, next level the users) has been generalized to a sequence of levels in the supervisor called "rings". The set of programs within a MULTICS process is organized in a hierarchical structure, the lower levels being known as the inner rings, and the higher levels being known as outer rings. Although the objects are programs, this relation is not the program hierarchy discussed in section 1. Calls occur in both directions and lower level programs may use higher level ones to get their work done [12].

Noting that certain data are much more crucial to operation of the system than other data, and that certain procedures are much more critical to the overall operation of the system than others, the designers have used this as the basis of their hierarchy. The data to which the system is most sensitive are controlled by the inner ring procedures, and transfers to those programs are very carefully controlled. Inner ring procedures have unrestricted access to programs and data in the outer rings. The outer rings contain data and procedures that effect a relatively small number of users and hence are less "sensitive". The hierarchy is most easily defined in terms of a relation "can be accessed by" since "sensitivity" in the sense used above is difficult to define. Low levels have unrestricted access to higher levels, but not vice versa.

It is clear that placing restrictions on the relation "can be accessed by" is important to system reliability and security.

It has, however, been suggested that by insisting that the relation "can be accessed by" be a hierarchy, we prevent certain accessibility patterns that might be desired. We might have three segments in which A requires access to B, B to C, and C to A. No other access rights are needed or desirable. If we insist that "can be accessed by" define a hierarchy, we must (in this case) use the trivial hierarchy in which A, B, C are considered one part.

In the view of the author, the member of pairs in the relation "can be accessed by" should be minimized, but he sees no advantage in insisting that it define a hierarchy [13] and [14].

The actual MULTICS restriction is even stronger than requiring a hierarchy. Within a process, the relation must be a complete ordering.

5. HIERARCHIES AND "TOP DOWN" DESIGN METHODOLOGY

About the time that the T.H.E. system work appeared, it became popular to discuss design methods using such terms as "top down" and "outside in"[15], [16], and [17]. The simultaneous appearance of papers suggesting how to design well and a well designed system led to the unfounded assumption that the T.H.E. system had been the result of a "top down" design process. Even in more recent work [18] top down design and structured programming are considered almost synonymous.

Actually "outside in" was a much better term for what was intended, than was "top down"! The intention was to begin with a description of the system's user interface, and work in small, verifiable steps towards the implementation. The "top" in that hierarchy consisted of those parts of the system that were visible to the user. In a system designed according to the "program hierarchy", the lower level functions will be used by the higher level functions, but some of them may also be visible to the user (store and load, for example). Some functions on higher levels may not be available to him (Restart system). Those participants in the design of the T.H.E. system with whom I have discussed the question [19], report that they did not proceed with the design of the higher levels first.

6. HIERARCHICAL STRUCTURE AND DECOMPOSITION INTO MODULES

Often one wants to view a system as divided into "modules" (e.g. with the purpose outlined in [5] and [2o]). This division defines a relation "part of". A group of sub-programs is collected into a module, groups of modules collected into bigger modules, etc. This process defines a relation "part of" whose graph is clearly loop-free. It remains loop-free even if we allow programs or modules to be part of several modules - the part never includes the whole.

Note that we may allow programs in one module to call programs in another module, so that the module hierarchy just defined need not have any connection with the program hierarchy. Even allowing recursive calls between modules does not defeat the purpose of the modular decomposition (e.g. flexibility) [5], provided that programs in one module do not assume much about the programs in another.

7. LEVELS OF LANGUAGE

It is so common to hear phrases such as "high level language", "low level language" and "linguistic level" that it is necessary to comment on the relation between the implied language hierarchy and the hierarchies discussed in the earlier sections of this paper. It would be nice, if, for example, the higher level languages were the languages of the higher level "abstract machines" in the program hierarchy. Unfortunately, this author can find no such relation and cannot define the hierarchy that is implied in the use of those phrases. In moments of scepticism one might suggest that the relation is "less efficient than" or "has a bigger grammar than" or "has a bigger compiler than", however, none of those phrases suggests an ordering, which is completely consistant with the use of the term. It would be nice, if the next person to use the phrase "higher level language" in a paper would define the hierarchy to which he refers.

SUMMARY

The computer system design literature now contains quite a number of valuable suggestions for improving the comprehensibility and predictability of computer systems by imposing a hierarchical structure on the programs. This paper has tried to demonstrate that, although these suggestions have been described in quite similar terms, the structures implied by those suggestions are not necessarily closely related. Each of the suggestions must be understood and evaluated (for its applicability to a particular system design problem) independantly. Further, we have tried to show that, while each of the suggestions offers some advantages over an "unstructured" design, there are also disadvantages, which must be considered. The main purpose of this paper has been to provide some guidance for those reading earlier literature and to suggest a way for future authors to include more precise definitions in their papers on design methods.

ACKNOWLEDGMENT

The Author acknowledges the valuable suggestions of Mr. W. Bartussek (Technische Hochschule Darmstadt) and Mr. John Shore (Naval Research Laboratory, Washington, D.C.). Both of these gentlemen have made substantial contributions to the more precise formulation of many of the concepts in this paper; neither should be held responsible for the fuzziness, which unfortunately remains.

REFERENCES

[1] Wulf, Cohen, Coowin, Jones, Levin, Pierson, Pollach, Hydra: The Kernel of a Multiprogramming System, Technical Report, Computer Science Department, Carnegie-Mellon University.

[2] David L. Parnas, Information Distribution Aspects of Design Methodology, Proceedings of the 1971 IFIP Congress, Booklet TA/3, 26-3o.

[3] E.W. Dijkstra, The Structure of the T.H.E. Multiprogramming System, Communications of the ACM, vol 11, no. 5, May 1968, 341-346.

[4] E.W. Dijkstra, Complexity controlled by Hierarchical Ordering of Function and Variability, Software Engineering, NATO.

[5] David L. Parnas, On the Criteria to be Used in Decomposing Systems into Modules, Communications of the ACM, vol. 15, no. 12, December 1972, 1o53-1o58.

[6] A.N. Habermann, On the Harmonious Cooperation of Abstract Machines, Doctoral Dissertation, Technische Hogeschool Eindhoven, The Netherlands.

[7] Edwin A. Abbott, Flatland, the Romance of Many Dimensions, Dover Publications, Inc., New York, 1952.

[8] Per Brinch Hansen, The Nucleus of a Multiprogramming System, Communications of the ACM, vol. 13, no. 4, April 197o, 238-25o.

[9] RC4000 Reference Manuals for the Operating System, Regnecentralen Denmark.

[1o] R.C. Varney, Process Selection in a Hierarchical Operating System, SIGOPS Operating Review, June 1972.

[11] W. Wulf, C. Pierson, Private Discussions.

[12] R.W. Graham, Protection in an Information Processing Utility, Communications of the ACM, May 1968.

[13] W.R. Price and David L. Parnas, The Design of the Virtual Memory Aspects of a Virtual Machine, Proceedings of the SIGARCH-SIGOPS Workshop on Virtual Machines, March 1973.

[14] W.R. Price, Doctoral Dissertation, Department of Computer Science, Carnegie-Mellon University, Pittsburgh, Pa., U.S.A.

[15] David L. Parnas and Darringer, SODAS and Methodology for System Design, Proceedings of 1967 FJCC.

[16] David L. Parnas, More on Design Methodology and Simulation, Proceedings of the 1969 SJCC.

[17] Zurcher and Randell, Iterative Multi-Level Modeling, Proceedings of the 1968 IFIP Congress.

[18] F.T. Baker, System Quality through Structured Programming, Proceedings of the 1972 FJCC.

[19] E.W. Dijkstra, A.N. Habermann, Private Discussions

[2o] David L. Parnas, Some Conclusions from an Experiment in Software Engineering, Proceedings of the 1972 FJCC.

[21] E.W. Dijkstra, A Short Introduction to the Art of Programming, in O.-J. Dahl, E.W. Dijkstra, and C.A.R. Hoare, Structured Programming, Academic Press, New York, 1972.

Algebraic Specification of Abstract Data Types

John Guttag
Professor and Head of Computer Science Department
Massachusetts Institute of Technology
guttag@mit.edu

Ph.D., University of Toronto

Co-leader,
MIT's Networks and
Mobile Systems Group

Member of the Board of Directors,
Empirix, Inc.

Member of the Board of Trustees,
MGH Institute of Health Professions

Major contributions:
abstract data types and specification techniques, the Larch family of specification languages and tools

Current interests: networking, wireless communication, medical computing

John V. Guttag

Abstract Data Types, Then and Now

Data abstraction has come to play an important role in software development. This paper presents one view of what data abstraction is, how it was viewed when it was introduced, and its long-term impact on programming.

1. Abstract Data Types in a Nutshell

The notion of an abstract data type is quite simple. It is a set of objects and the operations on those objects. The specification of those operations defines an interface between the abstract data type and the rest of the program. The interface defines the behavior of the operations – what they do, but not how they do it. The specification thus defines an abstraction barrier (Figure 1) that isolates the rest of the program from the data structures, algorithms, and code involved in providing a realization of the type abstraction.

Abstraction is all about forgetting. There are lots of ways to model this. I find it easiest to model it on my teenage children. They conveniently forget everything I say that they don't consider relevant. That's a pretty good abstraction process unless, as is often the case, their notion of relevance and mine conflict. The key to using abstraction effectively in programming is finding a notion of relevance that is appropriate for both the builder of an abstraction and the user of the abstraction. That is the true art of programming.

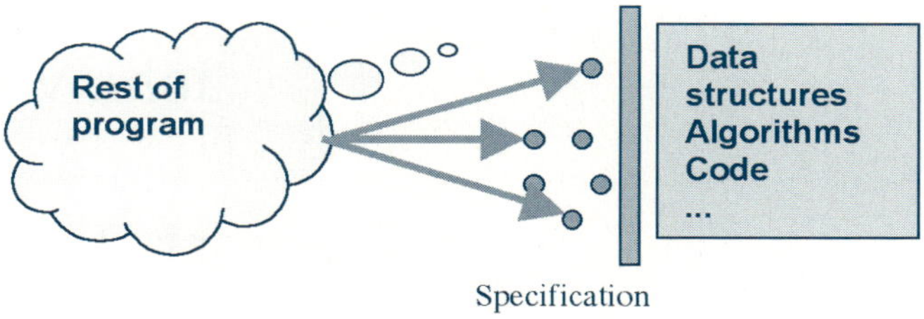

Fig. 1. A data abstraction hides information about how a type is implemented behind an interface.

2. The 1970s

Who invented abstract data types? No one, really. Like so many timely ideas it was "in the air." Parnas was preaching the virtues of information hiding [14]. Dahl and Nygaard were using Simula 67 [2] to create rich sets of abstractions. Wirth [15] and Liskov [10] were demonstrating how programming languages could be used to protect programmers from themselves. Hoare was showing how monitors [9] could be used to structure programs and axiomatic systems used to reason about them [7, 8], etc.

It was an exciting time to be thinking about how people should go about building programs. In 1975, under the guidance of Jim Horning, I completed a dissertation [4] that started with:

At most points in the program, one is concerned solely with the behavioral characteristics of the data object. What one can do with them; not with how the various operations on them are implemented. I will use the term "abstract data type" to refer to a class of objects defined by a representation independent specification.

Today this idea is commonplace, and it seems perfectly obvious that abstract data types are among the most useful kinds of abstractions.

So, abstract data types were an indisputably good idea with an impeccable pedigree. One would have assumed that in the mid-1970s everybody would have stood up and saluted. Wrong. There was significant resistance in both academia and industry.

There were other, apparently competing, ideas in the air. There was a popular (and good) book called *Algorithms + Data structures = Programs* [16]. Conferences and journals were full of papers on successive refinement. Thoughtful students of project management argued that code should belong to everyone on the project. Many interpretations of these ideas were in direct conflict with the notion of organizing programs around abstract data types and consequently led people to conclude that abstract data types were just a bad idea.

How did people working on the same problem reach such different conclusions? My guess is that they started from different sets of (often not clearly articulated) assumptions.

3. Assumptions Underlying Abstract Data Types

An important, and arguable, assumption is that software is not fractal. This is in contrast to the fractal view of software, "it's the same thing all the way down." The fractal view leads one to treat the process of creating software as the same at each level, and from there to development methods in which the same process is applied recursively all the way from the early design phases to executable code. An alternative view is that high-level design is qualitatively different from low-level design. In particular, the design of a system architecture is very different from the design of an algorithm or data structure. This non-fractal view motivated much of the work on abstract data types.

Another assumption underlying a lot of the early work on abstract data types is that knowledge is dangerous. "Drink deep, or taste not the Pierian Spring," is not necessarily good advice. Knowing too much is no better, and often worse, than knowing too little. People cannot assimilate very much information. Any programming method or approach that assumes that people will understand a lot is highly risky. More importantly, the more widely information is disseminated, the more likely it is that decisions believed by designers and implementers to be local will have global ramifications. This makes change challenging. If information hiding and data abstraction had been widely understood and practiced in the 1970s, there would have been no Y2K problem in the 1990s. It should have taken but a few minutes to change the number of digits in dates, but it is no mystery why it didn't. Programmers did not use data abstractions, and too many people knew how dates were implemented. A clear and concise discussion of how views on information hiding have evolved since the 1970s can be found in the 1995 edition of *The Mythical Man Month* [1].

An important attribute of programming with data abstraction is that it limits direct access to key data structures. A perceived risk of data abstraction is that preventing client programs from directly accessing critical data structures leads to unacceptable loss of efficiency. In the early days of data abstraction, many were concerned about the cost of introducing extraneous procedure calls. Modern compilation technology makes this concern moot. A more serious issue is that client programs will be forced to use inefficient algorithms. Consider a set of integers abstraction that supports only the operations ***createEmpty***, ***insert***, ***delete***, and ***elementOf***. There is no efficient way to write a client program that sums the elements of the set. One solution is to abandon the abstraction and allow the client program to directly access the data structure used to store the values of the set. A far better solution is to add an iterator operation [11] to the abstraction.

An assumption underlying approaches to programming that emphasize abstract data types is that performance is not primarily determined by local decisions. Building an efficient program rarely requires inventing clever data structures and algorithms. It more often depends upon developing a global understanding of the program that leads to a design that facilitates change. Certainly one needs to perform local optimization, but it is often difficult to know which things need to be optimized until relatively late in the day.

An apparent paradox of software is that it is so easy to replace that it is nearly immortal. It is not unusual for a programmer to be asked to maintain a program that originated before he or she was born. If one discovers that it is impossible to change some part of a program because the ramifications of that change cannot be ascertained, both the utility and the performance of that program will inevitably degrade over time – no matter how good the program was originally. Ultimately performance is determined by ease of repeated modification and optimization.

A final assumption underlying a belief in the usefulness of data abstraction is that inventing application-domain-specific types is a good thing, and a fixed set of type abstractions is not sufficient. That this assumption was not universally shared in the mid-1970s was driven home forcefully to me during my thesis defense. I vividly remember one of the examiners stating rather forcefully that he was quite certain that only a small set of types was needed. On another occasion, I recall someone saying something on the order of "I've looked at hundreds of thousands of lines of code and only eight types were ever used. Clearly that's all you need."

In 1980, I gave a short course at Softlab in Munich at the invitation of Ernst Denert. While preparing this paper, I uncovered the transparencies I used in that course. Apparently, I still found it necessary in 1980 to argue for the utility of inventing types. One of the (handwritten) transparencies contained the text, "Take a leap of faith, the programming process is facilitated by the introduction of new domains of discourse, it requires only imagination on the part of the designer."

Of course, it is possible to build too many highly specialized types. As it happens, many problems are not unique. They are similar to other problems, and one might just as well use the types that exist. Sets and numbers, vectors, and such can take one a long way.

4. From Assumptions to Abstract Data Types

From where did these assumptions arise? I think they came from trying to think about what's truly important in commercial software development. Time, cost, and quality matter. Truth and beauty are important, but they are means not ends. We would like our software to be elegant not because we are striving for elegance for its own sake, but because it helps us to build high-quality software on time and on budget. But, of course, even today software systems cost too much to build, take too long to build, and often do not do what people want.

Why is this? Most fundamentally, it is because programming is just plain hard. I don't think I understood this in the 1970s. At the time, I thought if only we could understand how best to go about building software it would become easy. I no longer believe that. I believe that we can improve our efficiency, but I don't think it will ever be easy to build interesting software systems. They are complex systems that people expect to be simultaneously reliable and infinitely malleable.

Programming is about managing complexity in a way that facilitates change. There are two powerful mechanisms available for accomplishing this, decomposition and abstraction. Decomposition creates structure in a program, and abstraction suppresses detail. The key is to suppress the appropriate details. This is where data abstraction hits the mark. One can create domain-specific types that provide a convenient decomposition. Ideally, these types capture concepts that will be relevant over the lifetime of a program. If, at the beginning, one devises types that will be relevant a decade later, one has a great leg up in maintaining that software.

Another thing that data abstraction achieves is representation independence. Think of the implementation of an abstract type as having several components:

- Implementations of the operations of the type,
- A data structure that holds values of the type, and
- Conventions about how the implementations of the operations are to use the data structure. The conventions are captured by the
    - Representation invariant and the
    - Abstraction function.

The representation invariant states which values of the data structure are valid representations of abstract values. Imagine implementing bounded sets of integers with lists of integers. The representation invariant might well restrict the lists that represent sets to those that contain no duplicates or perhaps to those in which the integers are in ascending order. The implementation of the operations would then be responsible for establishing and maintaining that invariant.

The abstraction function [8] maps values of the data structure implementing an abstract type onto the abstract values that the concrete value represents. For example, if one uses a sorted list to implement a set, the abstraction function might map NIL to { } and the list containing the elements <1, 2, 2, 3> to the set {1,2,3}.

The commutative diagram in Figure 2 captures the notion of representation independence. The box in the lower left corner stands for a representation of an abstract value. One can go across the bottom and apply the code implementing an operation of the abstract type to that data structure and get another concrete value. If one then applies the abstraction function, one gets an abstract value.

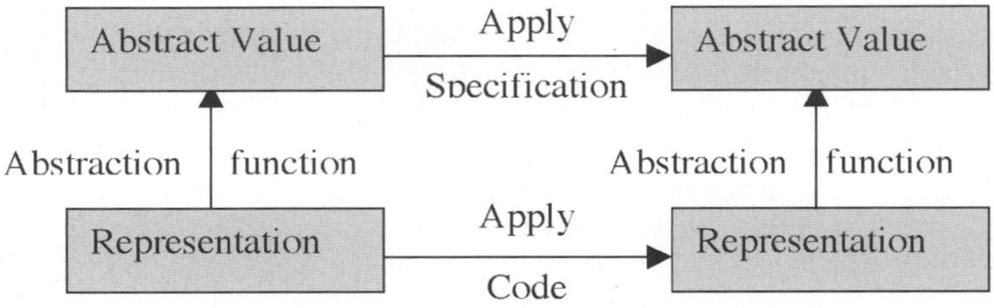

Fig. 2. A simplified commutative diagram. It is simplified in that it assumes that both specification and the code define a mapping (rather than a general relation).

The box in the upper left corner stands for the abstract value yielded by applying the abstraction function to the representation in the lower left corner. Think of applying an operation to that abstract value. The specification of the operation defines the abstract value that results. That each route yields the same abstract value indicates that the behavior of programs that rely on the interface provided by the abstract type is independent of the way the type is implemented. This is so because the interface to a module is characterized by the operations on that module.

Representation independence facilitates the construction of client programs, because those programming the clients need not be concerned with the representation of the abstract type. It also provides those implementing the abstract type with free reign to substitute one implementation for another.

5. Where Data Abstraction Has Led

Today, data abstraction is taken for granted. It is covered (more or less well) in many texts, e.g., [12], and supported (more or less well) by a number of programming languages, e.g., Java. I want to take this opportunity to speculate on some of the more indirect impacts of the work done on data abstraction. Cause and effect is rarely easy to prove, and I expect that some readers will see things quite differently.

The advent of data abstraction was a harbinger of a shift of emphasis in programming from decomposition, which dominated people's thinking for many years, to abstraction. This shift of emphasis appears in the literature on programming methods and in the design of programming languages.

A related development was an increased emphasis on interfaces; in particular on interfaces characterized by operations. The study of data abstraction led to a deeper understanding of the role of specification, including as an aid in designing and communicating interfaces [5].

The study of abstract data types led to a deeper understanding of types in general. In some sense, abstract data types existed from the very beginning of higher-level programming languages. FORTRAN programmers could ignore, at least some of the time, the way numbers were implemented. However, by better understanding such things as how to specify abstract types, how to support them in programming languages, and how to reason about them, we came to a deeper understanding of what types should be about. We came to understand the importance of things like representation invariants and abstraction functions. And, perhaps most importantly, we came to understand how to use induction to reason about types.

The data type induction principle [6] is quite simple, as illustrated in Figure 3.

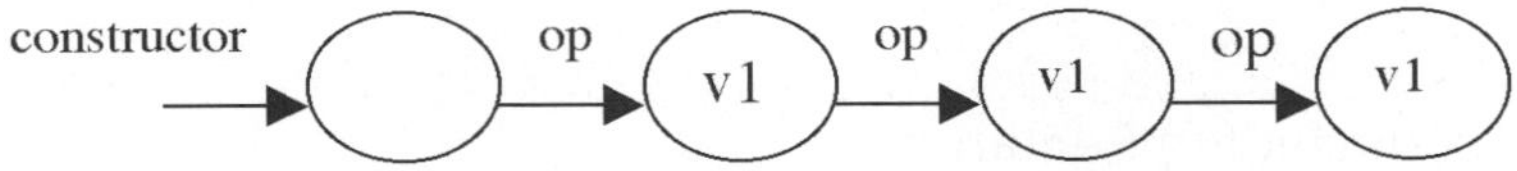

Fig. 3. Data type induction

Data type induction can be used to prove that something is always true about a type at the abstract level, i.e., an abstract invariant. It can also be used to prove representation invariants as part of the process of proving that an implementation of a type is correct. One starts with a constructor that creates some abstract value, *v1*, in this figure. One then applies an operation to get another abstract value, and so forth. The key is that the only abstract values that one can derive are those values that can be reached by applying a finite number of operations to the result of the constructor. This is so because the implementation is hidden, and (assuming the representation is not exposed through aliasing) there is no way to modify the data structures implementing the type except through the operations appearing in the interface. This yields an induction principle that allows one to reason about types. In short, data type induction provides for types what Floyd gave us for control structures. Floyd's inductive assertions method of reasoning [3] allows one to prove something about an infinite set of program executions in a finite number of proof steps. Data type induction allows one to do the same thing for data types.

Data abstraction encouraged program designers to focus on the centrality of data objects rather than procedures. That one should think about a program more as a collection of types than as a collection of procedures leads to a profoundly different organizing principle. It also encourages one to think about programming as a process of combining relatively large chunks, since data abstractions typically encompass much more functionality than do individual procedures. This, in turn, leads one to think of the essence of programming as a process not of writing individual lines of code, but of composing modules.

The reuse of relatively large modules is facilitated by flat program structures. The utility of flat structures was not obvious when abstract data types first came on the scene. As mentioned above, successive refinement was the rage. In some hands, it led to excessive levels of refinement. The problem with deep hierarchies is that, as one gets deeper in the hierarchy, there is a tendency to build increasingly specialized components. Flatter program structures typically lead one to build modules that assume relatively little about the context in which they are to be used. Such modules are more likely to be reusable.

The availability of reusable abstractions not only reduces development time, but also usually leads to more reliable programs because mature software is usually more reliable than new software. For many years, the only program libraries in common use were statistical or scientific. Today, however, there is a great range of useful program libraries, often based on a rich set of data abstractions.

6. Concluding Remarks

For the first 30 years of programming almost all of the real improvements in productivity were attributable to better tools, mostly better machines, and especially better programming languages. The movement from machine languages, to symbolic assembly languages, to higher-level programming languages produced huge improvements in programmer productivity.

In 1975, I thought that times were changing and that almost all future progress in software development would stem from innovations in the ways people thought about programs and programming and not from innovations in programming language technology.

I was not completely wrong. I was correctly pessimistic. There has not been much progress in programming languages since the late 1970s. If I look at the languages of today, I don't think that they are significantly better than some of the languages, e.g., CLU [11], that were available then. Certainly the compilers have gotten faster, they generate better code, and they do a bit more checking. But clearly nothing that has happened in the language domain has had anything like the impact of say FORTRAN, LISP, or Algol60. I was also correctly optimistic; there has indeed been considerable conceptual progress, e.g., the acceptance of abstract data types.

What I didn't anticipate was the enormous impact of improved hardware. This is a difficult thing for me to say as a computer scientist, but nothing in the last quarter century has had a comparable impact on programmer productivity and software utility. The vast memory and the fast processors in today's machines have done several things. Perhaps most importantly they have made it easier for us to use abstractions. The fact that we can so often ignore details of memory management has made it far easier to produce programs. In a similar vein, fast processors often allow one to use simple straightforward algorithms rather than the more subtle and efficient algorithms that were previously necessary.

What next? For one, we need to think about better ways to build embedded software. It seems likely that the majority of software will be part of embedded systems or at least connected to sensors or actuators. Such blue-collar software presents a number of challenges not usually present in white-collar software. Some of the difficulties are the difficulties of the past; embedded processors are often slow (by modern standards) and have little memory (again, by modern standards). The most fundamental aspect of this kind of software is that it will control systems that, with minimal or no human intervention, will take actions of critical importance. Such software will provide realtime control for automobiles, surgical tools, etc. The software will have to be immensely predictable and resilient. Static and dynamic analysis and principled testing of both designs and code will be essential.

Another challenge will be making everyone into a programmer. One of the reasons that programming is hard is that the programmer is often confused about what the user wants. The advent of spreadsheets solved this problem for one class of user. Tools like MatLab have helped to solve the problem for a different class of users. Many similar opportunities surely exist.

So where should you look for these things? Not in this volume. The software pioneers represented in this book did important work. For the most part, however, our day has passed. Look to the new pioneers.

Acknowledgements

Several people provided me with feedback both on earlier drafts of this paper and on a talk that preceded the paper. The overall shape and content of the paper owes much to the advice of Michael Ernst (one of the young pioneers to keep an eye on), Daniel Jackson (another young pioneer worth watching), Steve Garland, and John Ankcorn.

My thinking about data abstraction has been shaped by extensive interactions with some extraordinary colleagues. In particular, I owe much to Jim Horning, Dave Musser, and Steve Garland.

References

[1] F. Brooks, The Mythical-Man Month, Anniversary Edition: Essays on Software Engineering, Addison-Wesley (1995).

[2] O-J., Dahl, K. Nygaard, and B. Myhrhaug, "The SIMULA 67 Common Base Language," Norwegian Computing Centre, Forskningsveien 1B, Oslo (1968).

[3] R.W. Floyd, "Assigning Meaning to Programs," Proceedings of Symposium in Applied Mathematics, vol. IX, American Mathematical Society (1967).

[4] J.V. Guttag, The Specification and Application to Programming of Abstract Data Types, Ph.D. Thesis, Dept. of Computer Science, University of Toronto (1975).

[5] J.V. Guttag and J.J. Horning, "Formal Specification as a Design Tool," Seventh ACM Symposium on Principles of Programming Languages, Las Vegas (1980).

[6] J.V. Guttag, "Notes on Type Abstraction, Version 2," IEEE Transactions on Software Engineering vol. SE-6, no. 1 (1980).

[7] C.A.R. Hoare, "An Axiomatic Basis for Computer Programming," Communications of the ACM, vol. 12, no. 10 (1969).

[8] C.A.R. Hoare, "Proofs of Correctness of Data Representations," Acta Informatica, vol. 1, no. 4 (1972).

[9] C.A.R. Hoare, "Monitors: An Operating System Structuring Concept," Communications of the ACM, vol. 17, no. 10 (1974).

[10] B. Liskov and S.N. Zilles, "Programming with Abstract Data Types," Proceedings of ACM SIGPLAN Symposium on Very High Level Programming Languages, SIGPLAN Notices, vol. 9, no. 4 (1974).

[11] B. Liskov, A. Snyder, R. Atkinson, and C. Schaffert. "Abstraction Mechanisms in CLU," Communications of the ACM, vol. 20, no. 8 (1977).

[12] B Liskov and J. Guttag, Program Development in Java; Abstraction, Specification, and Object-Oriented Design, Addison-Wesley (2000).

[13] J.H. Morris, "Types Are Not Sets," ACM Symposium on the Principles of Programming Languages (1973).

[14] D.L. Parnas, "Information Distribution Aspects of Design Methodology," Proceedings of IFIP Congress 1971 (1971).

[15] N. Wirth, "The Programming Language Pascal," Acta Informatica, vol. 1, no. 1 (1972).

[16] N. Wirth, Algorithms + Data Structures = Programs, Prentice-Hall, Englewood Cliff (1976).

John V. Guttag

Abstract Data Types and the Development of Data Structures

Communications of the ACM,
Vol. 20 (6), 1977
pp. 396-404

Data: Abstraction, Definition, and Structure
B. Wegbreit Editor

Abstract Data Types and the Development of Data Structures

John Guttag
University of Southern California

This work was supported in part by the National Science Foundation under grant number MCS76-06089.

A version of this paper was presented as the SIGPLAN/SIGMOD Conference on Data: Abstraction, Definition, and Structure, Salt Lake City, Utah, March 22–24, 1976.

Author's address: Computer Science Department, University of Southern California, Los Angeles, CA 90007.

Abstract data types can play a significant role in the development of software that is reliable, efficient, and flexible. This paper presents and discusses the application of an algebraic technique for the specification of abstract data types. Among the examples presented is a top-down development of a symbol table for a block structured language; a discussion of the proof of its correctness is given. The paper also contains a brief discussion of the problems involved in constructing algebraic specifications that are both consistent and complete.

Key Words and Phrases: abstract data type, correctness proof, data type, data structure, specification, software specification

CR Categories: 4.34, 5.24

1. Introduction

Dijkstra [4] and many others have made the point that the amount of complexity that the human mind can cope with at any instant in time is considerably less than that embodied in much of the software that one might wish to build. Thus the key problem in the design and implementation of large software systems is reducing the amount of complexity or detail that must be considered at any one time. One way to do this is via the process of abstraction.

One of the most significant aids to abstraction used in programming is the self-contained subroutine. At the point where one decides to invoke a subroutine, one can (and most often should) treat it as a "black box." It performs a specific arbitrarily abstract function by means of an unprescribed algorithm. Thus, at the level where it is invoked, it separates the relevant detail of "what" from the irrelevant detail of "how." Similarly, at the level where it is implemented, it is usually unnecessary to complicate the "how" by considering the "why," i.e. the exact reasons for invoking a subroutine often need not be of concern to its implementor. By nesting subroutines, one may develop a hierarchy of abstractions.

Unfortunately, the nature of the abstractions that may be conveniently achieved through the use of subroutines is limited. Subroutines, while well suited to the description of abstract events (operations), are not particularly well suited to the description of abstract objects. This is a serious drawback, for in a great many applications the complexity of the data objects to be manipulated contributes substantially to the overall complexity of the problem.

2. The Abstraction of Data

The large knot of complexly interrelated attributes associated with a data object may be separated according to the nature of the information that the attributes convey regarding the data objects that they qualify. Two kinds of attributes, each of which may be studied in isolation, are:

(1) those that describe the representation of objects and the implementations of the operations associated with them in terms of other objects and operations, e.g. in terms of a physical store and a processor's order code;

(2) those that specify the names and define the abstract meanings of the operations associated with an object. Though these two kinds of attributes are in practice highly interdependent, they represent logically independent concepts.

The emphasis in this paper is on the second kind of attribute, i.e. on the specification of the operations associated with classes of data objects. At most points in a program one is concerned solely with the behavioral characteristics of a data object. One is interested in what one can do with it, not in how the various operations on it are implemented. The analogy with a closed procedure is exact. More often than not, one need be no more concerned with the underlying representation of the object being operated on than one is with the algorithm used to implement an invoked procedure.

If at a given level of refinement one is interested only in the behavioral characteristics of certain data objects, then any attempt to abstract data must be based upon those characteristics, and only those characteristics. The introduction of other attributes, e.g. a representation, can only serve to cloud the relevant issues. We use the term "abstract data type" to refer to a class of objects defined by a representation-independent specification.

The class construct of SIMULA 67 [3] has been used as the starting point for much of the more recent work on embedding abstract types in programming languages, e.g. [14, 16, 18]. While each of these offers a mechanism for binding together the operations and storage structures representing a type, they offer no representation-independent means for specifying the behavior of the operations. The only representation-independent information that one can supply are the domains and ranges of the various operations. One could, for example, define a type Queue (of Items) with the operations

NEW:	$\rightarrow$ Queue
ADD:	Queue $\times$ Item $\rightarrow$ Queue
FRONT:	Queue $\rightarrow$ Item
REMOVE:	Queue $\rightarrow$ Queue
IS_EMPTY?:	Queue $\rightarrow$ Boolean

Unfortunately, however, short of supplying a representation, the only mechanism for denoting what these operations "mean" is a judicious choice of names. Except for intuitions about the meaning of such words as Queue and FRONT, the operations might just as easily be defining type Stack as type Queue. The domain and range specifications for these two types are isomorphic. To rely on one's intuition about the meaning of names can be dangerous even when dealing with familiar types [19]. When dealing with unfamiliar types it is almost impossible. What is needed, therefore, is a mechanism for specifying the semantics of the operations of the type.

There are, of course, many possible approaches to the specification of the semantics of an abstract data type. Most, however, can be placed in one of two categories: operational or definitional. In an operational specification, instead of trying to describe the properties of the abstract data type, one gives a recipe for constructing it. One begins with some well-understood language or discipline and builds a model for the type in terms of that discipline. Wulf [24], for example, makes good use of sequences in modeling various data structures.

The operational approach to formal specification has many advantages. Most significantly, operational specifications seem to be relatively (compared to definitional specifications) easily constructed by those trained as programmers – chiefly because the construction of operational specifications so closely resembles programming. As the operations to be specified grow complex, however, operational specifications tend to get too long (see, for example, Batey [1]) to permit substantial confidence in their aptness. As the number of operations grows, problems arise because the relations among the operations are not explicitly stated, and inferring them becomes combinatorially harder.

The most serious problem associated with operational specifications is that they almost always force one to overspecify the abstraction. By introducing extraneous detail, they associate nonessential attributes with the type. This extraneous detail complicates the problem of proving the correctness of an implementation by introducing conditions that are irrelevant, yet nevertheless must be verified. More importantly, the introduction of extraneous detail places unnecessary constraints on the choice of an implementation and may potentially eliminate the best solutions to the problem.

Axiomatic definitions avoid this problem. The algebraic approach used here owes much to the work of Hoare [13] (which in turn owes much to Floyd [5]) and is closely related to Standish's "axiomatic specifications" [22] and Zilles' "algebraic specifications" [25]. Its formal basis stems from the heterogeneous algebras of Birkhoff and Lipson [2]. An algebraic specification of an abstract type consists of two pairs: a syntactic specification and a set of relations. The syntactic specification provides the syntactic information that many programming languages already require: the names, domains, and ranges of the operations associated with the type. The set of relations defines the meanings of the operations by stating their relationships to one another.

3. A Short Example

Consider type Queue (of Items) with the operations listed in the previous section. The syntactic specification is as above:

NEW:	$\rightarrow$ Queue
ADD:	Queue $\times$ Item $\rightarrow$ Queue
FRONT:	Queue $\rightarrow$ Item
REMOVE:	Queue $\rightarrow$ Queue
IS_EMPTY?:	Queue $\rightarrow$ Boolean

The distinguishing characteristic of a queue is that it is a first in–first out storage device. A good axiomatic definition of the above operations must therefore assert that and only that characteristic. The relations (or axioms) below comprise just such a definition. The meanings of the axioms should be relatively clear. ("=" has its standard meaning, "q" and "i" are typed free variables, and "error" is a distinguished value with the property that the value of any operation applied to an argument list containing error is error, e.g. $f_n(x_1, \ldots, x_i, \text{error}, x_{i+2}, \ldots, x_n) = \text{error}$.)

```
(1) IS_EMPTY? (NEW) = true
(2) IS_EMPTY? (ADD(q,i)) = false
(3) FRONT(NEW) = error
(4) FRONT (ADD(q,i)) = if IS_EMPTY? (q)
                          then i
                          else FRONT(q)
(5) REMOVE(NEW) = error
(6) REMOVE (ADD(q,i)) = if IS_EMPTY? (q)
                           then NEW
                           else ADD(REMOVE(q),i)
```

Note that this set of axioms involves no assumption about the attributes of type Item. In effect Item is a parameter of type Type, and the specification may be viewed as defining a type schema rather than a single type. This will be the case for many algebraic type specifications.

With some practice, one can become quite adept at reading algebraic axiomatizations. Practice also makes it easier to construct such specifications; see Guttag [11]. Unfortunately, it does not make it trivial. It is not always immediately clear how to attack the problem. Nor, once one has constructed an axiomatization, is it always easy to ascertain whether or not the axiomatization is consistent and sufficiently complete. The meaning of the operations is supplied by a set of individual statements of fact. If any two of these are contradictory, the axiomatization is inconsistent. If the combination of statements is not sufficient to convey all of the vital information regarding the meaning of the operations of the type, the axiomatization is not sufficiently complete.[1]

---

[1] Sufficiently complete is a technical notion first developed in Guttag [8]. It differs considerably from both the notion of completeness commonly used in logic and that used in Zilles [25].

Experience indicates that completeness is, in a practical sense, a more severe problem than consistency. If one has an intuitive understanding of the type being specified, one is unlikely to supply contradictory axioms. It is, on the other hand, extremely easy to overlook one or more cases. Boundary conditions, e.g. REMOVE(NEW), are particularly likely to be overlooked.

In an attempt to ameliorate this problem, we have devised heuristics to aid the user in the initial presentation of an axiomatic specification of the operations of an abstract type and a system to mechanically "verify" the sufficient-completeness of that specification. As the first step in defining a new type, the user would supply the system with the syntactic specification of the type and an axiomatization constructed with the aid of the heuristics mentioned above. Given this preliminary specification, the system would begin to prompt the user to supply the additional information necessary for the system to derive a sufficiently complete axiom set for the operations. A detailed look at sufficient-completeness is contained in Guttag [8, 9].

4. An Extended Example

A common data structuring problem is the design of the symbol table component of a compiler for a block structured language. Many sources contain good discussions of various symbol table organizations. Setting aside variations in form, the basic operations described vary little from source to source. They are:

INIT:	Allocate and initialize the symbol table.
ENTERBLOCK:	Prepare a new local naming scope.
LEAVEBLOCK:	Discard entries from the most recent scope entered, and reestablish the next outer scope.
IS_INBLOCK?:	Has a specified identifier already been declared in this scope? (Used to avoid duplicate declarations.)

ADD:	Add an identifier and its attributes to the symbol table.
RETRIEVE:	Return the attributes associated (in the most local scope in which it occurs) with a specified identifier.

Though many references provide insights into how these operations can be implemented, none presents a formal definition (other than implementations) of exactly what they mean. The abstract concept "symbol table" thus goes undefined. Those who attempt to write compilers in a top-down fashion suffer from a similar problem. Early refinements of parts of the compiler make use of the basic symbol table operations, but the "meaning" of these operations is provided only by subsequent levels of refinement. This is infelicitous in that the clear separation of levels of abstraction is lost and with it many of the advantages of top-down design. By providing axiomatic semantics for the operations, this problem can be avoided.

The thought of providing rigorous definitions for so many operations may, at first, seem a bit intimidating. Nevertheless, if one is to understand the refinement, one must know what each operation means. The following specification of abstract type Symboltable supplies these meanings.

Type: Symboltable

Operations:

INIT:	→ Symboltable
ENTERBLOCK:	Symboltable → Symboltable
LEAVEBLOCK:	Symboltable → Symboltable
ADD:	Symboltable × Identifier × Attributelist → Symboltable
IS_INBLOCK?:	Symboltable × Identifier → Boolean
RETRIEVE:	Symboltable × Identifier → Attributelist

Axioms:

(1) LEAVEBLOCK(INIT) = error

(2) LEAVEBLOCK(ENTERBLOCK(symtab)) = symtab

(3) LEAVEBLOCK(ADD(symtab, id, attrs)) = LEAVEBLOCK(symtab)

(4) IS_INBLOCK? (INIT, id) = false
(5) IS_INBLOCK? (ENTERBLOCK(symtab), id) = false
(6) IS_INBLOCK? (ADD(symtab, id, attrs), idl) =
    if IS_SAME? (id, idl)[2]
        then true
        else IS_INBLOCK? (symtab, id)
(7) RETRIEVE(INIT, id) = error
(8) RETRIEVE(ENTERBLOCK(symtab), id) =
    RETRIEVE(symtab, id)
(9) RETRIEVE(ADD(symtab, id, attrs), idl)=
    if IS_SAME? (id, idl)
        then attrs
        else RETRIEVE(symtab, idl)

This set of relations serves a dual purpose. Not only does it define an abstract type that can be used in the specification of various parts of the compiler, but it also provides a complete self-contained specification for a major subsystem of the compiler. If one wished to delegate the design and implementation of the symbol table subsystem, the algebraic characterization of the abstract type would (unlike the informal description in, say, McKeeman [15]) be a sufficient specification of the problem. In fact, the procedure discussed earlier can be used to formally prove the sufficient-completeness of this specification.

The next step in the design process is to further refine type Symboltable, i.e. to provide implementations of the operations of the type. These implementations will implicitly furnish representation for values of type Symboltable.

A representation of a type T consists of (i) any interpretation (implementation) of the operations of the type that is a model for the axioms of the specification of T, and (ii) a function Φ that maps terms in the model domain onto their representatives in the abstract domain. (This is basically the abstraction function of Hoare [12].)

---

[2] The definition of IS_SAME? is part of the specification of an independently defined type Identifier.

It is important to note that Φ may not have a proper inverse. Consider, for example, type Bounded Queue (with a maximum length of three). A reasonable representation of the values of this type might be based on a ring-buffer and top pointer. Given this representation, the program segment:

```
x := EMPTY.Q
x := ADD.Q(x, A)
x := ADD.Q(x, B)
x := ADD.Q(x, C)
x := REMOVE.Q(x)
x := ADD.Q(x, D)
```

would translate to a representation for *x* of the form:

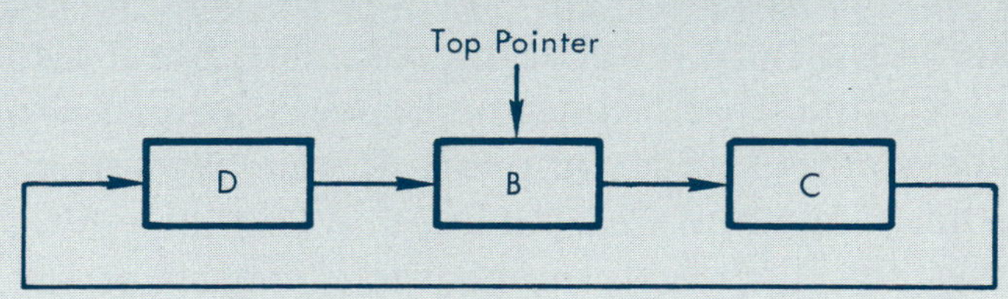

Similarly:

```
x := EMPTY.Q
x := ADD.Q(x, B)
x := ADD.Q(x, C)
x := ADD.Q(x, D)
```

would yield a representation for *x* of the form:

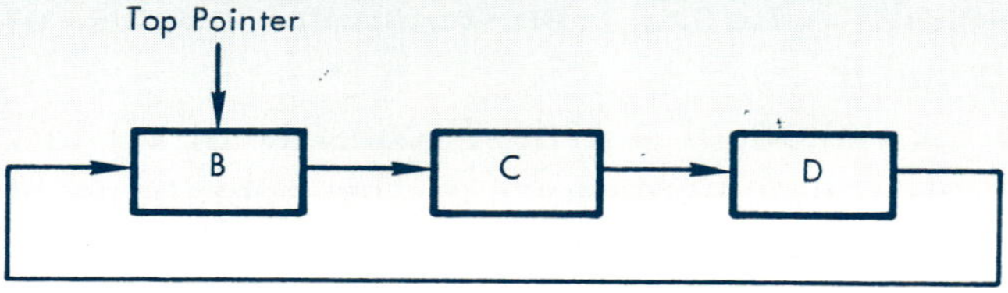

It is clear that these two representations though not identical, refer to the same abstract value. That is to say, the mapping from values to representations, Φ^{-1}, may be one-to-many.

The representation of type Symboltable will make use of the abstract data types Stack (of arrays) and Array (of attributelists) as defined below.

Type: Stack

Operations:

NEWSTACK:	→ Stack
PUSH:	Stack × Array → Stack
POP:	Stack → Stack
TOP:	Stack → Array
IS_NEWSTACK?:	Stack → Boolean
REPLACE:	Stack × Array → Stack

Axioms:

(10) IS_NEWSTACK? (NEWSTACK) = true
(11) IS_NEWSTACK? (PUSH(stk, arr)) = false
(12) POP(NEWSTACK) = error
(13) POP(PUSH(stk, arr)) = stk
(14) TOP(NEWSTACK) = error
(15) TOP(PUSH(stk, arr)) = arr
(16) REPLACE(stk, arr) = **if** IS_NEWSTACK? (stk)
then error
else PUSH(POP(stk), arr)

Type: Array

Operations:

EMPTY:	→ Array
ASSIGN:	Array × Identifier × Attributelist → Array
READ:	Array × Identifier → Attributelist
IS_UNDEFINED?:	Array × Identifier → Boolean

Axioms:

(17) IS_UNDEFINED? (EMPTY, id) = true
(18) IS_UNDEFINED? (ASSIGN(arr, id, attrs), idl) =
if IS_SAME? (id, idl)
then false
else IS_UNDEFINED? (arr, idl)
(19) READ(EMPTY, id) = error
(20) READ(ASSIGN(arr, id, attrs), idl) = **if** IS_SAME? (id, idl)
then attrs
else READ(arr, idl)

The general scheme of the representation of type Symboltable is to treat a value of the type as a stack of arrays (with index type Identifier), where each array contains the attributes for the identifiers declared in a single block. For every function f in the more abstract domain (e.g. type Symboltable), a function f' is defined in the lower-level domain; thus we have:

INIT':	→ Stack
ENTERBLOCK':	Stack → Stack
LEAVEBLOCK':	Stack → Stack
ADD':	Stack × Identifier × Attributelist → Stack
IS_INBLOCK?':	Stack × Identifier → Boolean
RETRIEVE':	Stack × Identifier → Attributelist

The "code" for each of these functions is ("::" means "is defined as"):

```
INIT' :: PUSH(NEWSTACK, EMPTY)
ENTERBLOCK'(stk) :: PUSH(stk, EMPTY)
LEAVEBLOCK'(stk) :: if IS_NEWSTACK? (POP(stk))
                      then error
                      else POP(stk)
ADD'(stk, id, attrs) :: REPLACE(stk, ASSIGN(TOP(stk), id,
   attrs))
IS_INBLOCK?'(stk, id) :: if IS_NEWSTACK? (stk)
                           then false
                           else ¬ IS_UNDEFINED? (TOP(stk),
                             id)
RETRIEVE'(stk, id) :: if IS_NEWSTACK? (stk)
                        then error
                        else ¬ IS_UNDEFINED? (TOP(stk), id)
                          then RETRIEVE'(POP(stk), id)
                          else READ(TOP(stk), id)
```

The interpretation function Φ is defined by:

(a) Φ(error) = error
(b) Φ(NEWSTACK) = error
(c) Φ(PUSH(stk, EMPTY)) = **if** IS_NEWSTACK? (stk)
  then INIT
  else ENTERBLOCK(Φ(stk))
(d) Φ(PUSH(stk, ASSIGN(arr, id, attrs))) = ADD(ΦPUSH(stk, arr)), id, attrs))

Before continuing to refine these operations, i.e. before supplying representations for types Array and Stack, let us consider the problem of proving that the above implementation of type Symboltable is correct.

In the course of such a proof two kinds of invariants may have to be verified: inherent invariants and representation invariants. The inherent invariants represent those invariant relationships that must be maintained by any representation of the type. They correspond to the axioms used in the specification of the type. A representation invariant, on the other hand, is peculiar to a particular representation of a type.

The basic procedure followed in verifying the inherent invariants is to take each axiom for type Symboltable and replace all instances of each function appearing in the axiomatization with its interpretation. Then, by using the axiomatizations of the operations used in constructing the representations, it is shown that the left-hand side of each axiom is equivalent to the right-hand side of that axiom. That is to say, they represent the same abstract value.

What must be shown therefore is that for every relation $f'(x^*) = z$ (where x^* is a list, possibly empty, of arguments), derived from the axiomatization of type Symboltable,

(a) if the range of f is the type being defined (i.e., Symboltable), $\Phi(f'(x^*)) = \Phi(z)$ for all legal assignments to the free variables of x^* and z, or

(b) if the range of f is a type other than that being defined, $f'(x^*) = z$ for all legal assignments to the free variables of x^* and z.

To show this, we have at our disposal a proof system consisting of the axioms and rules of inference of our programming language plus the axioms defining the abstract types used in the representation.

The proof depends upon the assumption that objects of type Symboltable are created and manipulated only via the operations defined in the specification of that type. (The use of classes as described in Palme [18] makes this assumption relatively easy to verify.) All that need be shown is that INIT′ establishes the invariants and that if on entry to an operation all invariants hold for all objects of type Symboltable to be manipulated by that operation, then all invariants on those objects hold upon completion of that operation. More complete discussions of how this may be done are contained in Guttag [8], Spitzen [21], and Wegbreit [23] (where it is called generator induction).

To verify that the implementation is consistent with Axioms 1 through 8 is quite straightforward. (It has, in fact, been done completely mechanically by David Musser [17] using the program verification system at the University of Southern California Information Sciences Institute [7]. Thus the proofs will not be presented here. Axiom 9, on the other hand, presents some problems that make the portion of the proof pertinent to that axiom worth examining.

The proof that the implementation satisfies Axiom 9 is based upon an assumption about the environment in which the operations of the type are to be used. In effect, the assumption asserts that an identifier is never added to an empty symbol table, i.e. a scope must have been established (on a more concrete level, an array must have been pushed onto the stack) before an identifier can be added. The concrete manifestation of this assumption is formally expressed:

Assumption 1. For any term, ADD′(symtab, id, attrs), IS_NEWSTACK? (symtab) = false.

The validity of the above assumption can be assured by adding to the implementation of ADD′ a check for this condition and having it execute an ENTERBLOCK′ if necessary. This would make it possible to construct a completely self-contained proof of the correctness of the representation. In most cases, however, it would also introduce needless inefficiency. The compiler must somewhere check for mismatched (i.e. extra) "end" statements. Any check in ADD′ would therefore be redundant.

This observation leads to a notion of conditional correctness, i.e. the representation of the abstract type is correct if the enclosing program obeys certain constraints. In practice, this is often an extremely useful notion of correctness, especially if the constraint is easily checked. If, on the other hand, the environment in which the abstract type is to be used is unknown (e.g. if the type is to be included in a library), this is probably unacceptably dangerous. Given the above assumption, the verification of Axiom 9 is straightforward but lengthy and will therefore not be presented here. It does appear in Guttag [8].

Now we know that, given implementations of types Stack and Array that are consistent with their specifications, the implementation of type Symboltable is "correct." Assuming PL/I-like based variables, pointers, and structures, the implementation of type Stack is trivial. The basic scheme is to represent a stack as a pointer to a list of structures of the form:

```
1. stack elem based,
   2. val Array,
   2. prev pointer.
```

The operations may be implemented as follows (PL/I keywords have been boldfaced):

```
NEWSTACK' :: null
PUSH'(symtab, newblock) ::
  procedure(symtab: pointer, newblock: Array)returns(pointer)
    declare elem_ptr pointer
    allocate(stack_elem) set(elem_ptr)
    elem_ptr → prev := symtab
    elem_ptr → val := newblock
    return(elem_ptr)
  end
POP'(symtab) ::
  procedure(symtab: pointer) returns(pointer)
    if symtab = null
      then return(error)
      else return(symtab → prev)
  end
TOP'(symtab) ::
  procedure(symtab: pointer) returns(Array)
```

```
    if symtab = null
      then return(error)
      else return(symtab → val)
  end
IS_NEWSTACK?'(symtab) :: symtab = null
REPLACE'(symtab, newblock) ::
  procedure(symtab: pointer, newblock: Array) returns(pointer)
    if symtab = null
      then return(error)
      else symtab → val := newblock
        return(symtab)
  end
```

Φ is defined by the mapping:

```
Φ(symtab) :: if symtab = null
               then NEWSTACK
               else PUSH(Φ(symtab → prev), symtab → val))
```

The implementation chosen for type Array is a bit more complicated. The basic scheme is to represent an array as a PL/I-like array, hash_tab, of *n* pointers to lists of structures of the form:

```
1. entry based,
  2. id Identifier
  2. attributes Attributelist,
  2. next pointer.
```

The correct element of hash_tab is selected by performing a hash on values of type Identifier. Therefore, in addition to the operations used in the code above, the implementation of type Array uses an operation

HASH:Identifier → {1, 2, . . . , n}

which is assumed to be defined in the type Identifier specification. The "code" implementing type Array is:

```
declare hash_tab(n) pointer based

EMPTY' ::
  procedure returns(pointer)
    declare new_hash_tab pointer
    allocate (hash_tab) set (new_hash_tab)
    do i := 1 to n
```

```
        new_hash_tab → hash_tab(i) := null
      end
      return(new_hash_tab)
    end
ASSIGN'(arr, indx, atr) ::
    procedure(arr: pointer, indx: Identifier, atr: Attributelist)
          returns(pointer)
      declare new_entry pointer
      allocate(entry) set (new_entry)
        new_entry → id := indx
        new_entry → attributes := atr
        new_entry → next := arr → hash_tab(HASH(indx))
        arr → hash_tab(HASH(indx)) := new_entry
      return(arr)
    end
READ'(arr, indx) ::
    procedure(arr: pointer, indx: Identifier) returns(Attributelist)
      declare bucket_ptr pointer
        bucket_ptr := arr → hash_tab(HASH(indx))
      do while(bucket_ptr ≠ null & ¬ IS_SAME?(bucket_ptr → id,
          indx))
        bucket_ptr := bucket_ptr → next
      end
      if bucket_ptr = null
        then return(error)
        else return (bucket_ptr → attributes)
    end
IS_UNDEFINED?'(arr, indx) ::
    procedure(arr: pointer, indx: Identifier) returns(Boolean)
      declare bucket_ptr pointer
        bucket_ptr := arr → hash_tab(HASH(indx))
      do while (bucket_ptr ≠ null & ¬ IS_SAME? (bucket_ptr → id,
          indx))
        bucket_ptr := bucket_ptr → next
      end
      return (bucket_ptr = null)
    end
```

As one might expect, Φ is a bit more complex for this representation. It is defined by using two intermediate functions: Φ1 to construct a union over all the entries in the hash table, and Φ2 to construct a union over the elements of an individual bucket.

```
(a) Φ(hash_tab_ptr) = Φ1(hash_tab_ptr, EMPTY, 1)
(b) Φ1(hash_tab_ptr, arr, i) =
      if i > n
        then arr
        else Φ1(hash_tab_ptr, Φ2(hash_tab_ptr → hash_tab(i), arr),
          i + 1)
(c) Φ2(bucket_ptr,arr) =
      if bucket_ptr = null
        then arr
        else ASSIGN(Φ2(bucket_ptr → next, arr), bucket_ptr → id,
          bucket_ptr → attributes)
```

The design of the symbol table subsystem of the compiler is now essentially complete. Given implementations of types Identifier and Attributelist and some obvious syntactic transformations, the above code could be compiled by a PL/I compiler. Before doing so, however, it would be wise to prove that the implementations of types Stack and Array are consistent with the specifications of those types. While such a proof would involve substantial issues related to the general program verification problem (e.g. vis à vis the integrity of the pointers and the question of modifying shared data structures), it would not shed further light on the role of abstract data types in program verification and is not presented in these pages.

The ease with which algebraic specifications can be adapted for different applications is one of the major strengths of the technique. Because the relationships among the various operations appear explicitly, the process of deciding which axioms must be altered to effect a change is straightforward. Let us consider a rather substantial change in the language to be compiled. Assume that the language permits the inheritance of global variables only if they appear in a "knows list," which lists, at block entry, all nonlocal variables to be used within the block [6]. The symbol table operations in a compiler for such a language would be much like those already discussed. The only difference visible to parts of the compiler other than the symbol table module would be in the ENTERBLOCK operation: It would have to be altered to include an argument of abstract type Knowlist. Within the specification of type Symboltable, all relations, and only those relations, that explicitly deal with the ENTERBLOCK operation would have to be altered. An appropriate set of axioms would be:

```
IS_INBLOCK?(ENTERBLOCK(symtab, klist), id) = false
LEAVEBLOCK(ENTERBLOCK(symtab, klist)) = symtab
RETRIEVE(ENTERBLOCK(symtab, klist), id) =
  if IS_IN?(klist, id)
    then RETRIEVE(symtab, id)
    else error
```

Note that the above relations are not well defined. The undefined symbol IS_IN?, an operation of the abstract type Knowlist, appears in the third axiom. The solution to this problem is simply to add another level to the specification by supplying an algebraic specification of the abstract type Knowlist. An appropriate set of operations might be:

CREATE:	→ Knowlist
APPEND:	Knowlist × Identifier → Knowlist
IS_IN?:	Knowlist × Identifier → Boolean

These operations could then be precisely defined by the following axioms:

```
IS_IN?(CREATE) = false
IS_IN?(APPEND(klist, id), idl) = if IS_SAME?(id, idl)
                                   then true
                                   else IS_IN?(klist, idl)
```


The implementation of abstract type Knowlist is trivial. The changes necessary to adapt the previously presented implementation of abstract type Symboltable would be more substantial. The kind of changes necessary can, however, be inferred from the changes made to the axiomatization.

5. Conclusions

We have not yet applied the techniques discussed in this paper to realistically large software projects. Nevertheless, there is reason to believe that the techniques demonstrated will "scale up." The size and complexity of a specification at any level of abstraction are essentially independent of both the size and complexity of the system being described and of the amount of mechanism ultimately used in the implementation. The independence springs in large measure from the ability to separate the precise meaning of a complex abstract data type from the details involved in its implementation. It is the ability to be precise without being detailed that encourages the belief that the approach outlined here can be applied even to "very large" systems can and perhaps reduce systems that were formerly "very large" (i.e. incomprehensible) to more manageable proportions.

Abstract types may thus play a vital role in the formulation and presentation of precise specifications for software. Many complex systems can be viewed as instances of an abstract type. A database management system, for example, might be completely characterized by an algebraic specification of the various operations available to users. For those systems that are not easily totally characterized in terms of algebraic relations, the use of algebraic type specifications to abstract various complex subsystems may still make a substantial contribution to the design process. The process of functional decomposition requires some means for specifying the communication among the various functions – data often fulfills this need. The use of algebraic specifications to provide abstract definitions of the operations used to establish communication among the various functions may thus play a significant role in simplifying the process of functional abstraction.

The extensive use of algebraic specifications of abstract types may also lead to better-designed data structures. The premature choice of a storage structure and set of access routines is a common cause of inefficiencies in software. Because they serve as the main means of communication among the various components of many systems, the data structures are often the first components designed. Unfortunately, the information required to make an intelligent choice among the various options is often not available at this stage of the design process. The designer may, for example, have poor insight into the relative frequency of the various operations to be performed on a data structure. By providing a representation-free, yet precise, description of the operations on a data structure, algebraic type definitions enable the designer to delay the moment at which a storage structure must be designed and frozen.

The second area in which we expect the algebraic specification of abstract types to have a substantial impact is on proofs of program properties. For verifications of programs that use abstract types, the algebraic specification of the types used provides a set of powerful rules of inference that can be used to demonstrate the consistency of the program and its specification. That is to say, the presence of axiomatic definitions of the abstract types provides a mechanism for proving a program to be consistent with its specifications, provided that the implementations of the abstract operations that it uses are consistent with their specifications. Thus a technique for factoring the proof is provided, for the algebraic type definitions serve as the specification of intent at a lower level of abstraction. For proofs of the correctness of representations of abstract types, the algebraic specification provides exactly those assertions that must be verified. The value of having such a set of assertions available should be apparent to any one who has attempted to construct, *a posteriori*, assertions appropriate to a correctness proof for a program. A detailed discussion of the use of algebraic specifications in a semiautomatic program verification system is contained in Guttag [10].

Given suitable restrictions on the form that axiomatizations may take, a system in which implementations and algebraic specifications of abstract types are interchangeable can be constructed. In the absence of an implementation, the operations of the algebra may be interpreted symbolically. Thus, except for a significant loss in efficiency, the lack of an implementation can be made completely transparent to the user. Such a system should prove valuable as a vehicle for facilitating the testing of software.

The ability to use specifications for testing is closely related to the policy of restricted information flow advocated in Parnas [20]. If a programmer is supplied with algebraic definitions of the abstract operations available to him and forced to write and test his module with only that information available to him, he is denied the opportunity to rely intentionally or accidentally upon information that should not be relied upon. This not only serves to localize the effect of implementation errors, but also to increase the ease with which one implementation may be replaced by another. This should, in general, serve to limit the danger of choosing a poor representation and becoming inextricably locked into it.

Before ending this paper, it seems fitting to mention some of the failings and problems associated with the work described. The specification technique presented here requires that all operations be specified as functions, i.e. as mappings from a cross product of values to a single value. Most programs, on the other hand, are laden with procedures that return several values (via parameters) or no value at all. (The latter kind of procedure is invoked purely for its side effects.) The inability to specify such procedures is a serious problem, but one that we believe can be solved with only minor changes to the specification techniques [10].

The value of abstraction in general and abstraction of data types in particular has been stressed throughout this paper. Nevertheless, the process is not without its dangers. It is all too easy to create abstractions that ignore crucial distinctions or attributes. The specification technique presented here, for example, provides no mechanism for specifying performance constraints and thus encourages one to ignore distinctions based on such criteria. In some environments, such considerations are crucial, and to abstract them out can be disastrous.

Another problem with algebraic specifications is that they supply little direction to implementors. Only experience will tell how easy it is to go from an algebraic specification to an implementation. It is clear, however, that the transition is less easy than from an operational specification.

Our most important reservation pertains to the ease with which algebraic specifications can be constructed and read. They should present no problem to those with formal training in computer science. At present, however, most people involved in the production of software have no such training. The extent to which the techniques described in this paper are generally applicable is thus somewhat open to conjecture.

Acknowledgment. The author is greatly indebted to J.J. Horning of the University of Toronto, who, as the author's thesis supervisor, provided three years of good advice.

References

1. Batey, M., Ed. Working Draft of ECMA/ANSI PL/I Standard Tenth Rev., ANSI, New York, (Sept. 1973).
2. Birkhoff, G., and Lipson, J.D. Heterogeneous algebras. *J. Combinatorial Theory 8* (1970), 115–133.
3. Dahl, O.-J., Nygaard, K., and Myhrhaug, B. The SIMULA 67 Common Base Language. Norwegian Comptng. Centre, Oslo, 1968.
4. Dijkstra, E.W. Notes on structured programming. In *Structured Programming*, Academic Press, New York, 1972.
5. Floyd, R.W. Assigning Meaning to Programs. Proc. Symp. in Applied Math., Vol. XIX, AMS, Providence, R.I., 1967, pp. 19–32.
6. Gannon, J.D. Language design to enhance programming reliability. Ph.D. Th., Comptr. Syst. Res. Group Tech. Rep. CSRG-47, Dept. Comptr. Sci., U. of Toronto, Ontario, 1975.
7. Good, D.I., London, R.L., and Bledsoe, W.W. An interactive program verification system. *IEEE Trans. on Software Engineering SE-1*, 1 (March 1975), 59–67.
8. Guttag, J.V. The specification and application to programming of abstract data types. Ph.D. Th., Comptr. Syst. Res. Group Tech. Rep. CSRG-59, Dept. Comptr. Sci. 1975, U. of Toronto, Ontario, 1975.
9. Guttag, J.V. and Horning, J.J., The algebraic specifications of abstract data types. *Acta Informatica* (to appear).
10. Guttag, J.V., Horowitz, E., and Musser, D.R. Abstract data types and software validation. Tech. Rep., Inform. Sci. Inst., U. of Southern California, Los Angeles, 1976.
11. Guttag, J.V., Horowitz, E., and Musser, D.R. The design of data type specifications. Proc. Second Int. Conf. on Software Eng., San Francisco, Oct. 1976, pp. 414–420.
12. Hoare, C.A.R., Proof of correctness of data representations. *Acta Informatica 1* (1972), 271–281.
13. Hoare, C.A.R., and Wirth, N. An axiomatic definition of the programming language PASCAL. *Acta Informatica 2* (1973), 335–355.
14. Liskov, B.H., and Zilles, S.N. Programming with abstract data types. Proc. ACM SIGPLAN Symp. on Very High Level Languages, SIGPLAN Notices (ACM) *9*, 4 (April 1974), 50–59.
15. McKeeman, W.M., Symbol Table Access. In *Compiler Construction, An Advanced Course*, T.L. Bauer, and J. Eichel, Eds., Springer-Verlag, New York, 1974.
16. Morris, J.H. Types are not sets. Conf. Rec. ACM Symp. on the Principles of Programming Languages, Boston, Mass., Oct. 1973, pp. 120–124.
17. Musser, D. Private communication, 1975.
18. Palme, J. Protected program modules in SIMULA 67. FOAP Rep. C8372-M3(E5), Res. Inst. of National Defense, Stockholm, 1973.
19. Parnas, D.L. A technique for the specification of software modules with examples. *Comm. ACM 15,* 5 (May 1973), 330–336.
20. Parnas, D.L. Information distribution aspects of design methodology. Information Processing 71, North Holland Pub. Co., Amsterdam, 1971, pp. 339–344.

21. Spitzen, J., and Wegbreit, B. The verification and synthesis of data structures. *Acta Informatica 4* (1975), 127–144.
22. Standish, T.A. Data structures: An axiomatic approach. BBN Rep. No. 2639, Bolt, Beranek and Newman, Cambridge, Mass., (1973).
23. Wegbreit, B., and Spitzen, J. Proving properties of complex data structures. *J. ACM 23,* 2 (April 1976), 389–396.
24. Wulf, W.A., London, R.L., and Shaw, M. Abstraction and verification in Alphard: Introduction to language and methodology. USC Inform. Sci. Tech. Rep., U. of Southern California, Los Angeles, 1976.
25. Zilles, S.N. Abstract specifications for data types. IBM Res. Lab., San Jose, Calif., 1975.

Data Structures Form Algorithms

Michael Jackson
Consultant
London
jacksonma@acm.org

Michael Jackson Systems Ltd.

Researcher,
AT&T Research Laboratories

Major contributions:
Jackson Structured Programming (JSP)
and Design (JSD)

Current interests:
analysis and structure of software
development problems

Michael Jackson

JSP in Perspective

Jackson Structured Programming (JSP) is a method of program design. Its origins lie in the data processing systems that grew up in the 1960s, when reliable, relatively cheap and adequately powerful computers first became generally available. The fundamental abstraction in JSP is the sequential data stream. Originally, this abstraction was inspired and motivated by the sequential tape files that characterised data processing in the 1960s, but it quickly became clear that it had a much wider applicability. Today the JSP design method is valuable for applications including embedded software, processing streams of EDI messages, handling network protocols and many others.

JSP arose from efforts by a small group of people in a data processing consultancy company to improve their programming practices, and to make their programs more reliable and easier to understand and to modify. In 1971 it became the core product of a very small new company, Michael Jackson Systems Limited, which offered development services, training courses, consultancy, and – from 1975 – software to support JSP design of COBOL programs. The name JSP was coined by the company's Swedish licensee in 1974. In the commercial world, IBM had appropriated the name 'Structured Programming' in the early 1970s, and Yourdon Inc. started offering courses in 'Structured Design' around 1974. A distinctive name was a commercial necessity. It was also technically appropriate to choose a distinctive and proprietary name: the JSP method was very different from its competitors.

1960s Data Processing Systems

Data processing systems of the early and middle 1960s were chiefly concerned with the processing of sequential files held on magnetic tape. Reliable tape drives had become widely available and were commonly used in the late 1950s; exchangeable disk drives first became available when IBM introduced the 1311 drive in 1963. Disk was a very limited and expensive medium compared to tape. At 1965 prices a 1311 disk pack cost about £200 and held 2 million characters; a 2400-foot tape reel cost about £7 and held between 20 million and 60 million characters. For large files, which might contain millions of records occupying many reels, tape was the only realistic choice. Most data processing systems had large files.

Because tape is an inherently sequential medium, updating a single record of a master file could be done only by reading the whole file and copying it, updated, to a new tape. This very slow process was economical only if many records were to be updated, so tape systems were almost inevitably batch systems. Transactions – for example, payments received – were recorded daily or weekly on a transaction tape file. The transaction file was then sorted into the same sequence as the file of master records – for example, customer accounts – to which the transactions were to be applied; it was then used in a batch update program to produce a new version of the master file whose records reflected the effects of the transactions.

It was always necessary to process the whole master file and to produce a complete new version, even if the batch contained transactions for very few master records. Processing a file that occupied one full tape might take an hour or more; some master files occupied dozens of tapes. Even worse, there might be several master files to be processed – for example, customers, orders, invoices, and products. The transaction file would then be sorted successively into the different sequences of the different master files, executing a batch update program for each master file and carrying partial results forward to the next update program in a transfer file that would also require sorting. To minimise processing time master files were amalgamated where possible. For example, the orders, instead of being held in a master file of their own, might be held within the customer master file, the order records for each customer following the customer record in the combined file. These choices resulted in a database with a hierarchical structure, held on magnetic tape: this was the kind of database for which IBM's database management system IMS was originally designed around 1966 in cooperation with North American Rockwell [4].

The Basic JSP Idea

A common design fault in batch update programs was failure to ensure that the program kept the files correctly synchronised as it traversed their hierarchical structures. A read operation performed at the wrong point in program execution might read beyond the record to which the next transaction should be applied, causing erroneous processing of that transaction and, often, of all the following transactions and master records. Another common design fault was failure to take account of empty sets – for example, of a customer with no outstanding orders. Such a fault could make irreparable nonsense of a complete update run that may have taken many hours. How could programs be designed to avoid such faults?

In commercial and industrial programming in the 1960s, the program design question was chiefly posed in terms of 'modular programming': What was the best decomposition for each particular program? The primary focus was on the decomposition structure, not on encapsulation. A 1968 conference [3] dedicated to modular programming attracted the participation of George Mealy, the computer scientist who gave his name to Mealy machines. The structured design ideas of coupling and cohesion [12, 13] took shape as an approach to modularity: it was claimed that a good design could be achieved by ensuring that the modules have high cohesion and low coupling.

The fundamental idea of JSP was very different: program structure should be dictated by the structures of its input and output data streams [7, 8]. If one of the sequential files processed by the program consisted of customer groups, each group consisting of a customer record followed by some number of order records, each of which is either a simple order or an urgent order, then the program should have the same structure: it should have a program part that processes the file, with a subpart to process each customer group; and that subpart should itself have one subpart that processes the customer record; and so on. The execution sequence of the parts should mirror the sequence of records and record groups in the file. Program parts could be very small and were not, in general, separately compiled.

The resulting structure can be represented in a JSP structure diagram, as in Figure 1. The diagram is a tree representation of the regular expression

(CustRecord (SimpleOrder | UrgentOrder)*)*

in which the expression and all of its subexpressions are labelled. Iteration is shown by the star in the iterated subexpression; selection is shown by the circle in each alternative.

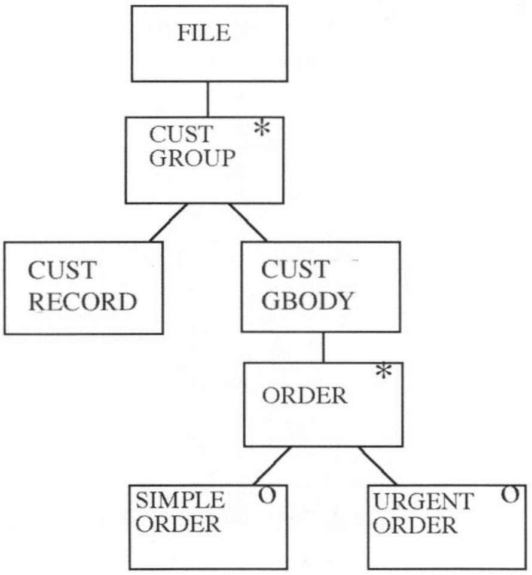

Fig. 1. Structure of a file and of a program

The structure is simultaneously the structure of the file and the structure of a program to process the file. As a data structure it may be verbalised like this:

"The *File* consists of zero or more *Customer Groups*. Each *Customer Group* consists of a *Customer Record* followed by a *Customer Group Body*. Each *Customer Group Body* consists of zero or more *Orders*. Each *Order* is either a *Simple Order* or an *Urgent Order*"

As a program structure it may be understood to mean:

```
program =
  { /* process file */
    while (another customer group) do
    { /* process customer group */
      process customer record;
      { /* process customer group body /*
        while (another order) do
        { /* process order */
          if simple order then
          { /* process simple order */ }
          else
          { /* process urgent order */ }

  } } } }
```

Multiple Streams

The JSP program design method insisted that the program structure should reflect all the stream data structures, not just one. Its first steps, then, are to identify the data structure of each file processed by the program, and to form a program structure that embodies them all. Such a program structure allows the designer to ensure easily that program execution will interleave all the file traversals correctly, keeping the files appropriately synchronised. Figure 2 shows an example of a program structure based on two data structures.

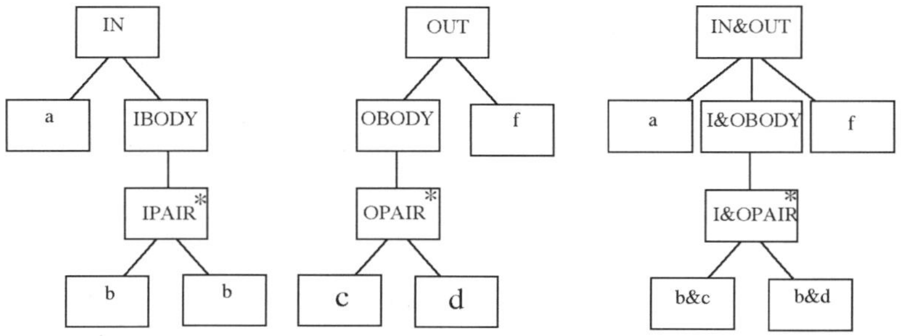

Fig. 2. Two file structures and a program structure

For brevity, the example is stylised and trivial. The program processes an input file *IN* and and output file *OUT*. *IN* contains an *a* record followed by an even number of *b* records; *OUT* contains *c* records, made from the odd-numbered *b* records, interleaved with *d* records made from the even-numbered *b* records, followed by a final *f* record. Thus, successive *OPAIRs* of *OUT* are constructed from successive *IPAIRs* of *IN*: that is, the *IPAIRs* and *OPAIRs* 'correspond functionally'. Similarly, the *c* and *d* records are computed from the first and second *b* records of each *IPAIR*: that is, the *c* and first *b* records correspond and the *d* and second *b* records correspond. In this trivial example it is easily seen that the program structure embodies both of the file structures exactly.

In more realistic examples, a program structure embodying all the file structures can be achieved by permissible rewritings of the file structures. Two such rewritings are shown in Figure 3.

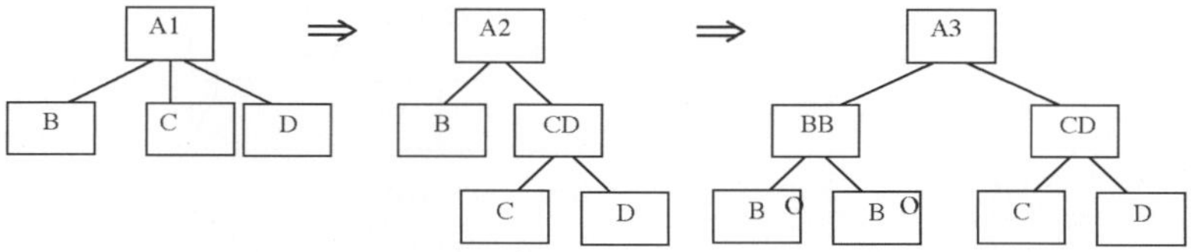

Fig. 3. Examples of regular expression rewritings

The data structure A1 can be rewritten as A2, and A2 as A3. Permissible rewritings preserve the set of leaf sequences defined by the structure: A1, A2 and A3 all define the sequence <B,C,D>. They must also preserve the intermediate nodes of each structure: A2 may not be rewritten as A1, because the node CD would be lost. The eventual program structure must have at least one component corresponding to each component of each data structure.

Operations and Conditions

The prime advantage of a program structure that embodies all the file data structures, respecting the correspondences among their parts, is that it provides an obviously correct place in the program text for each operation that must be executed. It also clarifies the conditions needed as guards in iteration (loop) and selection (if-else or case) constructs.

The operations to be executed are file traversal operations, such as *open*, *close*, *read* and *write*, and other operations that compute output record values from values of input records. For example, in the trivial illustration of Figure 2 the operations may be:

open IN, read IN, close IN, open OUT, write OUT record, close OUT, c := f(b), and so on. Each operation must appear in the program component that processes the operation's data. The *read* operations are a special case. Assuming that the input files can be parsed by looking ahead one record, there must be one *read* operation following the *open* at the beginning of the file, and one at the end of each component that completely processes one input record. The last read operation reads the notional EOF marker record, which is treated as a distinct record type.

For the example of Figure 2, the operations must be placed as shown in Figure 4.

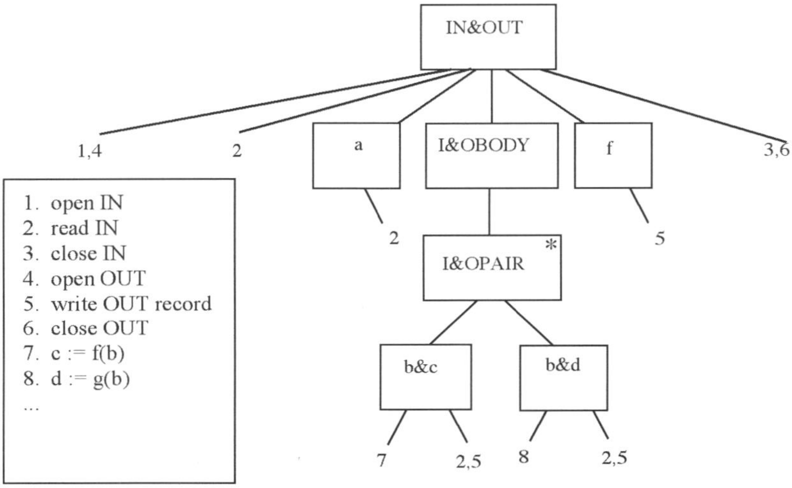

Fig. 4. Placing operations in program structure

The correspondence of program and data structures, together with the scheme of looking ahead one record, makes it easy to determine the iteration and selection conditions. For example, the condition on the iteration component *I&OBODY* is

while (another *I&OPAIR*)

which translates readily into

while (*IN* record is *b*)

in which '*IN* record' refers to the record that has been read ahead and is currently in the *IN* buffer. This condition is not satisfied by the *EOF* marker record.

Difficulties

The development procedures of a method should be closely matched to specific properties of the problems it can be used to solve. The development procedures of basic JSP, as they have been described here, require the problem to possess at least these two properties:

- The data structures of the input and output files, and the correspondences among their data components, are such that a single program structure can embody them all; and
- Each input file can be unambiguously parsed by looking ahead just one record.

Absence of a necessary property is immediately recognisable by difficulty in completing a part of the JSP design procedure. If the file structures do not correspond appropriately it is impossible to design a correct program structure: this difficulty is called a *structure clash*. If an input file cannot be parsed by single look-ahead it is impossible to write all the necessary conditions on the program's iterations and selections: this is a *recognition difficulty*.

Although these difficulties are detected during the basic JSP design procedure, they do not indicate a limitation of JSP. They indicate inherent complications in the problem itself, which cannot be ignored but must be dealt with somehow. In JSP they are dealt with by additional techniques within the JSP method.

Structure Clashes

There are three kinds of structure clash: *interleaving clash, ordering clash* and *boundary clash*.

In an interleaving clash, data groups that occur sequentially in one structure correspond functionally to groups that are interleaved in another structure. For example, the input file of a program may consist of chronologically ordered records of calls made at a telephone exchange; the program must produce a printed output report of the same calls arranged chronologically within subscriber. The 'subscriber groups' that occur successively in the printed report are interleaved in the input file.

In an ordering clash, corresponding data item instances are differently ordered in two structures. For example, an input file contains the elements of a matrix in row order, and the required output file contains the same elements in column order.

In a boundary clash, two structures have corresponding elements occurring in the same order, but the elements are differently grouped in the two structures. The boundaries of the two groupings are not synchronised.

Boundary clashes are surprisingly common. Here are three well-known examples:

- The calendar consists of a sequence of days. In one structure the days may be grouped by months, but in another structure by weeks. There is a boundary clash here: the weeks and months cannot be synchronised.

- A chapter of a printed book consists of text lines. In one structure the lines may be grouped by paragraphs, but in another structure by pages. There is a boundary clash because pages and paragraphs cannot be synchronised.
- A file in a low-level file handling system consists of variable-length records, each consisting of between 2 and 2000 bytes. The records must be stored sequentially in fixed blocks of 512 bytes. There is a boundary clash here: the boundaries of the records cannot be synchronised with the boundaries of the blocks.

The difficulty posed by a boundary clash is very real. The clash between weeks and months causes endless trouble in accounting: in 1923 the League of Nations set up a Special Committee of Enquiry into the Reform of the Calendar to determine whether the clash could be resolved by adopting a new calendar [1]. The clash between records and blocks affected the original IBM OS/360 file-handling software: the 'access method' that could handle the clash – software that supported 'spanned records' – proved the hardest to design and was the last to be delivered.

The JSP technique for dealing with a structure clash is to decompose the original program into two or more programs communicating by intermediate data structures. A boundary clash, for example, requires a decomposition into two programs communicating by an intermediate sequential stream. The structure of the intermediate stream is based on the 'highest common factor' of the two clashing structures. For example, an accounting program may have a boundary clash between an input file structured by months and an output file structured by weeks. Figure 5 shows the data structures.

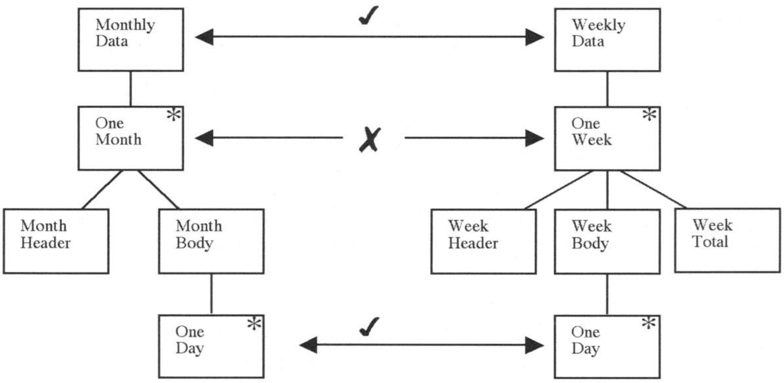

Fig. 5. Structures exhibiting a boundary clash

The solution to the difficulty is to decompose the program into two: one program handles the *Months*, producing an intermediate file structured by *Days*, which is input to a second program that handles the *Weeks*. For the second program the intermediate file structure must have no *Month* component: any necessary information from the *MonthHeader* record must therefore be encoded in the *OneDay* records of the intermediate file.

Recognition Difficulties

A recognition difficulty is present when an input file cannot be unambiguously parsed by single look-ahead. Sometimes the difficulty can be overcome by looking ahead two or more records; sometimes a more powerful technique is necessary.

The two cases are illustrated in Figure 6. The structure on the left can be parsed by looking ahead three records: the beginning of an *AGroup* is recognised when the third of the look-ahead records is an A. But the structure on the right cannot be parsed by any fixed look-ahead. The JSP technique needed for this structure is *backtracking*.

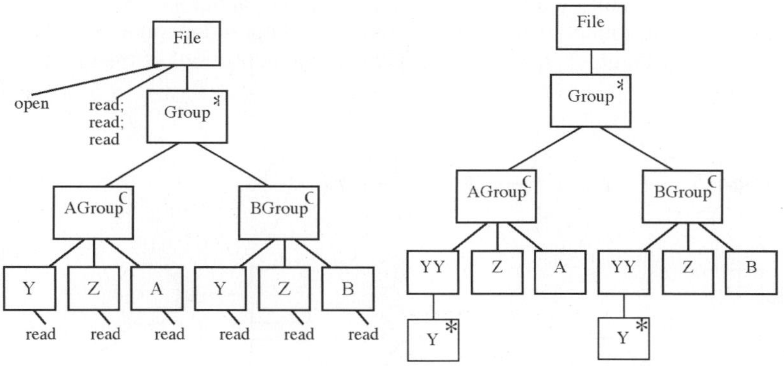

Fig. 6. Structures requiring multiple read-ahead and backtracking

The JSP procedure for the backtracking technique has three steps:

- First, the recognition difficulty is simply ignored. The program is designed, and the text for the *AGroup* and *BGroup* components is written, as usual. No condition is written on the Group selection component. The presence of the difficulty is marked only by using the keywords *posit* and *admit* in place of *if* and *else*.
- Second, a *quit* statement is inserted into the text of the *posit AGroup* component at each point at which it may be detected that the *Group* is, in fact, not an *AGroup*. In this example, the only such point is when the *B* record is encountered where an *A* record was expected. The *quit* statement is a tightly constrained form of *GO TO*: its meaning is that execution of the *AGroup* component is abandoned and control jumps to the beginning of the *admit BGroup* component.
- Third, the program text is modified to take account of side effects: that is, of the side effects of operations executed in *AGroup* before detecting that the *Group* was in fact a *BGroup*.

Program Inversion

The JSP solution to structure clash difficulties seems, at first sight, to add yet more sequential programs and intermediate files to systems already overburdened with time-consuming file processing. But JSP provides an implementation technique – *program inversion* – that both overcomes this obstacle and offers other important positive advantages.

The underlying idea of program inversion is that reading and writing sequential files on tape is only a specialised version of a more general form of communication. In the general form programs communicate by producing and consuming sequential streams of records, each stream being either unbuffered or buffered according to any of several possible regimes. The choice of buffering regime is, to a large extent, independent of the design of the communicating programs. But it is not independent of their scheduling. Figure 7 shows three possible implementations of the same system.

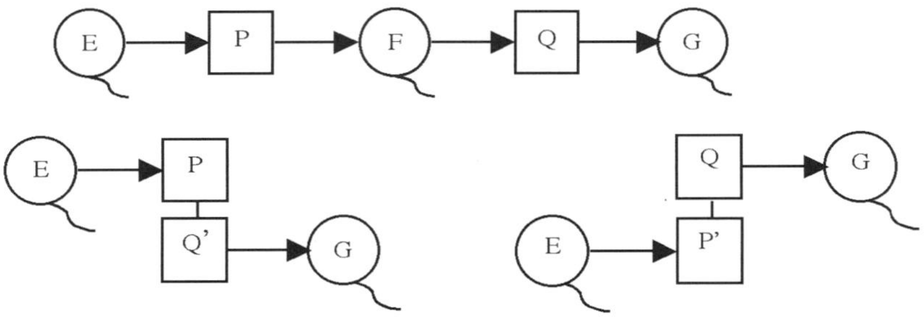

Fig. 7. Using inversion to eliminate an intermediate file

In the upper diagram two 'main' programs, P and Q, communicate by writing and reading an intermediate tape file F. First P is run to completion; then F is rewound; then Q is run to completion. In the lower left diagram the intermediate tape file has been eliminated: the 'main' program Q has been *inverted with respect to F.* This inversion converts Q into a subroutine Q', invoked by P whenever P requires to produce a record of F. Q' functions as a 'consume next F record' procedure. Similarly, in the lower right diagram the 'main' program P has been *inverted with respect to F.* This inversion converts P into a subroutine P', invoked by Q whenever Q requires to consume a record of F. P' functions as a 'produce next F record' procedure. Both inversions interleave execution of P and Q as tightly as possible: each F record is consumed as soon as it has been produced.

Two or more programs may be inverted in one system. Inverting Q with respect to F, and P with respect to E, gives a subroutine P" that uses the lower-level subroutine Q'; the function of the two together is to consume the next record of E, producing whatever records of G can then be constructed.

Inversion has an important effect on the efficiency of the system. First, it eliminates the storage space and device costs of the intermediate tape file. It eliminates the time required by the device to write and read each record of F, and also the 'rewind time' to reposition the newly written file for reading: for a magnetic tape file this may be many minutes for one reel. Second, it makes each successive record of G available with the least possible delay: each G record is produced as soon as P has consumed the relevant records of E.

Inversion also has an important effect on the program designer's view of a sequential program. The 'main program' P, the subroutine 'P inverted with respect to F', and the subroutine 'P inverted with respect to E', are seen as identical at the design level. This is a large economy of design effort. The JSP-COBOL preprocessor that supports JSP design for COBOL programs allows the three to differ only in their declarations of the files E and F.

The effect on the program designer's view goes deeper than an economy of design effort. An important example of the distinction between design and implementation is clarified; procedures with internal state can be designed as if they were main programs processing sequential message streams; and JSP design becomes applicable to all kinds of interactive systems, and even to interrupt handlers. Program inversion also suggests the modelling of real-world entities by objects with time-ordered behaviours: this is the basis of the eventual enlargement of JSP to handle system specification and design [9].

A Perspective View of JSP

Although JSP was originally rooted in mainframe data processing, it has been applied effectively in many environments. For applying JSP, the necessary problem characteristic is the presence of at least one external sequential data stream, to provide a given data structure that can and should be used as the basis of the program structure. Many programs have this characteristic. For example:

- A program that processes a stream of EDI messages;
- An automobile cruise control program that responds to the changing car state and the driver's actions;
- A program that justifies text for printing;
- A file handler that responds to invoked operations on the file;
- A program that generates HTML pages from database queries.

JSP was developed in the commercial world, often in ignorance of work elsewhere. Some of the JSP ideas were reinventions of what was already known, while others anticipated later research results. The JSP relationship between data structures and program structure is essentially the relationship exploited in parsing by recursive descent [2]. Some of the early detailed JSP discussion of recognition difficulties dealt with aspects that were well known to researchers in formal languages and parsing techniques. The idea of program inversion is closely related to the Simula [6] concept of semi-coroutines, and, of course, to the later Unix notion of pipes as a flexible implementation of sequential files. There was also one related program design approach in commercial use: the Warnier method [14] based program structure on the regular data structure of one file. Program decomposition into sequential processes communicating by a coroutine-like mechanism was discussed in a famous early paper [5]; it was also the basis of a little-known development method called Flow-Based Programming [11].

JSP has three central virtues. First, it provides a strongly systematic and prescriptive method for an important class of useful programs: independent JSP designers working on the same problem produce essentially the same solution. Second, it identifies major difficulties that can arise in such problems, and provides techniques for their recognition, classification and solution. Third, JSP keeps the program designer firmly in the world of static structures. Only in the last step of the backtracking technique, when dealing with side effects, is the JSP designer expected to consider the dynamic behaviour of the program. This restriction to designing in terms of static structures is a decisive contribution to program correctness for those problems to which JSP can be applied. It avoids the dynamic thinking – the mental stepping through the program execution – that has always proved so seductive and so fruitful a source of error.

Acknowledgements

The foundations of JSP were laid in the years 1966-1970, when the author was working with Barry Dwyer, a colleague in John Hoskyns and Company, an English data processing consultancy. Many of the underlying ideas can be traced back to Barry Dwyer's contributions in those years. Refining the techniques, and making JSP more systematic and more teachable in commercial courses, was the work of the following four years in Michael Jackson Systems Limited.

The JSP-COBOL preprocessor was designed by the author and Brian Boulter, a colleague in Michael Jackson Systems Limited. Brian Boulter was responsible for most of the implementation.

Many other people contributed to JSP over the years: John Cameron, Tony Debling, Leif Ingevaldsson, Ashley McNeile, Hans Naegeli, Dick Nelson (who introduced the name 'JSP'), Peter Savage, Mike Slavin and many others. A partial bibliography of JSP can be found in [10].

Daniel Jackson read a draft of this paper and suggested several improvements.

References

[1] E. Achelis; *The Calendar for the Modern Age*; Thomas Nelson, 1959.

[2] AV Aho and JD Ullman; *Principles of Compiler Design*; Addison-Wesley, 1977.

[3] Barnett and Constantine eds; *Modular Programming: Proceedings of a National Symposium*; Information and Systems Press, 1968.

[4] KR Blackman; *IMS celebrates thirty years as an IBM product*; IBM Systems Journal 37, 4, 1998.

[5] ME Conway; *Design of a Separable Transition-Diagram Compiler*; Communications of the ACM 6, 7, 1963.

[6] O-J Dahl, B Myhrhaug and K Nygaard; *SIMULA-67 Common Base Language*. Technical Report Number S-22, Norwegian Computer Centre, Oslo, 1970.

[7] MA Jackson; *Principles of Program Design*; Academic Press, 1975.

[8] MA Jackson; *Constructive Methods of Program Design*; in G Goos and J Hartmanis, eds; Proceedings of the 1st Conference of the European Cooperation in Informatics, Lecture Notes in Computer Science 44; Springer, 1976.

[9] MA Jackson; *System Development*; Prentice-Hall, 1983.

[10] M Jackson; *Jackson Development Methods: JSP and JSD*; in JJ Marciniak, ed; Encyclopaedia of Software Engineering, Volume I; Wiley, 1994.

[11] JP Morrison; *Flow-Based Programming: A New Approach to Application Development*; Van Nostrand Reinhold, 1994.

[12] GJ Myers; *Software Reliability: Principles & Practices*; Wiley, 1976.

[13] WP Stevens, GJ Myers, and LL Constantine; *Structured Design*; IBM Systems Journal 13, 2, 1974.

[14] JD Warnier; *Logical Construction of Programs*; HE Stenfert Kroese, 1974, and Van Nostrand Reinhold, 1976.

Michael Jackson

Constructive Methods of Program Design

Proceedings of the 1st Conference
of the European Cooperation in Informatics,
Amsterdam, August 9-12, 1976
Edited by K. Samelson
Lecture Notes in Computer Science, Vol. 44,
Springer-Verlag, Berlin, Heidelberg, New York
pp. 236-262

CONSTRUCTIVE METHODS OF PROGRAM DESIGN

M. A. Jackson
Michael Jackson Systems Limited
101 Hamilton Terrace, London NW8

Abstract

Correct programs cannot be obtained by attempts to test or to prove incorrect programs: the correctness of a program should be assured by the design procedure used to build it.

A suggestion for such a design procedure is presented and discussed. The procedure has been developed for use in data processing, and can be effectively taught to most practising programmers. It is based on correspondence between data and program structures, leading to a decomposition of the program into distinct processes. The model of a process is very simple, permitting use of simple techniques of communication, activation and suspension. Some wider implications and future possibilities are also mentioned.

1. Introduction

In this paper I would like to present and discuss what I believe to be a *more constructive method of program design*. The phrase itself is important; I am sure that no-one here will object if I use a LIFO discipline in briefly elucidating its intended meaning.

'Design' is primarily concerned with structure; the designer must say what parts there are to be and how they are to be arranged. The crucial importance of modular programming and structured programming (even in their narrowest and crudest manifestations) is that they provide some definition of what parts are permissible: a module is a separately compiled, parameterised subroutine; a structure component is a sequence, an iteration or a selection. With such definitions, inadequate though they may be, we can at least begin to think about design: what modules should make up that program, and how should they be arranged? should this program be an iteration of selections or a sequence of iterations? Without such definitions, design is meaningless. At the top level of a problem there are P^N possible designs, where P is the number of distinct types of permissible part and N is the number of parts needed to make up the whole. So, to preserve our sanity, both P and N must be small: modular programming, using tree or hierarchical structures, offers small values of N; structured programming offers, additionally, small values of P.

'Program' or, rather, 'programming' I would use in a narrow sense. Modelling the problem is 'analysis'; 'programming' is putting the model on a computer. Thus, for example, if we are asked to find a prime number in the range 10^{50} to 10^{60}, we need a number theorist for the analysis; if we are asked to program discounted cash flow, the analysis calls for a financial expert. One of the major ills in data processing stems from uncertainty about this distinction. In mathematical circles the distinction is often ignored altogether, to the detriment, I believe, of our understanding of programming. Programming is about computer programs, not about number theory, or financial planning, or production control.

'Method' is defined in the Shorter OED as a 'procedure for attaining an object'. The crucial word here is 'procedure'. The ultimate method, and the ultimate is doubtless unattainable, is a procedure embodying a precise and correct algorithm. To follow the method we need only execute the algorithm faithfully, and we will be led infallibly to the desired result. To the extent that a putative method falls short of this ideal it is less of a method.

To be 'constructive', a method must itself be decomposed into distinct steps, and correct execution of each step must assure correct execution of the whole method and thus the correctness of its product. The key requirement here is that the correctness of the execution of a step should be largely verifiable without reference to steps not yet executed by the designer. This is the central difficulty in stepwise refinement: we can judge the correctness of a refinement step only by reference to what is yet to come, and hence only by exercising a degree of foresight to which few people can lay claim.

Finally, we must recognise that design methods today are intended for use by human beings: in spite of what was said above about constructive methods, we need, now and for some time to come, a substantial ingredient of intuition and subjectivity. So what is presented below does not claim to be fully constructive - merely to be 'more constructive'. The reader must supply the other half of the comparison for himself, measuring the claim against the yardstick of his own favoured methods.

2. Basis of the Method

The basis of the method is described, in some detail, in (1). It is appropriate here only to illustrate it by a family of simple example problems.

Example 1

A cardfile of punched cards is sorted into ascending sequence of values of a key which appears in each card. Within this sequence, the first card for each group of cards with a common key value is a header card, while the others are detail cards. Each detail card carries an integer amount. It is required to produce a report showing the totals of amount for all keys.

Solution 1

The first step in applying the method is to describe the structure of the data. We use a graphic notation to represent the structures as trees:-

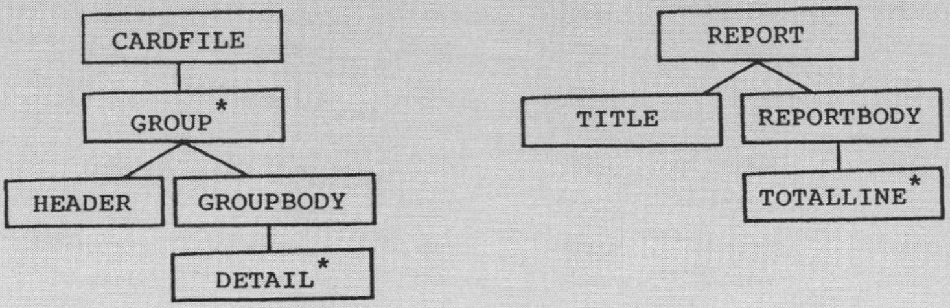

The above representations are equivalent to the following (in BNF with iteration instead of recursion):

<cardfile> ::= {<group>}$_{0}^{\infty}$
<group> ::= <header><groupbody>
<groupbody> ::= {<detail>}$_{0}^{\infty}$

<report> ::= <title><reportbody>
<reportbody> ::= {<totalline>}$_{0}^{\infty}$

The second step is to compose these data structures into a program structure:-

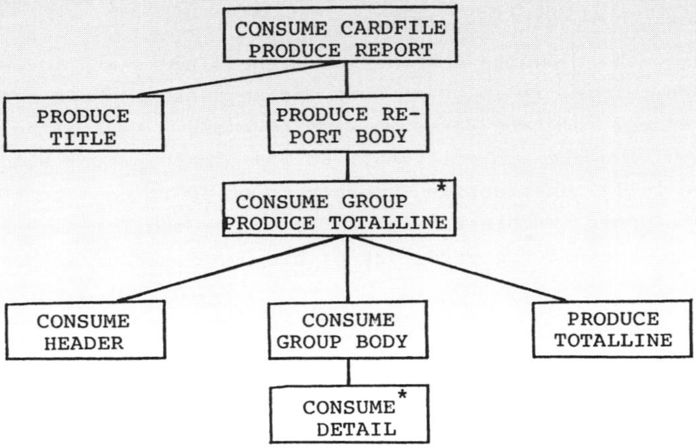

This structure has the following properties:

- It is related quite formally to each of the data structures. We may recover any one data structure from the program structure by first marking the leaves corresponding to leaves of the data structure, and then marking all nodes lying in a path from a marked node to the root.

- The correspondences (cardfile : report) and (group : totalline) are determined by the problem statement. One report is derivable from one cardfile; one totalline is derivable from one group, and the totallines are in the same order as the groups.
- The structure is vacuous, in the sense that it contains no executable statements: it is a program which does nothing; it is a tree without real leaves.

The third step in applying the method is to list the executable operations required and to allocate each to its right place in the program structure. The operations are elementary executable statements of the programming language, possibly after enhancement of the language by a bout of bottom-up design; they are enumerated, essentially, by working back from output to input along the obvious data-flow paths. Assuming a reasonably conventional machine and a line printer (rather than a character printer), we may obtain the list:

1. write title
2. write totalline (groupkey, total)
3. total := total + detail.amount
4. total := 0
5. groupkey := header.key
6. open cardfile
7. read cardfile
8. close cardfile

Note that every operation, or almost every operation, must have operands which are data objects. Allocation to a program structure is therefore a trivial task if the program structure is correctly based on the data structures. This triviality is a vital criterion of the success of the first two steps. The resulting program, in an obvious notation, is:

```
CARDFILE-REPORT  sequence
                 open cardfile; read cardfile; write title;
      REPORT-BODY  iteration until cardfile.eof
                   total := 0;  groupkey := header.key;
                   read cardfile;
         GROUP-BODY  iteration until cardfile.eof or
                                     detail.key ≠ groupkey
                     total := total + detail.amount;
                     read cardfile;
         GROUP-BODY  end
                   write totalline (groupkey, total);
      REPORT-BODY  end
                 close cardfile;
CARDFILE-REPORT  end
```

Clearly, this program may be transcribed without difficulty into any procedural programming language.

Comment

The solution has proceeded in three steps: first, we defined the data structures; second, we formed them into a program structure; third, we listed and allocated the executable operations. At each step we have criteria for the correctness of the step itself and an implicit check on the correctness of the steps already taken. For example, if at the first step we had wrongly described the structure of cardfile as

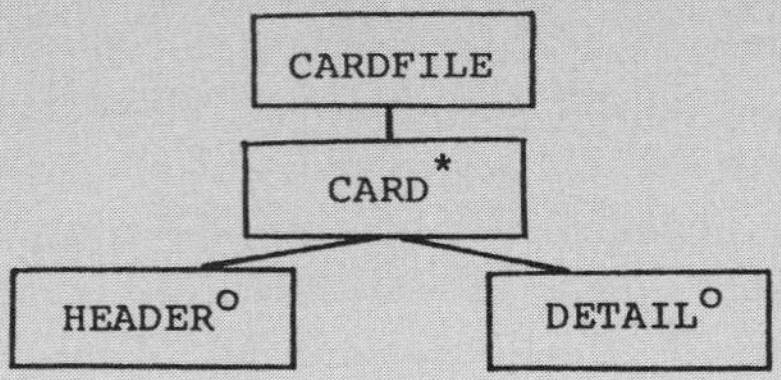

(that is: <cardfile> ::= {<card>}$_{0}^{\infty}$
<card> ::= <header>|<detail>), we should have been able to see at the first step that we had failed to represent everything we knew about the cardfile. If nonetheless we had persisted in error, we would have discovered it at the second step, when we would have been unable to form a program structure in the absence of a cardfile component corresponding to totalline in report.

The design has throughout concentrated on what we may think of as a static rather than a dynamic view of the problem: on maps, not on itineraries, on structures, not on logic flow. The logic flow of the finished program is a by-product of the data structures and the correct allocation of the 'read' operation. There is an obvious connection between what we have done and the design of a very simple syntax analysis phase in a compiler: the grammar of the input file determines the structure of the program which parses it. We may observe that the 'true' grammar of the cardfile is not context-free: within one group, the header and detail cards must all carry the same key value. It is because the explicit grammar cannot show this that we are forced to introduce the variable groupkey to deal with this stipulation.

Note that there is no error-checking. If we wish to check for errors in the input we must elaborate the structure of the input file to accommodate those errors explicitly. By defining a structure for an input file we define the domain of the program: if we wish to extend the domain, we must extend the input file structure accordingly. In a practical data processing system, we would always define the structure of primary input (such as decks of cards, keyboard messages, etc) to encompass all physically possible files: it would be absurd to construct a program whose operation is unspecified (and therefore, in principle, unpredictable) in the event of a card deck being dropped or a wrong key depressed.

Example 2

The cardfile of example 1 is modified so that each card contains a card-type indicator with possible values 'header', 'detail' and other. The program should take account of possible errors in the composition of a group: there may be no header card and/or there may be cards other than detail cards in the group body. Groups containing errors should be listed on an errorlist, but not totalled.

Solution 2

The structure of the report remains unchanged. The structure of the errorlist and of the new version of the cardfile are:

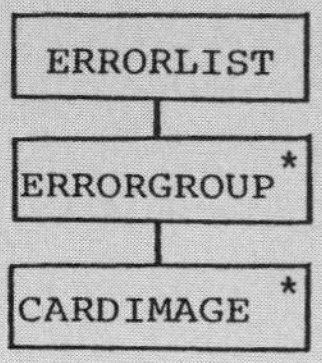

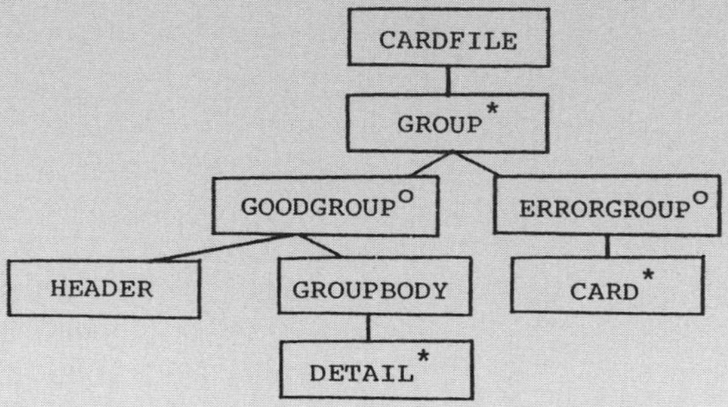

The structure of cardfile demands attention. Firstly, it is ambiguous: anything which is a goodgroup is also an errorgroup. We are forced into this ambiguity because it would be intolerably difficult - and quite unnecessary - to spell out all of the ways in which a group may be in error. The ambiguity is simply resolved by the conventions we use: the parts of a selection are considered to be ordered, and the first applicable part encountered in a left-to-right scan is chosen. So a group can be parsed as an errorgroup only if it has already been rejected as a goodgroup. Secondly, a goodgroup cannot be recognised by a left-to-right parse of the input file with any predetermined degree of lookahead. If we choose to read ahead R records, we may yet encounter a group containing an error only in the R+1'th card.

Recognition problems of this kind occur in many guises. Their essence is that we are forced to a choice during program execution at a time when we lack the evidence on which the choice must be based. Note that the difficulty is not structural but is confined to achieving a workable flow of control. We will call such problems 'backtracking' problems, and tackle them in three stages:-

a Ignore the recognition difficulty, imagining that a friendly demon will tell us infallibly which choice to make. In the present problem, he will tell us whether a group is a goodgroup or an errorgroup. Complete the design procedure in this blissful state of confidence, producing the full program text.

b Replace our belief in the demon's infallibility by a sceptical determination to verify each 'landmark' in the data which might prove him wrong. Whenever he is proved wrong we will execute a 'quit' statement which branches to the second part of the selection. These 'quit' statements are introduced into the program text created in stage a.

c Modify the program text resulting from stage b to ensure that side-effects are repealed where necessary.

The.result of stage a, in accordance with the design procedure used for example 1, is:

```
CFILE-REPT-ERR  sequence
                open cardfile;  read cardfile;  write title;
     REPORT-BODY  iteration until cardfile.eof
                  groupkey := card.key;
        GROUP-OUTG  select goodgroup
                    total := 0;
                    read cardfile;
          GOOD-GROUP  iteration until cardfile.eof or
                                      detail.key ≠ groupkey
                      total := total + detail.amount;
                      read cardfile;
          GOOD-GROUP  end
                    write totalline (groupkey, total);
        GROUP-OUTG  or errorgroup
          ERROR-GROUP  iteration until cardfile.eof or
                                       card.key ≠ groupkey
                       write errorline (card);

                       read cardfile;
          ERROR-GROUP  end
        GROUP-OUTG  end
     REPORT-BODY  end
                close cardfile;
CFILE-REPT-ERR  end
```

Note that we cannot completely transcribe this program into any programming language, because we cannot code an evaluable expression for the predicate goodgroup. However, we can readily verify the correctness of the program (assuming the infallibility of the demon). Indeed, if we are prepared to exert ourselves to punch an identifying character into the header card of each goodgroup - thus acting as our own demon - we can code and run the program as an informal demonstration of its acceptability.

We are now ready to proceed to stage b, in which we insert 'quit' statements into the first part of the selection GROUP-OUTG. Also, since quit statements are not present in a normal selection, we will replace the words 'select' and 'or' by 'posit' and 'admit' respectively, thus indicating the tentative nature of the initial choice. Clearly, the landmarks to be checked are the card-type indicators in the header and detail cards. We thus obtain the following program:

```
CFILE-REPT-ERR  sequence
                open cardfile; read cardfile; write title;
     REPORT-BODY  iteration until cardfile.eof
                  groupkey := card.key;
        GROUP-OUTG  posit goodgroup
                    total := 0;
                    quit GROUP-OUTG if card.type ≠ header;
                    read cardfile;
          GOOD-GROUP  iteration until cardfile.eof or
                                      card.key ≠ groupkey
                      quit GROUP-OUTG if card.type ≠ detail;
                      total := total + detail.amount;
                      read cardfile;
          GOOD-GROUP  end
                    write totalline (groupkey, total);
        GROUP-OUTG  admit errorgroup
         ERROR-GROUP  iteration until cardfile.eof or
                                      card.key ≠ groupkey;
                      write errorline (card);
                      read cardfile;
         ERROR-GROUP  end
        GROUP-OUTG  end
     REPORT-BODY  end
                close cardfile;
CFILE-REPT-ERR  end
```

The third stage, stage c, deals with the side-effects of partial execution of the first part of the selection. In this trivial example, the only significant side-effect is the reading of cardfile. In general, it will be found that the only troublesome side-effects are the reading and writing of serial files; the best and easiest way to handle them is to equip ourselves with input and output procedures capable of 'noting' and 'restoring' the state of the file and its associated buffers. Given the availability of such procedures, stage c can be completed by inserting a 'note' statement immediately following the 'posit' statement and a 'restore' statement immediately following the 'admit'. Sometimes side-effects will demand a more ad hoc treatment: when 'note' and 'restore' are unavailable there is no alternative to such cumbersome expedients as explicitly storing each record on disk or in main storage.

Comment

By breaking our treatment of the backtracking difficulty into three distinct stages, we are able to isolate distinct aspects of the problem. In stage a we ignore the backtracking difficulty entirely, and concentrate our efforts on obtaining a correct solution to the reduced problem. This solution is carried through the three main design steps, producing a completely specific program text: we are able to satisfy ourselves of the correctness of that text before going on to modify it in the second and third stages. In the second stage we deal only with the recognition difficulty: the difficulty is one of logic flow, and we handle it, appropriately, by modifying the logic flow of the program with quit statements. Each quit statement says, in effect, 'It is supposed (posited) that this is a goodgroup; but if, in fact, this card is not what it ought to be then this is not, after all, a goodgroup'. The required quit statements can be easily seen from the data structure definition, and their place is readily found in the program text because the program structure perfectly matches the data structure. The side-effects arise to be dealt with in stage 3 because of the quit statements inserted in stage b: the quit statements are truly 'go to' statements, producing discontinuities in the context of the computation and hence side-effects. The side-effects are readily identified from the program text resulting from stage b.

Note that it would be quite wrong to distort the data structures and the program structure in an attempt to avoid the dreaded four-letter word 'goto'. The data structures shown, and hence the program structure, are self-evidently the correct structures for the problem as stated: they must not be abandoned because of difficulties with the logic flow.

3. Simple Programs and Complex Programs

The design method, as described above, is severely constrained: it applies to a narrow class of serial file-processing programs. We may go further, and say that it defines such a class - the class of 'simple programs'. A 'simple program' has the following attributes:-

- The program has a fixed initial state; nothing is remembered from one execution to the next.
- Program inputs and outputs are serial files, which we may conveniently suppose to be held on magnetic tapes. There may be more than one input and more than one output file.
- Associated with the program is an explicit definition of the structure of each input and output file. These structures are tree structures, defined in the grammar used above. This grammar permits recursion in addition to the features shown above; it is not very different from a grammar of regular expressions.

- The input data structures define the domain of the program, the output data structures its range. Nothing is introduced into the program text which is not associated with the defined data structures.
- The data structures are compatible, in the sense that they can be combined into a program structure in the manner shown above.
- The program structure thus derived from the data structures is sufficient for a workable program. Elementary operations of the program language (possibly supplemented by more powerful or suitable operations resulting from bottom-up design) are allocated to components of the program structure without introducing any further 'program logic'.

A simple program may be designed and constructed with the minimum of difficulty, provided that we adhere rigorously to the design principles adumbrated here and eschew any temptation to pursue efficiency at the cost of distorting the structure. In fact, we should usually discount the benefits of efficiency, reminding ourselves of the mass of error-ridden programs which attest to its dangers.

Evidently, not all programs are simple programs. Sometimes we are presented with the task of constructing a program which operates on direct-access rather than on serial files, or which processes a single record at each execution, starting from a varying internal state. As we shall see later, a simple program may be clothed in various disguises which give it a misleading appearance without affecting its underlying nature. More significantly, we may find that the design procedure suggested cannot be applied to the problem given because the data structures are not compatible: that is, we are unable at the second step of the design procedure to form the program structure from the data structures.

<u>Example 3</u>

The input cardfile of example 1 is presented to the program in the form of a blocked file. Each block of this file contains a card count and a number of card images.

<u>Solution 3</u>

The structure of blockedfile is:

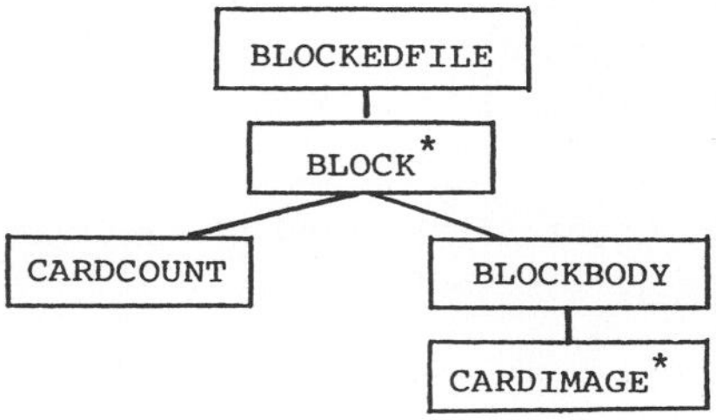

This structure does not, of course, show the arrangement of the cards in groups. It is impossible to show, in a single structure, both the arrangement in groups and the arrangement in blocks. But the structure of the report is still:

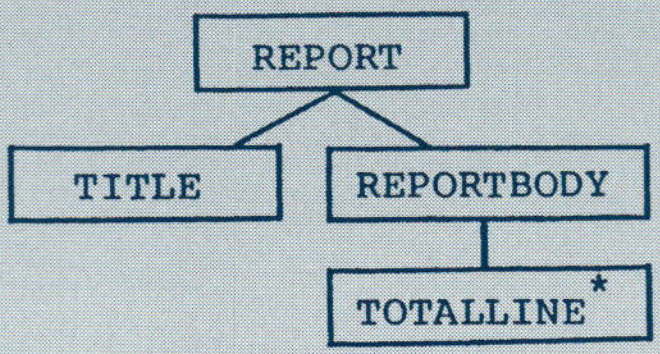

We cannot fit together the structures of report and blockedfile to form a program structure; nor would we be in better case if we were to ignore the arrangement in blocks. The essence of our difficulty is this: the program must contain operations to be executed once per block, and these must be allocated to a 'process block' component; it must also contain operations to be executed once per group, and these must be allocated to a 'process group' component; but it is impossible to form a single program structure containing both a 'process block' and a 'process group' component. We will call this difficulty a 'structure clash'.

The solution to the structure clash in the present example is obvious: more so because of the order in which the examples have been taken and because everyone knows about blocking and deblocking. But the solution can be derived more formally from the data structures. The clash is of a type we will call 'boundary clash': the boundaries of the blocks are not synchronised with the boundaries of the groups. The standard solution for a structure clash is to abandon the attempt to form a single program structure and instead decompose the problem into two or more simple programs. For a boundary clash the required decomposition is always of the form:

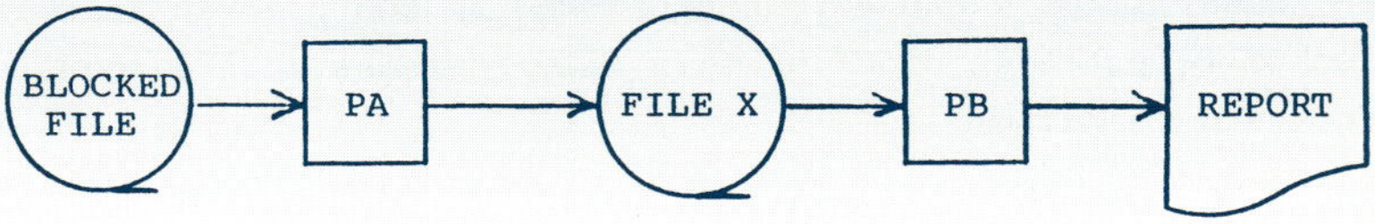

The intermediate file, file X, must be composed of records each of which is a cardimage, because cardimage is the highest common factor of the structures blockedfile and cardfile. The program PB is the program produced as a solution to example 1; the program PA is:

```
PA   sequence
     open blockedfile; open fileX; read blockedfile;
PABODY  iteration until blockedfile.eof
        cardpointer := 1;
```

```
    PBLOCK  iteration until cardpointer > block.cardcount
            write cardimage (cardpointer);
            cardpointer := cardpointer + 1;
    PBLOCK  end
          read blockedfile;
  PABODY  end
      close fileX; close blockedfile;
  PA  end
```

The program PB sees file X as having the structure of cardfile in example 1, while program PA sees its structure as:

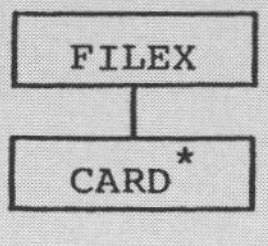

Comment

The decomposition into two simple programs achieves a perfect solution. Only the program PA is cognisant of the arrangement of cardimages in blocks; only the program PB of their arrangement in groups. The tape containing file X acts as a cordon sanitaire between the two, ensuring that no undesired interactions can occur: we need not concern ourselves at all with such questions as 'what if the header record of a group is the first cardimage in a block with only one cardimage?', or 'what if a group has no detail records and its header is the last cardimage in a block?'; in this respect our design is known to be correct.

There is an obvious inefficiency in our solution. By introducing the intermediate magnetic tape file we have, to a first approximation, doubled the elapsed time for program execution and increased the program's demand for backing store devices.

Example 4

The input cardfile of example 1 is incompletely sorted. The cards are partially ordered so that the header card of each group precedes any detail cards of that group, but no other ordering is imposed. The report has no title, and the totals may be produced in any order.

Solution 4

The best we can do for the structure of cardfile is:

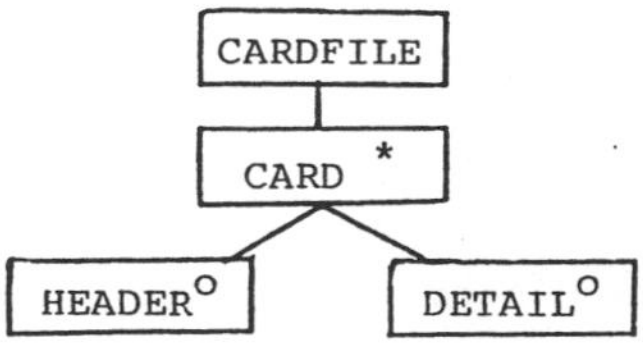

which is clearly incompatible with the structure of the report, since there is no component of cardfile corresponding to totalline in the report. Once again we have a structure clash, but this time of a different type. The cardfile consists of a number of groupfiles, each one of which has the form:

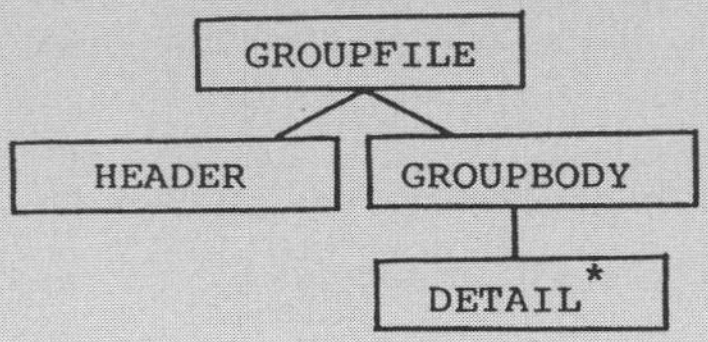

The cardfile is an arbitrary interleaving of these groupfiles. To resolve the clash (an 'interleaving clash') we must resolve cardfile into its constituent groupfiles:

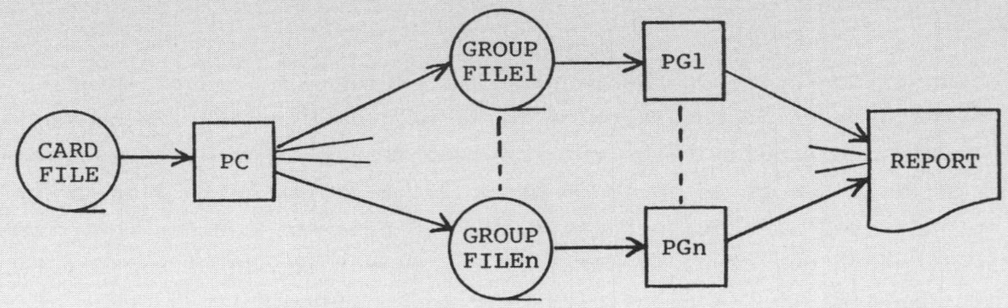

Allowing, for purposes of exposition, that a single report may be produced by the n programs PG1, ... PGn (each contributing one totalline), we have decomposed the problem into n+1 simple programs; of these, n are identical programs processing the n distinct groupfiles groupfile1, ... groupfilen; while the other, PC, resolves cardfile into its constituents.

Two possible versions of PC are:

```
PC1  sequence
     open cardfile; read cardfile;
     open all possible groupfiles;
PC1BODY  iteration until cardfile.eof
         write record to groupfile (record.key);
         read cardfile;
PC1BODY  end
     close all possible groupfiles;
     close cardfile;
PC1  end
```

and

```
PC2  sequence
     open cardfile; read cardfile;
```

```
PC2BODY  iteration until cardfile.eof
 REC-INIT  select new groupfile
           open groupfile (record.key);
 REC-INIT  end
         write record to groupfile (record.key);
         read cardfile;
PC2BODY  end
     close all opened groupfiles;
     close cardfile;
PC2  end
```

Both PC1 and PC2 present difficulties. In PC1 we must provide a groupfile for every possible key value, whether or not cardfile contains records for that key. Also, the programs PG1, ... PGn must be elaborated to handle the null groupfile:

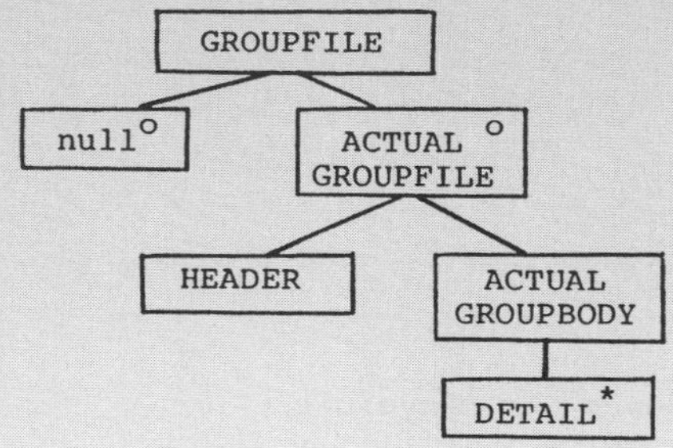

In PC2 we must provide a means of determining whether a groupfile already exists for a given key value. Note that it would be quite wrong to base the determination on the fact that a header must be the first record for a group: such a solution takes impermissible advantage of the structure of groupfile which, in principle, is unknown in the program PC; we would then have to make a drastic change to PC if, for example, the header card were made optional:

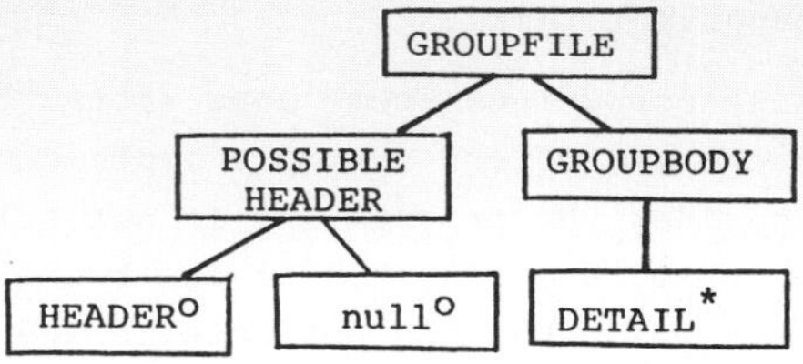

Further, in PC2 we must be able to run through all the actual key values in order to close all the groupfiles actually opened. This would still be necessary even if each group had a recognisable trailer record, for reasons similar to those given above concerning the header records.

Comment

The inefficiency of our solution to example 4 far outstrips the inefficiency of our solution to example 3. Indeed, our solution to example 4 is entirely impractical. Practical implementation of the designs will be considered below in the next section. For the moment, we may observe that the use of magnetic tapes for communication between simple programs enforces a very healthy discipline. We are led to use a very simple protocol: every serial file must be opened and closed. The physical medium encourages a complete decoupling of the programs: it is easy to imagine one program being run today, the tapes held overnight in a library, and a subsequent program being run tomorrow; the whole of the communication is visible in the defined structure of the files. Finally, we are strengthened in our resolve to think in terms of static structures, avoiding the notoriously error-prone activity of thinking about dynamic flow and execution-time events.

Taking a more global view of the design procedure, we may say that the simple program is a satisfactory high level component. It is a larger object than a sequence, iteration or selection; it has a more precise definition than a module; it is subject to restrictions which reveal to us clearly when we are trying to make a single program out of what should be two or more.

4. Programs, Procedures and Processes

Although from the design point of view we regard magnetic tapes as the canonical medium of communication between simple programs, they will not usually provide a practical implementation.

An obvious possibility for implementation in some environments is to replace each magnetic tape by a limited number of buffers in main storage, with a suitable regime for ensuring that the consumer program does not run ahead of the producer. Each simple program can then be treated as a distinct task or process, using whatever facilities are provided for the management of multiple concurrent tasks.

However, something more like coroutines seems more attractive (2). The standard procedure call mechanism offers a simple implementation of great flexibility and power. Consider the program PA, in our solution to example 3, which writes the intermediate file X. We can readily convert this program into a procedure PAX which has the characteristics of an input procedure for file X. That is, invocations of the procedure PAX will satisfactorily implement the operations 'open file X for reading', 'read file X' and 'close file X after reading'.

We will call this conversion of PA into PAX 'inversion of PA with respect to file X'. (Note that the situation in solution 3 is symmetrical: we could equally well decide to invert PB with respect to file X, obtaining an output procedure for file X.) The mechanics of inversion are a mere matter of generating the appropriate object coding from the text of the simple program: there is no need for any modification to that text. PA and PAX are the same program, not two different programs. Most practising programmers seem to be unaware of this identity of PA and PAX, and even those who are familiar with coroutines often program as if they supposed that PA and PAX were distinct things. This is partly due to the baleful influence of the stack as a storage allocation device: we cannot jump out of an inner block of PAX, return to the invoking procedure, and subsequently resume where we left off when we are next invoked. So we must either modify our compiler or modify our coding style, adopting the use of labels and go to statements as a standard in place of the now conventional compound statement of structured programming. It is common to find PAX, or an analogous program, designed as a selection or case statement: the mistake is on all fours with that of the kindergarten child who has been led to believe that the question 'what is 5 multiplied by 3?' is quite different from the question 'what is 3 multiplied by 5?'. At a stroke the poor child has doubled the difficulty of learning the multiplication tables.

The procedure PAX is, of course, a variable state procedure. The value of its state is held in a 'state vector' (or activation record), of which a vital part is the text pointer; the values of special significance are those associated with the suspension of PAX for operations on file X - open, write and close. The state vector is an 'own variable' par excellence, and should be clearly seen as such.

The minimum interface needed between PB and PAX is two parameters: a record of file X, and an additional bit to indicate whether the record is or is not the eof marker. This minimum interface suffices for example 3: there is no need for PB to pass an operation code to PAX (open read or close). It is important to understand that this minimum interface will not suffice for the general case. It is sufficient for example 3 only because the operation code is implicit in the ordering of operations. From the point of view of PAX, the first invocation must be 'open', and subsequent invocations must be 'read' until PAX has returned the eof marker to PB, after which the final invocation must be 'close'. This felicitous harmony is destroyed if, for example, PB is permitted to stop reading and close file X before reaching the eof marker. In such a case the interface must be elaborated with an operation code. Worse, the sequence of values of this operation code now constitutes a file in its own right: the solution becomes:

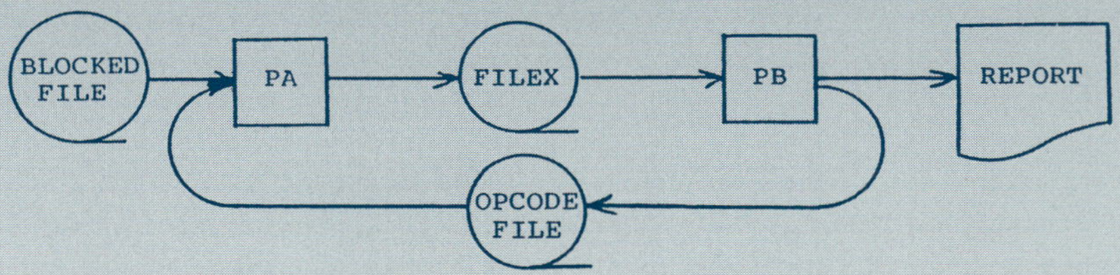

The design of PA is, potentially, considerably more complicated. The benefit we will obtain from treating this complication conscientiously is well worth the price: by making explicit the structure of the opcode file we define the problem exactly and simplify its solution. Failure to recognise the existence of the opcode file, or, just as culpable, failure to make its structure explicit, lies at the root of the errors and obscurities for which manufacturers' input-output software is deservedly infamous.

In solution 4 we created an intolerable multiplicity of files - groupfile1, ... groupfilen. We can rid ourselves of these by inverting the programs PG1, ... PGn with respect to their respective groupfiles: that is, we convert each of the programs PGi to an output procedure PGFi, which can be invoked by PC to execute operations on groupfilei. But we still have an intolerable multiplicity of output procedures, so a further step is required. The procedures are identical except for their names and the current values of their state vectors. So we separate out the pure procedure part - PGF - of which we need keep only one copy, and the named state vectors SVPGF1, ... SVPGFn. We must now provide a mechanism for storing and retrieving these state vectors and for associating the appropriate state vector with each invocation of PGF; many mechanisms are possible, from a fully-fledged direct-access file with serial read facilities to a simple arrangement of the state vectors in an array in main storage.

5. Design and Implementation

The model of a simple program and the decomposition of a problem into simple programs provides some unity of viewpoint. In particular, we may be able to see what is common to programs with widely different implementations. Some illustrations follow.

a A conversational program is a simple program of the form:

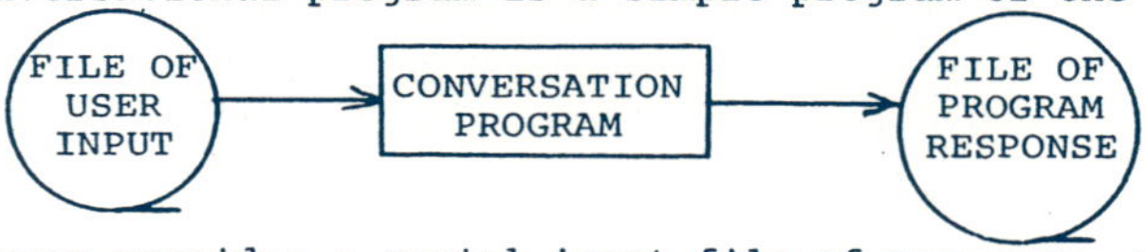

The user provides a serial input file of messages, ordered in time; the conversation program produces a serial file of responses. Inversion of the program with respect to the user input file gives an output procedure 'dispose of one message in a conversation'. The state vector of the inverted program must be preserved for the duration of the conversation: IBM's IMS provides the SPA (Scratchpad Area) for precisely this purpose. The conversation program must, of course, be designed and written as a single program: implementation restrictions may dictate segmentation of the object code.

b A 'sort-exit' allows the user of a generalised sorting program to introduce his own procedure at the point where each record is about to be written to the final output file. An interface is provided which permits 'insertion' and 'deletion' of records as well as 'updating'.

We should view the sort-exit procedure as a simple program:

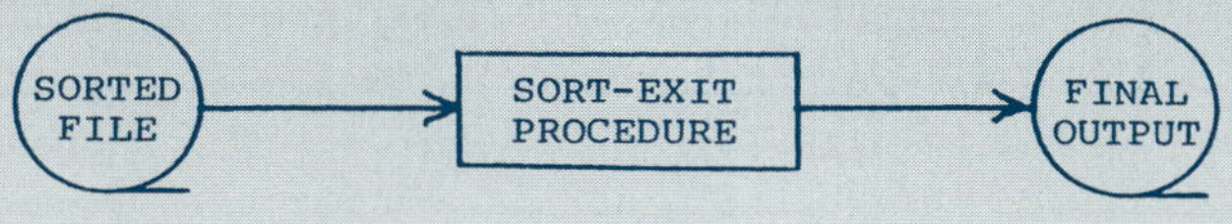

To fit it in with the sorting program we must invert it with respect to both the sortedfile and the finaloutput. The interface must provide an implementation of the basic operations: open sortedfile for reading; read sortedfile (distinguishing the eof marker); close sortedfile after reading; open finaloutput for writing; write finaloutput record; close finaloutput file after writing (including writing the eof marker).

Such concepts as 'insertion' and 'deletion' of records are pointless: at best, they serve the cause of efficiency, traducing clarity; at worst, they create difficulty and confusion where none need exist.

c Our solution to example 1 can be seen as an optimisation of the solution to the more general example 4. By sorting the cardfile we ensure that the groups do not overlap in time: the state vectors of the inverted programs PGF1, ... PGFn can therefore share a single area in main storage. The state vector consists only of the variable total; the variable groupkey is the name of the currently active group and hence of the current state vector. Because the records of a group are contiguous, the end of a group is recognisable at cardfile.eof or at the start of another group. The individual groupfile may therefore be closed, and the totalline written, at the earliest possible moment.

We may, perhaps, generalise so far as to say that an identifier is stored by a program only in order to give a unique name to the state vector of some process.

d A data processing system may be viewed as consisting of many simple programs, one for each independent entity in the real world model. By arranging the entities in sets we arrange the corresponding simple programs in equivalence classes. The 'master record' corresponding to an entity is the state vector of the simple program modelling that entity.

The serial files of the system are files of transactions ordered in time: some are primary transactions, communicating with the real world, some are secondary, passing between simple programs of the system. In general, the real world must be modelled as a network of entities or of entity sets; the data processing system is therefore a network of simple programs and transaction files.

Implementation of the system demands decisions in two major areas. First a scheduling algorithm must be decided; second, the representation and handling of state vectors. The extreme cases of the first are 'real-time' and 'serial batch'. In a pure 'real-time' system every primary transaction is dealt with as soon as it arrives, followed immediately by all of the secondary and consequent transactions, until the system as a whole becomes quiet. In a pure 'serial batch' system, each class (identifier set) of primary transactions is accumulated for a period (usually a day, week or month). Each simple program of that class is then activated (if there is a transaction present for it), giving rise to secondary transactions of various classes. These are then treated similarly, and so on until no more transactions remain to be processed.

Choosing a good implementation for a data processing system is difficult, because the network is usually large and many possible choices present themselves. This difficulty is compounded by the long-term nature of the simple programs: a typical entity, and hence a typical program, has a lifetime measured in years or even decades. During such a lifetime the system will inevitably undergo change: in effect, the programs are being rewritten while they are in course of execution.

e An interrupt handler is a program which processes a serial file of interrupts, ordered in time:

Inversion of the interrupt handler with respect to the interrupt file gives the required procedure 'dispose of one interrupt'. In general, the interrupt file will be composed of interleaved files for individual processes, devices, etc. Implementation is further complicated by the special nature of the invocation mechanism, by the fact that the records of the interrupt file are distributed in main storage, special registers and other places, and by the essentially recursive structure of the main interrupt file (unless the interrupt handler is permitted to mask off secondary interrupts).

f An input-output procedure (what IBM literature calls an 'access method') is a simple program which processes an input file of access requests and produces an output file of access responses. An access request consists of an operation code and, sometimes, a data record; an access response consists of a result code and, sometimes, a data record. For example, a direct-access method has the form:

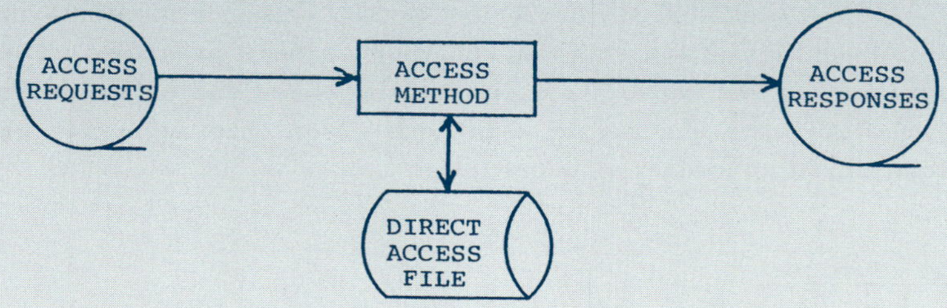

By inverting this simple program with respect to both the file of access requests and the file of access responses we obtain the desired procedure. This double inversion is always possible without difficulty, because each request must produce a response and that response must be calculable before the next request is presented.

The chief crime of access method designers is to conceal from their customers (and, doubtless, from themselves) the structure of the file of access requests. The user of the method is thus unable to determine what sequences of operations are permitted by the access method, and what their effect will be.

g Some aspects of a context-sensitive grammar may be regarded as interleaved context-free grammars. For example, in a grossly simplified version of the COBOL language we may wish to stipulate that any variable may appear as an operand of a MOVE statement, while only a variable declared as numeric may appear as an operand of an arithmetic (ADD, SUBTRACT, MULTIPLY or DIVIDE) statement. We may represent this stipulation as follows:

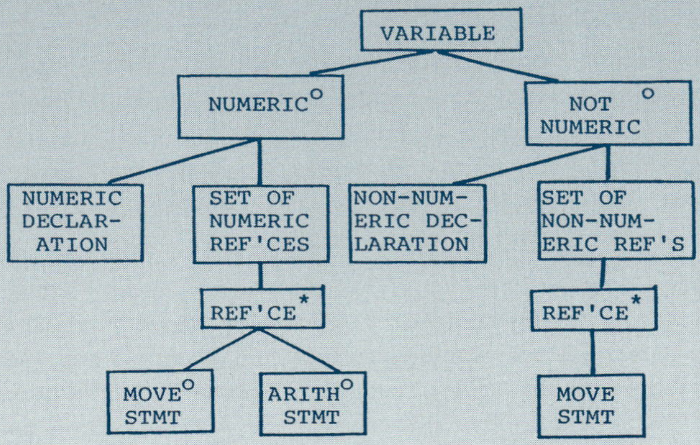

The syntax-checking part of the compiler consists, partly, of a simple program for each declared variable. The symbol table is the set of state vectors for these simple programs. The algorithm for activating and suspending these and other programs will determine the way in which one error interacts with another both for diagnosis and correction.

6. A Modest Proposal

It is one thing to propose a model to illuminate what has already been done, to clarify the sources of existing success or failure. It is quite another to show that the model is of practical value, and that it leads to the construction of acceptable programs. An excessive zeal in decomposition produces cumbersome interfaces and pointlessly redundant code. The "Shanley Principle" in civil engineering (3) requires that several functions be implemented in a single part; this is necessary for economy both in manufacturing and in operating the products of engineering design. It appears that a design approach which depends on decomposition runs counter to this principle: its main impetus is the separation of functions for implementation in distinct parts of the program.

But programs do not have the intractable nature of the physical objects which civil, mechanical or electrical engineers produce. They can be manipulated and transformed (for example, by compilers) in ways which preserve their vital qualities of correctness and modifiability while improving their efficiency both generally and in the specialised environment of a prticular machine. The extent to which a program can be manipulated and transformed is critically affected by two factors: the variety of forms it can take, and the semantic clarity of the text. Programs written using today's conventional techniques score poorly on both factors. There is a distressingly large variety of forms, and intelligibility is compromised or even destroyed by the introduction of implementation-orientated features. The justification for these techniques is, of course, efficiency. But in pursuing efficiency in this way we become caught in a vicious circle: because our languages are rich the compilers cannot understand, and hence cannot optimise, our programs so we need rich languages to allow us to obtain the efficiency which the compilers do not offer.

Decomposition into simple programs, as discussed above, seems to offer some hope of separating the considerations of correctness and modifiability from the considerations of efficiency. Ultimately, the objective is that the first should become largely trivial and the second largely automatic.

The first phase of design would produce the following documents:-

- a definition of each serial file structure for each simple program (including files of operation codes!);
- the text of each simple program;
- a statement of the communication between simple programs, perhaps in the form of identities such as

$$\text{output } (p_i, f_r) \equiv \text{input } (p_j, f_s).$$

It may then be possible to carry out some automatic checking of self-consistency in the design - for instance, to check that the inputs to a program are within its domain. We may observe, incidentally, that the 'inner' feature of Simula 67 (4) is a way of enforcing consistency of a file of operation codes between the consumer and producer processes in a very limited case. More ambitiously, it may be possible, if file-handling protocol is exactly observed, and read and write operations are allocated with a scrupulous regard to principle, to check the correctness of the simple programs in relation to the defined data structures.

In the second phase of design, the designer would specify, in greater or lesser detail:-

- the synchronisation of the simple programs;
- the handling of state vectors;
- the dissection and recombining of programs and state vectors to reduce interface overheads.

Synchronisation is already loosely constrained by the statements of program communication made in the first phase: the consumer can never run ahead of the producer. Within this constraint the designer may choose to impose additional constraints at compile time and/or at execution time. The weakest local constraint is to provide unlimited dynamic buffering at execution time, the consumer being allowed to lag behind the producer by anything from a single record to the whole file, depending on resource allocation elsewhere in the system. The strongest local constraints are use of coroutines or program inversion (enforcing a single record lag) and use of a physical magnetic tape (enforcing a whole file lag).

Dissection and recombining of programs becomes possible with coroutines or program inversion; its purpose is to reduce interface overheads by moving code between the invoking and invoked programs, thus avoiding some of the time and space costs of procedure calls and also, under certain circumstances, avoiding replication of program structure and hence of coding for sequencing control. It depends on being able to associate code in one program with code in another through the medium of the communicating data structure.

A trivial illustration is provided by solution 3, in which we chose to invert PA with respect to file X, giving an input procedure PAX for the file of cardimages. We may decide that the procedure call overhead is intolerable, and that we wish to dissect PAX and combine it with PB. This is achieved by taking the invocations of PAX in PB (that is, the statements 'open fileX', 'read fileX' and 'close fileX') and replacing those invocations by the code which PAX would execute in response to them. For example, in response to 'open fileX', PAX would execute the code 'open blockedfile'; therefore the 'open fileX' statement in PB can be replaced by the statement 'open blockedfile'.

A more substantial illustration is provided by the common practice of designers of 'real-time' data processing systems. Suppose that a primary transaction for a product gives rise to a secondary transaction for each open order item for that product, and that each of those in turn gives rise to a transaction for the open order of which it is a part, which then gives rise to a transaction for the customer who placed the order. Instead of having separate simple programs for the product, order item, order and customer, the designer will usually specify a 'transaction processing module': this consists of coding from each of those simple programs, the coding being that required to handle the relevant primary or secondary transaction.

Some interesting program transformations of a possibly relevant kind are discussed in a paper by Burstall and Darlington (5). I cannot end this paper better than by quoting from them:

> "The overall aim of our investigation has been to help people to write correct programs which are easy to alter. To produce such programs it seems advisable to adopt a lucid, mathematical and abstract programming style. If one takes this really seriously, attempting to free one's mind from considerations of computational efficiency, there may be a heavy penalty in program running time; in practice it is often necessary to adopt a more intricate version of the program, sacrificing comprehensibility for speed.

The question then arises as to how a lucid program can be transformed into a more intricate but efficient one in a systematic way, or indeed in a way which could be mechanised.

" ... We are interested in starting with programs having an extremely simple structure and only later introducing the complications which we usually take for granted even in high level language programs. These complications arise by introducing useful interactions between what were originally separate parts of the program, benefiting by what might be called 'economies of interaction'."

References

(1) Principles of Program Design; M A Jackson; Academic Press 1975.

(2) Hierarchical Program Structures; O-J Dahl; in Structured Programming; Academic Press 1972.

(3) Structured Programming with go to Statements; Donald E Knuth; in ACM Computing Surveys Vol 6 No 4 December 1974.

(4) A Structural Approach to Protection; C A R Hoare; 1975.

(5) Some Transformations for Developing Recursive Programs; R M Burstall & John Darlington; in Proceedings of 1975 Conference on Reliable Software; Sigplan Notices Vol 10 No 6 June 1975.

Structured Analysis

Tom DeMarco
Principal of the Atlantic Systems Guild
New York
tdemarco@atlsysguild.com

M.S., Columbia, Diplome,
University of Paris

Bell Labs: ESS-1 project

Manager of real-time projects,
distributed online banking systems

J.D. Warnier Prize, Stevens Prize

Fellow of IEEE

Major contributions:
Structured Analysis, Peopleware

Current interests: project management,
change facilitation, litigation
of software-intensive contracts

Tom DeMarco

Structured Analysis: Beginnings of a New Discipline

How It Happened

When I arrived at Bell Telephone Laboratories in the fall of 1963, I was immediately assigned to the ESS-1 project. This was a hardware/software endeavor to develop the world's first commercial stored program telephone switch (now installed in telephone offices all over the world). At the time, the project was made up of some 600 persons, divided about half-and-half between hardware and software. There was also a small simulation group (perhaps a dozen people) working to create an early prediction of system performance and robustness.

I was at first assigned to the hardware group. My assignment was to develop certain circuits that enabled Emergency Action, the reconfiguration of processors when one was judged to have failed. This was an intriguing assignment since each of the two processors would diagnose the other and then try to decide together and agree on which one was incapable of further processing – but somehow still capable to judge its own sanity.

To all of our surprise, the software for the project turned out to require a lot more effort than the hardware. By early 1964 an increasing number of hardware engineers were being switched into software to help that effort along. I was among them. By the end of that year I considered myself an experienced software engineer. I was 24 years old.

The simulation team under the management of Erna Hoover had by this time completed the bulk of its work. The findings were released in the form of a report and some internal documentation. The report (analysis of system performance and expected downtimes) was the major deliverable, but it was one aspect of the internal documentation that ended up getting more attention. Among the documents describing the simulation was a giant diagram that Ms. Hoover called a Petri Net. It was the first time I had ever seen such a diagram. It portrayed the system being simulated as a network of sub-component nodes with information flows connecting the nodes. In a rather elegant trick, some of the more complicated nodes were themselves portrayed as Petri Nets, sub-sub-component nodes with interconnecting information flows.

By now I was in full software construction mode, writing and testing programs to fulfill subsystem specifications handed down by the system architect. While these specifications were probably as good as any in the industry at the time, I found them almost incomprehensible. Each one was a lengthy narrative text describing what the subsystem had to do. I was not the only one having trouble with the specs. Among my fellow software craftsmen (out of deference to Dave Parnas, I shall henceforth not use the term 'software engineer' to describe myself or any of my 1960s colleagues), there was a general sense that the specs were probably correct but almost impossible to work from. The one document that we found ourselves using most was Erna's PetriNet. It showed how all the pieces of the puzzle were related and how they were obliged to interact. The lower-level networks gave us a useful pigeon-holing scheme for information from the subsystem specs. When all the elemental requirements from the spec had been slotted by node, it was relatively easy to begin implementation. One of my colleagues, Jut Kodner, observed that the diagram was a better spec than the spec.

When I left the Labs, I went to work for what today would be called a system integrator. I was assigned on contract to build first one and then another time-shared operating system for the then-new IBM 360. On both of these projects I created my own network diagrams. In place of a normal specification, I wrote one "mini-specification" per node. I used the same trick when I went to work for La CEGOS Informatique, a French consulting firm that had a contract to build a computerized conveyor system for the new merchandise mart at La Villette in Paris. (La Villette was the successor to the ancient market at Les Halles.) And I used the same trick again on a new telephone switch implemented for GTE System Telephone Companies. Note that all of these projects (two telephone switches, two time-shared executives, and a conveyor control system) were real time control systems, required to meet stringent time constraints on the order of a few milliseconds. All of my work up to this point was in the domain that I now call engineering systems. I had never participated in a single commercial data processing project.

By 1971 I went to work, for the first time in my life, outside the engineering sector. For the next four years I was involved building banking systems in Sweden, Holland, France, and finally New York. Here again I used my networking methods, though with a bit less success. What was different in financial applications was the presence of a database in the middle of my network. At first it could be treated as a simple file or repository of information. But over time these databases were to become more and more complex and my method gave me no particularly elegant way to deal with them. The truth is that the networks were an elegant and useful description of control systems, where data flow provides the best representation of overall system function, but a less useful tool for database systems where the structure of the repository itself is a better representation.

Though I have now come to believe that dataflow methods are ill-suited to business applications – at least compared to data modeling methods – the networks were as big a hit with my banking customers as they were with my engineering customers. Remember that in the early 1970s, the breakthroughs of data modeling including E-R diagrams and relational databases had yet to happen or were happening only in academia. For my customers, the network specifications that I was showing them were the only alternative to dreary and endless narrative specifications.

How It Became a Commercial Success

It was in 1974 that I first came across the work of Doug Ross and John Brackett of SofTech. Their tool, called SADT, was a much advanced and in some ways much more elegant variation on my network specifications. It also was the first time that I had seen the adjective 'structured' applied to leveled diagrams. Since all things structured were hot in 1974, this was good news to me. Imagine, something that I'd been doing for years now turned out to be 'structured'!

In 1975 I sought out my old friend Ed Yourdon and proposed to him to develop my network specification concept (much improved by my exposure to SADT) into a two-day training seminar for his new seminar company. Ed's little company had already dabbled in something called Structured Analysis, and though my concept and theirs were only marginally similar, he allowed me to use the term as if we'd really been talking about the same thing all along. Within a year, the original sense of what would constitute Structured Analysis was completely replaced by my concept of writing specifications in the form of leveled dataflow diagrams with complementary data dictionary and mini-specifications.

The course and its successor courses on Structured Analysis were a huge success. By the end of 1975, Ed had assigned a dozen instructors to teaching my courses, and with all the royalty income thus produced, I stole away for two months to write a book [*Structured Analysis and System Specification*, Prentice Hall, 1975] and a video training sequence which both subsequently became excellent royalty properties. A friend later observed – rather nastily, I thought – that if I had spent all my months as productively as those two months working on the video and the book, my present income would be several million dollars per month.

Over the next 20 years, the method prospered. It was implemented in virtually every CASE system ever produced. When CASE went away, the method persisted. Today dataflow representation is a component of virtually every current process, though few organizations are still using my 1975 prescription in its original form.

How My Own Perspective Has Changed Since 1975

While the market embraced the method called Structured Analysis, I myself have come to have some doubts about the whole approach. Remember that my own early experience applying the method had been in control systems, not commercial systems. When I first started advocating the method it was only to those who were building control systems. It was a great surprise to me that the commercial sector liked the method. In retrospect I believe that the whole dataflow approach is vastly more useful for control systems than typical commercial applications, and the appeal to the commercial users was mostly due to a complete lack of well-thought-out alternatives. Commercial IT ended up using a control system method because there were as yet no attractive commercial IT methods available.

When good alternatives did become available, many commercial organizations stuck loyally to my approach. I found myself regularly visiting clients who announced themselves to be 100% "DeMarcofied," though I myself would never have used my method on the projects they were doing. I came to the conclusion that these companies were using a method that was poorly suited to their real needs and the reason had little to do with anything purely technological. They were stuck on the method because it gave a comforting sense of completeness; it appeared to them to be The Answer to all of their problems. When it didn't solve their problems they blamed themselves and tried harder.

I now believe that my 1975 book was overly persuasive and that many in our industry were simply seduced by it. This is partly the result of my unconstrained enthusiasm for a method that had worked superbly for me (in a limited domain), and partly the result of the dismal state of the art of IT books at the time. Many people have told me that mine was the only IT book they ever got through awake, and the only one they ever enjoyed. I think they adopted its prescriptions thinking, "This guy may be dead wrong, but at least I understand what he's saying."

Bill of Particulars, Then and Now

Important parts of the Structured Analysis method were and are useful and practical, as much so today as ever. Other parts are too domain-specific to be generally applicable. And still other parts were simply wrong. In order to call attention to these different categories, I offer the following commented summary tables, the first showing what I thought I knew in 1975 and the second showing what I still believe to be true. As you will see, there are some substantial differences:

What I Thought I Knew in 1975

Principle	Commentary
1. Narrative specs are dumb	These "Victorian Novel" specifications neither specify nor inform
2. Four-stage modeling	A dataflow representation of a system is a model and the analysis life-cycle consists of building a sequence of these models showing four different stages
3. Dataflow is the essential view	The point of view of the data as it passes through the system is the most useful
4. Top-down partitioning	Top-down is good; bottom-up is evil
5. Loose connection criterion	The validity of any partitioning is a function of how thin the interfaces are
6. Defined process of analysis	System analysis always has the same well-defined steps
7. Pseudo-coded minispecs	The lowest level is defined in a formal way
8. Work at the user's desk	Analysts shouldn't hide in their own offices; the real work of analysis is at the user's desk
9. Philosophy of iteration	You can never get it right on the first try; success comes from numerous iterations, each one better than the last
10. The customer is king	The customer knows what the system has to be; the analyst's job is to listen and learn

There were other things that made up the discipline, but these ten were its essence. I felt strongly about all of these but at the time it was the top-down characteristic of the approach that most charmed me. After all, without the notion of top-down, the method could hardly be characterized as "structured," and that was an appellation that I coveted. (Remember that the structured disciplines were at their peak in 1975.)

I am writing this in the fall of 2001, and obviously much has changed. Not the least of what has changed is my own perception of the business of systems analysis and specification. In the next table I reproduce the ten principles showing in dark letters those that I think still apply in whole or in part. The gray shaded "ghosts" are just to remind you of the sense of those principles that didn't in my opinion survive the test of time:

What I Still Believe

Principle	Revised Commentary (as of 2001)
1. Narrative specs are dumb	Narrative specs are not the problem; a suitably partitioned spec with narrative text used at the bottom level makes a fine statement of work
2. Four-stage modeling	The four stages I proposed in 1975 were far too time consuming
3. **Dataflow** is the essential view	Dataflow is one of the essential views, not the only one
4. Top-down **partitioning**	Partitioning is essential in dealing with anything complex, but top-down partitioning is often far too difficult to achieve and not at all the great advantage it was touted to be
5. Loose connection criterion	This is an important truth: when you're attacking complexity by partitioning, the thinner the interface, the better the partitioning – if the interfaces are still thick, go back and partition again, searching for the natural seams of the domain
6. Defined process of analysis	Defined process is a holy grail that has never yet been found and probably never will be
7. Pseudo-coded **minispecs**	It's useful to partition the whole and then specify the pieces, but pseudo-code was an awful mistake (puts analysts into coding mode when they should be busy analyzing)
8. Work at the user's desk	Analysts have a tendency to hide at their own desks, but much of the action is in the business area and they need to venture out to find it
9. Philosophy of iteration	We never get it right the first time; the best we can do is improve from one iteration to the next; if we can continue to do this at each iteration, we can get arbitrarily close to a perfect product
10. The customer is king	See below . . .

If I'm right that the specification-by-network approach does not require and never did really benefit from being top-down, then the entire method never did justify the name 'structured.' That is what I believe today. We all profited by calling it structured, but it wasn't. To make matters worse, the attempt to achieve top-down representation sent projects off on a meaningless wild goose chase. These days I often encounter project teams working with enormous diagrams of connected software pieces. These diagrams take up a whole wall of a war room or are laid out on the floor with developers on their hands and knees crawling over them. Of course they are a pain to update, often hand-annotated, not reproducible, don't fit into anybody's documentation standard. And yet they are useful; that's why people use them. This use seems much more consistent with the early value I perceived in dataflow networks.

My final point (my loss of faith that "the customer is king") is not just a change in my own thinking, but a sign of the maturing of IT in specific and of the business climate in general. In 1975, the typical commercial system we built was a first-time automation of what had before been done manually. The customer, of course, was the only one who knew what this was all about and his/her sense of what the automated version would have to do was prime.

Today we are building third and fourth generation automated systems, and IT personnel are often as well or better informed about how the existing system works than their business partners. More important, the meaningful successes of IT today are no longer to be achieved by simple automation or re-automation of existing process. We have moved on to a new era: Our challenge today is to combine improved information technology and market opportunity in order to create a product that is radically different from what could have been achieved in a less-connected world. This tells us that the new king is neither client nor technologist, but their partnership: the tightly merged amalgam of business and technological expertise. Companies that achieve and maintain such a partnership are the ones who will prosper.

Tom DeMarco

Structured Analysis and System Specification

Yourdon, New York, 1978
pp. 1-7 and 37-44

STRUCTURED ANALYSIS

AND

SYSTEM SPECIFICATION

by

Tom DeMarco

Foreword by

P.J. Plauger

YOURDON inc.
1133 Avenue of the Americas
New York, New York 10036

First Printing, March 1978

Second Printing, June 1978

Revised, December 1978

 Library of Congress Catalog Card Number 78-51285.

ISBN: 0-917072-07-3

CONTENTS

PART 3: DATA DICTIONARY

PART 5: SYSTEM MODELING

PART 6: STRUCTURED ANALYSIS FOR A FUTURE SYSTEM

PART 1

BASIC CONCEPTS

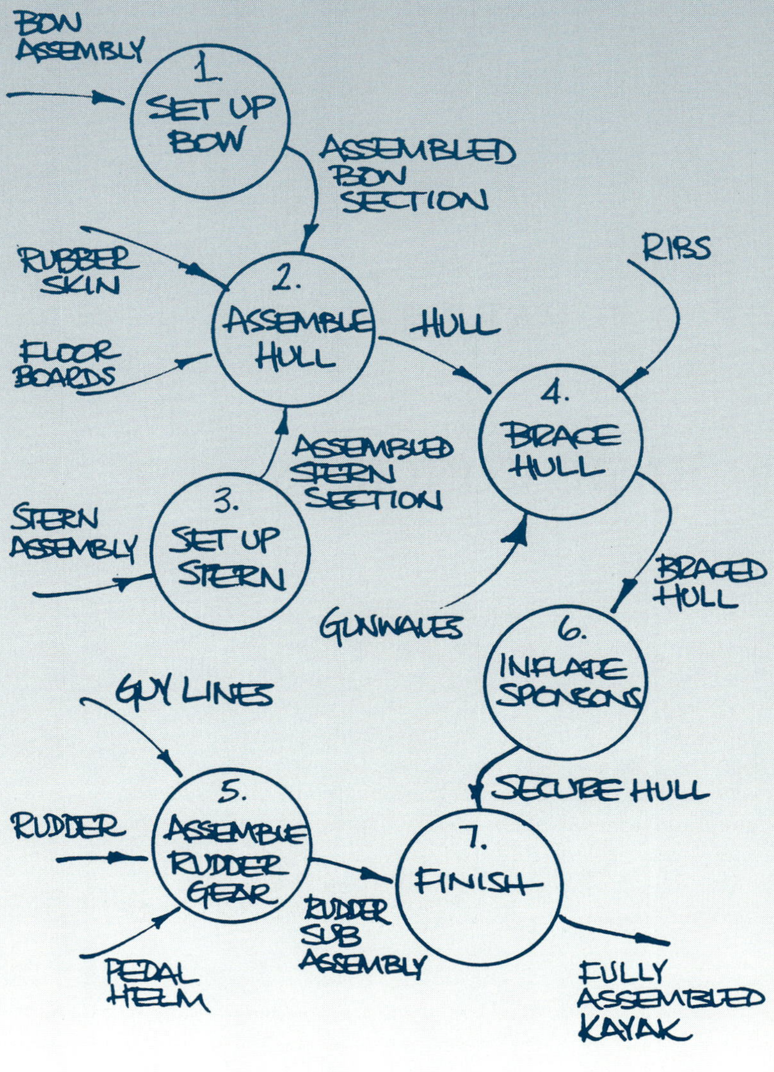
BOW ASSEMBLY
1. SET UP BOW
ASSEMBLED BOW SECTION
RUBBER SKIN
2. ASSEMBLE HULL
FLOOR BOARDS
HULL
RIBS
4. BRACE HULL
ASSEMBLED STERN SECTION
STERN ASSEMBLY
3. SET UP STERN
GUNWALES
BRACED HULL
6. INFLATE SPONSONS
GUY LINES
SECURE HULL
5. ASSEMBLE RUDDER GEAR
RUDDER
7. FINISH
RUDDER SUB ASSEMBLY
PEDAL HELM
FULLY ASSEMBLED KAYAK

Figure 1

1 THE MEANING OF STRUCTURED ANALYSIS

Let's get right to the point. This book is about Structured Analysis, and Structured Analysis is primarily concerned with a new kind of Functional Specification, the Structured Specification. Fig. 1 shows part of a Structured Specification.

The example I have chosen is a system of sorts, but obviously not a computer system. It is, in fact, a manual assembly procedure. Procedures like the one described in Fig. 1 are usually documented by a text narrative. Such descriptions have many of the characteristics of the classical Functional Specifications that we analysts have been producing for the last 20 years. (The Functional Specification describes *automated* procedures — that is the main difference between the two.) Take a look at a portion of the text that prompted me to draw Fig. 1.

Assembly Instructions for KLEPPER Folding Boats

1. Lay out hull in grass (or on carpet). Select a clean, level spot.
2. Take folded bow section (with red dot), lay it in grass, unfold 4 hinged gunwale boards. Kneel down, spread structure lightly with left hand near bow, place right hand on pullplate at *bottom* of hinged rib, and set up rib gently by pulling towards center of boat. Deckbar has a tongue-like fitting underneath which will connect with fitting on top of rib if you lift deckbar lightly, guide tongue to rib, press down on deckbar near bow to lock securely. Now lift whole bowsection using both arms wraparound style (to keep gunwales from flopping down) and slide into front of hull. Center seam of blue deck should rest on top of deckbar.
3. Take folded stern section (blue dot, 4 "horseshoes" attached), unfold 4 gunwales, set up rib by pulling on pullplate at *bottom* of rib. Deckbar locks to top of rib *from the side* by slipping a snaplock over a tongue attached to top of rib . . .

   And so forth.

The differences are fairly evident: The text plunges immediately into the details of the early assembly steps, while the structured variant tries to present the big picture first, with the intention of working smoothly from abstract to detailed. The Structured Specification is graphic and the text is not. The old-fashioned approach is one-dimensional (written narrative is always one-dimensional), and the structured variant is multidimensional. There are other differences as well; we'll get to those later. My intention here is only to give you an initial glimpse at a Structured Specification.

Now let's go back and define some terms.

1.1 What is analysis?

Analysis is the study of a problem, prior to taking some action. In the specific domain of computer systems development, analysis refers to the study of some business area or application, usually leading to the specification of a new system. The action we're going to be taking later on is the implementation of that system.

The most important product of systems analysis — of the analysis phase of the life cycle — is the specification document. Different organizations have different terms for this document: Functional Specification, External Specification, Design Specification, Memo of Rationale, Requirements Document. In order to avoid the slightly different connotations that these names carry, I would like to introduce a new term here: the Target Document. The Target Document establishes the goals for the rest of the project. It says what the project will have to deliver in order to be considered a success. The Target Document is the principal product of analysis.

Successful completion of the analysis phase involves all of the following:

1. selecting an optimal target
2. producing detailed documentation of that target in such a manner that subsequent implementation can be evaluated to see whether or not the target has been attained
3. producing accurate predictions of the important parameters associated with the target, including costs, benefits, schedules, and performance characteristics
4. obtaining concurrence on each of the items above from each of the affected parties

In carrying out this work, the analyst undertakes an incredibly large and diverse set of tasks. At the very minimum, analysts are responsible for: user liaison, specification, cost-benefit study, feasibility analysis, and estimating. We'll cover each of these in turn, but first an observation about some characteristics that are common to all the analyst's activities.

1.1.1 Characteristics of Analysis

Most of us come to analysis by way of the implementation disciplines — design, programming, and debugging. The reason for this is largely historical. In the past, the business areas being automated were the simpler ones, and the users were rather unsophisticated; it was more realistic to train computer people to understand the application than to train users to understand EDP technology. As we come to automate more and more complex areas, and as our users (as a result of prevalent computer training at the high school and college level) come to be more literate in automation technologies, this trend is reversing.

But for the moment, I'm sure you'll agree with me that most computer systems analysts are first of all computer people. That being the case, consider this observation: Whatever analysis is, it certainly is not very similar to the work of designing, programming, and debugging computer systems. Those kinds of activities have the following characteristics:

- The work is reasonably straightforward. Software sciences are relatively new and therefore not as highly specialized as more developed fields like medicine and physics.
- The interpersonal relationships are not very complicated nor are there very many of them. I consider the business of building computer systems and getting them to run a rather friendly activity, known for easy relationships.
- The work is very definite. A piece of code, for instance, is either right or wrong. When it's wrong, it lets you know in no uncertain terms by kicking and screaming and holding its breath, acting in obviously abnormal ways.
- The work is satisfying. A positive glow emanates from the programmer who has just found and routed out a bug. A friend of mine who is a doctor told me, after observing programmers in the debugging phase of a project, that most of them seemed "high as kites" much of the time. I think he was talking about the obvious satisfaction programmers take in their work.

The implementation disciplines are straightforward, friendly, definite, and satisfying. Analysis is none of these things:

- It certainly isn't easy. Negotiating a complex Target Document with a whole community of heterogeneous and conflicting users and getting them all to agree is a gargantuan task. In the largest systems for the most convoluted organizations, the diplomatic skills that the analyst must bring to bear are comparable to the skills of a Kissinger negotiating for peace in the Middle East.
- The interpersonal relationships of analysis, particularly those involving users, are complicated, sometimes even hostile.
- There is nothing definite about analysis. It is not even obvious when the analysis phase is done. For want of better termination criteria, the analysis phase is often considered to be over when the time allocated for it is up!
- Largely because it is so indefinite, analysis is not very satisfying. In the most complicated systems, there are so many compromises to be made that no one is ever completely happy with the result. Frequently, the various parties involved in the

negotiation of a Target Document are so rankled by their own concessions, they lose track of what a spectacular feat the analyst has achieved by getting them to agree at all.

So analysis is frustrating, full of complex interpersonal relationships, indefinite, and difficult. In a word, it is fascinating. Once you're hooked, the old easy pleasures of system building are never again enough to satisfy you.

1.1.2 The User Liaison

During the 1960's, our business community saw a rash of conglomerations in which huge corporate monoliths swallowed up smaller companies and tried to digest them. As part of this movement, many centralized computer systems were installed with an aim toward gathering up the reins of management, and thus allowing the conglomerate's directors to run the whole show. If you were an analyst on one of these large Management Information System (MIS) projects, you got to see the user-analyst relationship at its very worst. Users were dead set against their functions being conglomerated, and of course that's just what the MIS systems were trying to do. The philosophy of the 60's was that an adversary relationship between the analyst and the user could be very productive, that analysts could go in, as the representatives of upper management, and force the users to participate and comply.

Of course the record of such projects was absolutely dismal. I know of no conglomerate that made significant gains in centralization through a large Management Information System. The projects were often complete routs. Many conglomerates are now spinning off their acquisitions and finding it rather simple to do so because so little true conglomeration was ever achieved. Due to the experience of the 60's, the term Management Information System, even today, is likely to provoke stifled giggles in a group of computer people.

The lesson of the 60's is that no system is going to succeed without the active and willing participation of its users. Users have to be made aware of how the system will work and how they will make use of it. They have to be sold on the system. Their expertise in the business area must be made a key ingredient to system development. They must be kept aware of progress, and channels must be kept open for them to correct and tune system goals during development. All of this is the responsibility of the analyst. He is the users' teacher, translator, and advisor. This intermediary function is the most essential of all the analyst's activities.

1.1.3 Specification

The analyst is the middleman between the user, who decides what has to be done, and the development team, which does it. He bridges this gap with a Target Document. The business of putting this document together and getting it accepted by all parties is specification. Since the Target Document is the analyst's principal output, specification is the most visible of his activities.

If you visit the Royal Naval Museum at Greenwich, England, you will see the results of some of the world's most successful specification efforts, the admiralty models. Before any ship of the line was constructed, a perfect scale model had to be built and approved. The long hours of detail work were more than repaid by the clear understandings that come from studying and handling the models.

The success of the specification process depends on the product, the Target Document in our case, being able to serve as a model of the new system. To the extent that it helps you visualize the new system, the Target Document is the system model.

1.1.4 Cost-Benefit Analysis

The study of relative cost and benefits of potential systems is the feedback mechanism used by an analyst to select an optimal target. While Structured Analysis does not entail new methods for conduct of this study, it nonetheless has an important effect. An accurate and meaningful system model helps the user and the analyst perfect their vision of the new system and refine their estimates of its costs and benefits.

1.1.5 Feasibility Analysis

It is pointless to specify a system which defies successful implementation. Feasibility analysis refers to the continual testing process the analyst must go through to be sure that the system he is specifying can be implemented within a set of given constraints. Feasibility analysis is more akin to design than to the other analysis phase activities, since it involves building tentative physical models of the system and evaluating them for ease of implementation. Again, Structured Analysis does not prescribe new procedures for this activity. But its modeling tools will have some positive effect.

1.1.6 Estimating

Since analysis deals so heavily with a system which exists only on paper, it involves a large amount of estimating. The analyst is forever being called upon to estimate cost or duration of future activities, CPU load factors, core and disk extents, manpower allocation . . . almost anything. I have never heard of a project's success being credited to the fine estimates an analyst made; but the converse is frequently true — poor estimates often lead to a project's downfall, and in such cases, the analyst usually receives full credit.

Estimating is rather different from the other required analysis skills:

- *Nobody is an expert estimator.* You can't even take a course in estimating, because nobody is willing to set himself up as enough of an authority on the subject to teach it.
- *We don't build our estimating skills, because we don't collect any data about our past results.* At the end of a project we rarely go

back and carry out a thorough postmortem to see how the project proceeded. How many times have you seen project performance statistics published and compared to the original estimates? In my experience, this is done only in the very rare instance of a project that finishes precisely on time and on budget. In most cases, the original schedule has long since vanished from the record and will never be seen again.

- *None of this matters as much as it ought to anyway,* since most things we call "estimates" in computer system projects are not estimates at all. When your manager asks you to come up with a schedule showing project completion no later than June 1 and using no more than six people, you're not doing any real estimating. You are simply dividing up the time as best you can among the phases. And he probably didn't estimate either; chances are his dates and manpower loading were derived from budgetary figures, which were themselves based upon nothing more than Wishful Thinking.

All these factors aside, estimating plays a key part in analysis. There are some estimating heuristics that are a by-product of Structured Analysis; these will be discussed in a subsequent chapter. The key word here is *heuristic.* A heuristic is a cheap trick that often works well but makes no guarantee. It is not an algorithm, a process that leads to a guaranteed result.

1.1.7 The Defensive Nature of Analysis

In addition to the analysis phase activities presented above, there are many others; the analyst is often a project utility infielder, called upon to perform any number of odd jobs. As the project wears on, his roles may change. But the major activities, and the ones that will concern us most in this book, are: user liaison, specification, cost-benefit and feasibility analysis, and estimating.

In setting about these activities, the analyst should be guided by a rule which seems to apply almost universally: *The overriding concern of analysis is not to achieve success, but to avoid failure.* Analysis is essentially a defensive business.

This melancholy observation stems from the fact that the great flaming failures of the past have inevitably been attributable to analysis phase flaws. When a system goes disastrously wrong, it is the analyst's fault. When a system succeeds, the credit must be apportioned among many participants, but failure (at least the most dramatic kind) belongs completely to the analyst. If you think of a system project that was a true rout — years late, or orders of magnitude over budget, or totally unacceptable to the user, or utterly impossible to maintain — it almost certainly was an analysis phase problem that did the system in.

Computer system analysis is like child-rearing; you can do grievous damage, but you cannot ensure success.

My reason for presenting this concept here is to establish the following context for the rest of the book: The principal goal of Structured Analysis is to minimize the probability of critical analysis phase error. The tools of Structured Analysis are defensive means to cope with the most critical risk areas of analysis.

1.2 Problems of analysis

Projects can go wrong at many different points: The fact that we spend so much time, energy, and money on maintenance is an indication of our failures as designers; the fact that we spend so much on debugging is an indictment of our module design and coding and testing methods. But analysis failures fall into an entirely different class. When the analysis goes wrong, we don't just spend more money to come up with a desired result — we spend *much* more money, and often don't come up with any result.

That being the case, you might expect management to be super-conservative about the analysis phase of a project, to invest much more in doing the job correctly and thus avoid whole hosts of headaches downstream. Unfortunately, it is not as simple as that. Analysis is plagued with problems that are not going to be solved simply by throwing money at them. You may have experienced this yourself if you ever participated in a project where too much time was allocated to the analysis phase. What tends to happen in such cases is that work proceeds in a normal fashion until the major products of analysis are completed. In the remaining time, the project team spins its wheels, agonizing over what more it could do to avoid later difficulties. When the time is finally up, the team breathes a great sigh of relief and hurries on to design. Somehow the extra time is just wasted — the main result of slowing down the analysis phase and doing everything with exaggerated care is that you just get terribly bored. Such projects are usually every bit as subject to failures of analysis as others.

I offer this list of the major problems of analysis:

1. communication problems
2. the changing nature of computer system requirements
3. the lack of tools
4. problems of the Target Document
5. work allocation problems
6. politics

Before looking at these problems in more detail, we should note that none of them will be *solved* by Structured Analysis or by any other approach to analysis. The best we can hope for is some better means to grapple with them.

1.2.1 Communication Problems

A long-unsolved problem of choreography is the development of a rigorous notation to describe dance. Merce Cunningham, commenting on past failures to come up with a useful notation, has observed that the motor centers of the brain are separated from the reading and writing centers. This physical separation in the brain causes communication difficulties.

Computer systems analysis is faced with this kind of difficulty. The business of specification is, for the most part, involved in describing procedure. Procedure, like dance, resists description. (It is far easier to demonstrate procedure than to describe it, but that won't do for our purposes.) Structured Analysis attempts to overcome this difficulty through the use of graphics. When you use a picture instead of text to communicate, you switch mental gears. Instead of using one of the brain's serial processors, its reading facility, you use a parallel processor.

All of this is a highfalutin way to present a "lowfalutin" and very old idea: A picture is worth a thousand words. The reason I present it at all is that analysts seem to need some remedial work on this concept. When given a choice (in writing a Target Document, for instance) between a picture and a thousand words, most analysts opt unfailingly for the thousand words.

Communication problems are exacerbated in our case by the lack of a common language between user and analyst. The things we analysts work with — specifications, data format descriptions, flowcharts, code, disk and core maps — are totally inappropriate for most users. The one aspect of the system the user is most comfortable talking about is the set of human procedures that are his interface to the system, typically something we don't get around to discussing in great detail with him until well after analysis, when the user manuals are being written.

Finally, our communication problem is complicated by the fact that what we're describing is usually a system that exists only in our minds. There is no model for it. In our attempts to flesh out a picture of the system, we are inclined to fill in the physical details (CRT screens, report formats, and so forth) much too early.

To sum it up, the factors contributing to the communication problems of analysis are

1. the natural difficulty of describing procedure
2. the inappropriateness of our method (narrative text)
3. the lack of a common language between analyst and user
4. the lack of any usable early model for the system

1.2.2 The Changing Nature of Requirements

I sometimes think managers are sent to a special school where they are taught to talk about "freezing the specification" at least once a day during the analysis phase. The idea of freezing the specification is a sublime fiction. Changes won't go away and they can't be ignored. If a project lasts two years, you ought to expect as many legitimate changes (occasioned by changes in the way business is done) to occur during the project as would occur in the first two years after cutover. In addition to changes of this kind, an equal number of changes may arise from the user's increased understanding of the system. This type of change results from early, inevitable communication failures, failures which have since been corrected.

When we freeze a Target Document, we try to hold off or ignore change. But the Target Document is only an approximation of the true project target; therefore, by holding off and ignoring change, we are trying to proceed toward a target *without benefit of any feedback*.

There are two reasons why managers want to freeze the Target Document. First, they want to have a stable target to work toward, and second, an enormous amount of effort is involved in updating a specification. The first reason is understandable, but the second is ridiculous. *It is unacceptable to write specifications in such a way that they can't be modified.* Ease of modification has to be a requirement of the Target Document.

This represents a change of ground rules for analysis. In the past, it was expected that the Target Document would be frozen. It was a positive advantage that the document was impossible to change since that helped overcome resistance to the freeze. It was considered normal for an analyst to hold off a change by explaining that implementing it in the Target Document would require retyping every page. I even had one analyst tell me that the system, once built, was going to be highly flexible, so that it would be easier to put the requested change into the system itself rather than to put it into the specification!

Figures collected by GTE, IBM, and TRW over a large sample of system changes, some of them incorporated immediately and others deferred, indicate that the difference in cost can be staggering. It can cost two orders of magnitude more to implement a change after cutover than it would have cost to implement it during the analysis phase. As a rule of thumb, you should count on a 2:1 cost differential to result from deferring change until a subsequent project phase.[1]

My conclusion from all of this is that we must change our methods; we must begin building Target Documents that are highly maintainable. In fact, maintainability of the Target Document is every bit as essential as maintainability of the eventual system.

---

[1]See Barry Boehm's article, "Software Engineering," published in the *IEEE Transactions on Computers*, December 1976, for a further discussion of this topic.

1.2.3 The Lack of Tools

Analysts work with their wits plus paper and pencil. That's about it. The fact that you are reading this book implies that you are looking for some tools to work with. For the moment, my point is that most analysts don't have any.

As an indication of this, consider your ability to evaluate the products of each project phase. You would have little difficulty evaluating a piece of code: If it were highly readable, well submodularized, well commented, conformed to generally accepted programming practice, had no GOTO's, ALTER's, or other forms of pathology — you would probably be willing to call it a good piece of code. Evaluating a design is more difficult, and you would be somewhat less sure of your judgment. But suppose you were asked to evaluate a Target Document. Far from being able to judge its quality, you would probably be hard pressed to say whether it qualified as a Target Document at all. Our inability to evaluate any but the most incompetent efforts is a sign of the lack of analysis phase tools.

1.2.4 Problems of the Target Document

Obviously the larger the system, the more complex the analysis. There is little we can do to limit the size of a system; there are, however, intelligent and unintelligent ways to deal with size. An intelligent way to deal with size is to *partition*. That is exactly what designers do with a system that is too big to deal with conveniently — they break it down into component pieces (modules). Exactly the same approach is called for in analysis.

The main thing we have to partition is the Target Document. We have to stop writing Victorian novel specifications, enormous documents that can only be read from start to finish. Instead, we have to learn to develop dozens or even hundreds of "mini-specifications." And we have to organize them in such a way that the pieces can be dealt with selectively.

Besides its unwieldy size, the classical Target Document is subject to further problems:

- It is excessively redundant.
- It is excessively wordy.
- It is excessively physical.
- It is tedious to read and unbearable to write.

1.2.5 Work Allocation

Adding manpower to an analysis team is even more complicated than beefing up the implementation team. The more successful classical analyses are done by very small teams, often only one person. On rush projects, the analysis phase is sometimes shortchanged since people assume it will take forever, and there is no convenient way to divide it up.

I think it obvious that this, again, is a partitioning problem. Our failure to come up with an early partitioning of the subject matter (system or business area) means that we have no way to divide up the rest of the work.

1.2.6 Politics

Of course, analysis is an intensely political subject. Sometimes the analyst's political situation is complicated by communication failures or inadequacies of his methods. That kind of problem can be dealt with positively — the tools of Structured Analysis, in particular, will help.

But most political problems do not lend themselves to simple solutions. The underlying cause of political difficulty is usually the changing distribution of power and autonomy that accompanies the introduction of a new system. No new analysis procedures are going to make such an impending change less frightening.

Political problems aren't going to go away and they won't be "solved." The most we can hope for is to limit the effect of disruption due to politics. Structured Analysis approaches this objective by making analysis procedures more formal. To the extent that each of the analyst's tasks is clearly (and publicly) defined, and has clearly stated deliverables, the analyst can expect less political impact from them. Users understand the limited nature of his investigations and are less inclined to overreact. The analyst becomes less of a threat.

1.3 The user-analyst relationship

Since Structured Analysis introduces some changes into the user-analyst relationship, I think it is important to begin by examining this relationship in the classical environment. We need to look at the user's role, the analyst's role, and the division of responsibility between them.

1.3.1 What Is a User?

First of all, there is rarely just one user. In fact, the term "user" refers to at least three rather different roles:

- *The hands-on user*, the operator of the system. Taking an on-line banking system as an example, the hands-on users might include tellers and platform officers.
- *The responsible user*, the one who has direct business responsibility for the procedures being automated by the system. In the banking example, this might be the branch manager.
- *The system owner*, usually upper management. In the banking example, this might be the Vice President of Banking Operations.

Sometimes these roles are combined, but most often they involve distinctly different people. When multiple organizations are involved, you can expect the total number of users to be as much as three times the number of organizations.

The analyst must be responsible for communication with *all* of the users. I am continually amazed at how many development teams jeopardize their chances of success by failing to talk to one or more of their users. Often this takes the form of some person or organization being appointed "User Representative." This is done to spare the user the bother of the early system negotiations, and to spare the development team the bother of dealing with users. User Representatives would be fine if they also had authority to accept the system. Usually they do not. When it comes to acceptance, they step aside and let the real user come forward. When this happens, nobody has been spared any bother.

1.3.2 What Is an Analyst?

The analyst is the principal link between the user area and the implementation effort. He has to communicate the requirements to the implementors, and the details of how requirements are being satisfied back to the users. He may participate in the actual determination of what gets done: It is often the analyst who supplies the act of imagination that melds together applications and present-day technology. And, he may participate in the implementation. In doing this, he is assuming the role that an architect takes in guiding the construction of his building.

While the details may vary from one organization to the next, most analysts are required to be

- at ease with EDP concepts
- at ease with concepts particular to the business area
- able to communicate such concepts

1.3.3 Division of Responsibility Between Analyst and User

There is something terribly wrong with a user-analyst relationship in which the user specifies such physical details as hardware vendor, software vendor, programming language, and standards. Equally upsetting is the user who relies upon the analyst to decide how business ought to be conducted. What is the line that separates analyst functions from user functions?

I believe the analyst and the user ought to try to communicate across a "logical-physical" boundary that exists in any computer system project. Logical considerations include answers to the question, *What needs to be accomplished?* These fall naturally into the domain of the user. Physical considerations include answers to the question, *How shall we accomplish these things?* These are in the domain of the analyst.

1.4 What is Structured Analysis?

So far, most of what we have been discussing has been the classical analysis phase, its problems and failings. How is Structured Analysis different? To answer that question, we must consider

- New goals for analysis. While we're changing our methods, what new analysis phase requirements shall we consider?
- Structured tools for analysis. What is available and what can be adapted?

1.4.1 New Goals for Analysis

Looking back over the recognized problems and failings of the analysis phase, I suggest we need to make the following additions to our set of analysis phase goals:

- The products of analysis must be highly maintainable. This applies particularly to the Target Document.
- Problems of size must be dealt with using an effective method of partitioning. The Victorian novel specification is out.
- Graphics have to be used wherever possible.
- We have to differentiate between logical and physical considerations, and allocate responsibility, based on this differentiation, between the analyst and the user.
- We have to build a logical system model so the user can gain familiarity with system characteristics before implementation.

1.4.2 Structured Tools for Analysis

At the very least, we require three types of new analysis phase tools:

- Something to help us partition our requirement and document that partitioning before specification. For this I propose we use a *Data Flow Diagram,* a network of interrelated processes. Data Flow Diagrams are discussed in Chapters 4 through 10.
- Some means of keeping track of and evaluating interfaces without becoming unduly physical. Whatever method we select, it has to be able to deal with an enormous flood of detail — the more we partition, the more interfaces we have to expect. For our interface tool I propose that we adopt a set of *Data Dictionary* conventions, tailored to the analysis phase. Data Dictionary is discussed in Chapters 11 through 14.

- New tools to describe logic and policy, something better than narrative text. For this I propose three possibilities: *Structured English, Decision Tables,* and *Decision Trees.* These topics are discussed in Chapters 15 through 17.

1.4.3 Structured Analysis — A Definition

Now that we have laid all the groundwork, it is easy to give a working definition of Structured Analysis:

Structured Analysis is the use of these tools:

Data Flow Diagrams
Data Dictionary
Structured English
Decision Tables
Decision Trees

to build a new kind of Target Document, the Structured Specification.

Although the building of the Structured Specification is the most important aspect of Structured Analysis, there are some minor extras:

- estimating heuristics
- methods to facilitate the transition from analysis to design
- aids for acceptance test generation
- walkthrough techniques

1.4.4 What Structured Analysis Is Not

Structured Analysis deals mostly with a subset of analysis. There are many legitimate aspects of analysis to which Structured Analysis does not directly apply. For the record, I have listed items of this type below:

- cost-benefit analysis
- feasibility analysis
- project management
- performance analysis
- conceptual thinking (Structured Analysis might help you communicate better with the user; but if the user is just plain wrong, that might not be of much long-term benefit.)
- equipment selection
- personnel considerations
- politics

My treatment of these subjects is limited to showing how they fit in with the modified methods of Structured Analysis.

3 THE TOOLS OF STRUCTURED ANALYSIS

The purpose of this chapter is to give you a look at each one of the tools of Structured Analysis at work. Once you have a good idea of what they are and how they fit together, we can go back and discuss the details.

3.1 A sample situation

The first example I have chosen is a real one, involving the workings of our own company, Yourdon inc. To enhance your understanding of what follows, you ought to be aware of these facts:

1. Yourdon is a medium-sized computer consulting and training company that teaches public and inhouse sessions in major cities in North America and occasionally elsewhere.
2. People register for seminars by mail and by phone. Each registration results in a confirmation letter and invoice being sent back to the registrant.
3. Payments come in by mail. Each payment has to be matched up to its associated invoice to credit accounts receivable.
4. There is a mechanism for people to cancel their registrations if they should have to.
5. Once you have taken one of the company's courses, or even expressed interest in one, your name is added to a data base of people to be pursued relentlessly forever after. This data base contains entries for tens of thousands of people in nearly as many organizations.
6. In addition to the normal sales prompting usage of the data base, it has to support inquiries such as
   - When is the next Structured Design Programming Workshop in the state of California?
   - Who else from my organization has attended the Structured Analysis seminar? How did they rate it?
   - Which instructor is giving the Houston Structured Design and Programming Workshop next month?

In early 1976, Yourdon began a project to install a set of automated management and operational aids on a PDP-11/45, running under the UNIX operating system. Development of the system — which is now operational — first called for a study of sales and accounting functions. The study made use of the tools and techniques of Structured Analysis. The following subsections present some partial and interim products of our analysis.

3.2 A Data Flow Diagram example

An early model of the operations of the company is presented in Fig. 9. It is in the form of a Logical Data Flow Diagram. Refer to that figure now, and we'll walk through one of its paths. The rest should be clear by inference.

Input to the portrayed area comes in the form of Transactions ("Trans" in the figure). These are of five types: Cancellations, Enrollments, Payments, Inquiries, plus those that do not qualify as any of these, and are thus considered Rejects. Although there are no people or locations or departments shown on this figure (it is logical, not physical), I will fill some of these in for you, just as I would for a user to help him relate back to the physical situation that he knows. The receptionist (a physical consideration) handles all incoming transactions, whether they come by phone or by mail. He performs the initial edit, shown as Process 1 in the figure. People who want to take a course in Unmitigated Freelance Speleology, for example, are told to look elsewhere. Incomplete or improperly specified enrollment requests and inquiries, etc., are sent back to the originator with a note. Only clean transactions that fall into the four acceptable categories are passed on.

Enrollments go next to the registrar. His function (Process 2) is to use the information on the enrollment form to update three files: the People File, the Seminar File, and the Payments File. He then fills out an enrollment chit and passes it on to the accounting department. In our figure, the enrollment chit is called "E-Data," and the accounting process that receives it is Process 6.

Information on the chit is now transformed into an invoice. This process is partially automated, by the way — a ledger machine is used — but that information is not shown on a logical Data Flow Diagram.

The invoice passes on to the confirmation process (which happens to be done by the receptionist in this case). This task (Process 7) involves combining the invoice with a customized form letter, to be sent out together as a confirmation. The confirmation goes back to the customer.

3.2.1 Some Data Flow Diagram Conventions

If you have followed the narrative so far, you have already picked up the major Data Flow Diagram conventions:

- *The Data Flow Diagram shows flow of data, not of control.* This is the difference between Data Flow Diagrams and flowcharts. The Data Flow Diagram portrays a situation from the point of view of the data, while a flowchart portrays it from the point of view of those who act upon the data. For this reason, you almost never see a loop in a Data Flow Diagram. A loop is something that the data are unaware of; each datum typically goes through it once, and so from its point of view it is not a loop at all. Loops and decisions are control considerations and do not appear in Data Flow Diagrams.

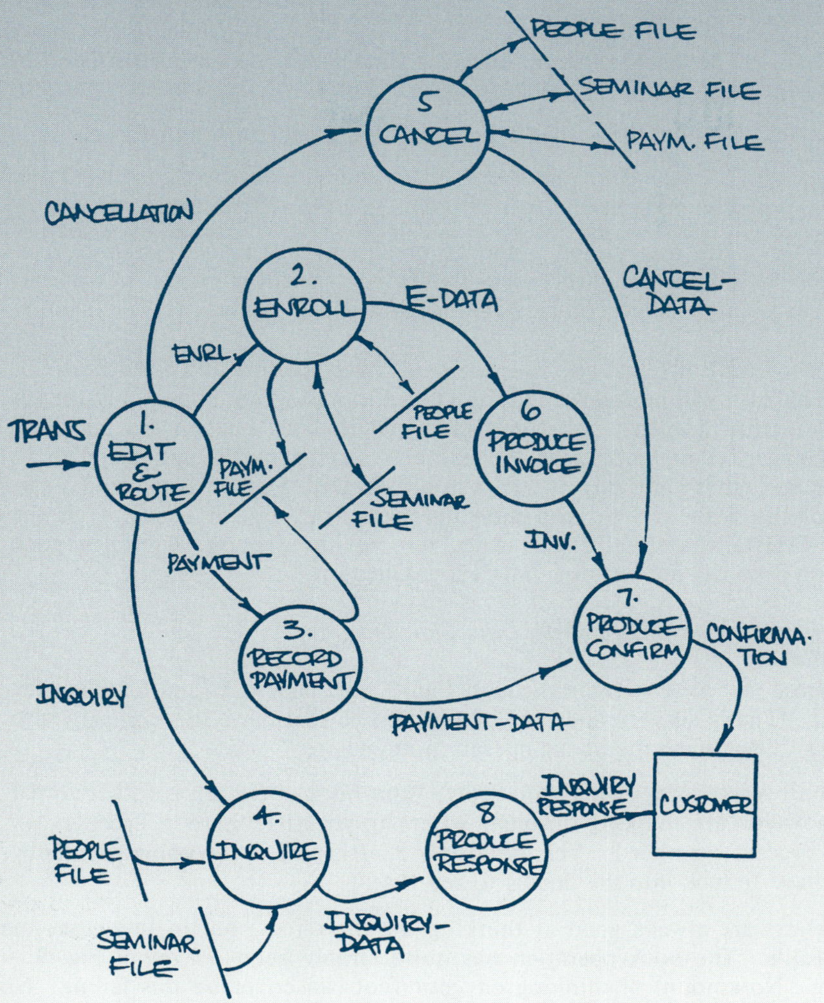
PEOPLE FILE
SEMINAR FILE
PAYM. FILE
5
CANCEL
CANCELLATION
CANCEL-
DATA
2.
ENROLL
E-DATA
ENRL.
PEOPLE
FILE
TRANS
1.
EDIT
&
ROUTE
PAYM.
FILE
SEMINAR
FILE
6
PRODUCE
INVOICE
INV.
PAYMENT
3.
RECORD
PAYMENT
7.
PRODUCE
CONFIRM
CONFIRMA-
TION
INQUIRY
PAYMENT-DATA
INQUIRY
RESPONSE
CUSTOMER
4.
INQUIRE
8
PRODUCE
RESPONSE
PEOPLE
FILE
SEMINAR
FILE
INQUIRY-
DATA

Figure 9

- *Four notational symbols are used.* These are:
  - The named vector (called a data flow), which portrays a data path.
  - The bubble (called a process), which portrays transformation of data.
  - The straight line, which portrays a file or data base.
  - The box (called a source or sink), which portrays a net originator or receiver of data — typically a person or an organization outside the domain of our study.

Since no control is shown, you can't tell from looking at a Data Flow Diagram which path will be followed. The Data Flow Diagram shows only the set of possible paths. Similarly, you can't tell what initiates a given process. You cannot assume, for instance, that Process 6 is started by the arrival of an E-Data — in fact, that's not how it works at all. E-Data's accumulate until a certain day of the week arrives, and then invoices all go out in a group. So the data flow E-Data indicates the data path, but not the prompt. The prompting information does not appear on a Data Flow Diagram.

3.2.2 An Important Advantage of the Data Flow Diagram

Suppose you were walking through Fig. 9 with your user and he made the comment: "That's all very fine, but in addition to seminars, this company also sells books. I don't see the books operation anywhere."

"Don't worry, Mr. User," you reply, "the book operation is fully covered here," (now you are thinking furiously where to stick it) "here in Process . . . um . . . Process Number 3. Yes, definitely 3. It's part of recording payments, only you have to look into the details to see that."

Analysts are always good at thinking on their feet, but in this case, the effort is futile. The book operation has quite simply been *left out* of Fig. 9 — it's wrong. No amount of thinking on your feet can cover up this failing. No books flow in or out, no inventory information is available, no reorder data flows are shown. Process 3 simply doesn't have access to the information it needs to carry out books functions. Neither do any of the others.

Your only option at this point is to admit the figure is wrong and fix it. While this might be galling when it happens, in the long run you are way ahead — making a similar change later on to the hard code would cost you considerably more grief.

I have seen this happen so many times: an analyst caught flat-footed with an incorrect Data Flow Diagram, trying to weasel his way out, but eventually being forced to admit that it is wrong and having to fix it. I conclude that it is a natural characteristic of the tool:

> *When a Data Flow Diagram is wrong, it is glaringly, demonstrably, indefensibly wrong.*

This seems to me to be an enormous advantage of using Data Flow Diagrams.

3.2.3 What Have We Accomplished With a Data Flow Diagram?

The Data Flow Diagram is documentation of a situation from the point of view of the data. This turns out to be a more useful viewpoint than that of any of the people or systems that process the data, because the data itself sees the big picture. So the first thing we have accomplished with the Data Flow Diagram is to come up with a meaningful portrayal of a system or a part of a system.

The Data Flow Diagram can also be used as a model of a real situation. You can try things out on it conveniently and get a good idea of how the real system will react when it is finally built.

Both the conceptual documentation and the modeling are valuable results of our Data Flow Diagramming effort. But something else, perhaps more important, has come about as a virtually free by-product of the effort: The Data Flow Diagram gives us a highly useful *partitioning* of a system. Fig. 9 shows an unhandily large operation conveniently broken down into eight pieces. It also shows all the interfaces among those eight pieces. (If any interface is left out, the diagram is simply wrong and has to be fixed.)

Notice that the use of a Data Flow Diagram causes us to go about our partitioning in a rather oblique way. If what we wanted to do was break things down, why didn't we just do that? Why didn't we concentrate on functions and subfunctions and just accomplish a brute-force partitioning? The reason for this is that a brute-force partitioning is too difficult. It is too difficult to say with any assurance that some task or group of tasks constitutes a "function." In fact, I'll bet you can't even define the word function except in a purely mathematical sense. Your dictionary won't do much better — it will give a long-winded definition that boils down to saying a function is a bunch of stuff to be done. The concept of function is just too imprecise for our purposes.

The oblique approach of partitioning by Data Flow Diagram gives us a "functional" partitioning, where this very special-purpose definition of the word functional applies:

> A partitioning may be considered *functional* when the interfaces among the pieces are minimized.

This kind of partitioning is ideal for our purposes.

3.3 A Data Dictionary example

Refer back to Fig. 9 for a moment. What is the interface between Process 3 and Process 7? As long as all that specifies the interface is the weak name "Payment-Data," we don't have a specification at all. "Payment-Data" could mean anything. We must state precisely what me mean by the data flow bearing that name in order for our Structured Specification to be anything more than a hazy sketch of the system. It is in the Data Dictionary that we state precisely what each of our data flows is made up of.

An entry from the sample project Data Dictionary might look like this:

```
Payment-Data        =  Customer-Name +
                       Customer-Address +
                       Invoice-Number +
                       Amount-of-Payment
```

In other words, the data flow called "Payment-Data" consists precisely of the items Customer-Name, Customer-Address, Invoice-Number, and Amount-of-Payment, concatenated together. They must appear in that order, and they must all be present. No other kind of data flow could qualify as a Payment-Data, even though the name might be applicable.

You may have to make several queries to the Data Dictionary in order to understand a term completely enough for your needs. (This also happens with conventional dictionaries — you might look up the term perspicacious, and find that it means sagacious; then you have to look up sagacious.) In the case of the example above, you may have to look further in the Data Dictionary to see exactly what an Invoice-Number is:

```
Invoice-Number      =  State-Code +
                       Customer-Account-Number +
                       Salesman-ID +
                       Sequential-Invoice-Count
```

Just as the Data Flow Diagram effects a partitioning of the area of our study, the Data Dictionary effects a top-down partitioning of our data. At the highest levels, data flows are defined as being made up of subordinate elements. Then the subordinate elements (also data flows) are themselves defined in terms of still more detailed subordinates.

Before our Structured Specification is complete, there will have to be a Data Dictionary entry for every single data flow on our Data Flow Diagram, and for all the subordinates used to define them. In the same fashion, we can use Data Dictionary entries to define our files.

3.4 A Structured English example

Partitioning is a great aid to specification, but you can't specify by partitioning alone. At some point you have to stop breaking things down into finer and finer pieces, and actually document the makeup of the pieces. In the terms of our Structured Specification, we have to state what it takes to do each of the data transformations indicated by a bubble on our Data Flow Diagram.

There are many ways we could go about this. Narrative text is certainly the most familiar of these. To the extent that we have partitioned sufficiently before beginning to specify, we may be spared the major difficulties of narrative description. However, we can do even better.

A tool that is becoming more and more common for process description is Structured English. Presented below is a Structured English example of a user's invoice handling policy from the sample analysis. It appears without clarification; if clarification is needed, it has failed in its intended purpose.

==

POLICY FOR INVOICE PROCESSING

If the amount of the invoice exceeds $500,
If the account has any invoice more than 60 days overdue,
hold the confirmation pending resolution of the debt.
Else (account is in good standing),
issue confirmation and invoice.
Else (invoice $500 or less),
If the account has any invoice more than 60 days overdue,
issue confirmation, invoice and write message on the credit action report.
Else (account is in good standing),
issue confirmation and invoice.

==

3.5 A Decision Table example

The same policy might be described as well by a Decision Table:

	RULES			
CONDITIONS	**1**	**2**	**3**	**4**
1. Invoice > $500	**Y**	**N**	**Y**	**N**
2. Account over-due by 60+ days	**Y**	**Y**	**N**	**N**
ACTIONS				
1. Issue Confirmation	**N**	**Y**	**Y**	**Y**
2. Issue Invoice	**N**	**Y**	**Y**	**Y**
3. Msg to C.A.R.	**N**	**Y**	**N**	**N**

3.6 A Decision Tree example

As a third alternative, you might describe the same policy with a Decision Tree. I have included the equivalent Decision Tree as Fig. 10.

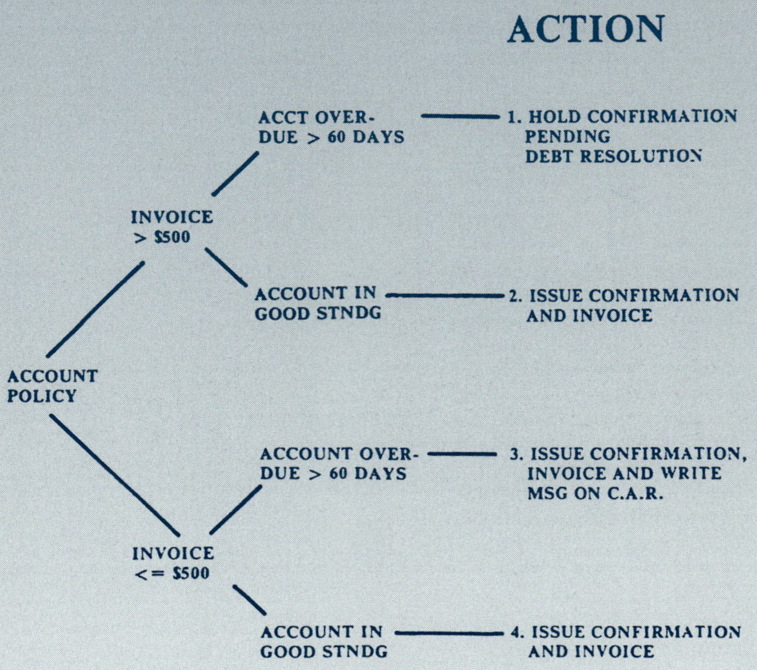
ACTION
ACCT OVER-
DUE > 60 DAYS
1. HOLD CONFIRMATION
PENDING
DEBT RESOLUTION
INVOICE
> $500
ACCOUNT IN
GOOD STNDG
2. ISSUE CONFIRMATION
AND INVOICE
ACCOUNT
POLICY
ACCOUNT OVER-
DUE > 60 DAYS
3. ISSUE CONFIRMATION,
INVOICE AND WRITE
MSG ON C.A.R.
INVOICE
<= $500
ACCOUNT IN
GOOD STNDG
4. ISSUE CONFIRMATION
AND INVOICE

Figure 10

Reviews and Inspections

Michael Fagan
President, Michael Fagan Associates
Palo Alto, California, USA
Michael@mfagan.com

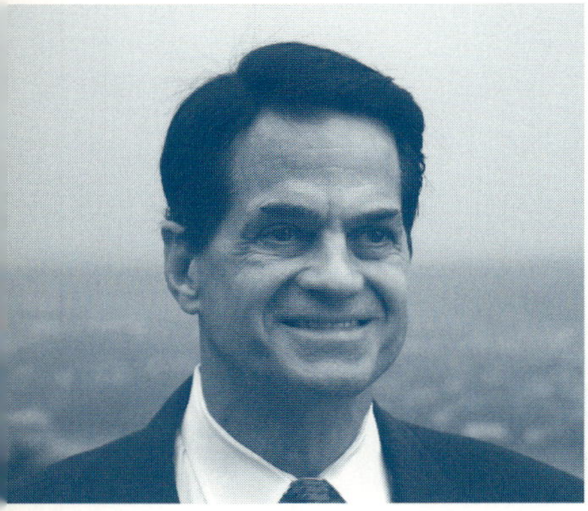

Senior technical staff member,
IBM T.J. Watson Research Lab

Co-founder,
IBM Quality Institute

Corporate Achievement Award
from IBM

Visiting Professor,
University of Maryland

Major contribution:
software inspection process

Current interests: improving
the Fagan Defect-Free Process

Michael Fagan

A History of Software Inspections

Inspections are now 30 years old and they continue to improve software quality and maintainability, reduce time to delivery, and lower development costs!

The concept of finding defects as early as possible in the software development process to save time, effort, and money seems intuitive in 2001. Many companies employ variations of the software inspections that I created as part of their development process in order to do just that. However, even 30 years after its creation, it is often not well understood and more often, poorly executed – yielding results that are positive, but well below their potential.

This paper will explore the history and creation of the software inspection process by Michael Fagan. Some readers will find a great deal of similarity between the development environment that led to the creation of this process and the one in which they are currently involved. The process itself has proven to be as effective and necessary today as when it was first created.

The Genesis of Inspections

My professional career began in hardware engineering and manufacturing. This shaped my thinking in terms of the cost of rework because when errors are cast in silicon, there is no recourse but to scrap the erroneous pieces, fix the design, and rerun the lot. In short, defect rework was obvious, measurable, and expensive. Every effort was made to find defects *before* production began.

In my area, at least, testing did not find a sufficient number of defects before release to production, so I formed teams of engineers to intellectually examine designs *after* exhaustive testing and before release to production. The results were startling: the engineers found many defects that testing had missed, significantly shortening the time needed to manufacture good products.

In 1971, on the advice of well-meaning friends, I made the switch from hardware development to the burgeoning world of software development. In this world, everything was new to me and posed quite a challenge. Development was chaotic, and, what was worse, no one seemed to have a way to get it under control.

There were no useful measurements that shed any light on what was going on or how to do better. There seemed to be an "everything is new under the sun" approach where each new project was undertaken as though no development had ever been done before. There was very little learning from project to project. I quickly realized that in order to be effective (and to survive), I had to really understand my new environment and then find a way to make order out of chaos.

It was obvious, even without measurement, that reworking defects was a very significant component of the software development effort. It was seen as being a completely natural accompaniment to creating design or writing code. Of course, it was and is. However, no one knew or focused attention on its relative size or how it may be examined in its own right as a consumer of vital software development resource.

It became clear to me that the magnitude of the rework effort was large enough that it needed to be better understood. I decided to measure it by applying a similar method that I had employed in hardware engineering. The idea was to intellectually examine software design and code in an organized manner and then measure the resultant effects on defect rates and defect rework during integration and testing. These human examinations reduced defect rework out of proportion to the effort expended on them. This sparked the development of software inspections.

Introducing Inspections into the Development Process

Creating a methodology and persuading others to use it are two very different things. Indeed, during the early years of inspections, even with results that looked convincing to me, I encountered everything from derision to exasperation, but very little acceptance from my peers. Even when third parties, who had no stake in the methodology, reported how it had helped them, some people were reluctant to try using the inspection process. Resistance to changing working habits was (and continues to be) an impediment to spreading the use of inspections.

Conventional software development wisdom at that time was to follow a traditional life-cycle model in which there were a series of steps that were approached in a logical, sequential order (later called the Waterfall Process, by Winston Royce). But this was not what I saw happening. Instead of one approach, there were several versions of the approach going on simultaneously. The product was developed in a series of cycles, similar to what we call "iterative development" or "spiral development" today.

Managing development of software (or anything) – making a plan, measuring accomplishment against the plan as the trackable elements were developed, and controlling the outcome to meet project objectives – as we tried to do in weekly "build meetings," required reconciling interpersonal communications among many people, who made perfect sense to themselves, but often not to one another.

I often left these meetings with more unanswered questions than answered ones: what were the actual requirements for the product we were trying to build, and were they clear and unambiguous to all the people who were responsible for satisfying them? Completion criteria for the requirements specification amounted to it being "signed off." However, being "signed off" did not necessarily signify that it was complete, correct, and non-ambiguous, because the signatories would often sign off when the deadline arrived, but rarely with a clear understanding of the contents of the specification. Development would proceed based upon these requirements nevertheless.

Was there really an accurate plan of how the components of a system really fit together *before* system integration? Did all the players have the same understanding of the definition of each of the components of a system? What were the trackable units in the system build plan, and was their relative granularity known? Measuring and tracking the degree of completion of design and code entities that comprised a function in the development phases between "start of design" and "end of integration" was simply infeasible.

Absence of commonly agreed upon, measurable completion criteria of the activities of requirements, design, code, and the other deliverables were a significant cause of misunderstandings, caused a good deal of rework, and did not provide any framework for managing software development. Without well-defined exit criteria for each stage of development, it was surprising how many different perspectives can be found within a development team about precisely when a particular type of work product (i.e., requirements or design) was "finished." For instance, are requirements finished when "signed off," when there is a testable product or feature defined, or when design, the next operation, accepts the document as valid input and begins work? This really was subject to a great deal of interpretation and was often more time (schedule) dependent than criteria dependent. Understandably, the confusion arising from this range of interpretations had to be resolved, and it was – in favor of meeting schedule dates, which are explicitly defined (even if it is unclear exactly what is to be delivered on a particular date!).

This brought up another question – how could the testing schedule be set when no one knew how much defect rework was involved during testing? Worse yet, immediately after a product was shipped, work would start on the next release. Unfortunately, developing the new release was frequently interrupted in order to fix the defects found by users in the previous release. Reworking defects in the previous release obviously subtracted from the effort committed to the new release and often caused it to ship late, or to ship on time with reduced functionality (i.e., fewer requirements satisfied), and/or to ship with more defects than it should. In addition, there was always a backlog of defects to be reworked and incorporated as a part of developing the new release. Supporting additional releases of legacy product compounded the problem.

[Note: the foregoing pattern is often found in young, start-up organizations. They get started with little or no formal process methodology and are successful until they must handle their own legacy software.]

However, it should be noted that if the percentage of total development effort that is devoted to fixing defects has an upward trend over the course of several releases, it is easy to extrapolate to a time when more effort will be spent on defect rework, with a diminishing percentage being available to develop new functionality. Since customers purchase products for their new functionality, continuation of this trend will lead to the point at which new functionality can only be added by employing more people – most of whom must rework defects in current and legacy products.

Answers to these and many other problems were not consistent and varied from case to case, without sufficient qualification of each case to enable extraction of any useful generalizations. The prevailing management climate was not sympathetic to dealing with these questions for the purpose of improving planning for future development. The thrust was to "**ship** this function **now**!" So, a lot of very good people worked very hard and wasted a lot of time dealing with miscommunications and crossed purposes.
We shipped products – time after time without much improvement to the development process between releases. We made software with (please forgive the expression) brute force – intellectual, of course – and heroic effort. Today, we would recognize this as an SEI CMM Level 1 organization (SEI = Software Engineering Institute, CMM = Capability Maturity Model.)

In short, I found myself in a software development world that I found largely unmanageable, with too high a potential for failure. (Many others confided the same concerns to me, so I did not feel too lonely.) Therefore, to help our ability to manage in the face of the prevailing problems endemic in creating innovative software, and improve the quality of software delivered to customers, I decided that I would have to modify the development process so as to make it more manageable – at least in my area.

Since I was a manager and it was my job to deliver high-quality software on time and within budget, I took the liberty and the attendant risk to implement an inspection process in my area. I did not receive much support – in fact, I was ridiculed, counseled and otherwise told to stop the nonsense and get on with the job of managing projects the way everyone else was doing it.

However, I was given just enough rope to hang myself with (many people were hoping I would) and I was able to implement inspections on my project – part of an operating system. Through perseverance and determination, my team and I learned that by applying inspections, we could improve the outcome of the project. Further, the more effective inspections were at finding defects, the more likely the project was to meet its schedule and cost objectives. This led me to ignore my critics and continue experiments and improve the inspection method and its application within the development process to get even better results.

One of the first steps taken was the creation of measurable exit criteria for all the operations in the development process. Explicit, measurable, succinct exit criteria for the operations which created the work products that would be inspected (i.e., requirements, design, code, etc.) were the first building blocks needed to support the inspection process. This eliminated many miscommunications and bad handoffs between development phases.

Another issue that needed to be addressed was the pressure to reduce the number of defects that reached customers. This dovetailed with the need I saw to minimize defect rework, which was such a large and uncontrolled component of development effort. Estimates of defect rework as a percentage of total development effort ranged from 30% to 80%. The two most obvious means to satisfy these needs were to reduce the number of defects injected during development, and to find and fix those that were injected as near to their point of origin in the process as possible. This led to formulation of the *dual objectives of the inspection process:*

- Find and fix all defects in the product, and
- Find and fix all defects in the development process that give rise to defects in the product (namely, remove the causes of defects in the product).

My development team and I modeled, measured, and implemented changes in the initial inspection process that we used for creating real live software products. The procedure we followed was to make changes in the inspection process, including adjusting the number of inspectors and their individual roles, measuring the results of executing the process with these changes installed, and then interviewing all the people involved in the inspection and in development. We analyzed the results of each change and made refinements and repeated the execution-refinement cycle up to the point when we experienced no further improvements. (This procedure was retained and periodically executed for monitoring and enabling continuous improvement of both the inspection and development processes.) After numerous cycles through this procedure, we reached a stage at which the inspection process was repeatable and efficiently found the highest number defects when it was executed by trained developers.

This initial work was done over a period of three and a half years. Many products and releases were involved, and several hundreds of developers and their managers participated – with varying degrees of willingness. The inspection methods employed did not delay any shipments, and customers noticed an improvement in the quality of products they received. Both the proponents and critics – and the latter abounded – were very helpful in making sure that we "turned over every stone." They helped us more than they know (or sometimes intended), so that by the time I wrote the paper, "Design and code inspections to reduce errors in program development," in the *IBM System Journal*, in 1976, I was confident that the process had been well wrung out in actual development and produced repeatable results. This paper provides a partial overview and useful highlights of the process, but was too short to fully describe the entire inspection process.

Over my next several years with IBM as a development manager of increasingly larger organizations, I continued measuring and repeating the execution-refinement cycle described above. After the process had been proven beyond a shadow of a doubt to be an effective means of reducing customer-reported defects and improving product quality as well as productivity, IBM asked me to promote this process within its other divisions. Again, I had to convince non-believers and change the work habits of development organizations. Eventually, I was even asked to share this process with a few large and key IBM customers.

For the creation of the software inspection process and the fact that it saved untold millions of dollars in development cost, IBM awarded me its largest corporate individual award at that time.

Implementation of Inspections in the Software Development World

The inspection process we created proved to be an extremely effective means of reducing the number of defects users find in software, as well as increasing development productivity. This was achieved by using inspections to find defects and remove them from all the work products that are created in the course of developing software, namely, in requirements specifications, design, coding, test plans and test cases, and user documentation. In addition, these inspections identified and caused to be fixed those defects in the development process that gave rise to defects in the products made using the process.

Over time, the list of noteworthy benefits has grown and now includes:

- Reduction in user-reported defects;
- Increased customer satisfaction;
- Increased development productivity, which materializes in shipping more function in a given time or reduction in time to delivery;
- Improvement in meeting committed schedules;
- Rapid cross-training of developers and maintainers on new products;
- Continuous process improvement through removal of systemic defects (which are the cause of defects in the product);
- Authors rapidly learned to avoid creating defects through participating in inspections that find defects in their own work products and in the work products of others;
- Team building; and
- Inspections have, in some cases, eliminated the need to implement unit test code.

With this list of benefits, it is hard to understand why the use of the Fagan Inspection Process is not "de rigueur" in all development organizations. In fact, supporters claim that all "world-class" product development groups are using inspections. This is not what I see today. There are still many organizations that are not using inspections. Additionally, there are precious few organizations who have been trained to execute the process fully and in a way that will provide the complete set of benefits of the process.

One reason that is expressed by developers and managers who are launched on tight delivery schedules for not employing the inspection process, is their dominant fear that inspections will take too long, lengthen the schedule, and delay shipment (this fear has stalked inspections throughout their history, and it continues). They note that inspections use 20–30% of the effort during the first half of product development and, they fear, this will add to the development cycle. Of course, what they overlook is the **net** reduction in development effort that results from inspections finding and fixing defects early in the life cycle at a very much lower cost (in terms of time and money) than the effort that is expended on fixing defects that are found during testing or in the field. This reduction in development effort often leads to shortening the time to delivery (Figure 1). Although experience has shown this concern to be a myth, fear of delaying shipment often causes managers to resist adopting inspections.

Additionally, mislabelling of process operations continues to this day in the software industry. If we insist on truth-in-advertising, "testing" would be called "defect rework" because in most cases much less than 50% of the effort expended during "testing" operations is actually used to verify that the product satisfies its requirements, while more than 50% of the effort is consumed doing defect rework. Exercising the product under test with one pass of the test cases is all that should be needed to verify that the product meets its requirements. This simple act of renaming would

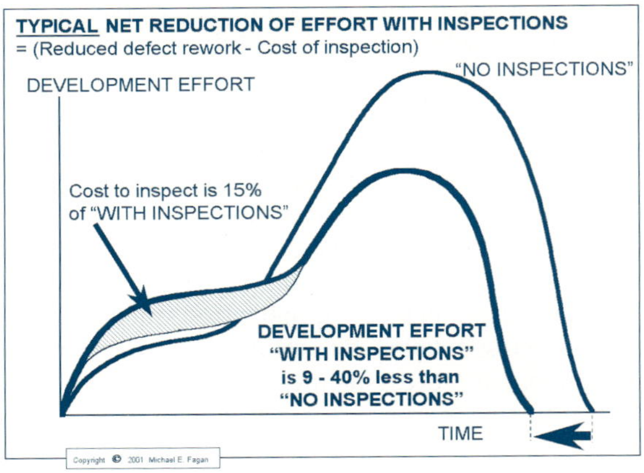

Fig. 1.

focus the attention of management (and financial auditors) on what is really going on, leaving testing to verify that the product satisfies its requirements.

Code inspections are more often discussed and receive a lot more attention than the inspection of requirement or design specifications, although the latter can produce even larger benefits. This could be because code is easier to inspect: since requirements are the foundation on which products are built, but are often improperly specified or reflect some other process deficiency, making them more difficult to inspect. Similarly, discussion of inspections of hardware and software systems design, logic design, test plans and cases, and user documentation tend to be under-reported, and under-utilized.

Since their inception, inspections have had the dual objectives of finding defects in the product and in the process used to create the product. Finding defects in the product is so immediate and satisfying that most often the added step of identifying defects in, and removing them from, the development process is overlooked. Skipping this second step is very shortsighted as it does not enable the reduction of defect injection in all future development, an essential element in "putting oneself out of the defect generation business."

Inspections are a team sport. There are two aspects to this: first, the organization of effort to accomplish the inspection meetings and, secondly, the conduct of the inspection meetings themselves. The team members, each playing their appointed role in the inspection process, include managers (all levels of line managers), project managers, requirements specifiers, hardware and software designers, coders, integrators, testers, quality assurance, writers of user documents, users of the product, and marketing. Like any team endeavor, successful inspections depend upon the individual capability of each team member and how well they work in inspection teams. Successful execution is an obvious, but frequently overlooked, determinant in effective inspections – see effectiveness factors, below.

There is often confusion about the relative roles of reviews, walkthroughs, and inspections. The fact is that both reviews and walkthroughs (taking these two activities as being essentially of the same type) and inspections are both important in the development of work products. Reviews/walkthroughs are conducted during development to demonstrate and refine approaches, solicit and compare viewpoints, develop improvements, and identify problems and create solutions to them. Inspections, on the other hand, are conducted when a work product is asserted to be complete. The objective of inspection is to find defects only, not to undertake the activities of reviews/walkthroughs. (A defect is defined as an instance in which the work product fails to satisfy the exit criteria of the operation in which it was created, including operational defects that could be reported by customers.)

The reason for maintaining this exclusive objective for inspections is that experimentation has shown that when review/walkthrough activities are included in the inspection or, indeed, even one more objective is included, there is significant degradation in the defect detection effectiveness of inspections. Occasionally, reviews and walkthroughs are spoken of as if they are competing activities, and at other times as if one has evolved into the other. In practice, combining their activities creates an amalgam of disparate objectives and appears to lower their individual and combined effectiveness. The fact is that reviews/walkthroughs and inspections come into play during different times in the development of a work product, and each adds unique value. When they are both used separately, each one contributes according to its unique objectives in the development of software that is well thought out and contains very few defects.

The Inspection Process Today

Since I left IBM, the incubator of the software inspection process, I have continued to study and make additional enhancements to the process through my company, Michael Fagan Associates.

The early inspection process included checklists of conditions to look for during preparation and the inspection meeting. Many of these checklists included items that could be found using tools and did not require the human brain. As the inspection process evolved, we eliminated checklists and learned to concentrate human intelligence on finding those defects that testing with computers could only detect with a lot of effort, and to let computers find mechanical types of defects. For example, compilers flag syntax errors and LINT-type compiler tools are good at finding procedure and standard violations. Using these tools relieves human inspectors from mechanical checklist tasks, and allows them to be much more effective by concentrating their attention on finding operational type defects – those that would otherwise be found in test or by the customer.

Other changes have included creating a set of "Effectiveness Factors" to quantify the relative quality of the inspections themselves. We know from history and analysis of compiled data which things have the largest positive and negative impact on the defect-finding effectiveness of an inspection. These factors can be used to manage the inspection process to produce its maximum set of benefits.

These and other recent enhancements have not been discussed in any other papers to date. They have been disseminated only to our clients through a course which provides training on the full inspection process.

Unfortunately, the terms "inspection," and even "Fagan Inspection," have become almost generalized to refer to things that do not produce the consistent and repeatable results that I consider necessary to good process management. Since the terms have come to mean something other than what I intended and implemented, I now speak of the "Fagan Defect-Free Process" which includes Fagan Inspections, as well as the other components needed to make this process successful and enduring.

The Fagan Defect-Free Process includes all of the necessary and interwoven components that are required to make software inspections successful. The three main ones include:

- Formal Process Definition – ensuring each member of the team is conversant in the objectives, function, and entry and exit criteria of each process phase;
- Inspection Process – the seven-step process used to find defects;
- Continuous Process Improvement – removing systemic defects from the development process.

The name "Fagan Inspection Process" has been misused to denote just the seven-step inspection process originally created by me. It has also been applied to the variations created and promoted by others without a real mastery of what makes the process work successfully. Unfortunately, many organizations have had comparatively modest to poor results without understanding that their implementation of the process was incomplete. Thus, while inspections have been very successful and have proven their value over many years and many projects, there are those who would rather not include inspections in their development processes because of mediocre results, due to a mediocre implementation.

As may be expected, cases of partial or improper implementation of the inspection process, which are incorrect executions of the process, often produced inferior results and confused the issue. Thus, although it should be obvious, experience has shown that it must be stated that to get the best results from the inspection process, it is necessary to execute this process precisely and consistently. Similarly, experience has shown that all variations of inspection processes do not produce similar results.

Summary

The "Fagan Inspection Process" is as relevant and important today as it was 30 years ago, perhaps even more so. With systems growing ever more complex and a shortage of resources with which to develop them, its benefits are even more attractive. Applied and executed as intended, it produces significant improvements to the software development process, including schedule and cost reduction, productivity improvements, and fewer customer-reported defects.

Nothing will be more satisfying in the future than to find the "Fagan Defect-Free Process," including Fagan Inspections, in increasingly widespread use not only in the software world, but also in other areas in which more diverse products and services are created. I have worked, and will continue to work, towards that goal.

Michael Fagan

Design and Code Inspections to Reduce Errors in Program Development

IBM Systems Journal,
Vol. 15 (3), 1976
pp. 182-211

Substantial net improvements in programming quality and productivity have been obtained through the use of formal inspections of design and of code. Improvements are made possible by a systematic and efficient design and code verification process, with well-defined roles for inspection participants. The manner in which inspection data is categorized and made suitable for process analysis is an important factor in attaining the improvements. It is shown that by using inspection results, a mechanism for initial error reduction followed by ever-improving error rates can be achieved.

Design and code inspections to reduce errors in program development

by M. E. Fagan

Successful management of any process requires planning, measurement, and control. In programming development, these requirements translate into defining the programming process in terms of a series of operations, each operation having its own exit criteria. Next there must be some means of measuring completeness of the product at any point of its development by inspections or testing. And finally, the measured data must be used for controlling the process. This approach is not only conceptually interesting, but has been applied successfully in several programming projects embracing systems and applications programming, both large and small. It has not been found to "get in the way" of programming, but has instead enabled higher predictability than other means, and the use of inspections has improved productivity and product quality. The purpose of this paper is to explain the planning, measurement, and control functions as they are affected by inspections in programming terms.

An ingredient that gives maximum play to the planning, measurement, and control elements is consistent and vigorous *discipline*. Variable rules and conventions are the usual indicators of a lack of discipline. An iron-clad discipline on all rules, which can stifle programming work, is not required but instead there should be a clear understanding of the flexibility (or nonflexibility) of each of the rules applied to various aspects of the project. An example of flexibility may be waiving the rule that all main paths will be tested for the case where repeated testing of a given path will logically do no more than add expense. An example of necessary inflexibility would be that *all* code must be inspected. A clear statement of the project rules and changes to these rules along with faithful adherence to the rules go a long way toward practicing the required project discipline.

A prerequisite of process management is a clearly defined series of operations in the process (Figure 1). The miniprocess within each operation must also be clearly described for closer management. A clear statement of the criteria that must be satisfied to exit each operation is mandatory. This statement and accurate data collection, with the data clearly tied to trackable units of known size and collected from specific points in the process, are some essential constituents of the information required for process management.

In order to move the form of process management from qualitative to more quantitative, process terms must be more specific, data collected must be appropriate, and the limits of accuracy of the data must be known. The effect is to provide more precise

Figure 1 Programming process

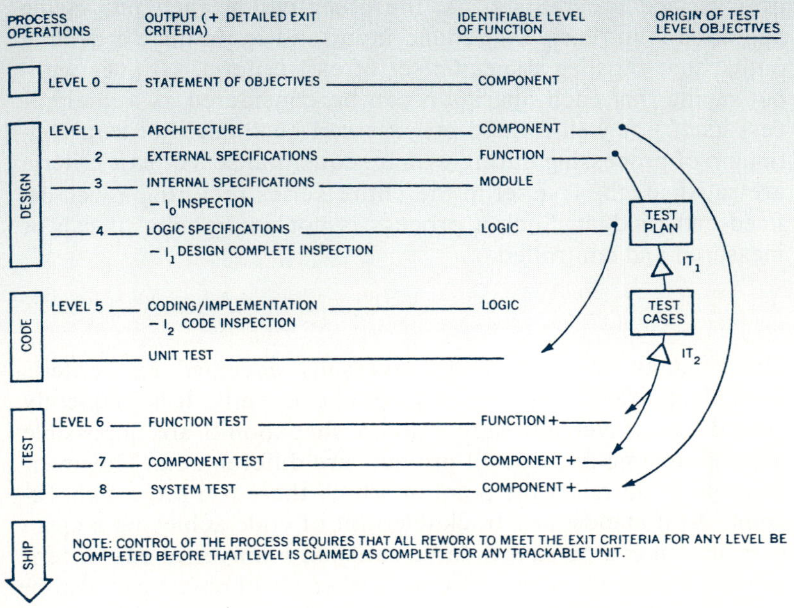

information in the correct process context for decision making by the process manager.

In this paper, we first describe the programming process and places at which inspections are important. Then we discuss factors that affect productivity and the operations involved with inspections. Finally, we compare inspections and walk-throughs on process control.

a manageable process A process may be described as a set of operations occurring in a definite sequence that operates on a given input and converts it to some desired output. A general statement of this kind is sufficient to convey the notion of the process. In a practical application, however, it is necessary to describe the input, output, internal processing, and processing times of a process in very specific terms if the process is to be executed and practical output is to be obtained.

In the programming development process, explicit requirement statements are necessary as input. The series of processing operations that act on this input must be placed in the correct sequence with one another, the output of each operation satisfying the input needs of the next operation. The output of the final operation is, of course, the explicitly required output in the form of a verified program. Thus, the objective of each processing operation is to receive a defined input and to produce a definite output that satisfies a specific set of exit criteria. (It goes without saying that each operation can be considered as a miniprocess itself.) A well-formed process can be thought of as a continuum of processing during which sequential sets of exit criteria are satisfied, the last set in the entire series requiring a well-defined end product. Such a process is not amorphous. It can be measured and controlled.

exit criteria Unambiguous, explicit, and universally accepted exit criteria would be perfect as process control checkpoints. It is frequently argued that universally agreed upon checkpoints are impossible in programming because all projects are different, etc. However, *all* projects do reach the point at which there is a project checkpoint. As it stands, any trackable unit of code achieving a clean compilation can be said to have satisfied a universal exit criterion or checkpoint in the process. Other checkpoints can also be selected, albeit on more arguable premises, but once the premises are agreed upon, the checkpoints become visible in most, if not all, projects. For example, there is a point at which the design of a program is considered complete. This point may be described as the level of detail to which a unit of design is reduced so that one design statement will materialize in an estimated three to 10 source code instructions (or, if desired, five to 20, for that matter). Whichever particular ratio is selected across a project, it provides a checkpoint for the process control of that project. In this way, suitable checkpoints may be selected throughout the development process and used in process management. (For more specific exit criteria see Reference 1.)

The cost of reworking errors in programs becomes higher the later they are reworked in the process, so every attempt should be made to find and fix errors as early in the process as possible. This cost has led to the use of the inspections described later and to the description of exit criteria which include assuring that all errors known at the end of the inspection of the new "clean-compilation" code, for example, have been correctly fixed. So, rework of all known errors up to a particular point must be complete before the associated checkpoint can be claimed to be met for any piece of code.

Where inspections are not used and errors are found during development or testing, the cost of rework as a fraction of overall development cost can be suprisingly high. For this reason, errors should be found and fixed as close to their place of origin as possible.

Production studies have validated the expected quality and productivity improvements and have provided estimates of standard productivity rates, percentage improvements due to inspections, and percentage improvements in error rates which are applicable in the context of large-scale operating system program production. (The data related to operating system development contained herein reflect results achieved by IBM in applying the subject processes and methods to representative samples. Since the results depend on many factors, they cannot be considered representative of every situation. They are furnished merely for the purpose of illustrating what has been achieved in sample testing.)

The purpose of the test plan inspection IT_1, shown in Figure 1, is to find voids in the functional variation coverage and other discrepancies in the test plan. IT_2, test case inspection of the test cases, which are based on the test plan, finds errors in the test cases. The total effects of IT_1 and IT_2 are to increase the integrity of testing and, hence, the quality of the completed product. And, because there are less errors in the test cases to be debugged during the testing phase, the overall project schedule is also improved.

A process of the kind depicted in Figure 1 installs all the intrinsic programming properties in the product as required in the statement of objectives (Level 0) by the time the coding operation (Level 5) has been completed—except for packaging and publications requirements. With these exceptions, all later work is of a verification nature. This verification of the product provides no contribution to the product during the essential development (Levels 1 to 5); it only adds error detection and elimination (frequently at one half of the development cost). I_0, I_1, and I_2 inspections were developed to measure and influence intrinsic quality (error content) in the early levels, where error rework can be most economically accomplished. Naturally, the beneficial effect on quality is also felt in later operations of the development process and at the end user's site.

An improvement in productivity is the most immediate effect of purging errors from the product by the I_0, I_1, and I_2 inspections. This purging allows rework of these errors very near their origin, early in the process. Rework done at these levels is 10 to 100 times less expensive than if it is done in the last half of the process. Since rework detracts from productive effort, it reduces productivity in proportion to the time taken to accomplish the rework. It follows, then, that finding errors by inspection and reworking them earlier in the process reduces the overall rework time and increases productivity even within the early operations and even more over the total process. Since less errors ship with the product, the time taken for the user to install programs is less, and his productivity is also increased.

The quality of documentation that describes the program is of as much importance as the program itself for poor quality can mislead the user, causing him to make errors quite as important as errors in the program. For this reason, the quality of program documentation is verified by publications inspections (PI_0, PI_1, and PI_2). Through a reduction of user-encountered errors, these inspections also have the effect of improving user productivity by reducing his rework time.

Figure 2 A study of coding productivity

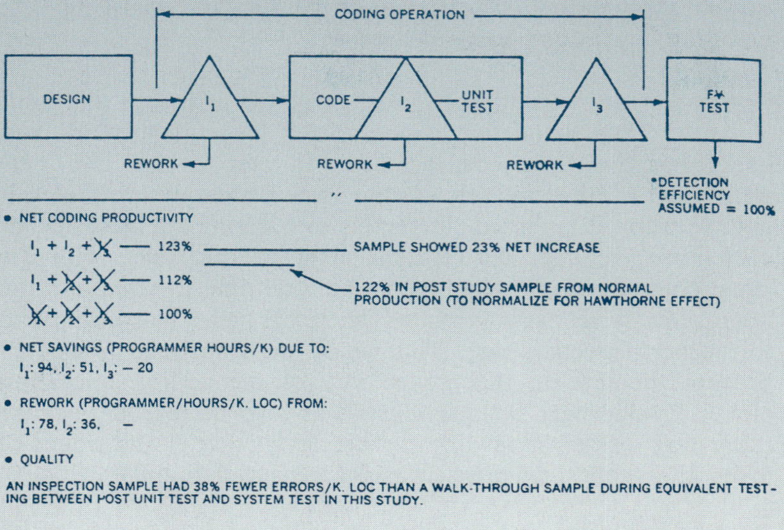

A study of coding productivity

A piece of the design of a large operating system component (all done in structured programming) was selected as a study sample (Figure 2). The sample was judged to be of moderate complexity. When the piece of design had been reduced to a level of detail sufficient to meet the Design Level 4 exit criteria[2] (a level of detail of design at which one design statement would ultimately appear as three to 10 code instructions), it was submitted to a design-complete inspection (100 percent), I_1. On conclusion of I_1, all error rework resulting from the inspection was completed, and the design was submitted for coding in PL/S. The coding was then done, and when the code was brought to the level of the first clean compilation,[2] it was subjected to a code inspection (100 percent), I_2. The resultant rework was completed and the code was subjected to unit test. After unit test, a unit test inspection, I_3, was done to see that the unit test plan had been fully executed. Some rework was required and the necessary changes were made. This step completed the coding operation. The study sample was then passed on to later process operations consisting of building and testing.

inspection sample

The inspection sample was considered of sufficient size and nature to be representative for study purposes. Three programmers designed it, and it was coded by 13 programmers. The inspection sample was in modular form, was structured, and was judged to be of moderate complexity on average.

coding operation productivity

Because errors were identified and corrected in groups at I_1 and I_2, rather than found one-by-one during subsequent work and handled at the higher cost incumbent in later rework, the overall amount of error rework was minimized, even within the coding operation. Expressed differently, considering the inclusion of *all* I_1 time, I_2 time, and resulting error rework time (with the usual coding and unit test time in the total time to complete the operation), a *net* saving resulted when this figure was compared to the no-inspection case. This net saving translated into a 23 percent increase in the productivity of the coding operation alone. Productivity in later levels was also increased because there was less error rework in these levels due to the effect of inspections, but the increase was not measured directly.

An important aspect to consider in any production experiment involving human beings is the Hawthorne Effect.[3] If this effect is not adequately handled, it is never clear whether the effect observed is due to the human bias of the Hawthorne Effect or due to the newly implemented change in process. In this case a *control sample* was selected at random from many pieces of work *after the I_1 and I_2 inspections were accepted as commonplace.* (Previous experience without I_1 and I_2 approximated the net coding productivity rate of 100 percent datum in Figure 2.) The difference in coding productivity between the experimental sample (with I_1 and I_2 for the first time) and the control sample was 0.9 percent. This difference is not considered significant. Therefore, the measured increase in coding productivity of 23 percent is considered to validly accrue from the only change in the process: addition of I_1 and I_2 inspections.

control sample

The control sample was also considered to be of representative size and was from the same operating system component as the study sample. It was designed by four programmers and was coded by seven programmers. And it was considered to be of moderate complexity on average.

net savings

Within the coding operation only, the net savings (including inspection and rework time) in programmer hours per 1000 Non-Commentary Source Statements (K.NCSS)[4] were I_1: 94, I_2: 51, and I_3: −20. As a consequence, I_3 is no longer in effect.

If personal fatigue and downtime of 15 percent are allowed in addition to the 145 programmer hours per K.NCSS, the saving approaches one programmer month per K.NCSS (assuming that our sample was truly representative of the rest of the work in the operating system component considered).

error rework

The error rework in programmer hours per K.NCSS found in this study due to I_1 was 78, and 36 for I_2 (24 hours for design errors and 12 for code errors). Time for error rework must be specifically scheduled. (For scheduling purposes it is best to develop rework hours per K.NCSS from history depending upon the particular project types and environments, but figures of 20 hours for I_1, and 16 hours for I_2 (*after the learning curve*) may be suitable to start with.)

quality

The only comparative measure of quality obtained was a comparison of the inspection study sample with a fully comparable piece of the operating system component that was produced similarly, except that walk-throughs were used in place of the I_1 and I_2 inspections. (Walk-throughs[5] were the practice before implementation of I_1 and I_2 inspections.) The process span in which the quality comparison was made was seven months of testing beyond unit test after which it was judged that both samples had been equally exercised. The results showed the inspection sample to contain 38 percent less errors than the walk-through sample.

Note that up to inspection I_2, no machine time has been used for debugging, and so machine time savings were not mentioned. Although substantial machine time is saved overall since there are less errors to test for in inspected code in later stages of the process, no actual measures were obtained.

Table 1 Error detection efficiency

Process Operations	*Errors Found per K.NCSS*	*Percent of Total Errors Found*
Design		
I_1 inspection		
Coding	38*	82
I_2 inspection		
Unit test		
Preparation for acceptance test	8	18
Acceptance test	0	
Actual usage (6 mo.)	0	
Total	46	100

*51% were logic errors, most of which were missing rather than due to incorrect design.

inspections in applications development

In the development of applications, inspections also make a significant impact. For example, an application program of eight modules was written in COBOL by Aetna Corporate Data Processing department, Aetna Life and Casualty, Hartford, Connecticut, in June 1975.[6] Two programmers developed the program. The number of inspection participants ranged between three and five. The only change introduced in the development process was the I_1 and I_2 inspections. The program size was 4,439 Non-Commentary Source Statements.

An automated estimating program, which is used to produce the normal program development time estimates for all the Corporate Data Processing department's projects, predicted that designing, coding, and unit testing this project would require 62 programmer days. In fact, the time actually taken was 46.5 programmer days including inspection meeting time. The resulting saving in programmer resources was 25 percent.

The inspections were obviously very thorough when judged by the inspection error detection efficiency of 82 percent and the later results during testing and usage as shown in Table 1.

The results achieved in Non-Commentary Source Statements per Elapsed Hour are shown in Table 2. These inspection rates are four to six times faster than for systems programming. If these rates are generally applicable, they would have the effect of making the inspection of applications programs much less expensive.

Table 2 Inspection rates in NCSS per hour

Operations	I_1	I_2
Preparation	898	709
Inspection	652	539

Inspections

Inspections are a *formal*, *efficient*, and *economical* method of finding errors in design and code. All instructions are addressed at least once in the conduct of inspections. Key aspects of inspections are exposed in the following text through describing the I_1 and I_2 inspection conduct and process. I_0, IT_1, IT_2, PI_0, PI_1, and PI_2 inspections retain the same essential properties as

Table 3. Inspection process and rate of progress

Process operations	Rate of progress* (loc/hr) Design I_1	Code I_2	Objectives of the operation
1. Overview	500	not necessary	Communication education
2. Preparation	100	125	Education
3. Inspection	130	150	*Find errors*
4. Rework	20 hrs/K.NCSS	16 hrs/K.NCSS	Rework and resolve errors found by inspection
5. Follow-up	–	–	See that all errors, problems, and concerns have been resolved

*These notes apply to systems programming and are conservative. Comparable rates for applications programming are much higher. Initial schedules may be started with these numbers and as project history that is keyed to unique environments evolves, the historical data may be used for future scheduling algorithms.

the I_1 and I_2 inspections but differ in materials inspected, number of participants, and some other minor points.

the people involved

The inspection team is best served when its members play their particular roles, assuming the particular vantage point of those roles. These roles are described below:

1. *Moderator*—The *key person* in a successful inspection. He must be a competent programmer but need *not* be a technical expert on the program being inspected. To preserve objectivity and to increase the integrity of the inspection, it is usually advantageous to use a moderator from an unrelated project. The moderator must manage the inspection team and offer leadership. Hence, he must use personal sensitivity, tact, and drive in balanced measure. His use of the strengths of team members should produce a synergistic effect larger than their number; in other words, *he is the coach.* The duties of moderator also include scheduling suitable meeting places, reporting inspection results within one day, and follow-up on rework. *For best results the moderator should be specially trained.* (This training is brief but very advantageous.)
2. *Designer*—The programmer responsible for producing the program design.
3. *Coder/Implementor*—The programmer responsible for translating the design into code.
4. *Tester*—The programmer responsible for writing and/or executing test cases or otherwise testing the product of the designer and coder.

If the coder of a piece of code also designed it, he will function in the designer role for the inspection process; a coder from some related or similar program will perform the role of the coder. If the same person designs, codes, and tests the product code, the coder role should be filled as described above, and another coder—preferably with testing experience—should fill the role of tester.

Four people constitute a good-sized inspection team, although circumstances may dictate otherwise. The team size should not be artificially increased over four, but if the subject code is involved in a number of interfaces, the programmers of code related to these interfaces may profitably be involved in inspection. Table 3 indicates the inspection process and rate of progress.

scheduling inspections and rework

The total time to complete the inspection process from overview through follow-up for I_1 or I_2 inspections with four people involved takes about 90 to 100 people-hours for systems programming. Again, these figures may be considered conservative but they will serve as a starting point. Comparable figures for applications programming tend to be much lower, implying lower cost per K.NCSS.

Because the error detection efficiency of most inspection teams tends to dwindle after two hours of inspection but then picks up after a period of different activity, it is advisable to schedule inspection sessions of no more than two hours at a time. Two two-hour sessions per day are acceptable.

The time to do inspections and resulting rework must be scheduled and managed with the same attention as other important project activities. (After all, as is noted later, for one case at least, it is possible to find approximately two thirds of the errors reported during an inspection.) If this is not done, the immediate work pressure has a tendency to push the inspections and/or rework into the background, postponing them or avoiding them altogether. The result of this short-term respite will obviously have a much more dramatic long-term negative effect since the finding and fixing of errors is delayed until later in the process (and after turnover to the user). Usually, the result of postponing early error detection is a lengthening of the overall schedule and increased product cost.

Scheduling inspection time for modified code may be based on the algorithms in Table 3 *and on judgment.*

I_1 inspection process

Keeping the objective of each operation in the forefront of team activity is of paramount importance. Here is presented an outline of the I_1 inspection process operations.

1. *Overview* (whole team) — The designer first describes the overall area being addressed and then the specific area he has designed in detail — logic, paths, dependencies, etc. Documentation of design is distributed to all inspection participants on conclusion of the overview. (For an I_2 inspection, no overview is necessary, but the participants should remain the same. Preparation, inspection, and follow-up proceed as for I_1 but, of course, using code listings *and* design specifications as inspection materials. Also, at I_2 the moderator should flag for special scrutiny those areas that were reworked since I_1 errors were found *and other design changes* made.)

2. *Preparation* (individual) — Participants, using the design documentation, literally do their homework to try to understand the design, its intent and logic. (Sometimes flagrant errors are found during this operation, but in general, the number of errors found is not nearly as high as in the inspection operation.) To increase their error detection in the inspection, the inspection team should first study the ranked distributions of error types found by recent inspections. This study will prompt them to concentrate on the most fruitful areas. (See examples in Figures 3 and 4.) Checklists of clues on finding these errors should also be studied. (See partial examples of these lists in Figures 5 and 6 and complete examples for I_0 in Reference 1 and for I_1 and I_2 in Reference 7.)

3. *Inspection* (whole team) — A "reader" chosen by the moderator (usually the coder) describes how he will implement the design. He is expected to paraphrase the design as expressed by the designer. Every piece of logic is covered at least once, and every branch is taken at least once. All higher-level documentation, high-level design specifications, logic specifications, etc., and macro and control block listings at I_2 must be available and present during the inspection.

   Now that the design is understood, *the objective is to find errors*. (Note that an error is defined as any condition that causes malfunction or that precludes the attainment of expected or previously specified results. Thus, deviations from specifications are clearly termed errors.) The finding of errors is actually done during the implementor/coder's dis-

Figure 3 Summary of design inspections by error type

VP	Individual Name	Inspection file Missing	Wrong	Extra	Errors	Error %
CD	CB Definition	16	2		18	3.5 } 10.4
CU	CB Usage	18	17	1	36	6.9 } 10.4
FS	FPFS	1			1	.2
IC	Interconnect Calls	18	9		27	5.2
IR	Interconnect Reqts	4	5	2	11	2.1
LO	Logic	126	57	24	207	39.8 ⬅
L3	Higher Lvl Docu	1		1	2	.4
MA	Mod Attributes	1			1	.2
MD	More Detail	24	6	2	32	6.2
MN	Maintainability	8	5	3	16	3.1
OT	Other	15	10	10	35	6.7
PD	Pass Data Areas		1		1	.2
PE	Performance	1	2	3	6	1.2
PR	Prologue/Prose	44	38	7	89	17.1 ⬅
RM	Return Code/Msg	5	7	2	14	2.7
RU	Register Usage	1	2		3	.6
ST	Standards					
TB	Test & Branch	12	7	2	21	4.0
		295	168	57	520	100.0
		57%	32%	11%		

Figure 4 Summary of code inspections by error type

VP	Individual Name	Inspection file Missing	Wrong	Extra	Errors	Error %
CC	Code Comments	5	17	1	23	6.6
CU	CB Usage	3	21	1	25	7.2
DE	Design Error	31	32	14	77	22.1 ⬅
F1			8		8	2.3
IR	Interconnect Calls	7	9	3	19	5.5
LO	Logic	33	49	10	92	26.4 ⬅
MN	Maintainability	5	7	2	14	4.0
OT	Other					
PE	Performance	3	2	5	10	2.9
PR	Prologue/Prose	25	24	3	52	14.9 ⬅
PU	PL/S or BAL Use	4	9	1	14	4.0
RU	Register Usage	4	2		6	1.7
SU	Storage Usage	1			1	.3
TB	Test & Branch	2	5		7	2.0
		123	185	40	348	100.0

course. Questions raised are pursued only to the point at which an error is recognized. It is noted by the moderator: its type is classified; severity (major or minor) is identified, and the inspection is continued. Often the solution of a problem is obvious. If so, it is noted, but no specific solution hunting is

Figure 5 Examples of what to examine when looking for errors at I_1

I_1 Logic

Missing

1. Are All Constants Defined?
2. Are All Unique Values Explicitly Tested on Input Parameters?
3. Are Values Stored after They Are Calculated?
4. Are All Defaults Checked Explicitly Tested on Input Parameters?
5. If Character Strings Are Created Are They Complete, Are All Delimiters Shown?
6. If a Keyword Has Many Unique Values, Are They All Checked?
7. If a Queue Is Being Manipulated, Can the Execution Be Interrupted; If So, Is Queue Protected by a Locking Structure; Can Queue Be Destroyed Over an Interrupt?
8. Are Registers Being Restored on Exits?
9. In Queuing/Dequeuing Should Any Value Be Decremented/Incremented?
10. Are All Keywords Tested in Macro?
11. Are All Keyword Related Parameters Tested in Service Routine?
12. Are Queues Being Held in Isolation So That Subsequent Interrupting Requestors Are Receiving Spurious Returns Regarding the Held Queue?
13. Should any Registers Be Saved on Entry?
14. Are All Increment Counts Properly Initialized (0 or 1)?

Wrong

1. Are Absolutes Shown Where There Should Be Symbolics?
2. On Comparison of Two Bytes, Should All Bits Be Compared?
3. On Built Data Strings, Should They Be Character or Hex?
4. Are Internal Variables Unique or Confusing If Concatenated?

Extra

1. Are All Blocks Shown in Design Necessary or Are They Extraneous?

to take place during inspection. (The inspection is *not* intended to redesign, evaluate alternate design solutions, or to find solutions to errors; it is intended just to find errors!) A team is most effective if it operates with only one objective at a time.

Within one day of conclusion of the inspection, the moderator should produce a written report of the inspection and its findings to ensure that all issues raised in the inspection will be addressed in the rework and follow-up operations. Examples of these reports are given as Figures 7A, 7B, and 7C.

4. *Rework*—All errors or problems noted in the inspection report are resolved by the designer or coder/implementor.

5. *Follow-Up*—It is imperative that every issue, concern, and error be entirely resolved at this level, or errors that result can be 10 to 100 times more expensive to fix if found later in

Figure 6 Examples of what to examine when looking for errors at I_2

INSPECTION SPECIFICATION

I_2 *Test Branch*

Is Correct Condition Tested (If X = ON vs. IF X = OFF)?
Is (Are) Correct Variable(s) Used for Test
(If X = ON vs. If Y = ON)?
Are Null THENs/ELSEs Included as Appropriate?
Is Each Branch Target Correct?
Is the Most Frequently Exercised Test Leg the THEN Clause?

I_2 *Interconnection (or Linkage) Calls*

For Each Interconnection Call to Either a Macro, SVC or Another Module:
Are All Required Parameters Passed Set Correctly?
If Register Parameters Are Used, Is the Correct Register Number Specified?
If Interconnection Is a Macro,
Does the Inline Expansion Contain All Required Code?
No Register or Storage Conflicts between Macro and Calling Module?
If the Interconnection Returns, Do All Returned Parameters Get Processed Correctly?

the process (programmer time only, machine time not included). It is the responsibility of the moderator to see that all issues, problems, and concerns discovered in the inspection operation have been resolved by the designer in the case of I_1, or the coder/implementor for I_2 inspections. If more than five percent of the material has been reworked, the team should reconvene and carry out a 100 percent reinspection. Where less than five percent of the material has been reworked, the moderator at his discretion may verify the quality of the rework himself or reconvene the team to reinspect either the complete work or just the rework.

commencing inspections

In Operation 3 above, it is one thing to direct people to find errors in design or code. It is quite another problem for them to find errors. Numerous experiences have shown that people have to be taught or prompted to find errors effectively. Therefore, it is prudent to condition them to seek the high-occurrence, high-cost error types (see example in Figures 3 and 4), and then describe the clues that usually betray the presence of each error type (see examples in Figures 5 and 6).

One approach to getting started may be to make a preliminary inspection of a design or code that is felt to be representative of the program to be inspected. Obtain a suitable quantity of errors, and analyze them by type and origin, cause, and salient indicative clues. With this information, an inspection specification may be constructed. This specification can be amended and improved in

Figure 7A Error list

1. PR/M/MIN	Line 3:	the statement of the prologue in the REMARKS section needs expansion.
2. DA/W/MAJ	Line 123:	ERR–RECORD–TYPE is out of sequence.
3. PU/W/MAJ	Line 147:	the wrong bytes of an 8-byte field (current–data) are moved into the 2-byte field (this year).
4. LO/W/MAJ	Line 169:	while counting the number of leading spaces in NAME, the wrong variable (I) is used to calculate "J".
5. LO/W/MAJ	Line 172:	NAME–CHECK is PERFORMED one time too few.
6. PU/E/MIN	Line 175:	In NAME–CHECK, the check for SPACE is redundant.
7. DE/W/MIN	Line 175:	the design should allow for the occurrence of a period in a last name.

Figure 7B Example of module detail report

DATE__________

CODE INSPECTION REPORT

MODULE DETAIL

MOD/MAC: CHECKER SUBCOMPONENT/APPLICATION__________

SEE NOTE BELOW

PROBLEM TYPE:	MAJOR*			MINOR		
	M	W	E	M	W	E
LO: LOGIC		9			1	
TB: TEST AND BRANCH						
EL: EXTERNAL LINKAGES						
RU: REGISTER USAGE						
SU: STORAGE USAGE						
DA: DATA AREA USAGE		2				
PU: PROGRAM LANGUAGE		2				1
PE: PERFORMANCE						
MN: MAINTAINABILITY					1	
DE: DESIGN ERROR					1	
PR: PROLOGUE				1		
CC: CODE COMMENTS						
OT: OTHER						
TOTAL:	13			5		

REINSPECTION REQUIRED? Y

*A PROBLEM WHICH WOULD CAUSE THE PROGRAM TO MALFUNCTION: A BUG. M = MISSING, W = WRONG, E = EXTRA.
NOTE: FOR MODIFIED MODULES, PROBLEMS IN THE CHANGED PORTION VERSUS PROBLEMS IN THE BASE SHOULD BE SHOWN IN THIS MANNER: 3(2), WHERE 3 IS THE NUMBER OF PROBLEMS IN THE CHANGED PORTION AND 2 IS THE NUMBER OF PROBLEMS IN THE BASE.

light of new experience and serve as an on-going directive to focus the attention and conduct of inspection teams. The objective of an inspection specification is to help maximize and make more consistent the error detection efficiency of inspections where

Figure 7C Example of code inspection summary report

CODE INSPECTION REPORT
SUMMARY

Date 11/20/-

To: Design Manager KRAUSS Development Manager GIOTTI
Subject: Inspection Report for CHECKER Inspection date 11 19 -
System/Application ____ Release ____ Build ____
Component ____ Subcomponents(s) ____

					ELOC Added, Modified, Deleted									Inspection People-hours (X.X)				
					Pre-insp			Est Post			Rework							
Mod/Mac Name	New or Mod	Full or Part Insp.	Programmer	Tester	A	M	D	A	M	D	A	M	D	Prep	Insp Meetg	Re-work	Follow-up	Sub-component
	N		McGINLEY	HALE	348			400			50			9.0	8.8	8.0	1.5	
				Totals														

Reinspection required? YES Length of inspection (clock hours and tenths) 2.2
Reinspection by (date) 11/25/- Additional modules/macros? NO
DCR #'s written C-2
Problem summary: Major 13 Minor 5 Total 18
Errors in changed code: Major ____ Minor ____ Errors in base code: Major ____ Minor ____

LARSON		McGINLEY		HALE	
Initial Desr	Detailed Dr	Programmer	Team Leader	Other	Moderator's Signature

Error detection efficiency

$$= \frac{\text{Errors found by an inspection}}{\text{Total errors in the product before inspection}} \times 100$$

reporting inspection results

The reporting forms and form completion instructions shown in the Appendix may be used for I_1 and I_2 inspections. Although these forms were constructed for use in systems programming development, they may be used for applications programming development with minor modification to suit particular environments.

The moderator will make hand-written notes recording errors found during inspection meetings. He will categorize the errors and then transcribe counts of the errors, by type, to the module detail form. By maintaining cumulative totals of the counts by error type, and dividing by the number of projected executable source lines of code inspected to date, he will be able to establish installation averages within a short time.

Figures 7A, 7B, and 7C are an example of a set of code inspection reports. Figure 7A is a partial list of errors found in code inspection. Notice that errors are described in detail and are classified by error type, whether due to something being missing, wrong, or extra as the cause, and according to major or minor severity. Figure 7B is a module level summary of the errors contained in the entire error list represented by Figure 7A. The code inspection summary report in Figure 7C is a summary of inspection results obtained on all modules inspected in a particular inspection session or in a subcomponent or application.

inspections and languages

Inspections have been successfully applied to designs that are specified in English prose, flowcharts, HIPO, (Hierarchy plus Input-Process-Output) and PIDGEON (an English prose-like meta language).

The first code inspections were conducted on PL/S and Assembler. Now, prompting checklists for inspections of Assembler, COBOL, FORTRAN, and PL/1 code are available.[7]

personnel considerations

One of the most significant benefits of inspections is the detailed feedback of results on a relatively real-time basis. The programmer finds out what error types he is most prone to make and their quantity and how to find them. This feedback takes place within a few days of writing the program. Because he gets early indications from the first few units of his work inspected, he is able to show improvement, and usually does, on later work even during the same project. In this way, feedback of results from inspections must be counted for the programmer's use and benefit: *they should not under any circumstances be used for programmer performance appraisal.*

Skeptics may argue that once inspection results are obtained, they will or even must count in performance appraisals, or at least cause strong bias in the appraisal process. The author can offer in response that inspections have been conducted over the past three years involving diverse projects and locations, hundreds of experienced programmers and tens of managers, and so far he has found no case in which inspection results have been used negatively against programmers. Evidently no manager has tried to "kill the goose that lays the golden eggs."

A preinspection opinion of some programmers is that they do not see the value of inspections because they have managed very well up to now, or because their projects are too small or somehow different. This opinion usually changes after a few inspections to a position of acceptance. The quality of acceptance is related to the success of the inspections they have experienced, the *conduct of the trained moderator*, and the *attitude demonstrated by management*. The acceptance of inspections by programmers and managers as a beneficial step in making programs is well-established amongst those who have tried them.

Process control using inspection and testing results

Obviously, the range of analysis possible using inspection results is enormous. Therefore, only a few aspects will be treated here, and they are elementary expositions.

most error-prone modules

A listing of either I_1, I_2, or combined $I_1 + I_2$ data as in Figure 8 immediately highlights which modules contained the highest error density on inspection. If the error detection efficiency of each of the inspections was fairly constant, the ranking of error-prone modules holds. Thus if the error detection efficiency of inspection is 50 percent, and the inspection found 10 errors in a module, then it can be estimated that there are 10 errors remaining in the module. This information can prompt many actions to control the process. For instance, in Figure 8, it may be decided to reinspect module "Echo" or to redesign and recode it entirely. Or, less drastically, it may be decided to test it "harder" than other modules and look especially for errors of the type found in the inspections.

distribution of error types

If a ranked distribution of error types is obtained for a group of "error-prone modules" (Figure 9), which were produced from the same Process A, for example, it is a short step to comparing this distribution with a "Normal/Usual Percentage Distribution." Large disparities between the sample and "standard" will lead to questions on why Process A, say, yields nearly twice as many internal interconnection errors as the "standard" process. If this analysis is done promptly on the first five percent of production, it may be possible to remedy the problem (if it is a problem) on the remaining 95 percent of modules for a particular shipment. Provision can be made to test the first five percent of the modules to remove the unusually high incidence of internal interconnection problems.

inspecting error-prone code

Analysis of the testing results, commencing as soon as testing errors are evident, is a vital step in controlling the process since future testing can be guided by early results.

Where testing reveals excessively error-prone code, it may be more economical and saving of schedule to select the most error-prone code and inspect it before continuing testing. (The business case will likely differ from project to project and case to case, but in many instances inspection will be indicated). The selection of the most error-prone code may be made with two considerations uppermost:

Figure 8 Example of most error-prone modules based on I_1 and I_2

Module name	Number of errors	Lines of code	Error density, Errors/K. Loc
Echo	4	128	31
Zulu	10	323	31
Foxtrot	3	71	28
Alpha	7	264	27←Average
Lima	2	106	19 Error
Delta	3	195	15 Rate
.	.	.	.
.	.	.	.
	67		

Figure 9 Example of distribution of error types

	Number of errors	%	Normal/usual distribution, %
Logic	23	35	44
Interconnection/Linkage (Internal)	21	31 ?	18
Control Blocks	6	9	13
–	.	8	10
–	.	7	7
–	.	6	6
–	.	4	2
		100%	100%

1. Which modules head a ranked list when the modules are rated by test errors per K.NCSS?
2. In the parts of the program in which test coverage is low, which modules or parts of modules are most suspect based on $(I_1 + I_2)$ errors per K.NCSS and programmer judgment?

From a condensed table of ranked "most error-prone" modules, a selection of modules to be inspected (or reinspected) may be made. Knowledge of the error types already found in these modules will better prepare an inspection team.

The reinspection itself should conform with the I_2 process, except that an overview may be necessary if the original overview was held too long ago or if new project members are involved.

Inspections and walk-throughs

Walk-throughs (or walk-thrus) are practiced in many different ways in different places, with varying regularity and thoroughness. This inconsistency causes the results of walk-throughs to vary widely and to be nonrepeatable. Inspections, however, having an established process and a formal procedure, tend to vary less and produce more repeatable results. Because of the variation in walk-throughs, a comparison between them and inspections is not simple. However, from Reference 8 and the walk-through procedures witnessed by the author and described to him by walk-through participants, as well as the inspection process described previously and in References 1 and 9, the comparison in Tables 4 and 5 is drawn.

effects on development process

Figure 10A describes the process in which a walk-through is applied. Clearly, the purging of errors from the product as it passes through the walk-through between Operations 1 and 2 is very beneficial to the product. In Figure 10B, the inspection process (and its feedback, feed-forward, and self-improvement) replaces the walk-through. The notes on the figure are self-explanatory.

Inspections are also an excellent means of measuring completeness of work against the exit criteria which must be satisfied to complete project checkpoints. (Each checkpoint should have a clearly defined set of exit criteria. Without exit criteria, a checkpoint is too negotiable to be useful for process control).

Inspections and process management

The most marked effects of inspections on the development process is to change the old adage that, "design is not complete until testing is completed," to a position where a very great deal must be known about the design before even the coding is begun. Although great discretion is still required in code implementation, more predictability and improvements in schedule, cost, and quality accrue. The old adage still holds true if one regards inspection as much a means of verification as testing.

percent of errors found

Observations in one case in systems programming show that approximately two thirds of all errors reported during development are found by I_1 and I_2 inspections prior to machine testing.

Table 4. Inspection and walk-through processes and objectives

Inspection		Walk-through	
Process Operations	*Objectives*	*Process Operations*	*Objectives*
1. Overview	Education (Group)	–	–
2. Preparation	Education (Individual)	1. Preparation	Education (Individual)
3. Inspection	Find errors! (Group)	2. Walk-through	Education (Group) Discuss design alternatives Find errors
4. Rework	Fix problems	–	
5. Follow-up	Ensure all fixes correctly installed	–	

Note the separation of objectives in the inspection process.

Table 5 Comparison of key properties of inspections and walk-throughs

Properties	*Inspection*	*Walk-Through*
1. Formal moderator training	Yes	No
2. Definite participant roles	Yes	No
3. Who "drives" the inspection or walk-through	Moderator	Owner of material (Designer or coder)
4. Use "How To Find Errors" checklists	Yes	No
5. Use distribution of error types to look for	Yes	No
6. Follow-up to reduce bad fixes	Yes	No
7. Less future errors because of detailed error feedback to individual programmer	Yes	Incidental
8. Improve inspection efficiency from analysis of results	Yes	No
9. Analysis of data → process problems → improvements	Yes	No

The error detection efficiencies of the I_1 and I_2 inspections separately are, of course, less than 66 percent. A similar observation of an application program development indicated an 82 percent find (Table 1). As more is learned and the error detection efficiency of inspection is increased, the burden of debugging on testing operations will be reduced, and testing will be more able to fulfill its prime objective of verifying quality.

Figure 10 (A) Walk-through process, (B) Inspection process

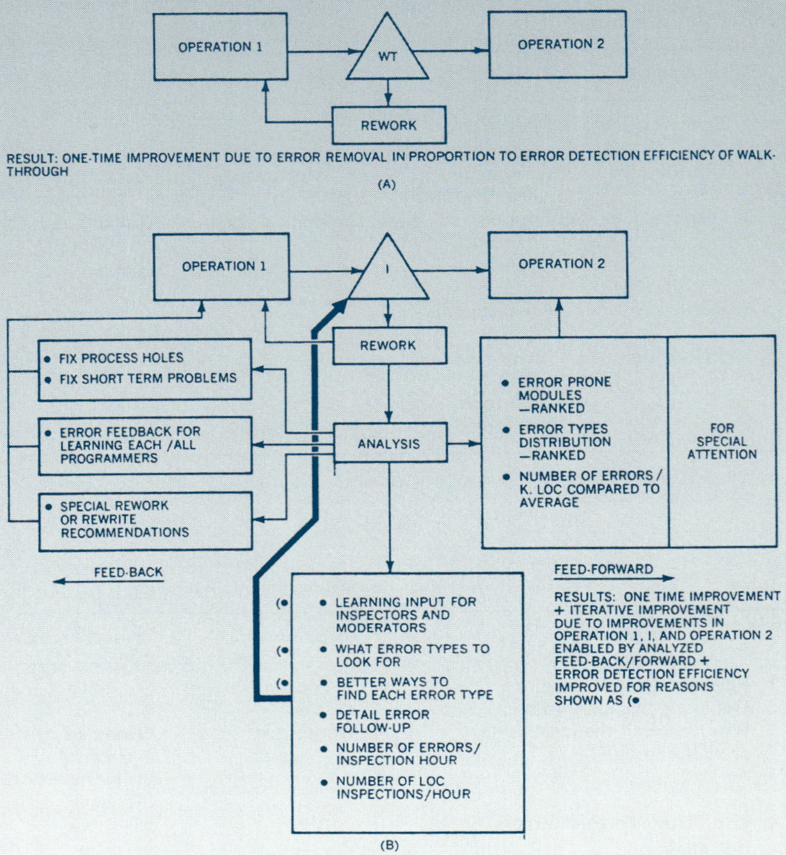

effect on cost and schedule

Comparing the "old" and "new" (with inspections) approaches to process management in Figure 11, we can see clearly that with the use of inspection results, error rework (which is a very significant variable in product cost) tends to be managed more during the first half of the schedule. This results in much lower cost than in the "old" approach, where the cost of error rework was 10 to 100 times higher and was accomplished in large part during the last half of the schedule.

process tracking

Inserting the I_1 and I_2 checkpoints in the development process enables assessment of project completeness and quality to be made early in the process (during the first half of the project instead of the latter half of the schedule, when recovery may be impossible without adjustments in schedule and cost). Since individually trackable modules of reasonably well-known size can

Figure 11 Effect of inspection on process management

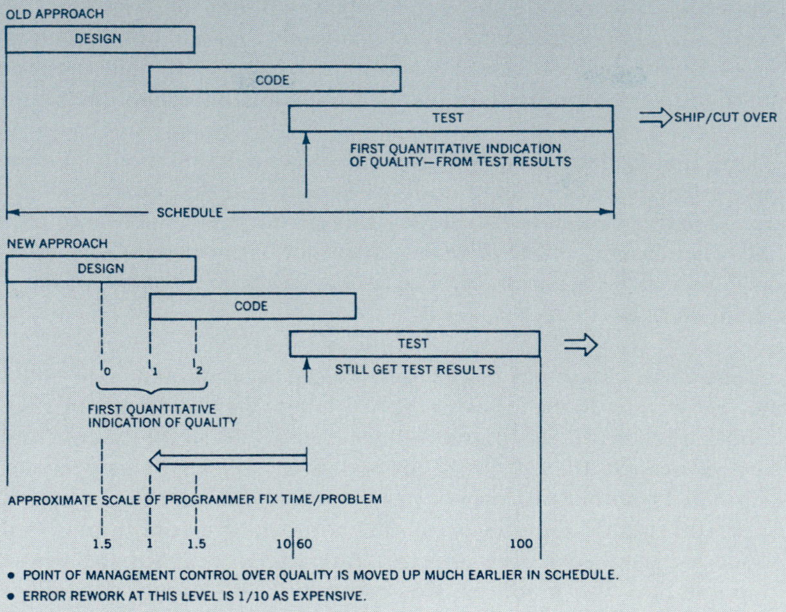

be counted as they pass through each of these checkpoints, the percentage completion of the project against schedule can be continuously and easily tracked.

effect on product knowledge

The overview, preparation, and inspection sequence of the operations of the inspection process give the inspection participants a high degree of product knowledge in a very short time. This important side benefit results in the participants being able to handle later development and testing with more certainty and less false starts. Naturally, this also contributes to productivity improvement.

An interesting sidelight is that because designers are asked at pre-I_1 inspection time for estimates of the number of lines of code (NCSS) that their designs will create, and they are present to count for themselves the actual lines of code at the I_2 inspection, the accuracy of design estimates has shown substantial improvement.

For this reason, an inspection is frequently a required event where responsibility for design or code is being transferred from one programmer to another. The complete inspection team is convened for such an inspection. (One-on-one reviews such as desk debugging are certainly worthwhile but do not approach the effectiveness of formal inspection.) Usually the side benefit of finding errors more than justifies the transfer inspection.

inspecting modified code

Code that is changed in, or inserted in, an existing module either in replacement of deleted code or simply inserted in the module is considered modified code. By this definition, a very large part of programming effort is devoted to modifying code. (The addition of entirely new modules to a system count as new, not modified, code.)

Some observations of errors per K.NCSS of modified code show its error rate to be considerably higher than is found in new code; (i.e., if 10.NCSS are replaced in a 100.NCSS module and errors against the 10.NCSS are counted, the error rate is described as number of errors per 10.NCSS, not number of errors per 100.NCSS). Obviously, if the number of errors in modified code are used to derive an error rate per K.NCSS for the whole module that was modified, this rate would be largely dependent upon the percentage of the module that is modified: this would provide a meaningless ratio. A useful measure is the number of errors per K.NCSS (modified) in which the higher error rates have been observed.

Since most modifications are small (e.g., 1 to 25 instructions), they are often erroneously regarded as trivially simple and are handled accordingly; the error rate goes up, and control is lost. In the author's experience, *all* modifications are well worth inspecting from an economic and a quality standpoint. A convenient method of handling changes is to group them to a module or set of modules and convene the inspection team to inspect as many changes as possible. But all changes must be inspected!

Inspections of modifications can range from inspecting the modified instructions and the surrounding instructions connecting it with its host module, to an inspection of the entire module. The choice of extent of inspection coverage is dependent upon the percentage of modification, pervasiveness of the modification, etc.

bad fixes

A very serious problem is the inclusion in the product of bad fixes. Human tendency is to consider the "fix," or correction, to a problem to be error-free itself. Unfortunately, this is all too frequently untrue in the case of fixes to errors found by inspections and by testing. The inspection process clearly has an operation called Follow-Up to try and minimize the bad-fix problem, but the fix process of testing errors very rarely requires scrutiny of fix quality before the fix is inserted. Then, if the fix is bad, the whole elaborate process of going from source fix to link edit, to test the fix, to regression test must be repeated at needlessly high cost. The number of bad fixes can be economically reduced by some simple inspection after clean compilation of the fix.

Summary

We can summarize the discussion of design and code inspections and process control in developing programs as follows:

1. Describe the program development process in terms of operations, and define exit criteria which must be satisfied for completion of each operation.
2. Separate the objectives of the inspection process operations to keep the inspection team focused on one objective at a time:

Operation	*Objective*
Overview	Communications/education
Preparation	Education
Inspection	Find errors
Rework	Fix errors
Follow-up	Ensure all fixes are applied correctly

3. Classify errors by type, and rank frequency of occurrence of types. Identify *which types* to spend most time looking for in the inspection.
4. Describe *how* to look for presence of error types.
5. Analyze inspection results and use for constant process improvement (until process averages are reached and then use for process control).

Some applications of inspections include function level inspections I_0, design-complete inspections I_1, code inspections I_2, test plan inspections IT_1, test case inspections IT_2, interconnections inspections IF, inspection of fixes/changes, inspection of publications, etc., and post testing inspection. Inspections can be applied to the development of system control programs, applications programs, and microcode in hardware.

We can conclude from experience that inspections increase productivity and improve final program quality. Furthermore, improvements in process control and project management are enabled by inspections.

ACKNOWLEDGMENTS

The author acknowledges, with thanks, the work of Mr. O. R. Kohli and Mr. R. A. Radice, who made considerable contributions in the development of inspection techniques applied to program design and code, and Mr. R. R. Larson, who adapted inspections to program testing.

CITED REFERENCES AND FOOTNOTES

1. O. R. Kohli, *High-Level Design Inspection Specification*, Technical Report TR 21.601, IBM Corporation, Kingston, New York (July 21, 1975).
2. It should be noted that the exit criteria for I_1 (design complete where one design statement is estimated to represent 3 to 10 code instructions) and I_2 (first clean code compilations) are checkpoints in the development process through which every programming project must pass.
3. The Hawthorne Effect is a psychological phenomenon usually experienced in human-involved productivity studies. The effect is manifested by participants producing above normal because they know they are being studied.
4. NCSS (Non-Commentary Source Statements), also referred to as "Lines of Code," are the sum of executable code instructions and declaratives. Instructions that invoke macros are counted once only. Expanded macroinstructions are also counted only once. Comments are not included.
5. Basically in a walk-through, program design or code is reviewed by a group of people gathered together at a structured meeting in which errors/issues pertaining to the material and proposed by the participants may be discussed in an effort to find errors. The group may consist of various participants but always includes the originator of the material being reviewed who usually plans the meeting and is responsible for correcting the errors. How it differs from an inspection is pointed out in Tables 2 and 3.
6. *Marketing Newsletter*, Cross Application Systems Marketing, "Program inspections at Aetna," MS-76-006, S2, IBM Corporation, Data Processing Division, White Plains, New York (March 29, 1976).
7. J. Ascoly, M. J. Cafferty, S. J. Gruen, and O. R. Kohli, *Code Inspection Specification*, Technical Report TR 21.630, IBM Corporation, Kingston, New York (1976).
8. N. S. Waldstein, *The Walk-Thru—A Method of Specification, Design and Review*, Technical Report TR 00.2536, IBM Corporation, Poughkeepsie, New York (June 4, 1974).
9. Independent study programs: *IBM Structured Programming Textbook*, SR20-7149-1, *IBM Structured Programming Workbook*, SR20-7150-0, IBM Corporation, Data Processing Division, White Plains, New York.

GENERAL REFERENCES

1. J. D. Aron, *The Program Development Process: Part 1: The Individual Programmer*, Structured Programs, 137–141, Addison-Wesley Publishing Co., Reading, Massachusetts (1974).

2. M. E. Fagan, *Design and Code Inspections and Process Control in the Development of Programs*, Technical Report TR 00.2763, IBM Corporation, Poughkeepsie, New York (June 10, 1976). This report is a revision of the author's *Design and Code Inspections and Process Control in the Development of Programs*, Technical Report TR 21.572, IBM Corporation, Kingston, New York (December 17, 1974).

3. O. R. Kohli and R. A. Radice, *Low-Level Design Inspection Specification*, Technical Report TR 21.629. IBM Corporation, Kingston, New York (1976).

4. R. R. Larson, *Test Plan and Test Case Inspection Specifications*, Technical Report TR 21.586, IBM Corporation, Kingston, New York (April 4, 1975).

Appendix: Reporting forms and form completion instructions

Instructions for Completing Design Inspection Module Detail Form

This form (Figure 12) should be completed for each module/macro that has valid problems against it. The problem-type information gathered in this report is important because a history of problem-type experience points out high-occurrence types. This knowledge can then be conveyed to inspectors so that they can concentrate on seeking the higher-occurrence types of problems.

Figure 12 Design inspection module detail form

DATE____________

DETAILED DESIGN INSPECTION REPORT

MODULE DETAIL

MOD/MAC:____________ SUBCOMPONENT/APPLICATION____________

SEE NOTE BELOW

	MAJOR*			MINOR		
PROBLEM TYPE:	M	W	E	M	W	E
LO: LOGIC						
TB: TEST AND BRANCH						
DA: DATA AREA USAGE						
RM: RETURN CODES/MESSAGES						
RU: REGISTER USAGE						
MA: MODULE ATTRIBUTES						
EL: EXTERNAL LINKAGES						
MD: MORE DETAIL						
ST: STANDARDS						
PR: PROLOGUE OR PROSE						
HL: HIGHER LEVEL DESIGN DOC.						
US: USER SPEC.						
MN: MAINTAINABILITY						
PE: PERFORMANCE						
OT: OTHER						
TOTAL:						

REINSPECTION REQUIRED?____________

*A PROBLEM WHICH WOULD CAUSE THE PROGRAM TO MALFUNCTION: A BUG. M = MISSING, W = WRONG, E = EXTRA.

NOTE: FOR MODIFIED MODULES, PROBLEMS IN THE CHANGED PORTION VERSUS PROBLEMS IN THE BASE SHOULD BE SHOWN IN THIS MANNER: 3(2), WHERE 3 IS THE NUMBER OF PROBLEMS IN THE CHANGED PORTION AND 2 IS THE NUMBER OF PROBLEMS IN THE BASE.

1. MOD/MAC: The module or macro name.
2. SUBCOMPONENT: The associated subcomponent.
3. PROBLEM TYPE: Summarize the number of problems by type (logic, etc.), severity (major/minor), and by category (missing, wrong, or extra). For modified modules, detail the number of problems in the changed design versus the number in the base design. (Problem types were developed in a systems programming environment. Appropriate changes, if desired, could be made for application development.)
4. REINSPECTION REQUIRED?: Indicate whether the module/macro requires a reinspection.

All valid problems found in the inspection should be listed and attached to the report. A brief description of each problem, its error type, and the rework time to fix it should be given (see Figure 7A, which describes errors in similar detail to that required but is at a coding level).

Instructions for Completing Design Inspection Summary Form

Following are detailed instructions for completing the form in Figure 13.

1. TO: The report is addressed to the respective design and development managers.
2. SUBJECT: The unit being inspected is identified.
3. MOD/MAC NAME: The name of each module and macro as it resides on the source library.

Figure 13 Design inspection summary form

DESIGN INSPECTION REPORT
SUMMARY

Date ________

To: Design Manager ________ Development Manager ________
Subject: Inspection Report for ________ Inspection date ________
System/Application ________ Release ________ Build ________
Component ________ Subcomponents(s) ________

Mod/Mac Name	New or Mod	Full or Part Insp.	Detailed Designer	Programmer	ELOC Added, Modified. Deleted									Inspection People-hours (X.X.)				Sub-component
					Est. Pre			Est. Post			Rework			Over-view & Prep.	Insp Meetg	Re-work	Follow-up	
					A	M	D	A	M	D	A	M	D					
				Totals														

Reinspection required? ________ Length of inspection (clock hours and tenths) ________
Reinspection by (date) ________ Additional modules/macros? ________
DCR #'s written ________
Problem summary: Major ________ Minor ________ Total ________
Errors in changed code: Major ____ Minor ____ Errors in base code: Major ____ Minor ____

Initial Desr | Detailed Dr | Programmer | Team Leader | Other | Moderator's Signature

4. NEW OR MOD: "N" if the module is new; "M" if the module is modified.
5. FULL OR PART INSP: If the module/macro is "modified," indicate "F" if the module/macro was fully inspected or "P" if partially inspected.
6. DETAILED DESIGNER: and PROGRAMMER: Identification of originators.
7. PRE-INSP EST ELOC: The estimated executable source lines of code (added, modified, deleted). Estimate made prior to the inspection by the designer.
8. POST-INSP EST ELOC: The estimated executable source lines of code. Estimate made after the inspection.
9. REWORK ELOC: The estimated executable source lines of code in rework as a result of the inspection.
10. OVERVIEW AND PREP: The number of people-hours (in tenths of hours) spent in preparing for the overview, in the overview meeting itself, and in preparing for the inspection meeting.
11. INSPECTION MEETING: The number of people-hours spent on the inspection meeting.
12. REWORK: The estimated number of people-hours spent to fix the problems found during the inspection.
13. FOLLOW-UP: The estimated number of people-hours spent by the moderator (and others if necessary) in verifying the correctness of changes made by the author as a result of the inspection.
14. SUBCOMPONENT: The subcomponent of which the module/macro is a part.
15. REINSPECTION REQUIRED?: Yes or no.
16. LENGTH OF INSPECTION: Clock hours spent in the inspection meeting.
17. REINSPECTION BY (DATE): Latest acceptable date for reinspection.
18. ADDITIONAL MODULES/MACROS?: For these subcomponents, are additional modules/macros yet to be inspected?
19. DCR #'S WRITTEN: The identification of Design Change Requests, DCR(s), written to cover problems in rework.
20. PROBLEM SUMMARY: Totals taken from Module Detail forms(s).
21. INITIAL DESIGNER, DETAILED DESIGNER, etc.: Identification of members of the inspection team.

Instructions for Completing Code Inspection Module Detail Form

This form (Figure 14) should be completed according to the instructions for completing the design inspection module detail form.

Instructions for Completing Code Inspection Summary Form

This form (Figure 15) should be completed according to the instructions for the design inspection summary form except for the following items.

1. PROGRAMMER AND TESTER: Identifications of original participants involved with code.
2. PRE-INSP. ELOC: The noncommentary source lines of code (added, modified, deleted). Count made prior to the inspection by the programmer.
3. POST-INSP EST ELOC: The estimated noncommentary source lines of code. Estimate made after the inspection.

Figure 14 Code inspection module detail form

DATE__________

CODE INSPECTION REPORT

MODULE DETAIL

MOD/MAC:__________ SUBCOMPONENT/APPLICATION__________

SEE NOTE BELOW

	MAJOR*			MINOR		
PROBLEM TYPE:	M	W	E	M	W	E
LO: LOGIC						
TB: TEST AND BRANCH						
EL: EXTERNAL LINKAGES						
RU: REGISTER USAGE						
SU: STORAGE USAGE						
DA: DATA AREA USAGE						
PU: PROGRAM LANGUAGE						
PE: PERFORMANCE						
MN: MAINTAINABILITY						
DE: DESIGN ERROR						
PR: PROLOGUE						
CC: CODE COMMENTS						
OT: OTHER						
TOTAL:						

REINSPECTION REQUIRED?__________

*A PROBLEM WHICH WOULD CAUSE THE PROGRAM TO MALFUNCTION: A BUG. M = MISSING, W = WRONG, E = EXTRA.

NOTE: FOR MODIFIED MODULES, PROBLEMS IN THE CHANGED PORTION VERSUS PROBLEMS IN THE BASE SHOULD BE SHOWN IN THIS MANNER: 3(2). WHERE 3 IS THE NUMBER OF PROBLEMS IN THE CHANGED PORTION AND 2 IS THE NUMBER OF PROBLEMS IN THE BASE.

Figure 15 Code inspection summary form

CODE INSPECTION REPORT
SUMMARY

Date______________

To: Design Manager____________________ Development Manager____________________

Subject: Inspection Report for____________________ Inspection date____________________

System/Application____________________ Release______________ Build__________

Component____________________ Subcomponents(s)____________________

					ELOC Added, Modified, Deleted									Inspection People-hours (X.X)				
					Pre-insp			Est Post			Rework							
Mod/Mac Name	New or Mod	Full or Part Insp.	Programmer	Tester	A	M	D	A	M	D	A	M	D	Prep	Insp Meetg	Re-work	Follow-up	Sub-component
				Totals														

Reinspection required?__________ Length of inspection (clock hours and tenths)__________

Reinspection by (date)__________ Additional modules/macros?__________

DCR #'s written__________

Problem summary: Major__________ Minor__________ Total__________

Errors in changed code: Major_____ Minor_____ Errors in base code: Major_____ Minor_____

Initial Desr | Detailed Dr | Programmer | Team Leader | Other | Moderator's Signature

4. REWORK ELOC: The estimated noncommentary source lines of code in rework as a result of the inspection.
5. PREP: The number of people hours (in tenths of hours) spent in preparing for the inspection meeting.

Reprint Order No. G321-5033.

Michael Fagan

Advances in Software Inspections

IEEE Transactions on Software Engineering,
Vol. SE-12 (7), 1986
pp. 744-751

Advances in Software Inspections

MICHAEL E. FAGAN, MEMBER, IEEE

Manuscript received September 30, 1985.
The author is with the IBM Thomas J. Watson Research Center, Yorktown Heights, NY 10598.
IEEE Log Number 8608192.

***Abstract*—This paper presents new studies and experiences that enhance the use of the inspection process and improve its contribution to development of defect-free software on time and at lower costs. Examples of benefits are cited followed by descriptions of the process and some methods of obtaining the enhanced results.**

Software inspection is a method of static testing to verify that software meets its requirements. It engages the developers and others in a formal process of investigation that usually detects more defects in the product—and at lower cost—than does machine testing. Users of the method report very significant improvements in quality that are accompanied by lower development costs and greatly reduced maintenance efforts. Excellent results have been obtained by small and large organizations in all aspects of new development as well as in maintenance. There is some evidence that developers who participate in the inspection of their own product actually create fewer defects in future work. Because inspections formalize the development process, productivity and quality enhancing tools can be adopted more easily and rapidly.

***Index Terms*—Defect detection, inspection, project management, quality assurance, software development, software engineering, software quality, testing, walkthru.**

Introduction

THE software inspection process was created in 1972, in IBM Kingston, NY, for the dual purposes of improving software quality and increasing programmer productivity. Its accelerating rate of adoption throughout the software development and maintenance industry is an acknowledgment of its effectiveness in meeting its goals. Outlined in this paper are some enhancements to the inspection process, and the experiences of some of the many companies and organizations that have contributed to its evolution. The author is indebted to and thanks the many people who have given their help so liberally.

Because of the clear structure the inspection process has brought to the development process, it has enabled study of both itself and the conduct of development. The latter has enabled process control to be applied from the point at which the requirements are inspected—a much earlier point in the process than ever before—and throughout development. Inspections provide data on the performance of individual development operations, thus providing a unique opportunity to evaluate new tools and techniques. At the same time, studies of inspections have isolated and fostered improvement of its key characteristics such that very high defect detection efficiency inspections may now be conducted *routinely*. This simultaneous study of development and design and code inspections prompted the adaptation of the principles of the inspection process to inspections of requirements, user information, and documentation, and test plans and test cases. In each instance, the new uses of inspection were found to improve product quality and to be cost effective, i.e., it saved more than it cost. Thus, as the effectiveness of inspections are improving, they are being applied in many new and different ways to improve software quality and reduce costs.

Benefits: Defect Reduction, Defect Prevention, and Cost Improvement

In March 1984, while addressing the IBM SHARE User Group on software service, L. H. Fenton, IBM Director of VM Programming Systems, made an important statement on quality improvement due to inspections [1]:

> ''Our goal is to provide defect free products and product information, and we believe the best way to do this is by refining and enhancing our existing software development process.
>
> Since we introduced the inspection process in 1974, we have achieved significant improvements in quality. IBM has nearly doubled the number of lines of code shipped for System/370 software products since 1976, while the number of defects per thousand lines of code has been reduced by *two-thirds*. Feedback from early MVS/XA and VM/SP Release

3 users indicates these products met and, in many cases, exceeded our ever increasing quality expectations."

Observation of a small sample of programmers suggested that early experience gained from inspections caused programmers to reduce the number of defects that were injected in the design and code of programs created later during the same project [3]. Preliminary analysis of a much larger study of data from recent inspections is providing similar results.

It should be noted that the improvements reported by IBM were made while many of the enhancements to inspections that are mentioned here were being developed. As these improvements are incorporated into everyday practice, it is probable that inspections will help bring further reductions in defect injection and detection rates.

Additional reports showing that inspections improve quality *and* reduce costs follow. (In all these cases, the cost of inspections is included in project cost. Typically, all design and code inspection costs amount to 15 percent of project cost.)

AETNA Life and Casualty. 4439 LOC [2]	—0 Defects in use. —25 percent reduction in development resource.
IBM RESPOND, U.K. 6271 LOC [3]	—0 Defects in use. —9 percent reduction in cost compared to walkthrus.
Standard Bank of South Africa. 143 000 LOC [4]	—0.15 Defects/KLOC in use. —95 percent reduction in corrective maintenance cost.
American Express, System code). 13 000 LOC	—0.3 Defects in use.

In the AETNA and IBM examples, inspections found 82 and 93 percent, respectively, of all defects (that would cause malfunction) detected over the life cycle of the products. The other two cases each found over 50 percent of all defects by inspection. While the Standard Bank of South Africa and American Express were unable to use trained inspection moderators, and the former conducted only code inspections, both obtained outstanding results. The tremendous reduction in corrective maintenance at the Standard Bank of South Africa would also bring impressive savings in life cycle costs.

Naturally, reduction in maintenance allows redirection of programmers to work off the application backlog, which is reputed to contain at least two years of work at most locations. Impressive cost savings and quality improvements have been realized by inspecting test plans and then the test cases that implement those test plans. For a product of about 20 000 LOC, R. Larson [5] reported that test inspections resulted in:

- modification of approximately 30 percent of the functional matrices representing test coverage,
- detection of 176 major defects in the test plans and test cases (i.e., in 176 instances testing would have missed testing critical function or tested it incorrectly), and
- savings of more than 85 percent in programmer time by detecting the major defects by inspection as opposed to finding them during functional variation testing.

There are those who would use inspections whether or not they are cost justified for defect removal because of the nonquantifiable benefits the technique supplies toward improving the service provided to users and toward creating a more professional application development environment [6].

Experience has shown that inspections have the effect of slightly front-end loading the committment of people resources in development, adding to requirements and design, while greatly reducing the effort required during testing and for rework of design and code. The result is an overall *net* reduction in development resource, and usually in schedule too. Fig. 1 is a pictorial description of the familiar ''snail'' shaped curve of software development resource versus the time schedule including and without inspections.

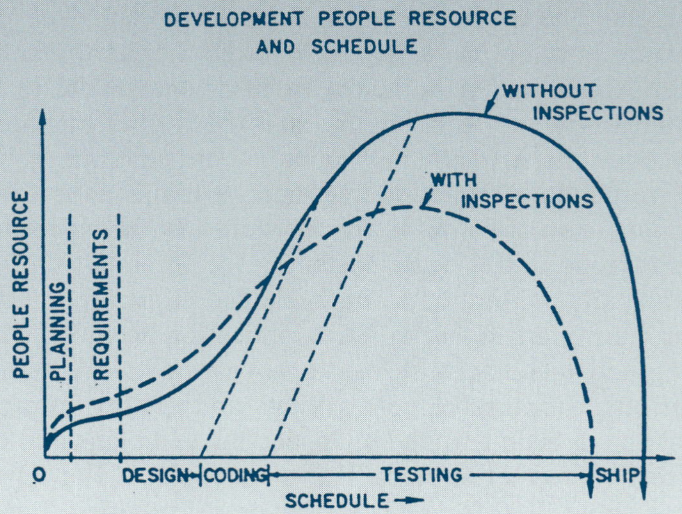

Fig. 1.

The Software Quality Problem

The software quality problem is the result of defects in code and documentation causing failure to satisfy user requirements. It also impedes the growth of the information processing industry. Validity of this statement is attested to by three of the many pieces of supporting evidence:

- The SHARE User Group Software Service Task Force Report, 1983 [1], that recommended an order of magnitude improvement in software quality over the next several years, with a like reduction in service. (Other manufacturers report similar recommendations from their users.)
- In 1979, 12 percent of programmer resource was consumed in post-shipment corrective maintenance alone and this figure was growing [8]. (Note that there is also a significant percentage of development and enhancement maintenance resource devoted to correcting defects. This is probably larger than the 12 percent expended in corrective maintenance, but there is no substantiating research.)
- The formal backlog of data processing tasks most quoted is three years [7].

At this point, a very important definition is in order:

A defect is an instance in which a requirement is not satisfied.

Here, it must be recognized that a requirement is any agreed upon commitment. It is not only the recognizable external product requirement, but can also include internal development requirements (e.g., the exit criteria of an operation) that must be met in order to satisfy the requirements of the end product. Examples of this would be the requirement that a test plan completely verifies that the product meets the agreed upon needs of the user, or that the code of a program must be complete before it is submitted to be tested.

While defects become manifest in the end product documentation or code, most of them are actually injected as the functional aspects of the product and its quality attributes are being created; during development of the requirements, the design and coding, or by insertion of changes. The author's research supports and supplements that of B. Boehm *et al.* [9] and indicates that there are eight attributes that must be considered when describing quality in a software product:

- intrinsic code quality,
- freedom from problems in operation,
- usability,
- installability,
- documentation for intended users,
- portability,
- maintainability and extendability, and "fitness for use"—that implicit conventional user needs are satisfied.

Inspections and the Software Quality Problem

Previously, each of these attributes of software quality were evaluated by testing and the end user. Now, some of them are being partly, and others entirely, verified against requirements by inspection. In fact, the product requirements themselves are often inspected to ascertain whether they meet user needs. In order to eliminate defects from the product it is necessary to address their prevention, or detection and resolution as soon as possible after their injection during development and maintenance. Prevention is the most desirable course to follow, and it is approached in many ways including the use of state machine representation of design, systematic programming, proof of correctness, process control, development standards, prototyping, and other methods. Defect detection, on the other hand, was once almost totally dependent upon testing during development and by the user. This has changed, and over the past decade walkthrus and inspections have assumed a large part of the defect detection burden; inspections finding from 60 to 90 percent defects. (See [2], [3], and other unpublished product experiences.) They are performed much nearer the point of injection of the defects than is testing, using less resource for rework and, thus, more than paying for themselves. In fact, inspections have been applied to most phases of development to verify that the key software attributes are present immediately after the point at which they should first be introduced into the product. They are also applied to test plans and test cases to improve the defect detection efficiency of testing. Thus, inspections have been instrumental in improving all aspects of software product quality, as well as the quality of logic design and code. In fact, inspections *supplement* defect prevention methods in improving quality.

Essential to the quality of inspection (or its defect detection efficiency) is proper definition of the development process. And, inspection quality is a direct contributor to product quality, as will be shown later.

Definition of the Development Process

The software development process is a series of operations so arranged that its execution will deliver the desired end product. Typically, these operations are: Requirements Definition, System Design, High Level Design, Low Level Design, Coding, Unit Testing, Component or Function Testing, System Testing, and then user support and Maintenance. In practice, some of these operations are repeated as the product is recycled through them to insert functional changes and fixes.

The attributes of software quality are invested along with the functional characteristics of the product during the early operations, when the cost to remedy defects is 10–100 times less than it would be during testing or maintenance [2]. Consequently, it is advantageous to find and correct defects as near to their point of origin as possible. This is accomplished by inspecting the output product of each operation to verify that it satisfies the output requirements or *exit criteria* of the operation. In most cases, these exit criteria are not specified with sufficient precision to allow go/no verification. Specification of exit criteria in unambiguous terms that are objective and preferably quantitative is an essential characteristic of any well defined process. Exit criteria are the standard against which inspections measure completion of the product at the end of an operation, and verify the presence or absence of quality attributes. (A deviation from exit criteria is a defect.)

Shown below are the essence of 4 key criteria taken from the full set of 15 exit criteria items for the Coding operation:

- The source code must be at the "first clean compilation" level. That means it must be properly compiled and be free of syntax errors.
- The code must accurately implement the low level design (which was the verified output of the preceding process operation).
- All design changes to date are included in the code.
- All rework resulting from the code inspection has been included and verified.

The code inspection, I2, must verify that all 15 of these exit criteria have been satisfied before a module or other entity of the product is considered to have completed the Coding operation. Explicit exit criteria for several of the other inspection types in use will be contained in the author's book in software inspections. However, there is no reason why a particular project could not define its own sets of exit criteria. What is important is that exit criteria should be as objective as possible, so as to be repeatable; they should completely describe what is required to exit each operation; and, *they must be observed by all those involved*.

The objective of process control is to measure completion of the product during stages of its development, to compare the measurement against the project plan, and then to remedy any deviations from plan. In this context, the quality of both exit criteria and inspections are of vital importance. And, they must both be properly described in the manageable development process, for such a process must be controllable by definition.

Development is often considered a subset of the maintenance process. Therefore, the maintenance process must be treated in the same manner to make it equally manageable.

Software Inspection Overview

This paper will only give an overview description of the inspection process that is sufficient to enable discussion of updates and enhancements. The author's original paper on the software inspections process [2] gives a brief description of the inspection process and what goes on in an inspection, and is the base to which the enhancements are added. His forthcoming companion books on this subject and on building defect-free software will provide an implementation level description and will include all the points addressed in this paper and more.

To convey the principles of software inspections, it is only really necessary to understand how they apply to design and code. A good grasp on this application allows tailoring of the process to enable inspection of virtually any operation in development or maintenance, and also allows inspection for any desired quality attribute. With this in mind, the main points of inspections will be exposed through discussing how they apply in design and code inspections.

There are three essential requirements for the implementation of inspections:

- definition of the DEVELOPMENT PROCESS in terms of operations and their EXIT CRITERIA,
- proper DESCRIPTION of the INSPECTION PROCESS, and

• CORRECT EXECUTION of the INSPECTION PROCESS. (Yes, correct execution of the process is vital.)

THE INSPECTION PROCESS

The inspection process follows any development operation whose product must be verified. As shown below, it consists of six operations, each with a specific objective:

Operation	Objectives
PLANNING	Materials to be inspected must meet inspection entry criteria.
	Arrange the availability of the right participants.
	Arrange suitable meeting place and time.
OVERVIEW	Group education of participants in what is to be inspected.
	Assign inspection roles to participants.
PREPARATION	Participants learn the material and prepare to fulfill their assigned roles.
INSPECTION	*Find defects.* (Solution hunting and discussion of design alternatives is discouraged.)
REWORK	The author reworks all defects.
FOLLOW-UP	Verification by the inspection moderator or the entire inspection team to assure that all fixes are effective and that no secondary defects have been introduced.

Evaluation of hundreds of inspections involving thousands of programmers in which alternatives to the above steps have been tried has shown that all these operations are really necessary. Omitting or combining operations has led to degraded inspection efficiency that outweighs the apparent short-term benefits. OVERVIEW is the only operation that under certain conditions can be omitted with slight risk. Even FOLLOW-UP is justified as study has shown that approximately one of every six fixes are themselves incorrect, or create other defects.

From observing scores of inspections, it is evident that participation in inspection teams is extremely taxing and should be limited to periods of 2 hours. Continuing beyond 2 hours, the defect detection ability of the team seems to diminish, but is restored after a break of 2 hours or so during which other work may be done. Accordingly, no more than two 2 hour sessions of inspection per day are recommended.

To assist the inspectors in finding defects, for not all inspectors start off being good detectives, a checklist of defect types is created to help them identify defects appropriate to the exit criteria of each operation whose product is to be inspected. It also serves as a guide to classification of defects found by inspection prior to their entry to the inspection and test defect data base of the project. (A database containing these and other data is necessary for quality control of development.)

People and Inspections

Inspection participants are usually programmers who are drawn from the project involved. The roles they play for design and code inspections are those of the *Author* (Designer or Coder), *Reader* (who paraphrases the design or code as if they will implement it), *Tester* (who views the product from the testing standpoint), and *Moderator*. These roles are described more fully in [2], but that level of detail is not required here. Some inspections types, for instance those of system structure, may require more participants, but it is advantageous to keep the number of people to a minimum. Involving the end users in those inspections in which they can truly participate is also very helpful.

The Inspection Moderator is a *key player* and *requires special training* to be able to conduct inspections that are optimally effective. Ideally, to preserve objectivity, the moderator should not be involved in development of the product that is to be inspected, but should come from another similar project. The moderator functions as a "player-coach" and is responsible for conducting the inspection so as to bring a peak of synergy from the group. This is a quickly learned ability by those with some interpersonal skill. In fact, when participants in the moderator training classes are questioned about their case studies, they invariably say that they sensed the presence of the "*Phantom Inspector*," who materialized as a feeling that there had been an additional presence contributed by the way the inspection team worked together. The moderator's task is to invite the Phantom Inspector.

When they are properly approached by management, programmers respond well to inspections. In fact, after they become familiar with them, many programmers have been known to complain when they were not allowed enough time or appropriate help to conduct inspections correctly.

Three separate classes of education have been recognized as a necessity for proper long lasting implementation of inspections. First, *Management* requires a class of one day to familiarize them with inspections and their benefits to management, and *their role* in making them successful. Next, the *Moderators* need three days of education. And, finally, the other *Participants* should receive one half day of training on inspections, the benefits, and their roles. Some organizations have started inspections without proper education and have achieved some success, but less than others who prepared their participants fully. This has caused some amount of start-over, which was frustrating to everyone involved.

Management and Inspections

A definite philosophy and set of attitudes regarding inspections and their results is essential. The management education class on inspections is one of the best ways found to gain the knowledge that must be built into day-to-day management behavior that is required to get the most from inspections on a continuing basis. For example, management must show encouragement for proper inspections. Requiring inspections and then asking for shortcuts will not do. And, people must be motivated to find defects by inspection. *Inspection results must never be used for personnel performance appraisal.* However, the results of testing should be used for performance appraisal. This promotes finding and reworking defects at the lowest cost, and allows testing for verification instead of debugging. In most situations programmers come to depend upon inspections; they prefer defect-free product. And, at those installations where management has taken and maintained a leadership role with inspections, they have been well accepted and very successful.

Inspection Results and Their Uses

The defects found by inspection are immediately recorded and classified by the moderator before being entered into the project data base. Here is an example:

In module: *XXX*, Line: *YYY*, NAME-CHECK is performed one less time than required—LO/W/MAJ

The description of the defect is obvious. The classification on the right means that this is a defect in Logic, that the logic is Wrong (as opposed to Missing or Extra), and that it is a Major defect. A MAJOR defect is one that would cause a malfunction or unexpected result if left uncorrected. Inspections also find MINOR defects. They will not cause malfunction, but are more of the nature of poor workmanship, like misspellings that do not lead to erroneous product performance.

Major defects are of the same type as defects found by testing. (One unpublished study of defects found by system testing showed that more than 87 percent could have been detected by inspection.) Because Major defects are equivalent to test defects, inspection results can be used to identify *defect prone design and code*. This is enabled because empirical data indicates a directly proportional relationship between the inspection detected defect rate in a piece of code and the defect rate found in it by subsequent testing. Using inspection results in this way, it is possible to identify defect prone code and correct it, in effect, performing real-time quality control of the product as it is being developed, *before it is shipped or put into use.*

There are, of course, many Process and Quality Control uses for inspection data including:

- Feedback to improve the development process by identification and correction of the root causes of systematic defects before more code is developed;
- *Feed-forward* to prepare the process ahead to handle problems *or to evaluate corrective action in advance* (e.g., handling defect prone code);
- Continuing improvement and control of inspections.

An outstanding benefit of feedback, as reported in [3] was that designers and coders through involvement in inspections of their own work learned to find defects they had created more easily. This enabled them to *avoid* causing these defects in future work, thus providing much higher quality product.

Various Applications of Inspections

The inspection process was originally applied to hardware logic, and then to software logic design and code. It was in the latter case that it first gained notice. Since then it has been very successfully applied to software test plans and test cases, user documentation, high level design, system structure design, design changes, requirements development, and microcode. It has also been employed for special purposes such as cleaning up defect prone code, and improving the quality of code that has already been tested. And, finally, it has been resurrected to produce defect-free hardware. It appears that virtually anything that is created by a development process and that can be made visible and readable can be inspected. All that is necessary for an inspection is to define the exit criteria of the process operation that will make the product to be inspected, tailor the inspection defect checklists to the particular product and exit criteria, and then to execute the inspection process.

What's in a Name?

In contrast to inspections, walkthrus, which can range anywhere from cursory peer reviews to inspections, do not usually practice a process that is repeatable or collect data (as with inspections), and hence this process cannot be reasonably studied and improved. Consequently, their defect detection efficiencies are usually quite variable and, when studied, were found to be much lower than those of inspections [2], [3]. However, the name "walkthru" (or "walkthrough") has a place, for in some management and national cultures it is more desirable than the term "inspection" and, in fact, the walkthrus in *some* of these situations are identical to formal inspections. (In almost all instances, however, the author's experience has been that the term walkthru has been accurately applied to the less efficient method—which process is actually in use can be readily determined by examining whether a formally defined development process with exit criteria is in effect, and by applying the criteria in [2, Table 5] to the activity. In addition, initiating walkthrus as a migration path to inspections has led to a lot of frustration in many organizations because once they start with the informal, they seem to have much more difficulty moving to the formal process than do those that introduce inspections from the start. And, programmers involved in inspections are usually more pleased with the results. In fact, their major complaints are generally to do with things that detract from inspection quality.) What is important is that the same results should not be expected of walkthrus as is required of inspections, *unless a close scrutiny proves the process and conduct of the "walkthru" is identical to that required for inspections*. Therefore, although walkthrus do serve very useful though limited functions, they are not discussed further in this paper.

Recognizing many of the abovementioned points, the IBM Information Systems Management Institute course on this subject is named: "Inspections: Formal Application Walkthroughs." They teach about inspection.

Contributors to Software Inspection Quality

Quality of inspection is defined as its ability to detect all instances in which the product does not meet its requirements. Studies, evaluations, and the observations of many people who have been involved in inspections over the past decade provide insights into the contributors to inspection quality. Listing contributors is of little value in trying to manage them as many have relationships with each other. These relationships must be understood in order to isolate and deal with initiating root causes of problems rather than to waste effort dealing with symptoms. The ISHIKAWA or FISHBONE CAUSE/EFFECT DIAGRAM [11], shown in Fig. 2, shows the contributors and their cause/effect relationships.

As depicted in Fig. 2, the main contributors, shown as main branches on the diagram, are: *PRODUCT INSPECTABILITY, INSPECTION PROCESS, MANAGERS,* and *PROGRAMMERS*. Subcontributors, like *INSPECTION MATERIALS* and *CONFORMS WITH STANDARDS*, which contribute to the *PRODUCT INSPECTABILITY*, are shown as twigs on these branches. Contributors to the subcontributors are handled similarly. Several of the relationships have been proven by objective statistical analysis, others are supported by empirical data, and some are evident from project experience. For example, one set of relationships very thoroughly established in a controlled study by F. O. Buck, in ''Indicators of Quality Inspections'' [10], are:

- excessive SIZE OF MATERIALS to be inspected leads to a PREPARATION RATE that is too high.
- PREPARATION RATE that is too high contributes to an excessive RATE OF INSPECTION, and
- Excessive RATE OF INSPECTION *causes fewer defects to be found*.

This study indicated that the following rates should be used in planning the I2 code inspection:

OVERVIEW:	500 Noncommentary Source Statements per Hour.
PREPARATION:	125 Noncommentary Source Statements per Hour.
INSPECTION:	90 Noncommentary Source Statements per Hour.
Maximum Inspection Rate:	125 Noncommentary Source Statements per Hour.

The rate of inspection seems tied to the thoroughness of the inspection, and there is evidence that defect detection efficiency diminishes at rates above 125 NCSS/h. (Many projects require reinspection if this maximum rate is exceeded, and the reinspection usually finds more defects.) Separate from this study, project data show that inspections conducted by trained moderators are very much more likely to approximate the permissible inspection rates, and yield higher quality product than moderators who have not been trained. Meeting this rate is not a direct conscious purpose of the moderator, but rather is the result of proper conduct of the inspection. In any event, as the study shows, requiring too much material to be inspected will induce insufficient PREPARATION which, in turn, will cause the INSPECTION to be conducted too fast. Therefore, it is the responsibility of management and the moderator to start off with a plan that will lead to successful inspection.

The planning rate for high level design inspection of *systems design* is approximately twice the rate for code inspection, and low level (Logic) design inspection is nearly the same (rates are based upon the designer's estimate of the number of source lines of code that will be needed to implement the design). Both these rates *may* depend upon the complexity of the material to be inspected and the manner in which it is prepared (e.g., unstructured code is more difficult to read and requires the inspection rate to be lowered. Faster inspection rates while retaining high defect detection efficiency *may* be feasible with highly structured, easy to understand material, *but further study is needed*). Inspections of requirements, test plans, and user documentation are governed by the same rules as for code inspection, although inspection rates are not as clear for them and are probably more product and project dependent than is the case of code.

With a good knowledge of and attention to the contributors to inspection quality, management can profoundly influence the quality, and the development and maintenance costs of the products for which they are responsible.

Summary

Experience over the past decade has shown software inspections to be a potent defect detection method, finding 60–90 percent of all defects, as well as providing feedback that enables programmers to avoid injecting defects in future work. As well as providing checkpoints to facilitate process management, inspections enable measurement of the performance of many tools and techniques in individual process operations. Because inspection engages similar skills to those used in creating the

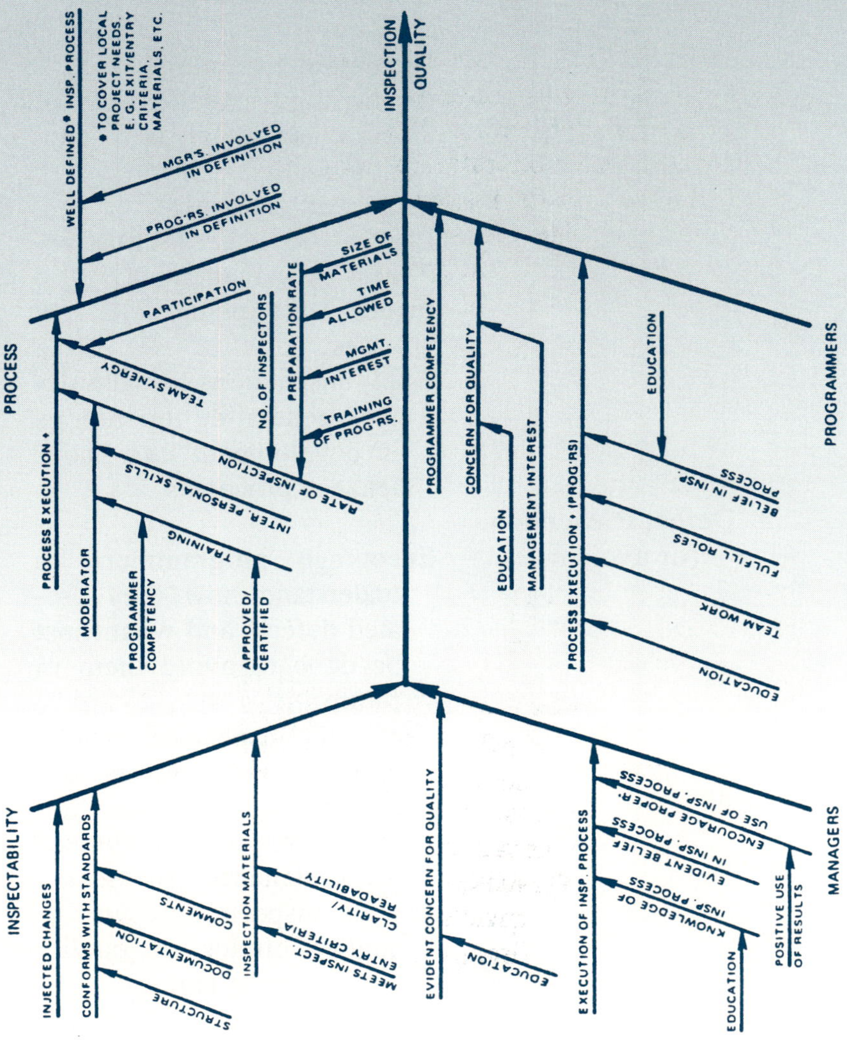

Fig. 2. Fishbone diagram of contributors to inspection quality.

product (and it has been applied to virtually every design technique and coding language), it appears that anything that can be created and described can also be inspected.

Study and observation have revealed the following key aspects that must be managed to take full advantage of the many benefits that inspections offer:

Capability	Action Needed to Enhance the Capability
• Defect Detection	— Management understanding and continuing support. This starts with education. — Inspection moderator training (3 days). — Programmer training. — Continuing management of the contributors to inspection quality. — Inspect all changes. — Periodic review of effectiveness by management. — Inspect test plans and test cases. — Apply inspections to main defect generating operations in development *and* maintenance processes.
• Defect Prevention (or avoidance)	— Encourage programmers to understand how they created defects and what must be done to avoid them in future. — Feedback inspection results promptly and removes root causes of systematic defects from the development or maintenance processes. — Provide inspection results to quality circles or quality improvement teams.

— Creation of requirements for expert system tools (for defect prevention) based upon analysis of inspection data.

- Process Management — Use inspection completions as checkpoints in the development plan and measure accomplishment against them.

References

[1] L. H. Fenton, "Response to the SHARE software service task force report," IBM Corp., Kingston, NY, Mar. 6, 1984.

[2] M. E. Fagan, "Design and code inspections to reduce errors in program development," *IBM Syst. J.*, vol. 15, no. 3, 1979.

[3] *IBM Technical Newsletter GN20-3814*, Base Publication GC20-2000-0, Aug. 15, 1978.

[4] T. D. Crossman, "Inspection teams, are they worth it?" in *Proc. 2nd Nat. Symp. EDP Quality Assurance*, Chicago, IL, Mar. 24–26, 1982.

[5] R. R. Larson, "Test plan and test case inspection specification," IBM Corp., Tech. Rep. TR21.585, Apr. 4, 1975.

[6] T. D. Crossman, "Some experiences in the use of inspection teams in application development," in *Proc. Applicat. Develop. Symp.*, Monterey, CA, 1979.

[7] G. D. Brown and D. H. Sefton, "The micro vs. the applications logjam," *Datamation*, Jan. 1984.

[8] J. H. Morrissey and L. S.-Y. Wu, "Software engineering: An economical perspective," in *Proc. IEEE Conf. Software Eng.*, Munich, West Germany, Sept. 14–19, 1979.

[9] B. Boehm *et al.*, *Characteristics of Software Quality.* New York: American Elsevier, 1978.

[10] F. O. Buck, "Indicators of quality inspections," IBM Corp., Tech. Rep. IBM TR21.802, Sept. 1981.

[11] K. Ishikawa, *Guide to Quality Control.* Tokyo, Japan: Asian Productivity Organization, 1982.

Michael E. Fagan (M'62) is a Senior Technical Staff Member at the IBM Corporation, Thomas J. Watson Research Center, Yorktown Heights, NY. While at IBM, he has had many interesting management and technical assignments in the fields of engineering, manufacturing, software development, and research. In 1972, he created the software inspection process, and has helped implement it within IBM and also promoted its use in the software industry. For this and other work, he has received IBM Outstanding Contribution and Corporate Achievement Awards. His area of interest is in studying and improving all the processes that comprise the software life cycle. For the past two years, he has been a Visiting Professor at, and is on the graduate council of, the University of Maryland.

Software Economics

Barry Boehm
TRW (Thomson Ramo Wooldridge, an American Aerospace and Technology corporation (www.trw.com)
Professor of Software Engineering
University of Southern California
boehm@sunset.usc.edu

Ph.D. in Mathematics, UCLA

Sc.D. (hon.) in Computer Science, University of Massachusetts

Chief Scientist, TRW

Director of the DARPA IT Office, DoD (Department of Defense)

Member of NASA Research and Technology Advisory Committee and of National Academy of Engineering

Fellow of ACM, IEEE, AIAA, INCOSE

Major contributions: Constructive Cost Model (COCOMO), Spiral Model

Current interests: model-based architecting, software engineering

Barry Boehm

Early Experiences in Software Economics

It Took Me a While to Appreciate Software Economics

My first exposure to software economics came on my first day in the software business, in June 1955 at General Dynamics in San Diego. My supervisor took me on a walking tour through the computer, an ERA 1103, which occupied most of a large room. His most memorable comment was,

"Now listen. We're paying this computer six hundred dollars an hour, and we're paying you two dollars an hour, and I want you to act accordingly."

This created some good habits for me, such as defensive programming, careful desk checking, test planning, and analyzing before coding. But it also created some bad habits – a preoccupation with saving microseconds, patching object code, etc. – which were hard to unlearn when the balance of hardware and software costs began to tip the other way. For example, when Fortran came to General Dynamics in 1957, there was a forward-looking bunch of programmers who came to work wearing Fortran T-shirts and bragged about how much faster they could develop programs. But I was still one of the recalcitrant assembly-language programmers, bragging about how we wasted less of those precious computer cycles.

Actually, at General Dynamics and at Rand Corporation between 1959 and 1968, I lived in a remarkably economics-independent work environment. It was the era of large cost-plus defense contracts, and performance at any reasonable cost was the top priority. I was jolted back into an appreciation of economics in attempting to transition some of my interactive-graphics computer-aided-design research into practice. I was involved in exploring the use of the Rand Tablet freehand-drawing input device described by Alan Kay for interactive computer-aided design and analysis of engineering and scientific applications such as rocketry, medicine, and networking.

The Rand engineers and analysts used the system enthusiastically, but I was surprised at how little interest there was in using it somewhere else. It was only then that I discovered how big a difference there was in using the system for free on a DARPA-purchased IBM 360/50 and paying $50/hour for the same capability elsewhere.

This and some related developments got me much more interested in economics. At the time, I was trying to get out of doing programming for others and into becoming a system analyst, which was the big thing to be at Rand. Rand was most appreciated by its major customer, the U.S. Air Force, for its large-scale cost-effectiveness analyses of defense force structures and operational concepts. It had also established a Domestic Studies division which did similar analyses of smaller-scale applications in health care, transportation, education, and public safety.

Learning and Doing Software Economics Analyses at Rand

It was an exciting time to be learning and practicing economics at Rand. At least two of the staff members were to become Nobel Prize economists: Harry Markowitz, also the originator of Simscript; and Bill Sharpe, also author of the seminal 1969 book, *The Economics of Computers* [1]. Rand mathematicians were also creating powerful economic analysis techniques, such as George Dantzig with linear programming and Richard Bellman with dynamic programming.

In 1968, I became head of a group at Rand called Computer Systems Analysis. Besides doing defense command-control and information systems analyses, we were involved in helping state and local governments make better use of computers and software. Several of these experiences convinced me that getting a software system's economic and systems analysis right was at least as important as getting its programming right. One of my role models at the time was Tom DeMarco, whose work convinced me that getting the people and culture factors right was also at least as important as getting the economics right.

In one case, I was involved in helping a state-government agency recover from a well-programmed solution to the wrong part of the problem. They had wanted to improve an inefficient and error-prone transaction processing system which had some cumbersome clerical and tabulator-equipment operations. They contracted a software house which had a software and computing solution: replacing the tabulator operations with a batch-processing computer operation. Although the software solution was well programmed, the end result was more inefficient and error-prone, and required more rather than fewer clerical people. The agency and its software contractor had neglected to analyze the clerical pre-processing steps, which were the main source of inefficiency and frustration, and these just got worse when the longer-turnaround batch-computing system went into operation.

In another case, an urban fire department dispatching system was overrun by 18 months and over one million dollars because they leaped into an overambitious solution without any preliminary feasibility analysis. This was an automated subsystem to determine the location of a fire from its street address. After trying to solve many vexing and tangled problems of discontinuous, curving, renamed streets with inconsistent numbering schemes, the beleaguered subsystem was dropped with little loss of effectiveness by having each dispatcher enter the fire location from a personalized map.

In a third case, a large state government agency had once brought a complex two-million-dollars second-generation mass storage system which rapidly became obsolete. Several years later, they were paying an extra four hundred thousand dollars per month in computer rental costs just to continue using and justifying the money sunk in their obsolete owned system. In each case, and in several others, the critical success factors involved the economic and people factors considerably more than the programming.

The Air Force CCIP-85 Study

In 1970 and 1971, Rand lent me to the Air Force to run a study called CCIP-85. This study involved a forecast of Air Force command and control information processing needs into the mid-1980s, and identification of the most critical information processing deficiencies, and a recommended Air Force R&D program to address deficiencies.

The study's sponsors and I expected that the main deficiencies would be computer speed and memory, large-screen displays, etc. But what we found in visiting Air Force command and control sites was that their biggest operational deficiencies came from shortfalls in the software. And we found that the balance of Air Force hardware-software costs was becoming dominated by software costs, as shown in Figure 1 [2].

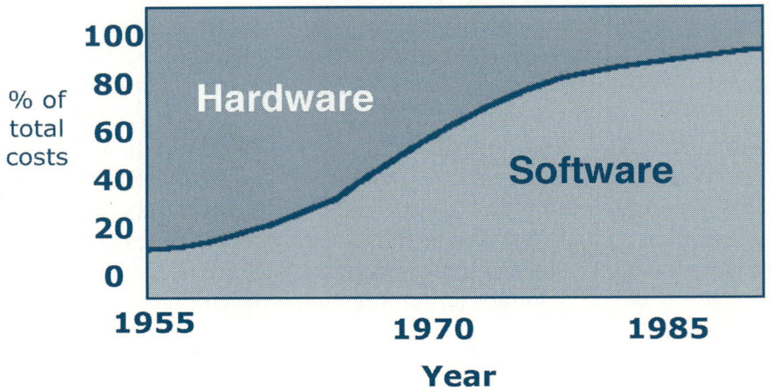

Fig. 1. Air Force computing cost trends – CCIP-85 study, 1971

The main CCIP-85 recommendations for Air Force investment in the software technology were for structured methods, software requirements and exercise technology, and software clarification technology. In structured methods one highlight learning experience I recall vividly was a conversation with Edsger W. Dijkstra, especially for the insight that software and structured methods would be increasingly concerned with concurrency issues in the future.

The emphasis on software requirements technology and its need to include exercise technology came partly from my experience during the study in attempting to lay out future airborne command post displays and asking Air Force generals to tell me where to improve them. Their main response was, "I can't tell you what I'd need for decision support, but once I try to exercise a prototype in a representative situation, I can tell you what needs improving." However, under certification technology, the study did also recommend research and development on requirements and design languages that enabled stronger early analysis of requirements and design completeness, consistency, traceability, and ambiguity problems. It also recommended extensions to programming language features that would eliminate certain classes of defects, such as those that Niklaus Wirth was putting into Pascal.

Besides finding that software costs were beginning to dominate hardware costs, the study also came up with results such as Figure 2, which showed that skimping on hardware costs actually increased software costs by requiring stricter algorithms and data structures. Besides helping me unlearn my 1955 conditioning about saving hardware costs, Figure 2 also helped computer companies sell a lot more hardware.

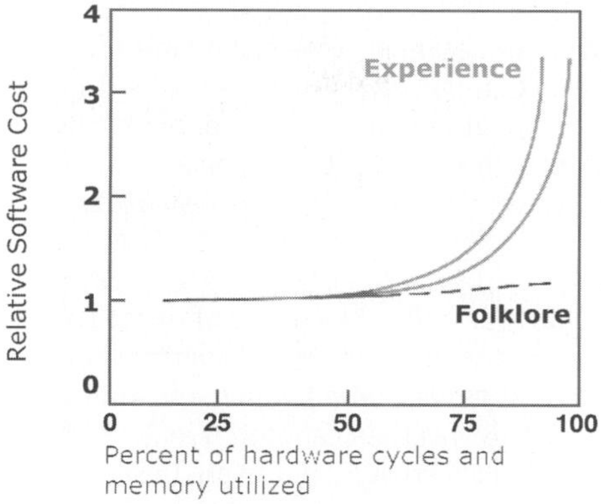

Fig. 2. Risk of minimizing hardware costs – CCIP-85 study, 1971

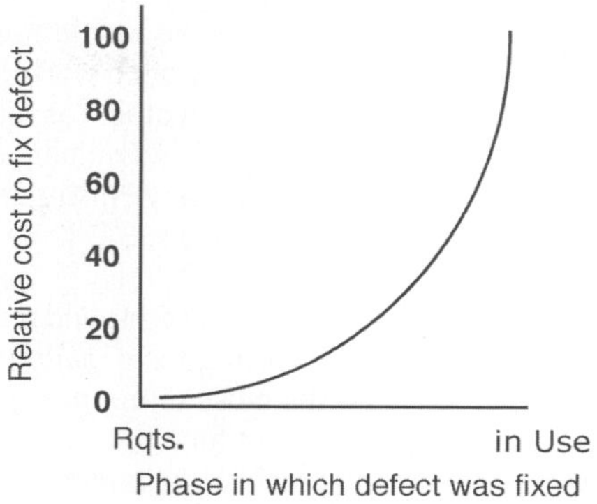

Fig. 3. Growth of software cost-to-fix by phase – TRW, 1976

Rejoining the Software Field at TRW

The CCIP-85 study led me to a career crossroads. I had successfully exited the software field and was enjoying doing systems analysis. But I had found that the software field was a source of a rich set of problems that could be helped by systems analysis approaches. However, there was a lot about understanding and reducing software costs that could not be done from a think tank such as Rand, so in 1973 I decided to join TRW as their Director of Software Research and Technology.

The first TRW software economics analysis I was involved in was the "Characteristics of Software Quality" study for NBS (*National Bureau of Standards*, now called *National Institute of Standards and Technology* (NIST, cf. http://www.100.nist.gov/directors.htm) in 1973 [3]. Its most significant conclusion for TRW was that the bulk of the serious software defects were inserted in the requirements and design phases, but that they weren't being eliminated until testing, when they were much more expensive to fix. This caused us to focus TRW's software technology investments on the front end of the software life cycle, leading to such tools as the Design Assertion Consistency Checker and the Software Requirements Engineering Methodology, now commercialized as RDD-100. Figure 3 summarizes our experience on the escalation of software cost-to-fix with time. The factor-of-100 escalation was similar to IBM's experience as discussed by Michael Fagan. Besides tools and inspections, this focused TRW software improvements on prototyping, architecting, and risk management.

At the time, TRW had a software cost estimation model developed by Ray Wolverton in 1972. It was a good match to TRW's needs at the time, but as TRW's software business expanded (eventually becoming number 2 in the world next to IBM in the Datamation 100 list), top management felt the need for a more general model, and this became my top-priority goal in 1976.

In surveying existing software cost estimation models at the time, I found some very good work being done by Claude Walston, Charles Felix, Capers Jones, and others at IBM [4], by Larry Putnam in U.S. Army applications [5], by Frank Freiman and Bob Park with RCA PRICE S [6], and in Montgomery Phister's monumental work, *Data Processing Technology and Economics* [7].

Based on this foundation of insights, we convened a group effort at TRW involving ten experienced software managers and analysts determining the most appropriate initial set of cost drivers, rating scales, and multiplier values for a provisional cost model. This was then calibrated to 20 TRW project data points to produce a model called SCEP (Software Cost Estimation Program). SCEP worked well enough for TRW that it was established as the company standard for use on all proposals and as the basis for software data collection. Its continued use led to a number of upgrades to add new cost drivers, phase and activity distributions, reuse and maintenance effects, etc. It also served as the quantitative basis for TRW's investment in productivity-enhancing tools, processes, personnel, and management practice in the 1980s such as the TRW Software Productivity System, rapid prototyping tools, risk management, and the Spiral Model [8].

Concurrently, I was gathering non-TRW data through the student term projects in my USC and UCLA courses, and through open sources such as the NASA Software Engineering Lab via Vic Basili. I tried fitting this data to the TRW SCEP model, but many of the data points did not fit well. In analyzing the patterns among the poorly fitting projects, I found that the differences could be explained fairly well by the concept of a "development mode." TRW's mode tended to be what became called the Embedded mode, characterized by tight, ambitious requirements and unprecedented applications. Batch, precedented business and scientific applications with easy-to-relax requirements and schedules fell into a different mode that became called the Organic mode. By fitting different size-to-cost coefficients and scaling exponents to these modes (and an intermediate mode), a three-mode model was able to explain the data fairly well (within 20% of the actuals, 70% of the time).

The resulting model was calibrated to 56 project data points in 1978, and produced comparably accurate estimates for another 7 project data points collected in 1979. It became the Constructive Cost Model (COCOMO) published in the book *Software Engineering Economics* in 1981 [9].

Where Are We Going?

At the time, I recall thinking that those of us in the software estimation field were like the Tycho Brahes in the 16th century, compiling observational data that later Keplers and Newtons would use to develop a quantitative science of software engineering. As we went from unprecedented to precedented software applications, our productivity would increase and our error in estimating software costs would continue to decrease, as on the left side of Figure 4.

However, this view rested on the assumption that, like the stars, planets, and satellites, software projects would continue to behave in the same way as time went on. But, due to the efforts of the software pioneers convened here and many others, this assumption turned out to be invalid. The software field is continually being reinvented via structured methods, abstract data types, information hiding, objects, patterns, reusable components, commercial packages, very high level languages, and rapid application development (RAD) processes.

With each reinvention, software cost estimation and other software engineering fields need to reinvent pieces of themselves just to keep up, resulting in the type of progress shown on the right side of Figure 4 [10]. And the rate of reinvention of the software field continues to increase.

The most encouraging thing in Figure 4, though, is that leading companies such as sd&m are able to build on this experience and continue to increase our relative productivity in delivering the huge masses of software our society increasingly depends on. Our biggest challenge for the future is to figure out how to selectively prune the parts of the software engineering experience base that become less relevant, and to conserve and build on the parts with lasting value for the future.

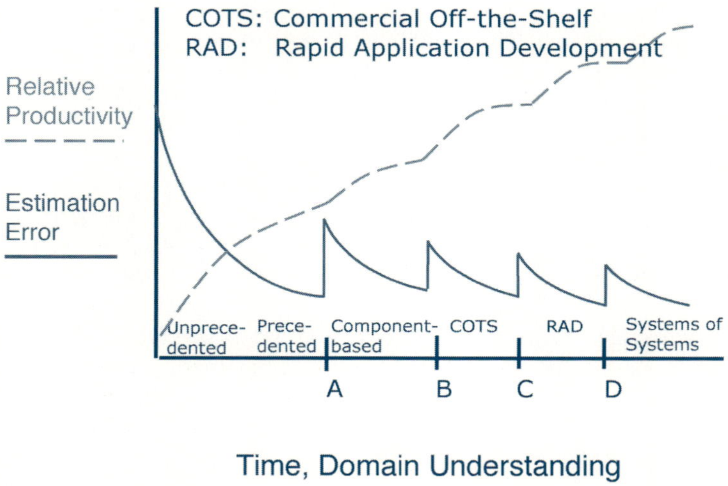

Fig. 4. Software estimation – The receding horizon

References

[1] W. Sharpe, The Economics of Computers, Columbia University Press, 1969.

[2] B. Boehm, "Software and Its Impact: A Quantitative Assessment," Datamation, May 1973, pp. 48-59.

[3] B. Boehm, J. Brown, H. Kaspar, M. Lipow, G. McLeod, and M. Merritt, Characteristics of Software Quality, NBS Report, 1973; North Holland (book), 1978.

[4] C. Walston and C. Felix, "A Method of Programming Measurement and Estimation," IBM Systems Journal 16(1), 1977, pp. 54-73.

[5] L. Putnam, "A General Empirical Solution to the Macro Software Sizing and Estimating Problem," IEEE Transactions, July 1978, pp. 345-361.

[6] F. Freiman and R. Park, "PRICE Software Model-Version 3: An Overview," Proc. IEEE-PINY Workshop on Quantitative Software Models, Oct. 1979, pp. 32-44.

[7] M. Phister, Data Processing Technology and Economics, Santa Monica Publishing Co., 1976.

[8] B. Boehm, Software Risk Management, IEEE-CS Press, 1989.

[9] B. Boehm, Software Engineering Economics, Prentice Hall, 1981.

[10] B. Boehm, C. Abts, A.W. Brown, S. Chulani, B. Clark, E. Horowitz, R. Madachy, D. Reifer, and B. Steece, Software Cost Estimation with COCOMO II, Prentice Hall, 2000.

Barry Boehm

Software Engineering Economics

IEEE Transactions on Software Engineering,
Vol. SE-10 (1),1984
pp. 4-21

Software Engineering Economics

BARRY W. BOEHM

Manuscript received April 26, 1983; revised June 28, 1983.
The author is with the Software Information Systems Division, TRW Defense Systems Group, Redondo Beach, CA 90278.

***Abstract*—This paper summarizes the current state of the art and recent trends in software engineering economics. It provides an overview of economic analysis techniques and their applicability to software engineering and management. It surveys the field of software cost estimation, including the major estimation techniques available, the state of the art in algorithmic cost models, and the outstanding research issues in software cost estimation.**

***Index Terms*—Computer programming costs, cost models, management decision aids, software cost estimation, software economics, software engineering, software management.**

I. Introduction

Definitions

The dictionary defines "economics" as "a social science concerned chiefly with description and analysis of the production, distribution, and consumption of goods and services." Here is another definition of economics which I think is more helpful in explaining how economics relates to software engineering.

> *Economics* is the study of how people make decisions in resource-limited situations.

This definition of economics fits the major branches of classical economics very well.

Macroeconomics is the study of how people make decisions in resource-limited situations on a national or global scale. It deals with the effects of decisions that national leaders make on such issues as tax rates, interest rates, foreign and trade policy.

Microeconomics is the study of how people make decisions in resource-limited situations on a more personal scale. It deals with the decisions that individuals and organizations make on such issues as how much insurance to buy, which word processor to buy, or what prices to charge for their products or services.

Economics and Software Engineering Management

If we look at the discipline of software engineering, we see that the microeconomics branch of economics deals more with the types of decisions we need to make as software engineers or managers.

Clearly, we deal with limited resources. There is never enough time or money to cover all the good features we would like to put into our software products. And even in these days of cheap hardware and virtual memory, our more significant software products must always operate within a world of limited computer power and main memory. If you have been in the software engineering field for any length of time, I am sure you can think of a number of decision situations in which you had to determine some key software product feature as a function of some limiting critical resource.

Throughout the software life cycle,[1] there are many decision situations involving limited resources in which software engineering economics techniques provide useful assistance. To provide a feel for the nature of these economic decision issues, an example is given below for each of the major phases in the software life cycle.

- *Feasibility Phase:* How much should we invest in information system analyses (user questionnaires and interviews, current-system analysis, workload characterizations, simulations, scenarios, prototypes) in order that we converge on an appropriate definition and concept of operation for the system we plan to implement?
- *Plans and Requirements Phase:* How rigorously should we specify requirements? How much should we invest in requirements validation activities (automated com-

1 Economic principles underlie the overall structure of the software life cycle, and its primary refinements of prototyping, incremental development, and advancemanship. The primary economic driver of the life-cycle structure is the significantly increasing cost of making a software change or fixing a software problem, as a function of the phase in which the change or fix is made. See [11, ch. 4].

pleteness, consistency, and traceability checks, analytic models, simulations, prototypes) before proceeding to design and develop a software system?

- *Product Design Phase:* Should we organize the software to make it possible to use a complex piece of existing software which generally but not completely meets our requirements?
- *Programming Phase:* Given a choice between three data storage and retrieval schemes which are primarily execution time-efficient, storage-efficient, and easy-to-modify, respectively; which of these should we choose to implement?
- *Integration and Test Phase:* How much testing and formal verification should we perform on a product before releasing it to users?
- *Maintenance Phase:* Given an extensive list of suggested product improvements, which ones should we implement first?
- *Phaseout:* Given an aging, hard-to-modify software product, should we replace it with a new product, restructure it, or leave it alone?

Outline of This Paper

The economics field has evolved a number of techniques (cost-benefit analysis, present value analysis, risk analysis, etc.) for dealing with decision issues such as the ones above. Section II of this paper provides an overview of these techniques and their applicability to software engineering.

One critical problem which underlies all applications of economic techniques to software engineering is the problem of estimating software costs. Section III contains three major sections which summarize this field:

III-A: Major Software Cost Estimation Techniques

III-B: Algorithmic Models for Software Cost Estimation

III-C: Outstanding Research Issues in Software Cost Estimation.

Section IV concludes by summarizing the major benefits of software engineering economics, and commenting on the major challenges awaiting the field.

II. Software Engineering Economics Analysis Techniques

Overview of Relevant Techniques

The microeconomics field provides a number of techniques for dealing with software life-cycle decision issues such as the ones given in the previous section. Fig. 1 presents an overall master key to these techniques and when to use them.[2]

As indicated in Fig. 1, standard optimization techniques can be used when we can find a single quantity such as dollars (or pounds, yen, cruzeiros, etc.) to serve as a "universal solvent" into which all of our decision variables can be converted. Or, if the nondollar objectives can be expressed as constraints (system availability must be at least 98 percent; throughput must be at least 150 transactions per second), then standard constrained optimization techniques can be used. And if cash flows occur at different times, then present-value techniques can be used to normalize them to a common point in time.

More frequently, some of the resulting benefits from the software system are not expressible in dollars. In such situations, one alternative solution will not necessarily dominate another solution.

An example situation is shown in Fig. 2, which compares the cost and benefits (here, in terms of throughput in transactions per second) of two alternative approaches to developing an operating system for a transaction processing system.

- *Option A:* Accept an available operating system. This will require only $80K in software costs, but will achieve a peak performance of 120 transactions per second, using five $10K minicomputer processors, because of a high multiprocessor overhead factor.
- *Option B:* Build a new operating system. This system would be more efficient and would support a higher peak throughput, but would require $180K in software costs.

2 The chapter numbers in Fig. 1 refer to the chapters in [11], in which those techniques are discussed in further detail.

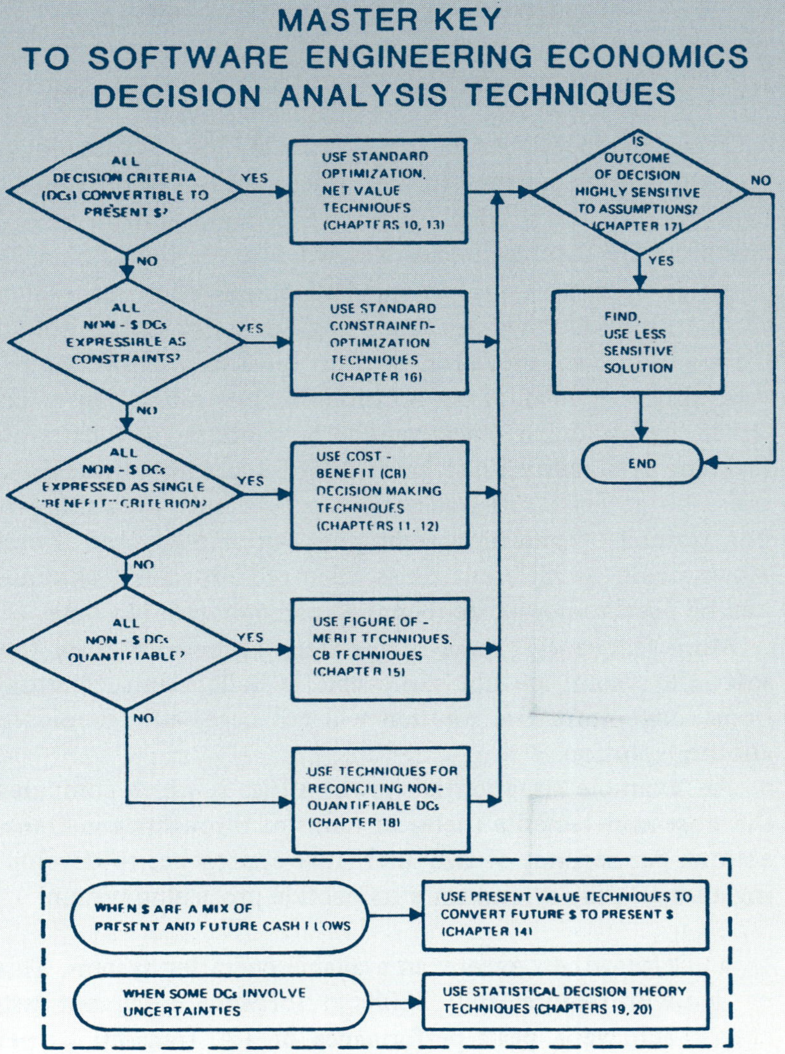

Fig. 1. Master key to software engineering economics decision analysis techniques.

The cost-versus-performance curve for these two options are shown in Fig. 2. Here, neither option dominates the other, and various cost-benefit decision-making techniques (maximum profit margin, cost/benefit ratio, return on investments, etc.) must be used to choose between Options A and B.

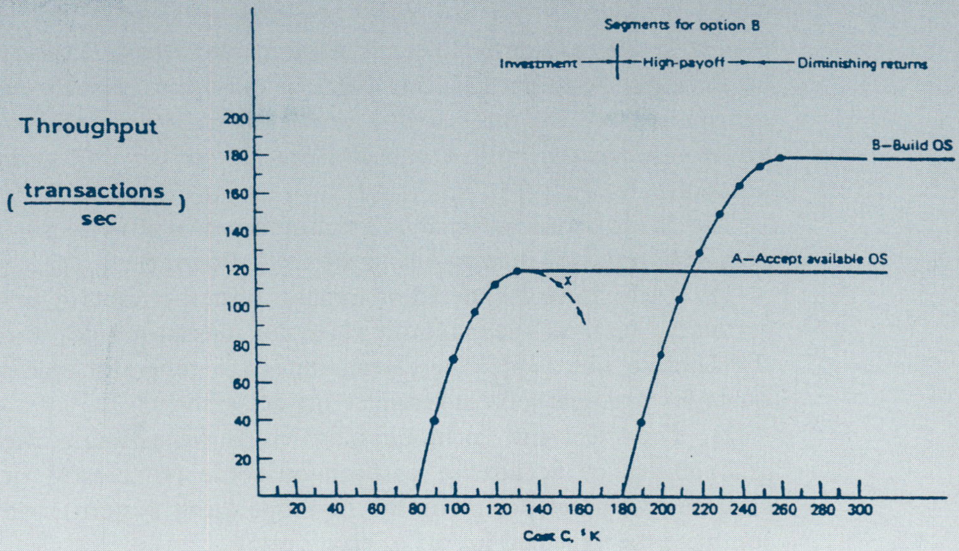

The cost-versus-performance curve for these two options are shown in Fig. 2. Here, neither option dominates the other, and various cost-benefit decision-making techniques (maximum profit margin, cost/benefit ratio, return on investments, etc.) must be used to choose between Options A and B.

In general, software engineering decision problems are even more complex than Fig. 2, as Options A and B will have several important criteria on which they differ (e.g., robustness, ease of tuning, ease of change, functional capability). If these criteria are quantifiable, then some type of figure of merit can be defined to support a comparative analysis of the preferability of one option over another. If some of the criteria are unquantifiable (user goodwill, programmer morale, etc.), then some techniques for comparing unquantifiable criteria need to be used. As indicated in Fig. 1, techniques for each of these situations are available, and discussed in [11].

Analyzing Risk, Uncertainty, and the Value of Information

In software engineering, our decision issues are generally even more complex than those discussed above. This is because the outcome of many of our options cannot be determined in advance. For example, building an operating system with a significantly lower multiprocessor overhead may be achievable, but on the other hand, it may not. In such circumstances, we are faced with a problem of *decision making under uncertainty,* with a considerable *risk* of an undesired outcome.

The main economic analysis techniques available to support us in resolving such problems are the following.

1) Techniques for decision making under complete uncertainty, such as the maximax rule, the maximin rule, and the Laplace rule [38]. These techniques are generally inadequate for practical software engineering decisions.

2) Expected-value techniques, in which we estimate the probabilities of occurrence of each outcome (successful or unsuccessful development of the new operating system) and complete the expected payoff of each option:

$$EV = \text{Prob(success)} * \text{Payoff(successful OS)}$$
$$+ \text{Prob(failure)} * \text{Payoff(unsuccessful OS)}.$$

These techniques are better than decision making under complete uncertainty, but they still involve a great deal of risk if the Prob(failure) is considerably higher than our estimate of it.

3) Techniques in which we reduce uncertainty by *buying information.* For example, *prototyping* is a way of buying information to reduce our uncertainty about the likely success or failure of a multiprocessor operating system; by developing a rapid prototype of its high-risk elements, we can get a clearer picture of our likelihood of successfully developing the full operating system.

In general, prototyping and other options for buying information[3] are most valuable aids for software engineering decisions. However, they always raise the following question: "how much information-buying is enough?"

In principle, this question can be answered via statistical decision theory techniques involving the use of Bayes' Law, which

[3] Other examples of options for buying information to support software engineering decisions include feasibility studies, user surveys, simulation, testing, and mathematical program verification techniques.

allows us to calculate the expected payoff from a software project as a function of our level of investment in a prototype or other information-buying option. (Some examples of the use of Bayes' Law to estimate the appropriate level of investment in a prototype are given in [11, ch. 20].)

In practice, the use of Bayes' Law involves the estimation of a number of conditional probabilities which are not easy to estimate accurately. However, the Bayes' Law approach can be translated into a number of *value-of-information guidelines*, or conditions under which it makes good sense to decide on investing in more information before committing ourselves to a particular course of action.

Condition 1: There exist attractive alternatives whose payoff varies greatly, depending on some critical states of nature. If not, we can commit ourselves to one of the attractive alternatives with no risk of significant loss.

Condition 2: The critical states of nature have an appreciable probability of occurring. If not, we can again commit ourselves without major risk. For situations with extremely high variations in payoff, the appreciable probability level is lower than in situations with smaller variations in payoff.

Condition 3: The investigations have a high probability of accurately identifying the occurrence of the critical states of nature. If not, the investigations will not do much to reduce our risk of loss due to making the wrong decision.

Condition 4: The required cost and schedule of the investigations do not overly curtail their net value. It does us little good to obtain results which cost more than they can save us, or which arrive too late to help us make a decision.

Condition 5: There exist significant side benefits derived from performing the investigations. Again, we may be able to justify an investigation solely on the basis of its value in training, team-building, customer relations, or design validation.

Some Pitfalls Avoided by Using the Value-of-Information Approach

The guideline conditions provided by the value-of-information approach provide us with a perspective which helps us avoid some serious software engineering pitfalls. The pitfalls below are expressed in terms of some frequently expressed but faulty pieces of software engineering advice.

Pitfall 1: Always use a simulation to investigate the feasibility of complex realtime software. Simulations are often extremely valuable in such situations. However, there have been a good many simulations developed which were largely an expensive waste of effort, frequently under conditions that would have been picked up by the guidelines above. Some have been relatively useless because, once they were built, nobody could tell whether a given set of inputs was realistic or not (picked up by Condition 3). Some have been taken so long to develop that they produced their first results the week after the proposal was sent out, or after the key design review was completed (picked up by Condition 4).

Pitfall 2: Always build the software twice. The guidelines indicate that the prototype (or build-it-twice) approach is often valuable, but not in all situations. Some prototypes have been built of software whose aspects were all straightforward and familiar, in which case nothing much was learned by building them (picked up by Conditions 1 and 2).

Pitfall 3: Build the software purely top-down. When interpreted too literally, the top-down approach does not concern itself with the design of low level modules until the higher levels have been fully developed. If an adverse state of nature makes such a low level module (automatically forecast sales volume, automatically discriminate one type of aircraft from another) impossible to develop, the subsequent redesign will generally require the expensive rework of much of the higher level design and code. Conditions 1 and 2 warn us to temper our top-down approach with a thorough top-to-bottom software risk analysis during the requirements and product design phases.

Pitfall 4: Every piece of code should be proved correct. Correctness proving is still an expensive way to get information on the fault-freedom of software, although it strongly satisfies Condition 3 by giving a very high assurance of a program's correctness. Conditions 1 and 2 recommend that proof techniques be used in situations where the operational cost of a software fault is very large, that is, loss of life, compromised national security, major financial losses. But if the operational cost of a software fault is small, the added information on fault-freedom provided by the proof will not be worth the investment (Condition 4).

Pitfall 5: Nominal-case testing is sufficient. This pitfall is just the opposite of Pitfall 4. If the operational cost of potential software faults is large, it is highly imprudent not to perform off-nominal testing.

Summary: The Economic Value of Information

Let us step back a bit from these guidelines and pitfalls. Put simply, we are saying that, as software engineers:

> "It is often worth paying for information because it helps us make better decisions."

If we look at the statement in a broader context, we can see that it is the primary reason why the software engineering field exists. It is what practically all of our software customers say when they decide to acquire one of our products: that it is worth paying for a management information system, a weather forecasting system, an air traffic control system, an inventory control system, etc., because it helps them make better decisions.

Usually, software engineers are *producers* of management information to be consumed by other people, but during the software life cycle we must also be *consumers* of management information to support our own decisions. As we come to appreciate the factors which make it attractive for us to pay for processed information which helps *us* make better decisions as software engineers, we will get a better appreciation for what our customers and users are looking for in the information processing systems we develop for *them.*

III. Software Cost Estimation

Introduction

All of the software engineering economics decision analysis techniques discussed above are only as good as the input data we can provide for them. For software decisions, the most critical and difficult of these inputs to provide are estimates of the cost of a proposed software project. In this section, we will summarize:

1) the major software cost estimation techniques available, and their relative strengths and difficulties;

2) algorithmic models for software cost estimation;

3) outstanding research issues in software cost estimation.

A. Major Software Cost Estimation Techniques

Table I summarizes the relative strengths and difficulties of the major software cost estimation methods in use today.

1) *Algorithmic Models:* These methods provide one or more algorithms which produce a software cost estimate as a function of a number of variables which are considered to be the major cost drivers.

2) *Expert Judgment:* This method involves consulting one or more experts, perhaps with the aid of an expert-consensus mechanism such as the Delphi technique.

3) *Analogy:* This method involves reasoning by analogy with one or more completed projects to relate their actual costs to an estimate of the cost of a similar new project.

4) *Parkinson:* A Parkinson principle ("work expands to fill the available volume") is invoked to equate the cost estimate to the available resources.

5) *Price-to-Win:* Here, the cost estimate is equated to the price believed necessary to win the job (or the schedule believed necessary to be first in the market with a new product, etc.).

6) *Top-Down:* An overall cost estimate for the project is derived from global properties of the software product. The total cost is then split up among the various components.

7) *Bottom-Up:* Each component of the software job is separately estimated, and the results aggregated to produce an estimate for the overall job.

The main conclusions that we can draw from Table I are the following.

- None of the alternatives is better than the others from all aspects.
- The Parkinson and price-to-win methods are unacceptable and do not produce satisfactory cost estimates.
- The strengths and weaknesses of the other techniques are complementary (particularly the algorithmic models versus expert judgment and top-down versus bottom-up).

TABLE I
STRENGTHS AND WEAKNESSES OF SOFTWARE COST-ESTIMATION METHODS

Method	Strengths	Weaknesses
Algorithmic model	• Objective, repeatable, analyzable formula • Efficient, good for sensitivity analysis • Objectively calibrated to experience	• Subjective inputs • Assessment of exceptional circumstances • Calibrated to past, not future
Expert judgment	• Assessment of representativeness, interactions, exceptional circumstances	• No better than participants • Biases, incomplete recall
Analogy	• Based on representative experience	• Representativeness of experience
Parkinson	• Correlates with some experience	• Reinforces poor practice
Price to win	• Often gets the contract	• Generally produces large overruns
Top-down	• System level focus • Efficient	• Less detailed basis • Less stable
Bottom-up	• More detailed basis • More stable • Fosters individual commitment	• May overlook system level costs • Requires more effort

- Thus, in practice, we should use combinations of the above techniques, compare their results, and iterate on them where they differ.

Fundamental Limitations of Software Cost Estimation Techniques

Whatever the strengths of a software cost estimation technique, there is really no way we can expect the technique to compensate for our lack of definition or understanding of the software job to be done. Until a software specification is fully defined, it actually represents a range of software products, and a corresponding range of software development costs.

This fundamental limitation of software cost estimation technology is illustrated in Fig. 3, which shows the accuracy within which software cost estimates can be made, as a function of the software life-cycle phase (the horizontal axis), or of the level of knowledge we have of what the software is intended to do. This level of uncertainty is illustrated in Fig. 3 with respect to a human-machine interface component of the software.

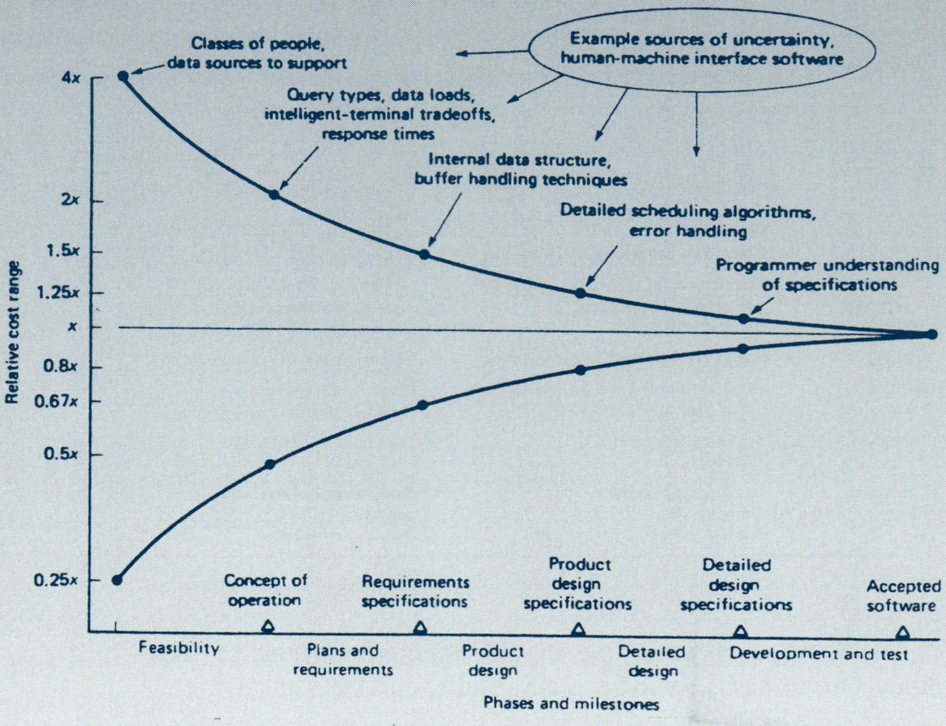

Fig. 3. Software cost estimation accuracy versus phase.

When we first begin to evaluate alternative concepts for a new software application, the relative range of our software cost estimates is roughly a factor of four on either the high or low side.[4] This range stems from the wide range of uncertainty we have at this time about the actual nature of the product. For the human-machine interface component, for example, we do not know at this time what classes of people (clerks, computer specialists, middle managers, etc.) or what classes of data (raw or pre-edited, numerical or text, digital or analog) the system will have to support. Until we pin down such uncertainties, a factor of four in either direction is not surprising as a range of estimates.

[4] These ranges have been determined subjectively, and are intended to represent 80 percent confidence limits, that is, "within a factor of four on either side, 80 percent of the time."

The above uncertainties are indeed pinned down once we complete the feasibility phase and settle on a particular concept of operation. At this stage, the range of our estimates diminishes to a factor of two in either direction. This range is reasonable because we still have not pinned down such issues as the specific types of user query to be supported, or the specific functions to be performed within the microprocessor in the intelligent terminal. These issues will be resolved by the time we have developed a software requirements specification, at which point, we will be able to estimate the software costs within a factor of 1.5 in either direction.

By the time we complete and validate a product design specification, we will have resolved such issues as the internal data structure of the software product and the specific techniques for handling the buffers between the terminal microprocessor and the central processors on one side, and between the microprocessor and the display driver on the other. At this point, our software estimate should be accurate to within a factor of 1.25, the discrepancies being caused by some remaining sources of uncertainty such as the specific algorithms to be used for task scheduling, error handling, abort processing, and the like. These will be resolved by the end of the detailed design phase, but there will still be a residual uncertainty about 10 percent based on how well the programmers really understand the specifications to which they are to code. (This factor also includes such consideration as personnel turnover uncertainties during the development and test phases.)

B. Algorithmic Models for Software Cost Estimation

Algorithmic Cost Models: Early Development

Since the earliest days of the software field, people have been trying to develop algorithmic models to estimate software costs. The earliest attempts were simple rules of thumb, such as:

- on a large project, each software performer will provide an average of one checked-out instruction per man-hour (or roughly 150 instructions per man-month);
- each software maintenance person can maintain four boxes of cards (a box of cards held 2000 cards, or roughly 2000 instructions in those days of few comment cards).

Somewhat later, some projects began collecting quantitative data on the effort involved in developing a software product, and its distribution across the software life cycle. One of the earliest of these analyses was documented in 1956 in [8]. It indicated that, for very large operational software products on the order of 100 000 delivered source instructions (100 KDSI), that the overall productivity was more like 64 DSI/man-month, that another 100 KDSI of support-software would be required; that about 15 000 pages of documentation would be produced and 3000 hours of computer time consumed; and that the distribution of effort would be as follows:

Program Specs:	10 percent
Coding Specs:	30 percent
Coding:	10 percent
Parameter Testing:	20 percent
Assembly Testing:	30 percent

with an additional 30 percent required to produce operational specs for the system. Unfortunately, such data did not become well known, and many subsequent software projects went through a painful process of rediscovering them.

During the late 1950's and early 1960's, relatively little progress was made in software cost estimation, while the frequency and magnitude of software cost overruns was becoming critical to many large systems employing computers. In 1964, the U.S. Air Force contracted with System Development Corporation for a landmark project in the software cost estimation field. This project collected 104 attributes of 169 software projects and treated them to extensive statistical analysis. One result was the 1965 SDC cost model [41] which was the best possible statistical 13-parameter linear estimation model for the sample data:

$$
\begin{aligned}
MM = & -33.63 \\
& +9.15 \text{ (Lack of Requirements) (0-2)} \\
& +10.73 \text{ (Stability of Design) (0-3)} \\
& +0.51 \text{ (Percent Math Instructions)} \\
& +0.46 \text{ (Percent Storage/Retrieval Instructions)}
\end{aligned}
$$

+0.40 (Number of Subprograms)

+7.28 (Programming Language) (0-1)

−21.45 (Business Application) (0-1)

+13.53 (Stand-Alone Program) (0.1)

+12.35 (First Program on Computer) (0-1)

+58.82 (Concurrent Hardware Development) (0-1)

+30.61 (Random Access Device Used) (0-1)

+29.55 (Difference Host, Target Hardware) (0-1)

+0.54 (Number of Personnel Trips)

−25.20 (Developed by Military Organization) (0-1).

The numbers in parentheses refer to ratings to be made by the estimator.

When applied to its database of 169 projects, this model produced a mean estimate of 40 MM and a standard deviation of 62 MM; not a very accurate predictor. Further, the application of the model is counterintuitive; a project with all zero ratings is estimated at minus 33 MM; changing language from a higher order language to assembly language adds 7 MM, independent of project size. The most conclusive result from the SDC study was that there were too many nonlinear aspects of software development for a linear cost-estimation model to work very well.

Still, the SDC effort provided a valuable base of information and insight for cost estimation and future models. Its cumulative distribution of productivity for 169 projects was a valuable aid for producing or checking cost estimates. The estimation rules of thumb for various phases and activities have been very helpful, and the data have been a major foundation for some subsequent cost models.

In the late 1960's and early 1970's, a number of cost models were developed which worked reasonably well for a certain restricted range of projects to which they were calibrated. Some of the more notable examples of such models are those described in [3], [54], [57].

The essence of the TRW Wolverton model [57] is shown in Fig. 4, which shows a number of curves of software cost per

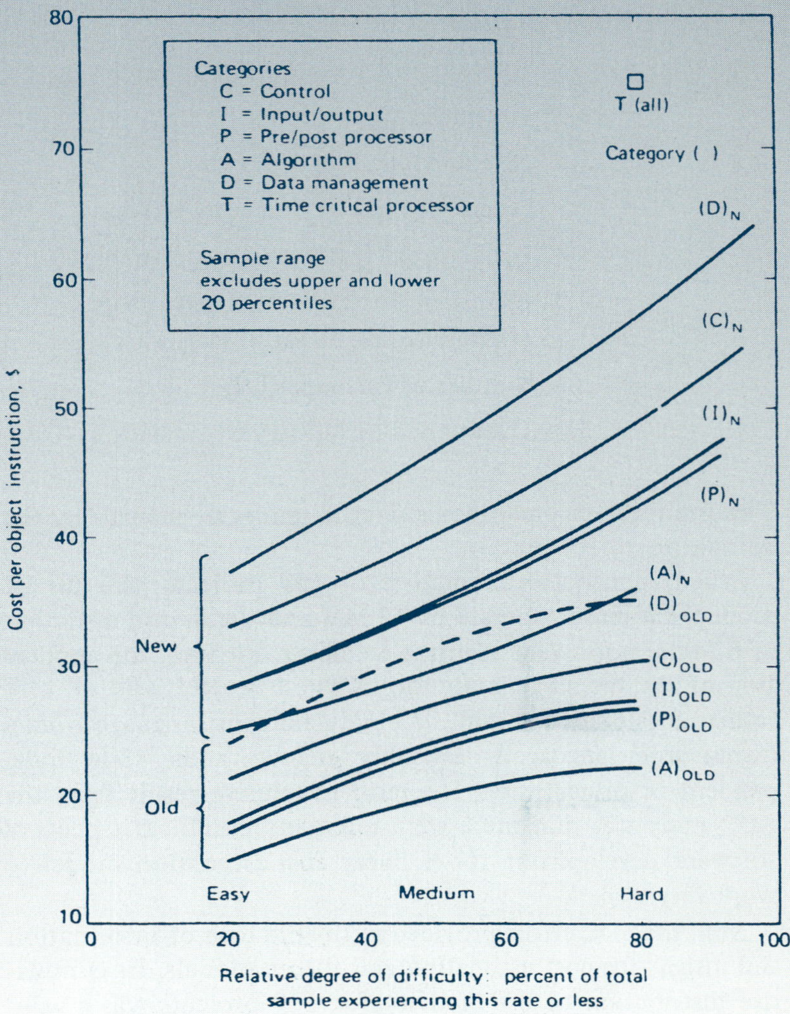

Fig. 4. TRW Wolverton model: Cost per object instruction versus relative degree of difficulty.

object instruction as a function of relative degree of difficulty (0 to 100), novelty of the application (new or old), and type of project. The best use of the model involves breaking the software into components and estimating their cost individually. This, a 1000 object-instruction module of new data management software of medium (50 percent) difficulty would be costed at $46/instruction, or $46 000.

This model is well-calibrated to a class of near-real-time government command and control projects, but is less accurate for some other classes of projects. In addition, the model provides a good breakdown of project effort by phase and activity.

In the late 1970's, several software cost estimation models were developed which established a significant advance in the state of the art. These included the Putnam SLIM Model [44], the Doty Model [27], the RCA PRICE S model [22], the COCOMO model [11], the IBM-FSD model [53], the Boeing model [9], and a series of models developed by GRC [15]. A summary of these models, and the earlier SDC and Wolverton models, is shown in Table II, in terms of the size, program, computer, personnel, and project attributes used by each model to determine software costs. The first four of these models are discussed below.

The Putnam SLIM Model [44], [45]

The Putnam SLIM Model is a commercially available (from Quantitative Software Management, Inc.) software product based on Putnam's analysis of the software life cycle in terms of the Rayleigh distribution of project personnel level versus time. The basic effort macro-estimation model used in SLIM is

$$S_s = C_k K^{1/3} t_d^{4/3}$$

where

S_s = number of delivered source instructions
K = life-cycle effort in man-years
t_d = development time in years
C_k = a "technology constant."

Values of C_k typically range between 610 and 57 314. The current version of SLIM allows one to calibrate C_k to past projects or to past projects or to estimate it as a function of a project's use of modern programming practices, hardware constraints, personnel experience, interactive development, and other factors. The required development effort, DE, is estimated as roughly 40 percent of the life-cycle effort for large

TABLE II
FACTORS USED IN VARIOUS COST MODELS

GROUP	FACTOR	SDC, 1965	TRW, 1972	PUTNAM, SLIM	DOTY	RCA, PRICE S	IBM	BOEING, 1977	GRC, 1979	COCOMO	SOFCOST	DSN	JENSEN
SIZE ATTRIBUTES	SOURCE INSTRUCTIONS			X	X		X	X		X	X	X	X
	OBJECT INSTRUCTIONS	X	X		X	X							
	NUMBER OF ROUTINES	X				X					X		
	NUMBER OF DATA ITEMS						X			X	X		
	NUMBER OF OUTPUT FORMATS								X			X	
	DOCUMENTATION				X		X				X		X
	NUMBER OF PERSONNEL			X			X	X			X		X
PROGRAM ATTRIBUTES	TYPE	X	X	X	X	X	X	X			X		
	COMPLEXITY		X	X		X	X			X	X	X	X
	LANGUAGE	X		X				X	X		X	X	
	REUSE			X		X		X	X	X	X	X	X
	REQUIRED RELIABILITY			X		X				X	X		X
	DISPLAY REQUIREMENTS				X						X		X
COMPUTER ATTRIBUTES	TIME CONSTRAINT		X	X	X	X }	X	X	X	X	X	X	X
	STORAGE CONSTRAINT			X	X	X }	X		X	X	X	X	X
	HARDWARE CONFIGURATION	X				X							
	CONCURRENT HARDWARE DEVELOPMENT	X			X	X	X			X	X	X	X
	INTERFACING EQUIPMENT, S/W										X	X	
PERSONNEL ATTRIBUTES	PERSONNEL CAPABILITY			X		X	X			X	X	X	X
	PERSONNEL CONTINUITY						X					X	
	HARDWARE EXPERIENCE	X		X	X	X	X		X }	X	X	X	X
	APPLICATIONS EXPERIENCE		X	X		X	X	X	X }	X	X	X	X
	LANGUAGE EXPERIENCE			X		X	X		X }	X	X	X	X
PROJECT ATTRIBUTES	TOOLS AND TECHNIQUES			X		X	X	X		X	X	X	X
	CUSTOMER INTERFACE	X					X				X	X	
	REQUIREMENTS DEFINITION	X			X		X				X	X	X
	REQUIREMENTS VOLATILITY	X			X	X	X		X	X	X	X	X
	SCHEDULE			X		X				X	X	X	X
	SECURITY						X				X	X	
	COMPUTER ACCESS			X	X		X	X		X	X	X	X
	TRAVEL/REHOSTING/MULTI-SITE	X			X	X					X	X	X
	SUPPORT SOFTWARE MATURITY									X		X	
CALIBRATION FACTOR				X		X				X			
EFFORT EQUATION	$MM_{NOM} = C(DSI)^X$, X =		1.0		1.047		0.91	1.0		1.05 - 1.2		1.0	1.2
SCHEDULE EQUATION	$t_D = C\,(MM)^X$, X =						0.35			0.32 - 0.38		0.356	0.333

systems. For smaller systems, the percentage varies as a function of system size.

The SLIM model includes a number of useful extensions to estimate such quantities as manpower distribution, cash flow, major-milestone schedules, reliability levels, computer time, and documentation costs.

The most controversial aspect of the SLIM model is its tradeoff relationship between development effort K and between development time t_d. For a software product of a given size, the SLIM software equation above gives

$$K = \frac{\text{constant}}{t_d^4}.$$

For example, this relationship says that one can cut the cost of a software project in half, simply by increasing its de-

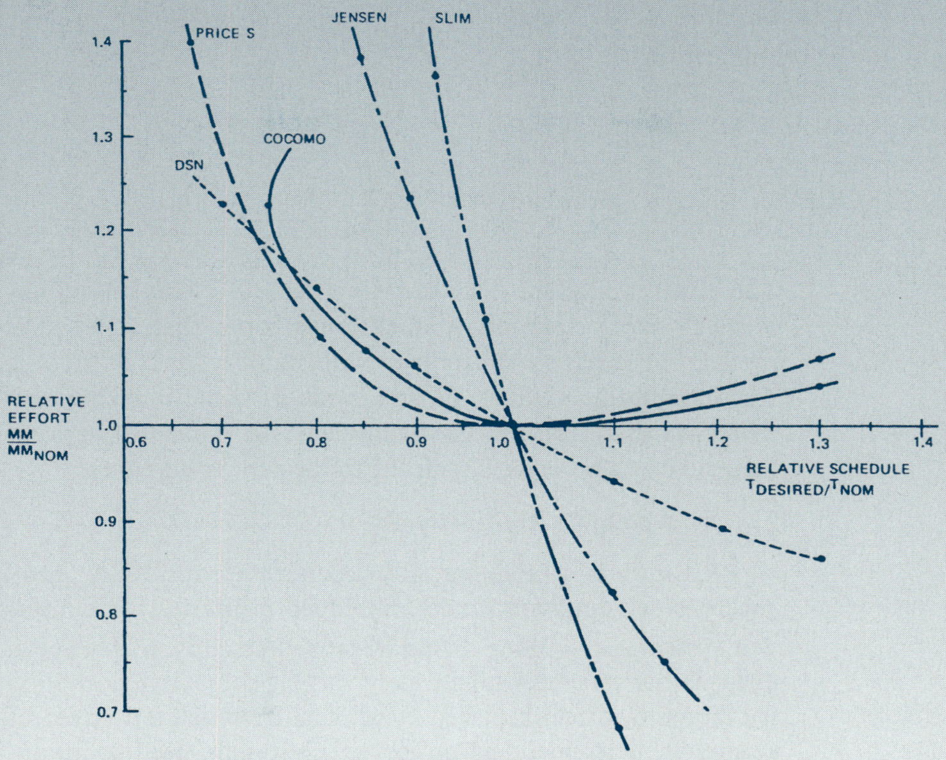

Fig. 5. Comparative effort-schedule tradeoff relationships.

velopment time by 19 percent (e.g., from 10 months to 12 months). Fig. 5 shows how the SLIM tradeoff relationship compares with those of other models; see [11, ch. 27] for further discussion of this issue.

On balance, the SLIM approach has provided a number of useful insights into software cost estimation, such as the Rayleigh-curve distribution for one-shot software efforts, the explicit treatment of estimation risk and uncertainty, and the cube-root relationship defining the minimum development time achievable for a project requiring a given amount of effort.

The Doty Model [27]

This model is the result of an extensive data analysis activity, including many of the data points from the SDC sample. A number of models of similar form were developed for different application areas. As an example, the model for general application is

$$MM = 5.288\,(\mathrm{KDSI})^{1.047}, \qquad \text{for KDSI} \geqslant 10$$

$$MM = 2.060\,(\mathrm{KDSI})^{1.047}\left(\prod_{j=1}^{14} f_j\right), \qquad \text{for KDSI} < 10.$$

The effort multipliers f_i are shown in Table III. This model has a much more appropriate functional form than the SDC model, but it has some problems with stability, as it exhibits a discontinuity at KDSI = 10, and produces widely varying estimates via the f factors (answering "yes" to "first software developed on CPU" adds 92 percent to the estimated cost).

The RCA PRICE S Model [22]

PRICE S is a commercially available (from RCA, Inc.) macro cost-estimation model developed primarily for embedded system applications. It has improved steadily with experience; earlier versions with a widely varying subjective complexity factor have been replaced by versions in which a number of computer, personnel, and project attributes are used to modulate the complexity rating.

TABLE III
DOTY MODEL FOR SMALL PROGRAMS*

$$MM = 2.060\, I^{1.047} \prod_{j=1}^{j=14} f_j$$

Factor	f_j	Yes	No
Special display	f_1	1.11	1.00
Detailed definition of operational requirements	f_2	1.00	1.11
Change to operational requirements	f_3	1.05	1.00
Real-time operation	f_4	1.33	1.00
CPU memory constraint	f_5	1.43	1.00
CPU time constraint	f_6	1.33	1.00
First software developed on CPU	f_7	1.92	1.00
Concurrent development of ADP hardware	f_8	1.82	1.00
Timeshare versus batch processing, in development	f_9	0.83	1.00
Developer using computer at another facility	f_{10}	1.43	1.00
Development at operational site	f_{11}	1.39	1.00
Development computer different than target computer	f_{12}	1.25	1.00
Development at more than one site	f_{13}	1.25	1.00
Programmer access to computer	f_{14}	Limited	1.00
		Unlimited	0.90

* Less than 10,000 source instructions

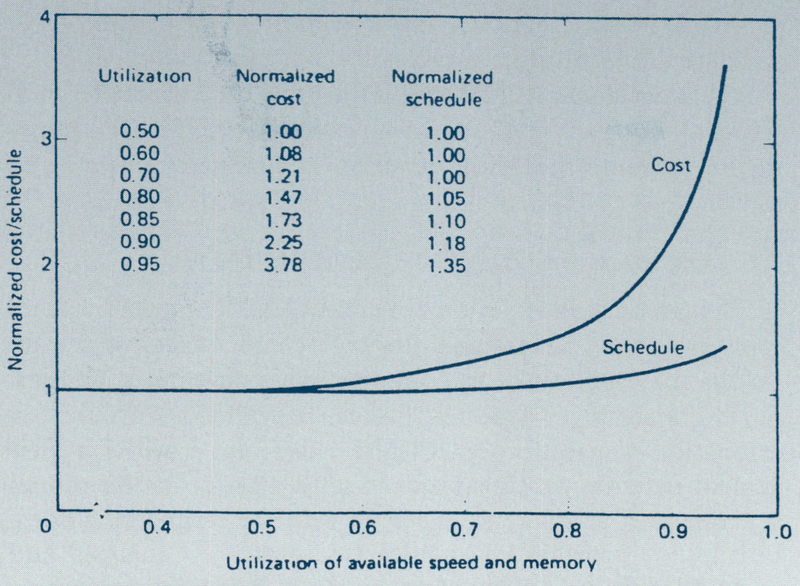

Fig. 6. RCA PRICE S model: Effect of hardware constraints.

PRICE S has extended a number of cost-estimating relationships developed in the early 1970's such as the hardware constraint function shown in Fig. 6 [10]. It was primarily developed to handle military software projects, but now also includes rating levels to cover business applications.

PRICE S also provides a wide range of useful outputs on gross phase and activity distributions analyses, and monthly project cost-schedule-expected progress forecasts. Price S uses a two-parameter beta distribution rather than a Rayleigh curve to calculate development effort distribution versus calendar time.

PRICE S has recently added a software life-cycle support cost estimation capability called PRICE SL [34]. It involves the definition of three categories of support activities.

- *Growth:* The estimator specifies the amount of code to be added to the product. PRICE SL then uses its standard techniques to estimate the resulting life-cycle-effort distribution.
- *Enhancement:* PRICE SL estimates the fraction of the existing product which will be modified (the estimator may

provide his own fraction), and uses its standard techniques to estimate the resulting life-cycle effort distribution.

- *Maintenance:* The estimator provides a parameter indicating the quality level of the developed code. PRICE SL uses this to estimate the effort required to eliminate remaining errors.

The COnstructive COst MOdel (COCOMO) [11]

The primary motivation for the COCOMO model has been to help people understand the cost consequences of the decisions they will make in commissioning, developing, and supporting a software product. Besides providing a software cost estimation capability, COCOMO therefore provides a great deal of material which explains exactly what costs the model is estimating, and why it comes up with the estimates it does. Further, it provides capabilities for sensitivity analysis and tradeoff analysis of many of the common software engineering decision issues.

COCOMO is actually a hierarchy of three increasingly detailed models which range from a single macro-estimation scaling model as a function of product size to a micro-estimation model with a three-level work breakdown structure and a set of phase-sensitive multipliers for each cost driver attribute. To provide a reasonably concise example of a current state of the art cost estimation model, the intermediate level of COCOMO is described below.

Intermediate COCOMO estimates the cost of a proposed software product in the following way.

1) A nominal development effort is estimated as a function of the product's size in delivered source instructions in thousands (KDSI) and the project's development mode.

2) A set of effort multipliers are determined from the product's ratings on a set of 15 cost driver attributes.

3) The estimated development effort is obtained by multiplying the nominal effort estimate by all of the product's effort multipliers.

4) Additional factors can be used to determine dollar costs, development schedules, phase and activity distributions, computer costs, annual maintenance costs, and other elements from the development effort estimate.

TABLE IV
COCOMO SOFTWARE DEVELOPMENT MODES

Feature	Mode: Organic	Semidetached	Embedded
Organizational understanding of product objectives	Thorough	Considerable	General
Experience in working with related software systems	Extensive	Considerable	Moderate
Need for software conformance with pre-established requirements	Basic	Considerable	Full
Need for software conformance with external interface specifications	Basic	Considerable	Full
Concurrent development of associated new hardware and operational procedures	Some	Moderate	Extensive
Need for innovative data processing architectures, algorithms	Minimal	Some	Considerable
Premium on early completion	Low	Medium	High
Product size range	<50 KDSI	<300 KDSI	All sizes
Examples	Batch data reduction Scientific models Business models Familiar OS, compiler Simple inventory, production control	Most transaction processing systems New OS, DBMS Ambitious inventory, production control Simple command-control	Large, complex transaction processing systems Ambitious, very large OS Avionics Ambitious command-control

Step 1–Nominal Effort Estimation: First, Table IV is used to determine the project's development mode. Organic-mode projects typically come from stable, familiar, forgiving, relatively unconstrained environments, and were found in the COCOMO data analysis of 63 projects have a different scaling equation from the more ambitious, unfamiliar, unforgiving, tightly constrained embedded mode. The resulting scaling equations for each mode are given in Table V; these are used to determine the nominal development effort for the project in man-months as a function of the project's size in KDSI and the project's development mode.

For example, suppose we are estimating the cost to develop the microprocessor-based communications processing software for a highly ambitious new electronic funds transfer network with high reliability, performance, development schedule, and interface requirements. From Table IV, we determine

TABLE V
COCOMO NOMINAL EFFORT AND SCHEDULE EQUATIONS

DEVELOPMENT MODE	NOMINAL EFFORT	SCHEDULE
Organic	$(MM)_{NOM} = 3.2(KDSI)^{1.05}$	$TDEV = 2.5(MM_{DEV})^{0.38}$
Semidetached	$(MM)_{NOM} = 3.0(KDSI)^{1.12}$	$TDEV = 2.5(MM_{DEV})^{0.35}$
Embedded	$(MM)_{NOM} = 2.8(KDSI)^{1.20}$	$TDEV = 2.5(MM_{DEV})^{0.32}$

(KDSI = thousands of delivered source instructions)

that these characteristics best fit the profile of an embedded-mode project.

We next estimate the size of the product as 10 000 delivered source instructions, or 10 KDSI. From Table V, we then determine that the nominal development effort for this Embedded-mode project is

$$2.8(10)^{1.20} = 44 \text{ man-months (MM)}.$$

Step 2–Determine Effort Multipliers: Each of the 15 cost driver attributes in COCOMO has a rating scale and a set of effort multipliers which indicate by how much the nominal effort estimate must be multiplied to account for the project's having to work at its rating level for the attribute.

These cost driver attributes and their corresponding effort multipliers are shown in Table VI. The summary rating scales for each cost driver attribute are shown in Table VII, except for the complexity rating scale which is shown in Table VIII (expanded rating scales for the other attributes are provided in [11]).

The results of applying these tables to our microprocessor communications software example are shown in Table IX. The effect of a software fault in the electronic fund transfer system could be a serious financial loss; therefore, the project's RELY rating from Table VII is High. Then, from Table VI, the effort multiplier for achieving a High level of required reliability is 1.15, or 15 percent more effort than it would take to develop the software to a nominal level of required reliability.

TABLE VI
INTERMEDIATE COCOMO SOFTWARE DEVELOPMENT EFFORT MULTIPLIERS

Cost Drivers	Ratings					
	Very Low	Low	Nominal	High	Very High	Extra High
Product Attributes						
RELY Required software reliability	.75	.88	1.00	1.15	1.40	
DATA Data base size		.94	1.00	1.08	1.16	
CPLX Product complexity	.70	.85	1.00	1.15	1.30	1.65
Computer Attributes						
TIME Execution time constraint			1.00	1.11	1.30	1.66
STOR Main storage constraint			1.00	1.06	1.21	1.56
VIRT Virtual machine volatility*		.87	1.00	1.15	1.30	
TURN Computer turnaround time		.87	1.00	1.07	1.15	
Personnel Attributes						
ACAP Analyst capability	1.46	1.19	1.00	.86	.71	
AEXP Applications experience	1.29	1.13	1.00	.91	.82	
PCAP Programmer capability	1.42	1.17	1.00	.86	.70	
VEXP Virtual machine experience*	1.21	1.10	1.00	.90		
LEXP Programming language experience	1.14	1.07	1.00	.95		
Project Attributes						
MODP Use of modern programming practices	1.24	1.10	1.00	.91	.82	
TOOL Use of software tools	1.24	1.10	1.00	.91	.83	
SCED Required development schedule	1.23	1.08	1.00	1.04	1.10	

* For a given software product, the underlying virtual machine is the complex of hardware and software (OS, DBMS, etc.) it calls on to accomplish its tasks.

The effort multipliers for the other cost driver attributes are obtained similarly, except for the Complexity attribute, which is obtained via Table VIII. Here, we first determine that communications processing is best classified under device-dependent operations (column 3 in Table VIII). From this column, we determine that communication line handling typically has a complexity rating of Very High; from Table VI, then, we determine that its corresponding effort multiplier is 1.30.

Step 3–Estimate Development Effort: We then compute the estimated development effort for the microprocessor communications software as the nominal development effort (44 MM) times the product of the effort multipliers for the 15 cost driver attributes in Table IX (1.35, in Table IX). The resulting estimated effort for the project is then

(44 MM) (1.35) = 59 MM.

Step 4–Estimate Related Project Factors: COCOMO has additional cost estimating relationships for computing the re-

TABLE VII
COCOMO SOFTWARE COST DRIVER RATINGS

Cost Driver	Ratings					
	Very Low	Low	Nominal	High	Very High	Extra High
Product attributes						
RELY	Effect: slight inconvenience	Low, easily recoverable losses	Moderate, recoverable losses	High financial loss	Risk to human life	
DATA		$\frac{\text{DB bytes}}{\text{Prog. DSI}} < 10$	$10 \leq \frac{D}{P} < 100$	$100 \leq \frac{D}{P} < 1000$	$\frac{D}{P} \geq 1000$	
CPLX	See Table 8					
Computer attributes						
TIME			≤ 50% use of available execution time	70%	85%	95%
STOR			≤ 50% use of available storage	70%	85%	95%
VIRT		Major change every 12 months Minor: 1 month	Major: 6 months Minor: 2 weeks	Major: 2 months Minor: 1 week	Major: 2 weeks Minor: 2 days	
TURN		Interactive	Average turnaround <4 hours	4–12 hours	>12 hours	
Personnel attributes						
ACAP	15th percentile*	35th percentile	55th percentile	75th percentile	90th percentile	
AEXP	≤4 months experience	1 year	3 years	6 years	12 years	
PCAP	15th percentile*	35th percentile	55th percentile	75th percentile	90th percentile	
VEXP	≤1 month experience	4 months	1 year	3 years		
LEXP	≤1 month experience	4 months	1 year	3 years		
Project attributes						
MODP	No use	Beginning use	Some use	General use	Routine use	
TOOL	Basic microprocessor tools	Basic mini tools	Basic midi/maxi tools	Strong maxi programming, test tools	Add requirements, design, management, documentation tools	
SCED	75% of nominal	85%	100%	130%	160%	

* Team rating criteria: analysis (programming) ability, efficiency, ability to communicate and cooperate

sulting dollar cost of the project and for the breakdown of cost and effort by life-cycle phase (requirements, design, etc.) and by type of project activity (programming, test planning, management, etc.). Further relationships support the estimation of the project's schedule and its phase distribution. For example, the recommended development schedule can be obtained from the estimated development man-months via the embedded-mode schedule equation in Table V:

$$T_{DEV} = 2.5(59)^{0.32} = 9 \text{ months.}$$

As mentioned above, COCOMO also supports the most common types of sensitivity analysis and tradeoff analysis involved in scoping a software project. For example, from Tables VI and VII, we can see that providing the software developers with an interactive computer access capability (Low turnaround time) reduces the TURN effort multiplier from 1.00 to

TABLE VIII
COCOMO MODULE COMPLEXITY RATINGS VERSUS TYPE OF MODULE

Rating	Control Operations	Computational Operations	Device-dependent Operations	Data Management Operations
Very low	Straightline code with a few non-nested SP[a] operators: DOs, CASEs, IFTHENELSEs. Simple predicates	Evaluation of simple expressions: e.g., A = B + C * (D − E)	Simple read, write statements with simple formats	Simple arrays in main memory
Low	Straightforward nesting of SP operators. Mostly simple predicates	Evaluation of moderate-level expressions, e.g., D = SQRT (B**2−4.*A*C)	No cognizance needed of particular processor or I/O device characteristics. I/O done at GET/PUT level. No cognizance of overlap	Single file subsetting with no data structure changes, no edits, no intermediate files
Nominal	Mostly simple nesting. Some intermodule control. Decision tables	Use of standard math and statistical routines. Basic matrix/vector operations	I/O processing includes device selection, status checking and error processing	Multi-file input and single file output. Simple structural changes, simple edits
High	Highly nested SP operators with many compound predicates. Queue and stack control. Considerable intermodule control.	Basic numerical analysis: multivariate interpolation, ordinary differential equations. Basic truncation, roundoff concerns	Operations at physical I/O level (physical storage address translations; seeks, reads, etc). Optimized I/O overlap	Special purpose subroutines activated by data stream contents. Complex data restructuring at record level
Very high	Reentrant and recursive coding. Fixed-priority interrupt handling	Difficult but structured N.A.: near-singular matrix equations, partial differential equations	Routines for interrupt diagnosis, servicing, masking. Communication line handling	A generalized, parameter-driven file structuring routine. File building, command processing, search optimization
Extra high	Multiple resource scheduling with dynamically changing priorities. Microcode-level control	Difficult and unstructured N.A.: highly accurate analysis of noisy, stochastic data	Device timing-dependent coding, micro-programmed operations	Highly coupled, dynamic relational structures. Natural language data management

[a] SP = structured programming

TABLE IX
COCOMO COST DRIVER RATINGS: MICROPROCESSOR COMMUNICATIONS SOFTWARE

Cost Driver	Situation	Rating	Effort Multiplier
RELY	Serious financial consequences of software faults	High	1.15
DATA	20,000 bytes	Low	0.94
CPLX	Communications processing	Very High	1.30
TIME	Will use 70% of available time	High	1.11
STOR	45K of 64K store (70%)	High	1.06
VIRT	Based on commercial microprocessor hardware	Nominal	1.00
TURN	Two-hour average turnaround time	Nominal	1.00
ACAP	Good senior analysts	High	0.86
AEXP	Three years	Nominal	1.00
PCAP	Good senior programmers	High	0.86
VEXP	Six months	Low	1.10
LEXP	Twelve months	Nominal	1.00
MODP	Most techniques in use over one year	High	0.91
TOOL	At basic minicomputer tool level	Low	1.10
SCED	Nine months	Nominal	1.00
	Effort adjustment factor (product of effort multipliers)		1.35

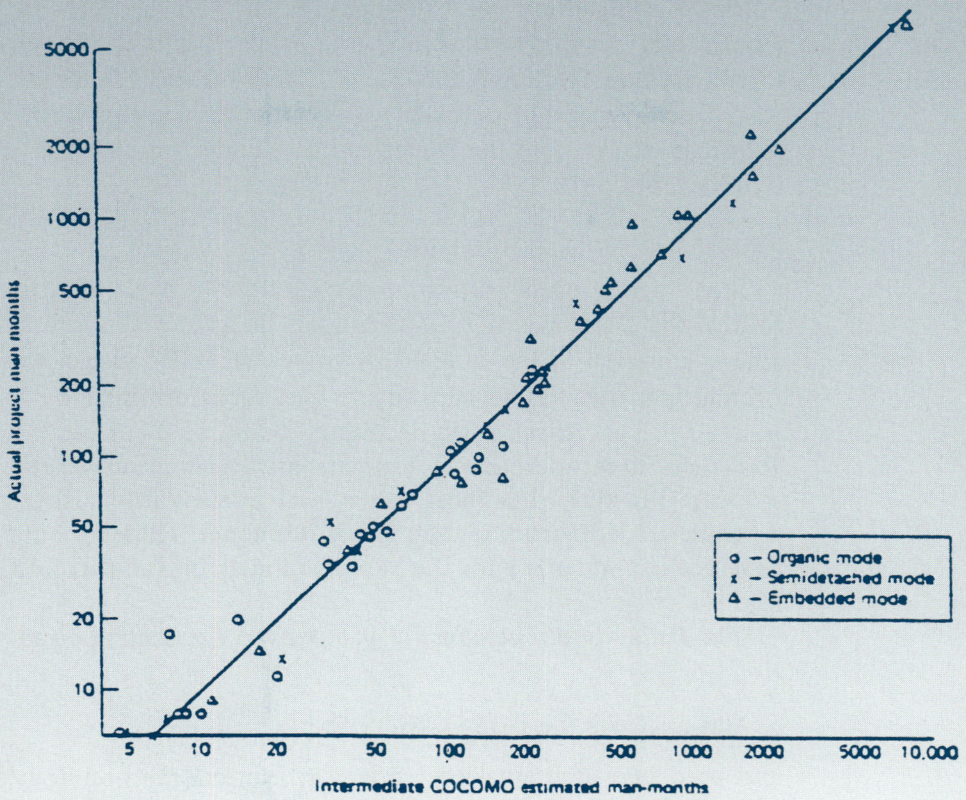

Fig. 7. Intermediate COCOMO estimates versus project actuals.

0.87, and thus reduces the estimated project effort from 59 MM to

(59 MM) (0.87) = 51 MM.

The COCOMO model has been validated with respect to a sample of 63 projects representing a wide variety of business, scientific, systems, real-time, and support software projects. For this sample, Intermediate COCOMO estimates come within 20 percent of the actuals about 68 percent of the time (see Fig. 7). Since the residuals roughly follow a normal distribution, this is equivalent to a standard deviation of roughly 20 percent of the project actuals. This level of accuracy is representative of the current state of the art in software cost models. One can do somewhat better with the aid of a calibration coefficient (also a COCOMO option), or within a limited applications context, but it is difficult to improve significantly on this level of accuracy while the accuracy of software data collection remains in the "±20 percent" range.

A Pascal version of COCOMO is available for a nominal distribution charge from the Wang Institute, under the name WICOMO [18].

Recent Software Cost Estimation Models

Most of the recent software cost estimation models tend to follow the Doty and COCOMO models in having a nominal scaling equation of the form $MM_{NOM} = c(KDSI)^x$ and a set of multiplicative effort adjustment factors determined by a number of cost driver attribute ratings. Some of them use the Rayleigh curve approach to estimate distribution across the software life-cycle, but most use a more conservative effort/schedule tradeoff relation than the SLIM model. These aspects have been summarized for the various models in Table II and Fig. 5.

The Bailey-Basili meta-model [4] derived the scaling equation

$$MM_{NOM} = 3.5 + 0.73\,(KDSI)^{1.16}$$

and used two additional cost driver attributes (methodology level and complexity) to model the development effort of 18 projects in the NASA-Goddard Software Engineering Laboratory to within a standard deviation of 15 percent. Its accuracy for other project situations has not been determined.

The Grumman SOFCOST Model [19] uses a similar but unpublished nominal effort scaling equation, modified by 30 multiplicative cost driver variables rated on a scale of 0 to 10. Table II includes a summary of these variables.

The Tausworthe Deep Space Network (DSN) model [50] uses a linear scaling equation ($MM_{NOM} = a(KDSI)^{1.0}$) and a similar set of cost driver attributes, also summarized in Table II. It also has a well-considered approach for determining the equivalent KDSI involved in adapting existing software within a new product. It uses the Rayleigh curve to determine the phase distribution of effort, but uses a considerably more conservative version of the SLIM effort-schedule tradeoff relationship (see Fig. 5).

The Jensen model [30], [31] is a commercially available model with a similar nominal scaling equation, and a set of cost driver attributes very similar to the Doty and COCOMO models (but with different effort multiplier ranges); see Table II. Some of the multiplier ranges in the Jensen model vary as functions of other factors; e.g., increasing access to computer resources widens the multiplier ranges on such cost drivers as personnel capability and use of software tools. It uses the Rayleigh curve for effort distribution, and a somewhat more conservative effort-schedule tradeoff relation than SLIM (see Fig. 5). As with the other commercial models, the Jensen model produces a number of useful outputs on resource expenditure rates, probability distributions on costs and schedules, etc.

C. Outstanding Research Issues in Software Cost Estimation

Although a good deal of progress has been made in software cost estimation, a great deal remains to be done. This section updates the state-of-the-art review published in [11], and summarizes the outstanding issues needing further research:

1) Software size estimation;
2) Software size and complexity metrics;
3) Software cost driver attributes and their effects;
4) Software cost model analysis and refinement;
5) Quantitative models of software project dynamics;
6) Quantitative models of software life-cycle evolution;
7) Software data collection.

1) Software Size Estimation: The biggest difficulty in using today's algorithmic software cost models is the problem of providing sound sizing estimates. Virtually every model requires an estimate of the number of source or object instructions to be developed, and this is an extremely difficult quantity to determine in advance. It would be most useful to have some formula for determining the size of a software product in terms of quantities known early in the software life cycle, such as the number and/or size of the files, input formats, reports, displays, requirements specification elements, or design specification elements.

Some useful steps in this direction are the function-point approach in [2] and the sizing estimation model of [29], both of which have given reasonably good results for small-to-medium sized business programs within a single data processing organization. Another more general approach is given by DeMarco in [17]. It has the advantage of basing its sizing estimates on the properties of specifications developed in conformance with DeMarco's paradigm models for software specifications and designs: number of functional primitives, data elements, input elements, output elements, states, transitions between states, relations, modules, data tokens, control tokens, etc. To date, however, there has been relatively little calibration of the formulas to project data. A recent IBM study [14] shows some correlation between the number of variables defined in a state-machine design representation and the product size in source instructions.

Although some useful results can be obtained on the software sizing problem, one should not expect too much. A wide range of functionality can be implemented beneath any given specification element or I/O element, leading to a wide range of sizes (recall the uncertainty ranges of this nature in Fig. 3). For example, two experiments, involving the use of several teams developing a software program to the same overall functional specification, yielded size ranges of factors of 3 to 5 between programs (see Table X).

TABLE X
SIZE RANGES OF SOFTWARE PRODUCTS PERFORMING SAME FUNCTION

Experiment	Product	No. of Teams	Size Range (source-instr.)
Weinberg & Schulman [55]	Simultaneous linear equations	6	33-165
Boehm, Gray, & Seewaldt [13]	Interactive cost model	7	1514-4606

The primary implication of this situation for practical software sizing and cost estimation is that *there is no royal road to software sizing*. This is no magic formula that will provide an easy and accurate substitute for the process of thinking through and fully understanding the nature of the software product to be developed. There are still a number of useful things that one can do to improve the situation, including the following.

- Use techniques which explicitly recognize the ranges of variability in software sizing. The PERT estimation technique [56] is a good example.
- Understand the primary sources of bias in software sizing estimates. See [11, ch. 21].
- Develop and use a corporate memory on the nature and size of previous software products.

2) Software Size and Complexity Metrics: Delivered source instructions (DSI) can be faulted for being too low-level a metric for use in early sizing estimation. On the other hand, DSI can also be faulted for being too high-level a metric for precise software cost estimation. Various complexity metrics have been formulated to more accurately capture the relative information content of a program's instructions, such as the Halstead Software Science metrics [24], or to capture the relative control complexity of a program, such as the metrics formulated by McCabe in [39]. A number of variations of these metrics have been developed; a good recent survey of them is given in [26].

However, these metrics have yet to exhibit any practical superiority to DSI as a predictor of the relative effort required to develop software. Most recent studies [48], [32] show a reasonable correlation between these complexity metrics and development effort, but no better a correlation than that between DSI and development effort.

Further, the recent [25] analysis of the software science results indicates that many of the published software science "successes" were not as successful as they were previously considered. It indicates that much of the apparent agreement between software science formulas and project data was due to factors overlooked in the data analysis: inconsistent definitions and interpretations of software science quantities, unrealistic or inconsistent assumptions about the nature of the projects analyzed, overinterpretation of the significance of statistical measures such as the correlation coefficient, and lack of investigation of alternative explanations for the data. The software science use of psychological concepts such as the Stroud number have also been seriously questioned in [16].

The overall strengths and difficulties of software science are summarized in [47]. Despite the difficulties, some of the software science metrics have been useful in such areas as identifying error-prone modules. In general, there is a strong intuitive argument that more definitive complexity metrics will eventually serve as better bases for definitive software cost estimation than will DSI. Thus, the area continues to be an attractive one for further research.

3) Software Cost Driver Attributes and Their Effects: Most of the software cost models discussed above contain a selection of cost driver attributes and a set of coefficients, functions, or tables representing the effect of the attribute on software cost (see Table II). Chapters 24–28 of [11] contain summaries of the research to date on about 20 of the most significant cost driver attributes, plus statements of nearly 100 outstanding research issues in the area.

Since the publication of [11] in 1981, a few new results have appeared. Lawrence [35] provides an analysis of 278 business data processing programs which indicate a fairly uniform development rate in procedure lines of code per hour, some significant effects on programming rate due to batch turnaround time and level of experience, and relatively little effect due to use of interactive operation and modern programming practices (due, perhaps, to the relatively repetitive nature of the software jobs sampled). Okada and Azuma [42] analyzed 30 CAD/CAM programs and found some significant effects due to type of software, complexity, personnel skill level, and requirements volatility.

4) Software Cost Model Analysis and Refinement: The most useful comparative analysis of software cost models to date is the Thibodeau [52] study performed for the U.S. Air Force. This study compared the results of several models (the Wolverton, Doty, PRICE S, and SLIM models discussed earlier, plus models from the Boeing, SDC, Tecolote, and Aerospace corporations) with respect to 45 project data points from three sources.

Some generally useful comparative results were obtained, but the results were not definitive, as models were evaluated with respect to larger and smaller subsets of the data. Not too surprisingly, the best results were generally obtained using models with calibration coefficients against data sets with few points. In general, the study concluded that the models with calibration coefficients achieved better results, but that none of the models evaluated were sufficiently accurate to be used as a definitive Air Force software cost estimation model.

Some further comparative analyses are currently being conducted by various organizations, using the database of 63 software projects in [11], but to date none of these have been published.

In general, such evaluations play a useful role in model refinement. As certain models are found to be inaccurate in certain situations, efforts are made to determine the causes, and to refine the model to eliminate the sources of inaccuracy.

Relatively less activity has been devoted to the formulation, evaluation, and refinement of models to cover the effects of more advanced methods of software development (prototyping, incremental development, use of application generators, etc.) or to estimate other software-related life-cycle costs (conversion, maintenance, installation, training, etc.). An exception is the excellent work on software conversion cost estimation performed by the Federal Conversion Support Center [28]. An extensive model to estimate avionics software support costs using a weighted-multiplier technique has recently been developed [49]. Also, some initial experimental results have been obtained on the quantitative impact of prototyping in [13] and on the impact of very high level nonprocedural languages in [58]. In both studies, projects using prototyping and VHLL's were completed with significantly less effort.

5) Quantitative Models of Software Project Dynamics: Current software cost estimation models are limited in their ability to represent the internal dynamics of a software project, and to estimate how the project's phase distribution of effort and schedule will be affected by environmental or project management factors. For example, it would be valuable to have a model which would accurately predict the effort and schedule distribution effects of investing in more thorough design verification, of pursuing an incremental development strategy, of varying the staffing rate or experience mix, of reducing module size, etc.

Some current models assume a universal effort distribution, such as the Rayleigh curve [44] or the activity distributions in [57], which are assumed to hold for any type of project situation. Somewhat more realistic, but still limited are models with phase-sensitive effort multipliers such as PRICE S [22] and Detailed COCOMO [11].

Recently, some more realistic models of software project dynamics have begun to appear, although to date none of them have been calibrated to software project data. The Phister phase-by-phase model in [43] estimates the effort and schedule required to design, code, and test a software product as a function of such variables as the staffing level during each phase, the size of the average module to be developed, and such factors as interpersonal communications overhead rates and error detection rates. The Abdel Hamid–Madnick model [1], based on Forrester's System Dynamics world-view, estimates the time distribution of effort, schedule, and residual defects as a function of such factors as staffing rates, experience mix, training rates, personnel turnover, defect introduction rates, and initial estimation errors. Tausworthe [51] derives and calibrates alternative versions of the SLIM effort-schedule tradeoff relationship, using an intercommunication-overhead model of project dynamics. Some other recent models of software project dynamics are the Mitre SWAP model and the Duclos [21] total software life-cycle model.

6) Quantitative Models of Software Life-Cycle Evolution: Although most of the software effort is devoted to the software maintenance (or life-cycle support) phase, only a few significant results have been obtained to date in formulating quantitative models of the software life-cycle evolution process. Some basic studies by Belady and Lehman analyzed data on several projects and derived a set of fairly general "laws of program evolution" [7], [37]. For example, the first of these laws states:

> "A program that is used and that as an implementation of its specification reflects some other reality, undergoes continual change or becomes progressively less useful. The change or decay process continues until it is judged more cost effective to replace the system with a re-created version."

Some general quantitative support for these laws was obtained in several studies during the 1970's, and in more recent studies such as [33]. However, efforts to refine these general laws into a set of testable hypotheses have met with mixed results. For example, the Lawrence [36] statistical analysis of the Belady-Lahman data showed that the data supported an even stronger form of the first law ("systems grow in size over their useful life"); that one of the laws could not be formulated precisely enough to be tested by the data; and that the other three laws did not lead to hypotheses that were supported by the data.

However, it is likely that variant hypotheses can be found that are supported by the data (for example, the operating system data supports some of the hypotheses better than does the applications data). Further research is needed to clarify this important area.

7) Software Data Collection: A fundamental limitation to significant progress in software cost estimation is the lack of unambiguous, widely-used standard definitions for software data. For example, if an organization reports its "software development man-months," do these include the effort devoted to requirements analysis, to training, to secretaries, to quality assurance, to technical writers, to uncompensated overtime? Depending on one's interpretations, one can easily cause variations of over 20 percent (and often over a factor of 2) in the meaning of reported "software development man-months" between organizations (and similarly for "delivered instructions," "complexity," "storage constraint," etc.) Given such uncertainties in the ground data, it is not surprising that software cost estimation models cannot do much better than "within 20 percent of the actuals, 70 percent of the time."

Some progress towards clear software data definitions has been made. The IBM FSD database used in [53] was carefully collected using thorough data definitions, but the detailed data and definitions are not generally available. The NASA-Goddard Software Engineering Laboratory database [5], [6], [40] and the COCOMO database [11] provide both clear data definitions and an associated project database which are available for general use (and reasonably compatible). The recent Mitre SARE report [59] provides a good set of data definitions.

But there is still no commitment across organizations to establish and use a set of clear and uniform software data definitions. Until this happens, our progress in developing more precise software cost estimation methods will be severely limited.

IV. Software Engineering Economics Benefits and Challenges

This final section summarizes the benefits to software engineering and software management provided by a software engineering economics perspective in general and by software cost estimation technology in particular. It concludes with some observations on the major challenges awaiting the field.

Benefits of a Software Engineering Economics Perspective

The major benefit of an economic perspective on software engineering is that it provides a balanced view of candidate software engineering solutions, and an evaluation framework which takes account not only of the programming aspects of a situation, but also of the human problems of providing the best possible information processing service within a resource-limited environment. Thus, for example, the software engineering economics approach does not say, "we should use these structured structures because they are mathematically elegant" or "because they run like the wind" or "because they are part of the structured revolution." Instead, it says "we should use these structured structures because they provide people with more benefits in relation to their costs than do other approaches." And besides the framework, of course, it also provides the techniques which help us to arrive at this conclusion.

Benefits of Software Cost Estimation Technology

The major benefit of a good software cost estimation model is that it provides a clear and consistent universe of discourse within which to address a good many of the software engineering issues which arise throughout the software life cycle. It can help people get together to discuss such issues as the following.

- Which and how many features should we put into the software product?
- Which features should we put in first?
- How much hardware should we acquire to support the software product's development, operation, and maintenance?
- How much money and how much calendar time should we allow for software development?

- How much of the product should we adapt from existing software?
- How much should we invest in tools and training?

Further, a well-defined software cost estimation model can help avoid the frequent misinterpretations, underestimates, overexpectations, and outright buy-ins which still plague the software field. In a good cost-estimation model, there is no way of reducing the estimated software cost without changing some objectively verifiable property of the software project. This does not make it impossible to create an unachievable buy-in, but it significantly raises the threshold of credibility.

A related benefit of software cost estimation technology is that it provides a powerful set of insights on how a software organization can improve its productivity. Many of a software cost model's cost-driver attributes are management controllables: use of software tools and modern programming practices, personnel capability and experience, available computer speed, memory, and turnaround time, software reuse. The cost model helps us determine how to adjust these management controllables to increase productivity, and further provides an estimate of how much of a productivity increase we are likely to achieve with a given level of investment. For more information on this topic, see [11, ch. 33], [12] and the recent plan for the U.S. Department of Defense Software Initiative [20].

Finally, software cost estimation technology provides an absolutely essential foundation for software project planning and control. Unless a software project has clear definitions of its key milestones and realistic estimates of the time and money it will take to achieve them, there is no way that a project manager can tell whether his project is under control or not. A good set of cost and schedule estimates can provide realistic data for the PERT charts, work breakdown structures, manpower schedules, earned value increments, etc., necessary to establish management visibility and control.

Note that this opportunity to improve management visibility and control requires a complementary management commitment to define and control the reporting of data on software progress and expenditures. The resulting data are therefore worth collecting simply for their management value in comparing plans versus achievements, but they can serve another valuable function as well: they provide a continuing stream of calibration data for evolving a more accurate and refined software cost estimation models.

Software Engineering Economics Challenges

The opportunity to improve software project management decision making through improved software cost estimation, planning, data collection, and control brings us back full-circle to the original objectives of software engineering economics: to provide a better quantitative understanding of how software people make decisions in resource-limited situations.

The more clearly we as software engineers can understand the quantitative and economic aspects of our decision situations, the more quickly we can progress from a pure seat-of-the-pants approach on software decisions to a more rational approach which puts all of the human and economic decision variables into clear perspective. Once these decision situations are more clearly illuminated, we can then study them in more detail to address the deeper challenge: achieving a quantitative understanding of how people work together in the software engineering process.

Given the rather scattered and imprecise data currently available in the software engineering field, it is remarkable how much progress has been made on the software cost estimation problem so far. But, there is not much further we can go until better data becomes available. The software field cannot hope to have its Kepler or its Newton until it has had its army of Tycho Brahes, carefully preparing the well-defined observational data from which a deeper set of scientific insights may be derived.

References

[1] T. K. Abdel-Hamid and S. E. Madnick, "A model of software project management dynamics," in *Proc. IEEE COMPSAC 82*, Nov. 1982, pp. 539–554.

[2] A. J. Albrecht, "Measuring Application Development Productivity," in *SHARE-GUIDE*, 1979, pp. 83–92.

[3] J. D. Aron, "Estimating resources for large programming systems." NATO Sci. Committee, Rome, Italy, Oct. 1969.

[4] J. J. Bailey and V. R. Basili, "A meta-model for software development resource expenditures," in *Proc. 5th Int. Conf. Software Eng.*, IEEE/ACM/NBS, Mar. 1981, pp. 107–116.

[5] V. R. Basili, "Tutorial on models and metrics for software and engineering," IEEE Cat. EHO-167-7, Oct. 1980.

[6] V. R. Basili and D. M. Weiss, "A methodology for collecting valid software engineering data," Univ. Maryland Technol. Rep. TR-1235, Dec. 1982.

[7] L. A. Belady and M. M. Lehman, "Characteristics of large systems," in *Research Directions in Software Technology*, P. Wegner, Ed. Cambridge, MA: MIT Press, 1979.

[8] H. D. Benington, "Production of large computer programs," in *Proc. ONR Symp. Advanced Programming Methods for Digital Computers*, June 1956, pp. 15–27.

[9] R. K. D. Black, R. P. Curnow, R. Katz, and M. D. Gray, "BCS software production data," Boeing Comput. Services, Inc., Final Tech. Rep., RADC-TR-77-116, NTIS AD-A039852, Mar. 1977.

[10] B. W. Boehm, "Software and its impact: A quantitative assessment," *Datamation*, pp. 48–59, May 1973.

[11] ——, *Software Engineering Economics*. Englewood Cliffs, NJ: Prentice-Hall, 1981.

[12] B. W. Boehm, J. F. Elwell, A. B. Pyster, E. D. Stuckle, and R. D. Williams, "The TRW software productivity system," in *Proc. IEEE 6th Int. Conf. Software Eng.*, Sept. 1982.

[13] B. W. Boehm, T. E. Gray, and T. Seewaldt, "Prototyping vs. specifying: A multi-project experiment," *IEEE Trans. Software Eng.*, to be published.

[14] R. N. Britcher and J. E. Gaffney, "Estimates of software size from state machine designs," in *Proc. NASA-Goddard Software Eng. Workshop*, Dec. 1982.

[15] W. M. Carriere and R. Thibodeau, "Development of a logistics software cost estimating technique for foreign military sales," General Res. Corp., Rep. CR-3-839, June 1979.

[16] N. S. Coulter, "Software science and cognitive psychology," *IEEE Trans. Software Eng.*, pp. 166–171, Mar. 1983.

[17] T. DeMarco, *Controlling Software Projects*. New York: Yourdon, 1982.

[18] M. Demshki, D. Ligett, B. Linn, G. McCluskey, and R. Miller, "Wang Institute cost model (WICOMO) tool user's manual," Wang Inst. Graduate Studies, Tyngsboro, MA, June 1982.

[19] H. F. Dircks, "SOFCOST: Grumman's software cost eliminating model," in *IEEE NAECON 1981*, May 1981.

[20] L. E. Druffel, "Strategy for DoD software initiative," RADC/DACS, Griffiss AFB, NY, Oct. 1982.

[21] L. C. Duclos, "Simulation model for the life-cycle of a software product: A quality assurance approach," Ph.D. dissertation, Dep. Industrial and Syst. Eng., Univ. Southern California, Dec. 1982.

[22] F. R. Freiman and R. D. Park, "PRICE software model—Version 3: An overview," in *Proc. IEEE-PINY Workshop on Quantitative Software Models*, IEEE Cat. TH0067-9, Oct. 1979, pp. 32–41.

[23] R. Goldberg and H. Lorin, *The Economics of Information Processing*. New York: Wiley, 1982.

[24] M. H. Halstead, *Elements of Software Science*. New York: Elsevier, 1977.

[25] P. G. Hamer and G. D. Frewin, "M. H. Halstead's software science—A critical examination," in *Proc. IEEE 6th Int. Conf. Software Eng.*, Sept. 1982, pp. 197–205.

[26] W. Harrison, K. Magel, R. Kluczney, and A. DeKock, "Applying software complexity metrics to program maintenance," *Computer*, pp. 65–79, Sept. 1982.

[27] J. R. Herd, J. N. Postak, W. E. Russell, and K. R. Stewart, "Software cost estimation study—Study results," Doty Associates, Inc., Rockville, MD, Final Tech. Rep. RADC-TR-77-220, vol. 1 (of two), June 1977.

[28] C. Houtz and T. Buschbach, "Review and analysis of conversion cost-estimating techniques," GSA Federal Conversion Support Center, Falls Church, VA, Rep. GSA/FCSC-81/001, Mar. 1981.

[29] M. Itakura and A. Takayanagi, " A model for estimating program size and its evaluation," in *Proc. IEEE 6th Software Eng.*, Sept. 1982, pp. 104–109.

[30] R. W. Jensen, "An improved macrolevel software development resource estimation model," in *Proc. 5th ISPA Conf.*, Apr. 1983, pp. 88–92.

[31] R. W. Jensen and S. Lucas, "Sensitivity analysis of the Jensen software model," in *Proc. 5th ISPA Conf.*, Apr. 1983, pp. 384–389.

[32] B. A. Kitchenham, "Measures of programming complexity," *ICL Tech. J.*, pp. 298–316, May 1981.

[33] ——, "Systems evolution dynamics of VME/B," *ICL Tech. J.*, pp. 43–57, May 1982.

[34] W. W. Kuhn, "A software lifecycle case study using the PRICE model," in *Proc. IEEE NAECON*, May 1982.

[35] M. J. Lawrence, "Programming methodology, organizational environment,and programming productivity," *J. Syst. Software*, pp. 257–270, Sept. 1981.

[36] ——, "An examination of evolution dynamics," in *Proc. IEEE 6th Int. Conf. Software Eng.*, Sept. 1982, pp. 188–196.

[37] M. M. Lehman, "Programs, life cycles, and laws of software evolution," *Proc. IEEE*, pp. 1060–1076, Sept. 1980.

[38] R. D. Luce and H. Raiffa, *Games and Decisions*. New York: Wiley, 1957.

[39] T. J. McCabe, "A complexity measure," *IEEE Trans. Software Eng.*, pp. 308–320, Dec. 1976.

[40] F. E. McGarry, "Measuring software development technology: What have we learned in six years," in *Proc. NASA-Goddard Software Eng. Workshop*, Dec. 1982.

[41] E. A. Nelson, "Management handbook for the estimation of computer programming costs," Syst. Develop. Corp., AD-A648750, Oct. 31, 1966.

[42] M. Okada and M. Azuma, "Software development estimation study—A model from CAD/CAM system development experiences," in *Proc. IEEE COMPSAC 82*, Nov. 1982, pp. 555–564.

[43] M. Phister, Jr., "A model of the software development process," *J. Syst. Software*, pp. 237–256, Sept. 1981.

[44] L. H. Putnam, "A general empirical solution to the macro software sizing and estimating problem," *IEEE Trans. Software Eng.*, pp. 345–361, July 1978.

[45] L. H. Putnam and A. Fitzsimmons, "Estimating software costs," *Datamation*, pp. 189–198, Sept. 1979; continued in *Datamation*, pp. 171–178, Oct. 1979 and pp. 137–140, Nov. 1979.

[46] L.H. Putnam, "The real economics of software development," in *The Economics of Information Processing*, R. Goldberg and H. Lorin. New York: Wiley, 1982.

[47] V. Y. Shen, S. D. Conte, and H. E. Dunsmore, "Software science revisited: A critical analysis of the theory and its empirical support," *IEEE Trans. Software Eng.*, pp. 155–165, Mar. 1983.

[48] T. Sunohara, A. Takano, K. Uehara, and T. Ohkawa, "Program complexity measure for software development management," in *Proc. IEEE 5th Int. Conf. Software Eng.*, Mar. 1981, pp. 100–106.

[49] SYSCON Corp., "Avionics software support cost model," USAF Avionics Lab., AFWAL-TR-1173, Feb. 1, 1983.

[50] R. C. Tausworthe, "Deep space network software cost estimation model," Jet Propulsion Lab., Pasadena, CA, 1981.

[51] ——, "Staffing implications of software productivity models," in *Proc. 7th Annu. Software Eng. Workshop*, NASA/Goddard, Greenbelt, MD, Dec. 1982.

[52] R. Thibodeau, "An evaluation of software cost estimating models," General Res. Corp., Rep. T10-2670, Apr. 1981.

[53] C. E. Walston and C. P. Felix, "A method of programming measurement and estimation," *IBM Syst. J.*, vol. 16, no. 1, pp. 54–73, 1977.

[54] G. F. Weinwurm, Ed., *On the Management of Computer Programming*. New York: Auerbach, 1970.

[55] G. M. Weinberg and E. L. Schulman, "Goals and performance in computer programming," *Human Factors*, vol. 16, no. 1, pp. 70–77, 1974.

[56] J. D. Wiest and F. K. Levy, *A Management Guide to PERT/CPM*. Englewood Cliffs, NJ: Prentice-Hall, 1977.

[57] R. W. Wolverton, "The cost of developing large-scale software," *IEEE Trans. Comput.*, pp. 615–636, June 1974.

[58] E. Harel and E. R. McLean, "The effects of using a nonprocedural computer language on programmer productivity," UCLA Inform. Sci. Working Paper 3-83, Nov. 1982.

[59] R. L. Dumas, "Final report: Software acquisition resource expenditure (SARE) data collection methodology," MITRE Corp., MTR 9031, Sept. 1983.

Barry W. Boehm received the B.A. degree in mathematics from Harvard University, Cambridge, MA, in 1957 and the M.A. and Ph.D. degrees from the University of California, Los Angeles, in 1961 and 1964, respectively.

From 1978 to 1979 he was a Visiting Professor of Computer Science at the University of Southern California. He is currently a Visiting Professor at the University of California, Los Angeles, and Chief Engineer of TRW's Software Information Systems Division. He was previously Head of the Information Sciences Department at The Rand Corporation, and Director of the 1971 Air Force CCIP-85 study. His responsibilities at TRW include direction of TRW's internal software R&D program, of contract software technology projects, of the TRW software development policy and standards program, of the TRW Software Cost Methodology Program, and the TRW Software Productivity Program. His most recent book is *Software Engineering Economics*, by Prentice-Hall.

Dr. Boehm is a member of the IEEE Computer Society and the Association for Computing Machinery, and an Associate Fellow of the American Institute of Aeronautics and Astronautics.

Design Patterns

Erich Gamma
Technical Director,
Object Technology International, Zürich
erich_gamma@oti.com

Doctorate in Informatics,
University of Zürich

Union Bank of Switzerland

Taligent

Major contribution: design patterns

Current interests:
Java development tools, frameworks,
building large object-oriented systems

Erich Gamma

Design Patterns – Ten Years Later

Design patterns have changed the way software developers design object-oriented systems. Rather than rediscovering solutions to recurring design problems over and over again, developers can now refer to a body of literature that captures THE best practices of system design. This article looks back to where design patterns came from, shows design patterns in action, and provides an overview of where patterns are today.

Some History

Looking back, we can see that patterns came to us via two routes: the pattern language route and the micro-architecture route.

The pattern language route came first, thanks to Christopher Alexander, author of several books on building architecture [2]. Alexander's goal was to create architectures that improve quality of life. His theory begins by identifying what makes a building great: freedom, wholeness, completeness, comfort, harmony, habitability, durability, openness, resilience, variability, and adaptability. But he concludes that greatness lies in something even more fundamental, something he calls the "Quality Without a Name."

To impart this quality systematically, Alexander and his colleagues developed a "pattern language"– a set of patterns that build on each other. Each pattern captures a proven solution to a problem that arises during design. Alexander characterizes them this way: "Each pattern describes a problem which occurs over and over again in our environment and then describes the core of the solution to that problem, in such a way that you can use this solution a million times over, without ever doing it the same way twice." A pattern language guides a designer's application of individual patterns to the entire design.

Ward Cunningham and Kent Beck had Alexander's books on their shelves long before they recognized the applicability of his ideas to software. By 1987 they began introducing those ideas to the object-oriented programming community [25]. Ward and Kent were successful in generating interest in Alexander's ideas, but it all sounded rather theoretical to software developers. What was lacking were concrete examples for software development.

Coincidently, the micro-architecture route got started. I was in a doctoral program working with André Weinand on ET++ [31], a framework for interactive graphical applications written in C++. You can't publish C++ code as a dissertation, of course, so I planned to write about research issues in object-oriented design based on my experience with ET++.

As I reflected on ET++, it became apparent that a mature framework contains recurring design structures that give "frameworkers" warm and fuzzy feelings. These structures promote qualities like decoupling, abstraction, openness, and flexibility. They can be considered micro-architectures that contribute to an overall system architecture. I ended up documenting a dozen such micro-architectures in my thesis [13]. This approach to patterns differed from pattern languages: rather than coming up with a set of interwoven patterns top-down, micro-architectures constitute isolated patterns that eventually relate to each other bottom-up.

I "outed" my thesis work at an informal OOPSLA '90 session organized by Bruce Anderson titled, "Towards an Architecture Handbook." Richard Helm immediately understood my intent, and we found that we had similar ideas on good object-oriented design. This motivated us to meet for a day in 1991 in Zurich. Fueled by liters of ice tea, we came up with the idea of a catalogue of design patterns. We wanted to move away from prose towards a more structured format, which we hoped would make it easy to compare, categorize, and finally catalogue the different patterns. We ended up with a rough sketch of the pattern format and skeletal descriptions of several patterns adhering to it.

We were not shy about showing our catalogue to people. Surprisingly, most of the feedback we got from experts was along the lines of, "Oh, I've done that." Such feedback encouraged us to continue, but we also realized that we should leverage insights from some of these experts to move forward faster.

As a first step, we infected John Vlissides with the idea. John was a contributor to InterViews [20], a C++ library that was popular at the time. InterViews and ET++ were competitors, giving rise to debates over which system's way of solving problems was best. A lesson of these discussions was the importance of documenting not just one solution but multiple alternatives, along with their trade-offs.

Around the same time, another contributor was making patterns more concrete. Jim Coplien collected and published a catalogue of C++ "idioms" – small-scale patterns applicable to a particular programming language [9]. As useful as idioms are, though, one of our goals was language neutrality. We did not want to limit ourselves to one language, no matter how popular; we wanted to capture problems and solutions that applied across object-oriented languages.

To that end, we sought to broaden our group skills to include Smalltalk expertise, and we eventually convinced Ralph Johnson to join us. Ralph was particularly interested in using patterns to document frameworks [18]. The group, quickly dubbed "Gang of Four" or just "GoF" by onlookers, was complete. Much as some of us disliked that moniker, it became a convenient shorthand for the Internet-wide discussions that ensued.

A summary of the group's initial work was presented at ECOOP [14]. That effort convinced us that we couldn't do justice to the topic with anything less than a book-length work. We also quickly learned that you can't get a pattern right the first time. It took us more than two years and several rewrites, requiring over 2000 sometimes-heated e-mail exchanges, to arrive at the catalog of 23 patterns we finally published [15].

While we GoF worked on our patterns, Kent Beck began forming the Hillside Group, which was instrumental in the pattern community's formation. In particular, Hillside was the catalyst for the various Pattern Languages of Programs (PLoP) conferences. The PLoP charter was to create a new literature of what is explicitly not new, but works in practice. The centerpiece of a PLoP conference is its Writers' Workshops. Instead of one or more authors presenting their work to a large audience, they listen silently as a small group discusses each pattern and debates its strengths and weaknesses. The products of PLoPs are carefully reviewed and revised patterns [10, 16, 21, 30] from many different domains.

Several other pattern books appeared around the time of the first few PLoPs, and they did much to popularize patterns. Examples include *Pattern-Oriented Software Architecture* [7], *Smalltalk Best Practice Patterns* [5], or the seminal work on analysis patterns [11].

This short history would be incomplete without mention of the GoF's 1999 indictment by the International Tribunal for the Prosecution of Crimes Against Computer Science for crimes too numerous to list here [28]. A show trial held at OOPSLA that year had a predictable outcome, though only one of the defendants received a jail sentence. He subsequently escaped and remains at large.

Conceptions and Misconceptions

The popularity of design patterns has given rise to many different opinions about what they are and how they should be written. I won't try to come up with the ultimate definition here. I will focus instead on conceptions and misconceptions about patterns.

A design pattern has three key characteristics:

1. *It promotes design reuse and provides guidance.* The whole point of a pattern is to reuse, rather than reinvent, someone else's successful design. It represents a bite-sized chunk of engineering knowledge. A pattern can help you ask the right questions at the right time. In particular, it can help you avoid obvious but flawed solutions.
2. *It conforms to a literary style.* A pattern articulates design decisions, discusses their trade-offs, and provides implementation hints using a specific style of presentation. Many different pattern styles are in use today.
3. *Together, they define a vocabulary for discussing design.* Designers can use pattern names to communicate concisely. They can design at a higher level of abstraction.

The following, meanwhile, are misconceptions:

- *A pattern is a Unified Modeling Language (UML) diagram with a name.* Name aside, the most memorable part of a design pattern is its structure diagram. But it is wrong to reduce a pattern to a diagram. A pattern will have variations that one diagram can't capture. So while the diagram is important, equally important is a pattern's discussion of trade-offs, variant implementations, and examples.
- *Patterns make software more reusable.* Design patterns are the fruit of studying reusable object-oriented software. But just because a system was designed using patterns doesn't necessarily mean it's reusable. As a matter of fact, the same is true of object-oriented programming: objects merely enable reuse; they don't guarantee it.
- *The more patterns, the better the system.* Patterns make it easy to make a system complex. They achieve flexibility and variability by introducing levels of indirection, which can complicate a design. It's better to start simple and evolve a design as needed. This is in contrast to the over-enthusiasm evident in the following quote from a newsgroup: "In this project I tried to use all the GoF patterns, but I was able to use only 21 of them."

- *A pattern is a new way of solving a problem.* A pattern never suggests a new way to solve a problem; it's necessarily a proven solution found in existing systems. Experienced designers shouldn't be surprised by a pattern. The rule we followed in our catalogue was that unless we could point to at least two uses in existing systems, we didn't consider it a pattern. The fact that a pattern is a proven solution lends confidence when it is applied, since others arrived at the same solution.
- *Design patterns are all you need to know.* Today there is a wide range of pattern domains and styles. To understand the full breadth of patterns, you have to learn more than just our own. The Patterns Home Page [26] is a good starting point for pattern-related material.

Frameworks

Design patterns have their roots in object-oriented *frameworks*. A framework is a set of customizable, cooperating classes that implements a reusable solution to a given problem. To make a framework reusable, it's imperative to design-in flexibility and extensibility. Frameworks are therefore design-dense, making them a rich source of design solutions. Many pattern authors have experience in framework development.

When we wrote *Design Patterns*, we were excited about frameworks and forecast that there would be lots of them in the near future. Time has shown, however, that frameworks aren't for everyone. Developing a good framework is hard and requires a large up-front investment that doesn't always pay off. It is not uncommon for a framework to be deemed a success on the supplier side, only to produce frustration ("frameworkitis") on the consumer side thanks to a steep learning curve. I was actively involved in one such framework [29], and so I know what I'm talking about. But this is not a diatribe against frameworks; if you have a framework that matches your problem, then by all means use it. However, be careful when developing a new framework, even if you know all about patterns.

Design patterns are interesting in this context because they let you apply insights gained in framework design to day-to-day application development. Design patterns are more abstract than frameworks; they are mainly prose descriptions with few opportunities for code reuse. As a consequence, patterns are highly "portable" and can be adapted and tweaked when they are applied. Patterns are also smaller architectural building blocks than frameworks. You can combine them in myriad ways.

Mature frameworks use many patterns. The same isn't typically true of applications, because they do not require the flexibility of a framework. An application's pattern density is generally lower than a framework's.

Design Patterns in Action

The best way to understand design patterns is by seeing them in action. The example I'll use is the Java JUnit testing framework [27]. JUnit supports writing and running tests. At the press of a button, the developer can run a collection of tests to gain confidence in the current state of development. I had the pleasure of pair-programming JUnit with Kent Beck. JUnit is a compelling example – simple, yet with several patterns working together in a handful of classes.

Modeling Tests

Let's examine the design of modeling tests using patterns as a guide. First we need an abstraction for the tests themselves. JUnit should be able to run a test without knowing any details about it. Scanning through our patterns' intents, we encounter this one: "Encapsulate a request as an object, thereby letting you parameterize clients with different requests, queue or log requests, and support undoable operations." That's the intent of the Command pattern. While we're not particularly interested in the undo aspect, the idea of representing a "run test" request as an object will effectively hide details of the test from JUnit – and make the tests extensible as well.

We apply the Command pattern by introducing a **TestCase** class with a single method, **run**. Figure 1 illustrates the result of this first design step.

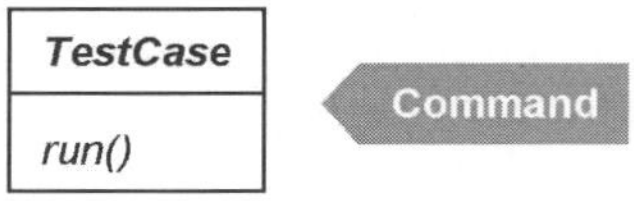

Fig. 1. Applying Command

We use UML class diagrams to show snapshots of our design as it evolves. We annotate the UML with shaded boxes identifying the pattern being applied.

Breaking Down run

At this stage, implementing a test requires subclassing **TestCase** and overriding the run method. We can do better – once we observe that the execution of a test has a common structure: first the test setup is created, then the test is run, and finally the test setup is cleaned-up. The framework can codify this common structure.

The Template Method pattern can do just that. Its intent is to, "Define the skeleton of an algorithm in an operation, deferring some steps to subclasses. Template Method lets subclasses redefine certain steps of an algorithm without changing the algorithm's structure." The pattern shows us how to turn **run** into a template method made up of primitive methods **setUp**, **runTest**, and **tearDown**. Figure 2 shows the result.

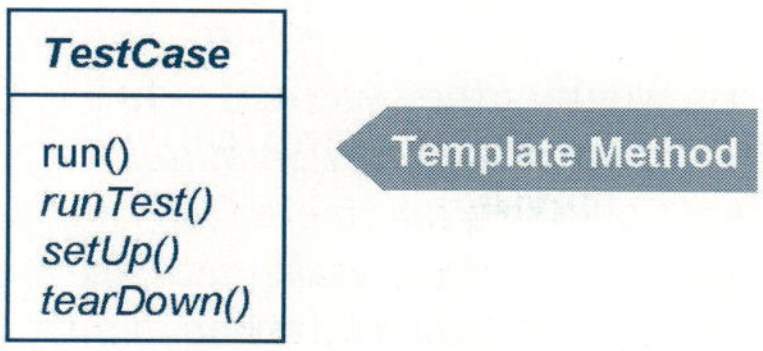

Fig. 2. Applying Template Method

Modeling Test Suites

The next design problem has to do with how JUnit supports suites of tests. Tests should be composable with other tests for two reasons. First, you might want to test scattered modifications all at once. Second, you might want to combine your tests with those of other developers.

Composite is a powerful pattern for assembling objects. Its intent reads, "Compose objects into tree structures to represent part-whole hierarchies. Composite lets clients treat individual objects and compositions of objects uniformly." To that end, we introduce **TestSuite**, a composite class for assembling tests.

The power of the Composite pattern stems from the conformance of the TestSuite and TestCase interfaces – a TestSuite is a TestCase. How you express this conformance depends on your programming language. In Java, you introduce a common interface; in Smalltalk, you simply define the same method selectors.

This conformance lets you assemble TestSuite objects recursively to form larger suites. In other words, clients no longer have to distinguish between a single test and many tests. Figure 3 illustrates the introduction of Composite.

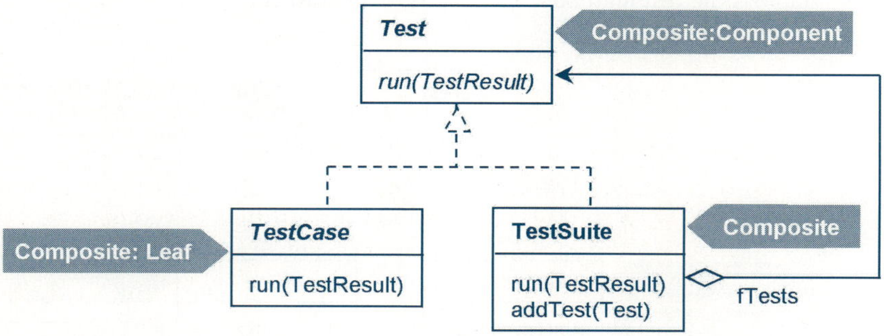

Fig. 3. Applying Composite

Collecting Test Results

At the end of the day, you run tests to see whether they succeed or fail. To do that, we need to collect the results of executing tests. *Smalltalk Best Practice Patterns* [5] has a pattern called "Collecting Parameter" for this purpose. When you need to collect results over several methods, the pattern suggests adding a parameter to each method and passing in an object in which to hold the results. In our design, then, we create a new object, **TestResult**, that holds the results of running tests (Figure 4).

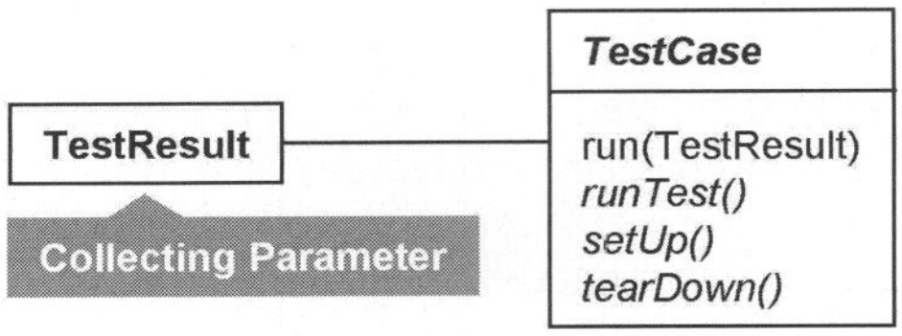

Fig. 4. Applying Collecting Parameter

Cutting Down on Classes

Our design is coming along nicely, but the framework remains harder to use than it could be. Developers must create a new subclass of TestCase for every test. Loads of classes, each implementing a lone method, are a maintenance burden.

We can always put multiple tests into the class to reduce subclassing. But we still have a problem: JUnit offers just one method, **runTest**, for carrying out a test.

The Adapter pattern comes in handy here. Its intent is to "convert the interface of a class into another interface." Of particular interest is an implementation variation called *pluggable* adapters. A pluggable adapter builds interface adaptation into a class. The aspect we want to make pluggable is the test method that TestCase executes.

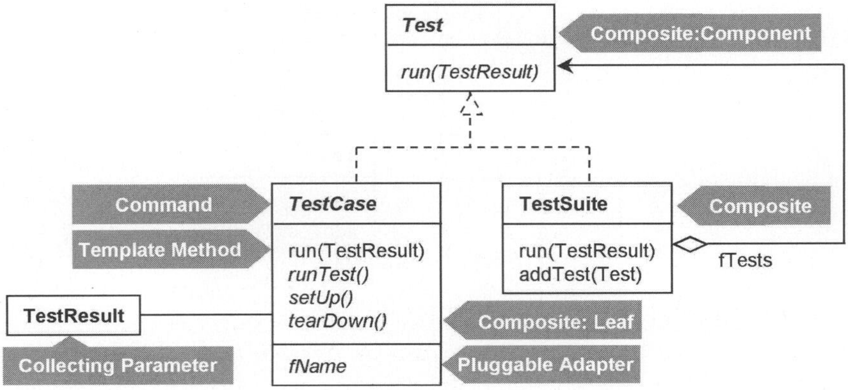

Fig. 5. Framework summary

Accordingly, JUnit lets you specify a test method name when creating a TestCase. The **runTest** method can then invoke that method using Java reflection. Figure 5 illustrates the final design of the JUnit framework.

Postmortem

Some observations are in order on our use of design patterns in this example:

- Design patterns provide a vocabulary for communicating design concisely. A reader can immediately understand a design problem, its solutions, and the consequences thereof by referring to a pattern.
- Patterns work in teams. Note how TestCase, the central abstraction in the framework, embodies four patterns. Pictures of mature object designs show similar "pattern density." As another example, consider a diagram (Figure 6) showing the key classes in Java's AWT toolkit [17], annotated with patterns.

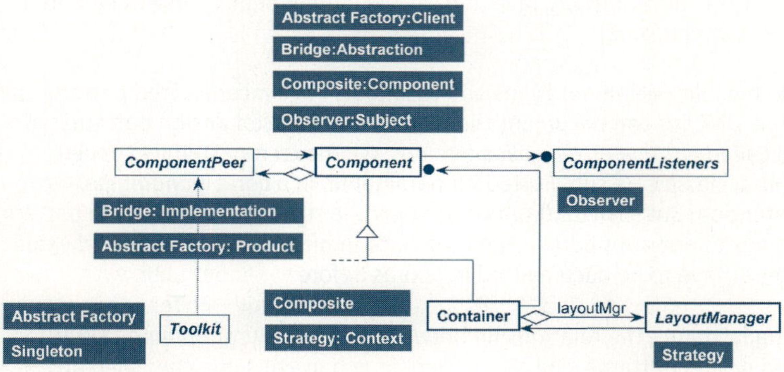

Fig. 6. The Java AWT toolkit

- Patterns speed up your understanding of a design. They make design motives and elements explicit. In particular, patterns help you understand how classes relate and interact. Since you understand the designer's intent better and faster, you're less likely to struggle with the design itself.

Design Patterns Today

The patterns literature has expanded greatly in the past few years. Published patterns and pattern languages cover a wide range of domains. Many contributions come from the PLoP conferences held throughout the world. In fact, the ever-growing body of patterns is increasingly difficult to manage.

This is a good problem to have, as it attests to patterns' popularity. But at the same time, the pattern community must work to bring structure to the morass. A good first step in this direction is *The Pattern Almanac* [23], an index and cross-reference of most of the patterns published through 1999.

It takes time, but design patterns eventually emerge from challenging new problem domains where design expertise is especially appreciated. For example, concurrent programming has many degrees of freedom beyond conventional sequential programming. This relatively new and complicated problem domain is already covered by patterns for concurrent Java programming [19] and for programming networked objects [24]. Another recent example is the work on J2EE patterns for Java enterprise application development [1]. You can expect this trend to continue and new pattern catalogs to be developed as new technologies emerge.

Many people have reported anecdotal success with patterns in various domains [4]. Quantitative results from controlled experiments tend to back them up [22]. They have corroborated, for example, that documenting the patterns in a system does indeed make the system easier to understand.

Patterns are now used to teach software design and programming at many universities. They have found their way into the APIs of today's software platforms, most notably Java and .NET. (Figure 6 depicts an example from the Java platform.)

Meanwhile, design notations and case tools have incorporated pattern support. UML [8] can document the structural aspects of design patterns with so-called *parameterized collaborations*. Whenever a pattern is applied, actual classes are substituted for parameters in a corresponding pattern definition. But such diagrams cover only the structural aspects of a pattern; its most important parts – the trade-offs, implementation hints, and examples – have to be documented in text as before.

Finally, there's the relationship between Extreme Programming (XP) [6] and design patterns. The XP methodology balks at some cherished attitudes of software development. In particular, it de-emphasizes *up-front* design in favor of evolutionary design. In *evolutionary* design, you start with the simplest design that could possibly work and continually evolve it through refactoring [12]. Patterns are about building flexibility into a design up front. So the question is, can design patterns be ignored in XP?

It's true that pattern over-enthusiasm is in strong conflict with XP practices. However, XP emphasizes refactoring to improve a system's design, and that's where patterns come into play. Design patterns serve as targets for refactoring; they show you how to improve existing code, particularly how to make it simpler. For example, the State and Strategy patterns show you how to transform conditional code into polymorphic code.

Given the centrality of refactoring in XP, I have no doubt that patterns belong in an XP developer's skill set. By the way, as Kent (the father of XP) and I pair-programmed JUnit, the pattern names I've mentioned peppered our conversation as we coded along.

Conclusion

Design patterns have been successful and influential because they tackle design problems in a way many developers can appreciate. Here are my nominees for the four biggest benefits of design patterns:

- They accelerate discourse by providing a vocabulary for design.
- They let you repeat successful designs rather than rediscover them.
- As targets for refactoring, they show you how to improve an existing system.
- A system that documents its patterns is easier to understand than one that doesn't.

Design patterns are tools, not rules. Applying a pattern involves tweaking and adapting it to your needs. You can't turn off your brain when you use patterns. You have to choose them wisely and then decide *how* to use them or – perhaps more importantly – how *not* to.

The most powerful patterns are the ones you've internalized, and they're not easy to come by. You can't absorb patterns solely by reading a book or by attending a course. Unless you have experienced a design problem firsthand, you cannot appreciate the solution a pattern describes.

Acknowledgements

Thanks to John Vlissides for carefully reading and many suggesting corrections to improve the readability of this article.

References

[1] Deepak, A., J. Crupi, D. Malks. Core J2EE Patterns: Best Practices and Design Strategies. Prentice Hall, 2001.

[2] Alexander, C., S. Ishikawa, M. Silverstein, et al. A Pattern Language. Oxford University Press, New York, 1977

[3] Alexander, C. The Timeless Way of Building. Oxford University Press, New York, 1979

[4] Beck, K., J.O. Coplien, R. Crocker, et al. Industrial experience with design patterns. Proceedings of the 18th International Conference on Software Engineering (pp. 103-114), Berlin, Germany, March 1996.

[5] Beck, K. Smalltalk Best Practice Patterns. Prentice Hall, 1996

[6] Beck, K. Extreme Programming Explained: Embrace Change. Addison-Wesley, Reading, MA, 1999.

[7] Buschmann, F., R. Menunier, H. Rohnert, M. Stal. Pattern-Oriented Software Architecture - A System of Patterns. Wiley, Chichester, England, 1996.

[8] Booch, G., I. Jacobson, J. Rumbaugh. The Unified Modeling Language User Guide. Addison-Wesley, Reading, MA, 1998.

[9] Coplien, J. Advanced C++ Programming Styles and Idioms. Addison-Wesley, Reading, MA, 1992.

[10] Coplien, J.O., D. C. Schmidt (Eds.). Pattern Languages of Program Design. Addison-Wesley (Software Patterns Series), Reading, MA, 1995

[11] Fowler, M. Analysis Patterns: Reusable Object Models. Addison-Wesley, Reading, MA, 1997.

[12] Fowler, M. Refactoring: Improving the Design of Existing Code. Addison-Wesley, Reading, MA, 1999.

[13] Gamma, E. Object-Oriented Software Development Based on ET++: Design Patterns, Class Library, Tools (in German). Doctoral thesis, University of Zurich, Institut für Informatik, 1991.

[14] Gamma, E., R. Helm, R. Helm, R. Johnson, and J. Vlissides. Design Patterns: Abstraction and Reuse of Object-Oriented Design ECOOP '93, Lecture Notes in Computer Science 707. Springer, Berlin Heidelberg New York, p. 406 ff.

[15] Gamma, E., R. Helm, R. Johnson, J. Vlissides. Design Patterns: Elements of Reusable Object-Oriented Software. Addison-Wesley, Reading, MA, 1995.

[16] Harrison, N., B. Foote, H. Rohnert (Eds.). Pattern Languages of Program Design 4. Addison-Wesley (Software Patterns Series), Reading, MA, 2000

[17] JavaSoft, Inc. Java Development Kit. Mountain View, CA, 1997

[18] Johnson, R. Documenting Frameworks Using Patterns. In Proc. OOPSLA '92, pages 63-76, Vancouver BC, Oct. 1992.

[19] Lea, D. Concurrent Programming in Java, Design Principles and Patterns, 2nd edition. Addison-Wesley, Reading, MA, 1999

[20] Linton, M., J. Vlissides, and P. Calder. Composing User Interfaces with InterViews. Computer 22(2):8-22, 1989.

[21] Martin, R., D. Riehle, and F. Buschmann (Eds.). Pattern Languages of Program Design 3. Addison-Wesley, Reading, MA, 1998.

[22] Prechelt, L. An experiment on the usefulness of design patterns: Detailed description and evaluation. Technical Report 9/1997, University of Karlsruhe, Germany, June 1997.

[23] Rising, L. The Pattern Almanac 2000. Addison-Wesley, Reading, MA, 2000.

[24] Schmidt, D., M. Stal, H. Rohnert, F. Buschmann. Pattern-Oriented Software Architecture - Patterns for Concurrent and Networked Objects. Wiley and Sons Ltd., Chichester, England, 2000.

[25] http://c2.com/doc/oopsla87.html

[26] http://hillside.net/patterns

[27] http://www.junit.org

[28] http://www.acm.org/sigplan/oopsla/oopsla99/2_ap/tech/2d1a_gang.html

[29] Cotter, S., M. Potel. Inside Taligent Technology. Addison-Wesley, Reading, MA, 1995.

[30] Vlissides, J., J. Coplien, and N. Kerth (Eds.). Pattern Languages of Program Design 2. Addison-Wesley, Reading, MA, 1996.

[31] Weinand, A., E. Gamma, and R. Marty. ET++–An object-oriented application framework in C++. In Object-Oriented Programming Systems, Languages, and Applications Conference Proceedings, pages 46–57, San Diego, CA. ACM Press, 1988.

Erich Gamma

Erich Gamma, Richard Helm,
Ralph Johnson, John Vlissides

Design Patterns: Abstraction and Reuse of Object-Oriented Design

Proceedings ECOOP'93, Kaiserslautern,
July 1993, ed. by Oscar Nierstrasz
Lecture Notes in Computer Science, Vol. 707
Springer-Verlag, Heidelberg
pp. 406-431

Design Patterns: Abstraction and Reuse of Object-Oriented Design

Erich Gamma[1*], Richard Helm[2], Ralph Johnson[3], John Vlissides[2]

[1] Taligent, Inc.
10725 N. De Anza Blvd., Cupertino, CA 95014-2000 USA

[2] I.B.M. Thomas J. Watson Research Center
P.O. Box 704, Yorktown Heights, NY 10598 USA

[3] Department of Computer Science
University of Illinois at Urbana-Champaign
1034 W. Springfield Ave., Urbana, IL 61801 USA

Abstract. We propose **design patterns** as a new mechanism for expressing object-oriented design experience. Design patterns identify, name, and abstract common themes in object-oriented design. They capture the intent behind a design by identifying objects, their collaborations, and the distribution of responsibilities. Design patterns play many roles in the object-oriented development process: they provide a common vocabulary for design, they reduce system complexity by naming and defining abstractions, they constitute a base of experience for building reusable software, and they act as building blocks from which more complex designs can be built. Design patterns can be considered reusable micro-architectures that contribute to an overall system architecture. We describe how to express and organize design patterns and introduce a catalog of design patterns. We also describe our experience in applying design patterns to the design of object-oriented systems.

1 Introduction

Design methods are supposed to promote good design, to teach new designers how to design well, and to standardize the way designs are developed. Typically, a design method comprises a set of syntactic notations (usually graphical) and a set of rules that govern how and when to use each notation. It will also describe problems that occur in a design, how to fix them, and how to evaluate a design. Studies of expert programmers for conventional languages, however, have shown that knowledge is not organized simply around syntax, but in larger conceptual structures such as algorithms, data structures and idioms [1, 7, 9, 27], and plans that indicate steps necessary to fulfill a particular goal [26]. It is likely that designers do not think about the notation they are using for recording the design. Rather, they look for patterns to match against plans, algorithms, data structures, and idioms they have learned in the past. Good designers, it appears, rely

* Work performed while at UBILAB, Union Bank of Switzerland, Zurich, Switzerland.

on large amounts of design experience, and this experience is just as important as the notations for recording designs and the rules for using those notations.

Our experience with the design of object-oriented systems and frameworks [15, 17, 22, 30, 31] bears out this observation. We have found that there exist idiomatic class and object structures that help make designs more flexible, reusable, and elegant. For example, the Model-View-Controller (MVC) paradigm from Smalltalk [19] is a design structure that separates representation from presentation. MVC promotes flexibility in the choice of views, independent of the model. Abstract factories [10] hide concrete subclasses from the applications that use them so that class names are not hard-wired into an application.

Well-defined design structures like these have a positive impact on software development. A software architect who is familiar with a good set of design structures can apply them immediately to design problems without having to rediscover them. Design structures also facilitate the reuse of successful architectures—expressing proven techniques as design structures makes them more readily accessible to developers of new systems. Design structures can even improve the documentation and maintenance of existing systems by furnishing an explicit specification of class and object interactions and their underlying intent.

To this end we propose **design patterns**, a new mechanism for expressing design structures. Design patterns identify, name, and abstract common themes in object-oriented design. They preserve design information by capturing the intent behind a design. They identify classes, instances, their roles, collaborations, and the distribution of responsibilities. Design patterns have many uses in the object-oriented development process:

- Design patterns provide a common vocabulary for designers to communicate, document, and explore design alternatives. They reduce system complexity by naming and defining abstractions that are above classes and instances. A good set of design patterns effectively raises the level at which one programs.
- Design patterns constitute a reusable base of experience for building reusable software. They distill and provide a means to reuse the design knowledge gained by experienced practitioners. Design patterns act as building blocks for constructing more complex designs; they can be considered **micro-architectures** that contribute to overall system architecture.
- Design patterns help reduce the learning time for a class library. Once a library consumer has learned the design patterns in one library, he can reuse this experience when learning a new class library. Design patterns help a novice perform more like an expert.
- Design patterns provide a target for the reorganization or refactoring of class hierarchies [23]. Moreover, by using design patterns early in the lifecycle, one can avert refactoring at later stages of design.

The major contributions of this paper are: a definition of design patterns, a means to describe them, a system for their classification, and most importantly, a catalog containing patterns we have discovered while building our own class libraries and patterns we have collected from the literature. This work has its roots in Gamma's thesis [11], which abstracted design patterns from the ET++ framework. Since then the work has been refined and extended based on our collective experience. Our thinking has also been influenced and inspired by discussions within the Architecture Handbook Workshops at recent OOPSLA conferences [3, 4].

This paper has two parts. The first introduces design patterns and explains techniques to describe them. Next we present a classification system that characterizes common aspects of patterns. This classification will serve to structure the catalog of patterns presented in the second part of this paper. We discuss how design patterns impact object-oriented programming and design. We also review related work.

The second part of this paper (the Appendix) describes our current catalog of design patterns. As we cannot include the complete catalog in this paper (it currently runs over 90 pages [12]), we give instead a brief summary and include a few abridged patterns. Each pattern in this catalog is representative of what we judge to be good object-oriented design. We have tried to reduce the subjectivity of this judgment by including only design patterns that have seen practical application. Every design pattern we have included works—most have been used at least twice and have either been discovered independently or have been used in a variety of application domains.

2 Design Patterns

A design pattern consists of three essential parts:

1. An abstract description of a class or object collaboration and its structure. The description is abstract because it concerns abstract design, not a particular design.
2. The issue in system design addressed by the abstract structure. This determines the circumstances in which the design pattern is applicable.
3. The consequences of applying the abstract structure to a system's architecture. These determine if the pattern should be applied in view of other design constraints.

Design patterns are defined in terms of object-oriented concepts. They are sufficiently abstract to avoid specifying implementation details, thereby ensuring wide applicability, but a pattern may provide hints about potential implementation issues.

We can think of a design pattern as a micro-architecture. It is an architecture in that it serves as a blueprint that may have several realizations. It is "micro" in that it defines something less than a complete application or library. To be useful, a design pattern should be applicable to more than a few problem domains; thus design patterns tend to be relatively small in size and scope. A design pattern can also be considered a transformation of system structure. It defines the context for the transformation, the change to be made, and the consequences of this transformation.

To help readers understand patterns, each entry in the catalog also includes detailed descriptions and examples. We use a template (Figure 1) to structure our descriptions and to ensure uniformity between entries in the catalog. This template also explains the motivation behind its structure. The Appendix contains three design patterns that use the template. We urge readers to study the patterns in the Appendix as they are referenced in the following text.

3 Categorizing Design Patterns

Design patterns vary in their granularity and level of abstraction. They are numerous and have common properties. Because there are many design patterns, we need a way to organize them. This section introduces a classification system for design patterns. This classification makes it easy to refer to families of related patterns, to learn the patterns in the catalog, and to find new patterns.

		Characterization		
		Creational	**Structural**	**Behavioral**
Jurisdiction	**Class**	Factory Method	Adapter (class) Bridge (class)	Template Method
	Object	Abstract Factory Prototype Solitaire	Adapter (object) Bridge (object) Flyweight Glue Proxy	Chain of Responsibility Command Iterator (object) Mediator Memento Observer State Strategy
	Compound	Builder	Composite Wrapper	Interpreter Iterator (compound) Walker

Table 1. Design Pattern Space

We can think of the set of all design patterns in terms of two orthogonal criteria, **jurisdiction** and **characterization**. Table 1 organizes our current set of patterns according to these criteria.

Jurisdiction is the domain over which a pattern applies. Patterns having **class** jurisdiction deal with relationships between base classes and their subclasses;

Design Pattern Name | **Jurisdiction Characterization**

What is the pattern's name and classification? The name should convey the pattern's essence succinctly. A good name is vital, as it will become part of the design vocabulary.

Intent

What does the design pattern do? What is its rationale and intent? What particular design issue or problem does it address?

Motivation

A scenario in which the pattern is applicable, the particular design problem or issue the pattern addresses, and the class and object structures that address this issue. This information will help the reader understand the more abstract description of the pattern that follows.

Applicability

What are the situations in which the design pattern can be applied? What are examples of poor designs that the pattern can address? How can one recognize these situations?

Participants

Describe the classes and/or objects participating in the design pattern and their responsibilities using CRC conventions [5].

Collaborations

Describe how the participants collaborate to carry out their responsibilities.

Diagram

A graphical representation of the pattern using a notation based on the Object Modeling Technique (OMT) [25], to which we have added method pseudo-code.

Consequences

How does the pattern support its objectives? What are the trade-offs and results of using the pattern? What does the design pattern objectify? What aspect of system structure does it allow to be varied independently?

Implementation

What pitfalls, hints, or techniques should one be aware of when implementing the pattern? Are there language-specific issues?

Examples

This section presents examples from real systems. We try to include at least two examples from different domains.

See Also

What design patterns have closely related intent? What are the important differences? With which other patterns should this one be used?

Fig. 1. **Basic Design Pattern Template**

class jurisdiction covers static semantics. The **object** jurisdiction concerns relationships between peer objects. Patterns having **compound** jurisdiction deal with recursive object structures. Some patterns capture concepts that span jurisdictions. For example, iteration applies both to collections of objects (i.e., object jurisdiction) and to recursive object structures (compound jurisdiction). Thus there are both object and compound versions of the Iterator pattern.

Characterization reflects what a pattern does. Patterns can be characterized as either **creational**, **structural**, or **behavioral**. Creational patterns concern the process of object creation. Structural patterns deal with the composition of classes or objects. Behavioral patterns characterize the ways in which classes or objects interact and distribute responsibility.

The following sections describe pattern jurisdictions in greater detail for each characterization using examples from our catalog.

3.1 Class Jurisdiction

Class Creational. Creational patterns abstract how objects are instantiated by hiding the specifics of the creation process. They are useful because it is often undesirable to specify a class name explicitly when instantiating an object. Doing so limits flexibility; it forces the programmer to commit to a particular class instead of a particular protocol. If one avoids hard-coding the class, then it becomes possible to defer class selection to run-time.

Creational class patterns in particular defer some part of object creation to subclasses. An example is the Factory Method, an abstract method that is called by a base class but defined in subclasses. The subclass methods create instances whose type depends on the subclass in which each method is implemented. In this way the base class does not hard-code the class name of the created object. Factory Methods are commonly used to instantiate members in base classes with objects created by subclasses.

For example, an abstract Application class needs to create application-specific documents that conform to the Document type. Application instantiates these Document objects by calling the factory method DoMakeDocument. This method is overridden in classes derived from Application. The subclass DrawApplication, say, overrides DoMakeDocument to return a DrawDocument object.

Class Structural. Structural class patterns use inheritance to compose protocols or code. As a simple example, consider using multiple inheritance to mix two or more classes into one. The result is an amalgam class that unites the semantics of the base classes. This trivial pattern is quite useful in making independently-developed class libraries work together [15].

Another example is the class-jurisdictional form of the Adapter pattern. In general, an Adapter makes one interface (the adaptee's) conform to another, thereby providing a uniform abstraction of different interfaces. A class Adapter accomplishes this by inheriting privately from an adaptee class. The Adapter then expresses its interface in terms of the adaptee's.

Class Behavioral. Behavioral class patterns capture how classes cooperate with their subclasses to fulfill their semantics. Template Method is a simple and well-known behavioral class pattern [32]. Template methods define algorithms step by step. Each step can invoke an abstract method (which the subclass must define) or a base method. The purpose of a template method is to provide an abstract definition of an algorithm. The subclass must implement specific behavior to provide the services required by the algorithm.

3.2 Object Jurisdiction

Object patterns all apply various forms of non-recursive object composition. Object composition represents the most powerful form of reusability—a collection of objects are most easily reused through variations on how they are composed rather than how they are subclassed.

Object Creational. Creational object patterns abstract how sets of objects are created. The Abstract Factory pattern (page 18) is a creational object pattern. It describes how to create "product" objects through an generic interface. Subclasses may manufacture specialized versions or compositions of objects as permitted by this interface. In turn, clients can use abstract factories to avoid making assumptions about what classes to instantiate. Factories can be composed to create larger factories whose structure can be modified at run-time to change the semantics of object creation. The factory may manufacture a custom composition of instances, a shared or one-of-a-kind instance, or anything else that can be computed at run-time, so long as it conforms to the abstract creation protocol.

For example, consider a user interface toolkit that provides two types of scroll bars, one for Motif and another for Open Look. An application programmer may not want to hard-code one or the other into the application—the choice of scroll bar will be determined by, say, an environment variable. The code that creates the scroll bar can be encapsulated in the class **Kit**, an abstract factory that abstracts the specific type of scroll bar to instantiate. Kit defines a protocol for creating scroll bars and other user interface elements. Subclasses of Kit redefine operations in the protocol to return specialized types of scroll bars. A MotifKit's scroll bar operation would instantiate and return a Motif scroll bar, while the corresponding OpenLookKit operation would return an Open Look scroll bar.

Object Structural. Structural object patterns describe ways to assemble objects to realize new functionality. The added flexibility inherent in object composition stems from the ability to change the composition at run-time, which is impossible with static class composition[4].

Proxy is an example of a structural object pattern. A proxy acts as a convenient surrogate or placeholder for another object. A proxy can be used as a

[4] However, object models that support dynamic inheritance, most notably Self [29], are as flexible as object composition in theory.

local representative for an object in a different address space (remote proxy), to represent a large object that should be loaded on demand (virtual proxy), or to protect access to the original object (protected proxy). Proxies provide a level of indirection to particular properties of objects. Thus they can restrict, enhance, or alter an object's properties.

The Flyweight pattern is concerned with object sharing. Objects are shared for at least two reasons: efficiency and consistency. Applications that use large quantities of objects must pay careful attention to the cost of each object. Substantial savings can accrue by sharing objects instead of replicating them. However, objects can only be shared if they do not define context-dependent state. Flyweights have no context-dependent state. Any additional information they need to perform their task is passed to them when needed. With no context-dependent state, flyweights may be shared freely. Moreover, it may be necessary to ensure that all copies of an object stay consistent when one of the copies changes. Sharing provides an automatic way to maintain this consistency.

Object Behavioral. Behavioral object patterns describe how a group of peer objects cooperate to perform a task that no single object can carry out by itself. For example, patterns such as Mediator and Chain of Responsibility abstract control flow. They call for objects that exist solely to redirect the flow of messages. The redirection may simply notify another object, or it may involve complex computation and buffering. The Observer pattern abstracts the synchronization of state or behavior. Entities that are co-dependent to the extent that their state must remain synchronized may exploit Observer. The classic example is the model-view pattern, in which multiple views of the model are notified whenever the model's state changes.

The Strategy pattern (page 21) objectifies an algorithm. For example, a text composition object may need to support different line breaking algorithms. It is infeasible to hard-wire all such algorithms into the text composition class and subclasses. An alternative is to objectify different algorithms and provide them as **Compositor** subclasses. The interface for Compositors is defined by the abstract Compositor class, and its derived classes provide different layout strategies, such as simple line breaks or full page justification. Instances of the Compositor subclasses can be coupled with the text composition at run-time to provide the appropriate text layout. Whenever a text composition has to find line breaks, it forwards this responsibility to its current Compositor object.

3.3 Compound Jurisdiction

In contrast to patterns having object jurisdiction, which concern peer objects, patterns with compound jurisdiction affect recursive object structures.

Compound Creational. Creational compound patterns are concerned with the creation of recursive object structures. An example is the Builder pattern. A Builder base class defines a generic interface for incrementally constructing recursive object structures. The Builder hides details of how objects in the structure are created, represented, and composed so that changing or adding a new representation only requires defining a new Builder class. Clients will be unaffected by changes to Builder.

Consider a parser for the RTF (Rich Text Format) document exchange format that should be able to perform multiple format conversions. The parser might convert RTF documents into (1) plain ASCII text and (2) a text object that can be edited in a text viewer object. The problem is how to make the parser independent of these different conversions.

The solution is to create an RTFReader class that takes a Builder object as an argument. The RTFReader knows how to parse the RTF format and notifies the Builder whenever it recognizes text or an RTF control word. The builder is responsible for creating the corresponding data structure. It separates the parsing algorithm from the creation of the structure that results from the parsing process. The parsing algorithm can then be reused to create any number of different data representations. For example, an ASCII builder ignores all notifications except plain text, while a Text builder uses the notifications to create a more complex text structure.

Compound Structural. Structural compound patterns capture techniques for structuring recursive object structures. A simple example is the Composite pattern. A Composite is a recursive composition of one or more other Composites. A Composite treats multiple, recursively composed objects as a single object.

The Wrapper pattern (page 24) describes how to flexibly attach additional properties and services to an object. Wrappers can be nested recursively and can therefore be used to compose more complex object structures. For example, a Wrapper containing a single user interface component can add decorations such as borders, shadows, scroll bars, or services like scrolling and zooming. To do this, the Wrapper must conform to the interface of its wrapped component and forward messages to it. The Wrapper can perform additional actions (such as drawing a border around the component) either before or after forwarding a message.

Compound Behavioral. Finally, behavioral compound patterns deal with behavior in recursive object structures. Iteration over a recursive structure is a common activity captured by the Iterator pattern. Rather than encoding and distributing the traversal strategy in each class in the structure, it can be extracted and implemented in an Iterator class. Iterators objectify traversal algorithms over recursive structures. Different iterators can implement pre-order, in-order, or post-order traversals. All that is required is that nodes in the structure provide services to enumerate their sub-structures. This avoids hard-wiring traversal algorithms throughout the classes of objects in a composite structure. Iterators may be replaced at run-time to provide alternative traversals.

4 Experience with Design Patterns

We have applied design patterns to the design and construction of a several systems. We briefly describe two of these systems and our experience.

4.1 ET++SwapsManager

The ET++SwapsManager [10] is a highly interactive tool that lets traders value, price, and perform what-if analyses for a financial instrument called a **swap**. During this project the developers had to first learn the ET++ class library, then implement the tool, and finally design a framework for creating "calculation engines" for different financial instruments. While teaching ET++ we emphasized not only learning the class library but also describing the applied design patterns. We noticed that design patterns reduced the effort required to learn ET++. Patterns also proved helpful during development in design and code reviews. Patterns provided a common vocabulary to discuss a design. Whenever we encountered problems in the design, patterns helped us explore design alternatives and find solutions.

4.2 QOCA: A Constraint Solving Toolkit

QOCA (Quadratic Optimization Constraint Architecture) [14, 15] is a new object-oriented constraint-solving toolkit developed at IBM Research. QOCA leverages recent results in symbolic computation and geometry to support efficient incremental and interactive constraint manipulation. QOCA's architecture is designed to be flexible. It permits experimentation with different classes of constraints and domains (e.g., reals, booleans, etc.), different constraint-solving algorithms for these domains, and different representations (doubles, infinite precision) for objects in these domains. QOCA's object-oriented design allows parts of the system to be varied independently of others. This flexibility was achieved, for example, by using Strategy patterns to factor out constraint solving algorithms and Bridges to factor out domains and representations of variables. In addition, the Observable pattern is used to propagate notifications when variables change their values.

4.3 Summary of Observations

The following points summarize the major observations we have made while applying design patterns:

– Design patterns motivate developers to go beyond concrete objects; that is, they objectify concepts that are not immediately apparent as objects in the problem domain.
– Choosing intuitive class names is important but also difficult. We have found that design patterns can help name classes. In the ET++SwapsManager's calculation engine framework we encoded the name of the design pattern

in the class name (for example CalculationStrategy or TableAdaptor). This convention results in longer class names, but it gives clients of these classes a hint about their purpose.

- We often apply design patterns *after* the first implementation of an architecture to improve its design. For example, it is easier to apply the Strategy pattern after the initial implementation to create objects for more abstract notions like a calculation engine or constraint solver. Patterns were also used as targets for class refactorings. We often find ourselves saying, "Make this part of a class into a Strategy," or, "Let's split the implementation portion of this class into a Bridge."
- Presenting design patterns together with examples of their application turned out to be an effective way to teach object-oriented design by example.
- An important issue with any reuse technology is how a reusable component can be adapted to create a problem-specific component. Design patterns are particularly suited to reuse because they are abstract. Though a concrete class structure may not be reusable, the design pattern underlying it often is.
- Design patterns also reduce the effort required to learn a class library. Each class library has a certain design "culture" characterized by the set of patterns used implicitly by its developers. A specific design pattern is typically reused in different places in the library. A client should therefore learn these patterns as a first step in learning the library. Once they are familiar with the patterns, they can reuse this understanding. Moreover, because some patterns appear in other class libraries, it is possible to reuse the knowledge about patterns when learning other libraries as well.

5 Related Work

Design patterns are an approach to software reuse. Krueger [20] introduces the following taxonomy to characterize different reuse approaches: software component reuse, software schemas, application generators, transformation systems, and software architectures. Design patterns are related to both software schemas and reusable software architectures. Software schemas emphasize reusing abstract algorithms and data structures. These abstractions are represented formally so they can be instantiated automatically. The Paris system [18] is representative of schema technology. Design patterns are higher-level than schemas; they focus on design structures at the level of collaborating classes and not at the algorithmic level. In addition, design patterns are not formal descriptions and cannot be instantiated directly. We therefore prefer to view design patterns as reusable software architectures. However, the examples Krueger lists in this category (blackboard architectures for expert systems, adaptable database subsystems) are all coarse-grained architectures. Design patterns are finer-grained and therefore can be characterized as reusable micro-architectures.

Most research into patterns in the software engineering community has been geared towards building knowledge-based assistants for automating the application of patterns for synthesis (that is, to write programs) and analysis (in debugging, for example) [13, 24]. The major difference between our work and that of the knowledge-based assistant community is that design patterns encode higher-level expertise. Their work has tended to focus on patterns like enumeration and selection, which can be expressed directly as reusable components in most existing object-oriented languages. We believe that characterizing and cataloging higher-level patterns that designers already use informally has an immediate benefit in teaching and communicating designs.

A common approach for reusing object-oriented software architectures are object-oriented frameworks [32]. A framework is a codified architecture for a problem domain that can be adapted to solve specific problems. A framework makes it possible to reuse an architecture together with a partial concrete implementation. In contrast to frameworks, design patterns allow only the reuse of abstract micro-architectures without a concrete implementation. However, design patterns can help define and develop frameworks. Mature frameworks usually reuse several design patterns. An important distinction between frameworks and design patterns is that frameworks are implemented in a programming language. Our patterns are ways of *using* a programming language. In this sense frameworks are more concrete than design patterns.

Design patterns are also related to the idioms introduced by Coplien [7]. These idioms are concrete design solutions in the context of C++. Coplien "focuses on idioms that make C++ programs more expressive." In contrast, design patterns are more abstract and higher-level than idioms. Patterns try to abstract design rather than programming techniques. Moreover, design patterns are usually independent of the implementation language.

There has been interest recently within the object-oriented community [8] in pattern languages for the architecture of buildings and communities as advocated by Christopher Alexander in *The Timeless Way of Building* [2]. Alexander's patterns consist of three parts:

- A context that describes when a pattern is applicable.
- The problem (or "system of conflicting forces") that the pattern resolves in that context.
- A configuration that describes physical relationships that solve the problem.

Both design patterns and Alexander's patterns share the notion of context/problem/configuration, but our patterns currently do not form a complete system of patterns and so do not strictly define a pattern language. This may be because object-oriented design is still a young technology—we may not have had enough experience in what constitutes good design to extract design patterns that cover all phases of the design process. Or this may be simply because the problems encountered in software design are different from those found in architecture and are not amenable to solution by pattern languages.

Recently, Johnson has advocated pattern languages to describe how to use use object-oriented frameworks [16]. Johnson uses a pattern language to explain how to extend and customize the Hotdraw drawing editor framework. However, these patterns are not design patterns; they are more descriptions of how to reuse existing components and frameworks instead of rules for generating new designs.

Coad's recent paper on object-oriented patterns [6] is also motivated by Alexander's work but is more closely related to our work. The paper has seven patterns: "Broadcast" is the same as Observer, but the other patterns are different from ours. In general, Coad's patterns seem to be more closely related to analysis than design. Design patterns like Wrapper and Flyweight are unlikely to be generated naturally during analysis unless the analyst knows these patterns well and thinks in terms of them. Coad's patterns could naturally arise from a simple attempt to model a problem. In fact, it is hard to see how any large model could avoid using patterns like "State Across a Collection" (which explains how to use aggregation) or "Behavior Across a Collection" (which describes how to distribute responsibility among objects in an aggregate). The patterns in our catalog are typical of a mature object-oriented design, one that has departed from the original analysis model in an attempt to make a system of reusable objects. In practice, both types of patterns are probably useful.

6 Conclusion

Design patterns have revolutionized the way we think about, design, and teach object-oriented systems. We have found them applicable in many stages of the design process—initial design, reuse, refactoring. They have given us a new level of abstraction for system design.

New levels of abstraction often afford opportunities for increased automation. We are investigating how interactive tools can take advantage of design patterns. One of these tools lets a user explore the space of objects in a running program and watch their interaction. Through observation the user may discover existing or entirely new patterns; the tool lets the user record and catalog his observations. The user may thus gain a better understanding of the application, the libraries on which it is based, and design in general.

Design patterns may have an even more profound impact on how object-oriented systems are designed than we have discussed. Common to most patterns is that they permit certain aspects of a system to be varied independently. This leads to thinking about design in terms of "What aspect of a design should be variable?" Answers to this question lead to certain applicable design patterns, and their application leads subsequently to modification of a design. We refer to this design activity as **variation-oriented design** and discuss it more fully in the catalog of patterns [12].

But some caveats are in order. Design patterns should not be applied indiscriminately. They typically achieve flexibility and variability by introducing additional levels of indirection and can therefore complicate a design. A design pattern should only be applied when the flexibility it affords is actually needed. The consequences described in a pattern help determine this. Moreover, one is often tempted to brand any new programming trick a new design pattern. A true design pattern will be non-trivial and will have had more than one application.

We hope that the design patterns described in this paper and in the companion catalog will provide the object-oriented community both a common design terminology and a repertoire of reusable designs. Moreover, we hope the catalog will motivate others to describe their systems in terms of design patterns and develop their own design patterns for others to reuse.

7 Acknowledgements

The authors wish to thank Doug Lea and Kent Beck for detailed comments and discussions about this work, and Bruce Anderson and the participants of the Architecture Handbook workshops at OOPSLA '91 and '92.

References

1. B. Adelson and Soloway E. The role of domain experience in software design. *IEEE Transactions on Software Engineering*, 11(11):1351–1360, 1985.
2. Christopher Alexander. *The Timeless Way of Building*. Oxford University Press, New York, 1979.
3. Association for Computing Machinery. *Addendum to the Proceedings, Object-Oriented Programming Systems, Languages, and Applications Conference*, Phoenix, AZ, October 1991.
4. Association for Computing Machinery. *Addendum to the Proceedings, Object-Oriented Programming Systems, Languages, and Applications Conference*, Vancouver, British Columbia, October 1992.
5. Kent Beck and Ward Cunningham. A laboratory for teaching object-oriented thinking. In *Object-Oriented Programming Systems, Languages, and Applications Conference Proceedings*, pages 1–6, New Orleans, LA, October 1989.
6. Peter Coad. Object-oriented patterns. *Communications of the ACM*, 35(9):152–159, September 1992.
7. James O. Coplien. *Advanced C++ Programming Styles and Idioms*. Addison-Wesley, Reading, Massechusetts, 1992.
8. Ward Cunningham and Kent Beck. Constructing abstractions for object-oriented applications. Technical Report CR-87-25, Computer Research Laboratory, Tektronix, Inc., 1987.
9. Bill Curtis. Cognitive issues in reusing software artifacts. In Ted J. Biggerstaff and Alan J. Perlis, editors, *Software Reusability, Volume II*, pages 269–287. Addison-Wesley, 1989.
10. Thomas Eggenschwiler and Erich Gamma. The ET++SwapsManager: Using object technology in the financial engineering domain. In *Object-Oriented Programming Systems, Languages, and Applications Conference Proceedings*, pages 166–178, Vancouver, British Columbia, October 1992.
11. Erich Gamma. *Objektorientierte Software-Entwicklung am Beispiel von ET++: Design-Muster, Klassenbibliothek, Werkzeuge*. Springer-Verlag, Berlin, 1992.
12. Erich Gamma, Richard Helm, Ralph Johnson, and John Vlissides. A catalog of object-oriented design patterns. Technical Report in preparation, IBM Research Division, 1992.

13. Mehdi T. Harandi and Frank H. Young. Software design using reusable algorithm abstraction. In *In Proc. 2nd IEEE/BCS Conf. on Software Engineering*, pages 94–97, 1985.
14. Richard Helm, Tien Huynh, Catherine Lassez, and Kim Marriott. A linear constraint technology for user interfaces. In *Graphics Interface*, pages 301–309, Vancouver, British Columbia, 1992.
15. Richard Helm, Tien Huynh, Kim Marriott, and John Vlissides. An object-oriented architecture for constraint-based graphical editing. In *Proceedings of the Third Eurographics Workshop on Object-Oriented Graphics*, pages 1–22, Champéry, Switzerland, October 1992. Also available as IBM Research Division Technical Report RC 18524 (79392).
16. Ralph Johnson. Documenting frameworks using patterns. In *Object-Oriented Programming Systems, Languages, and Applications Conference Proceedings*, pages 63–76, Vancouver, BC, October 1992.
17. Ralph E. Johnson, Carl McConnell, and J. Michael Lake. The RTL system: A framework for code optimization. In Robert Giegerich and Susan L. Graham, editors, *Code Generation—Concepts, Tools, Techniques. Proceedings of the International Workshop on Code Generation*, pages 255–274, Dagstuhl, Germany, 1992. Springer-Verlag.
18. S. Katz, C.A. Richter, and K.-S. The. Paris: A system for reusing partially interpreted schemas. In *Proc. of the Ninth International Conference on Software Engineering*, 1987.
19. Glenn E. Krasner and Stephen T. Pope. A cookbook for using the model-view controller user interface paradigm in Smalltalk-80. *Journal of Object-Oriented Programming*, 1(3):26–49, August/September 1988.
20. Charles W. Krueger. Software reuse. *ACM Computing Surveys*, 24(2), June 1992.
21. Mark A. Linton. Encapsulating a C++ library. In *Proceedings of the 1992 USENIX C++ Conference*, pages 57–66, Portland, OR, August 1992.
22. Mark A. Linton, John M. Vlissides, and Paul R. Calder. Composing user interfaces with InterViews. *Computer*, 22(2):8–22, February 1989.
23. William F. Opdyke and Ralph E. Johnson. Refactoring: An aid in designing application frameworks and evolving object-oriented systems. In *SOOPPA Conference Proceedings*, pages 145–161, Marist College, Poughkeepsie, NY, September 1990.
24. Charles Rich and Richard C. Waters. Formalizing reusable software components in the programmer's apprentice. In Ted J. Biggerstaff and Alan J. Perlis, editors, *Software Reusability, Volume II*, pages 313–343. Addison-Wesley, 1989.
25. James Rumbaugh, Michael Blaha, William Premerlani, Frederick Eddy, and William Lorenson. *Object-Oriented Modeling and Design*. Prentice Hall, Englewood Cliffs, New Jersey, 1991.
26. Elliot Soloway and Kate Ehrlich. Empirical studies of programming knowledge. *IEEE Transactions on Software Engineering*, 10(5), September 1984.
27. James C. Spohrer and Elliot Soloway. Novice mistakes: Are the folk wisdoms correct? *Communications of the ACM*, 29(7):624–632, July 1992.
28. ParcPlace Systems. *ParcPlace Systems, Objectworks/Smalltalk Release 4 Users Guide*. Mountain View, California, 1990.
29. David Ungar and Randall B. Smith. Self: The power of simplicity. In *Object-Oriented Programming Systems, Languages, and Applications Conference Proceedings*, pages 227–242, Orlando, Florida, October 1987.

30. John M. Vlissides and Mark A. Linton. Unidraw: A framework for building domain-specific graphical editors. *ACM Transactions on Information Systems*, 8(3):237–268, July 1990.
31. André Weinand, Erich Gamma, and Rudolf Marty. ET++—An object-oriented application framework in C++. In *Object-Oriented Programming Systems, Languages, and Applications Conference Proceedings*, pages 46–57, San Diego, CA, September 1988.
32. Rebecca Wirfs-Brock and Ralph E. Johnson. A survey of current research in object-oriented design. *Communications of the ACM*, 33(9):104–124, 1990.

A Catalog Overview

The following summarizes the patterns in our current catalog.

Abstract Factory provides an interface for creating generic product objects. It removes dependencies on concrete product classes from clients that create product objects.

Adapter makes the protocol of one class conform to the protocol of another.

Bridge separates an abstraction from its implementation. The abstraction may vary its implementations transparently and dynamically.

Builder provides a generic interface for incrementally constructing aggregate objects. A Builder hides details of how objects in the aggregate are created, represented, and composed.

Command objectifies the request for a service. It decouples the creator of the request for a service from the executor of that service.

Composite treats multiple, recursively-composed objects as a single object.

Chain of Responsibility defines a hierarchy of objects, typically arranged from more specific to more general, having responsibility for handling a request.

Factory Method lets base classes create instances of subclass-dependent objects.

Flyweight defines how objects can be shared. Flyweights support object abstraction at the finest granularity.

Glue defines a single point of access to objects in a subsystem. It provides a higher level of encapsulation for objects in the subsystem.

Interpreter defines how to represent the grammar, abstract syntax tree, and interpreter for simple languages.

Iterator objectifies traversal algorithms over object structures.

Mediator decouples and manages the collaboration between objects.

Memento opaquely encapsulates a snapshot of the internal state of an object and is used to restore the object to its original state.

Observer enforces synchronization, coordination, and consistency constraints between objects.

Prototype creates new objects by cloning a prototypical instance. Prototypes permit clients to install and configure dynamically the instances of particular classes they need to instantiate.

Proxy acts as a convenient surrogate or placeholder for another object. Proxies can restrict, enhance, or alter an object's properties.

Solitaire defines a one-of-a-kind object that provides access to unique or well-known services and variables.

State lets an object change its behavior when its internal state changes, effectively changing its class.

Strategy objectifies an algorithm or behavior.

Template Method implements an abstract algorithm, deferring specific steps to subclass methods.

Walker centralizes operations on object structures in one class so that these operations can be changed independently of the classes defining the structure.

Wrapper attaches additional services, properties, or behavior to objects. Wrappers can be nested recursively to attach multiple properties to objects.

Abstract Factory

Object Creational

Intent

Abstract Factory provides an interface for creating generic product objects. It removes dependencies on concrete product classes from clients that create product objects.

Motivation

Consider a user interface toolkit that supports multiple standard look-and-feels, say, Motif and Open Look, and provides different scroll bars for each. It is undesirable to hard-code dependencies on either standard into the application—the choice of look-and-feel and hence scroll bar may be deferred until run-time. Specifying the class of scroll bar limits flexibility and reusability by forcing a commitment to a particular class instead of a particular protocol. An Abstract Factory avoids this commitment.

An abstract base class WindowKit declares services for creating scroll bars and other controls. Controls for Motif and Open Look are derived from common abstract classes. For each look-and-feel there is a concrete subclass of WindowKit that defines services to create the appropriate control. For example, the CreateScrollBar() operation on the MotifKit would instantiate and return a Motif scroll bar, while the corresponding operation on the OpenLookKit returns an Open Look scroll bar. Clients access a specific kit through the interface declared by the WindowKit class, and they access the controls created by a kit only by their generic interface.

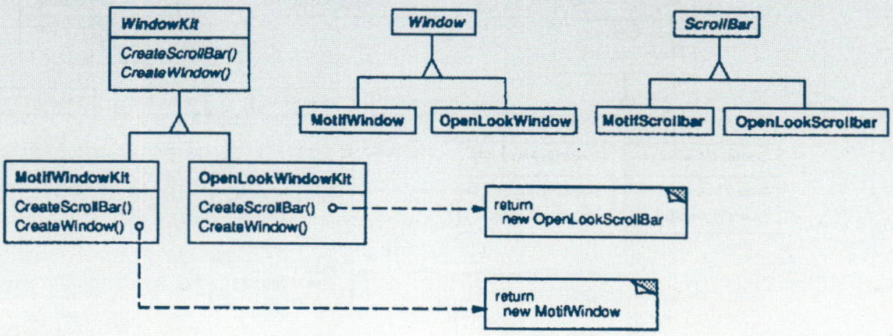

Applicability

When the classes of the product objects are variable, and dependencies on these classes must be removed from a client application.

When variations on the creation, composition, or representation of aggregate objects or subsystems must be removed from a client application. Differences in configuration can be obtained by using different concrete factories. Clients do not explicitly create and configure the aggregate or subsystem but defer this responsibility to an AbstractFactory class. Clients instead call a method of the AbstractFactory that returns an object providing access to the aggregate or subsystem.

Participants

- **AbstractFactory**
  - declares a generic interface for operations that create generic product objects.
- **ConcreteFactory**
  - defines the operations that create specific product objects.
- **GenericProduct**
  - declares a generic interface for product objects.
- **SpecificProduct**
  - defines a product object created by the corresponding concrete factory.
  - all product classes must conform to the generic product interface.

Collaborations

- Usually a single instance of a ConcreteFactory class is created at run-time. This concrete factory creates product objects having a particular implementation. To use different product objects, clients must be configured to use a different concrete factory.
- AbstractFactory defers creation of product objects to its ConcreteFactory subclasses.

Diagram

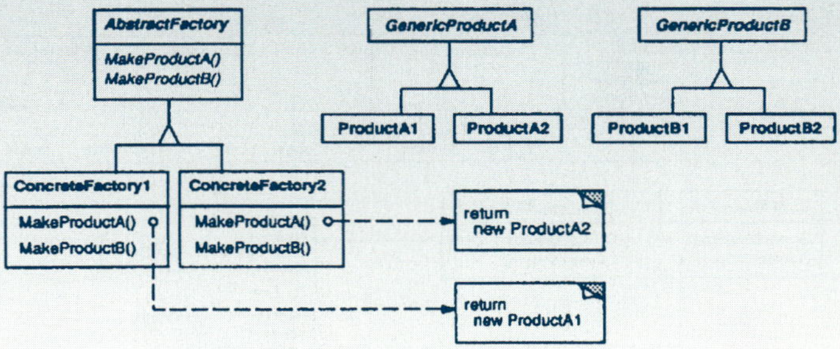

Consequences

Abstract Factory provides a focus during development for changing and controlling the types of objects created by clients. Because a factory objectifies the responsibility for and the process of creating product objects, it isolates clients from implementation classes. Only generic interfaces are visible to clients. Implementation class names do not appear in client code. Clients can be defined and implemented solely in terms of protocols instead of classes.

Abstract factories that encode class names in operation signatures can be difficult to extend with new kinds of product objects. This can require redeclaring the AbstractFactory and all ConcreteFactories. Abstract factories can be composed with subordinate factory objects. Responsibility for creating objects is delegated to these sub-factories. Composition of abstract factories provides a simple way to extend the kinds of objects a factory is responsible for creating.

Examples

InterViews uses the "Kit" suffix [21] to denote abstract factory classes. It defines WidgetKit and DialogKit abstract factories for generating look-and-feel-specific user interface objects. InterViews also includes a LayoutKit that generates different composition objects depending on the layout desired.

ET++ [31] employs the Abstract Factory pattern to achieve portability across different window systems (X Windows and SunView, for example). The WindowSystem abstract base class defines the interface for creating objects representing window system resources (for example, MakeWindow, MakeFont, MakeColor). Concrete subclasses implement the interfaces for a specific window system. At runtime ET++ creates an instance of a concrete WindowSystem subclass that creates system resource objects.

Implementation

A novel implementation is possible in Smalltalk. Because classes are first-class objects, it is not necessary to have distinct ConcreteFactory subclasses to create the variations in products. Instead, it is possible to store classes that create these products in variables inside a concrete factory. These classes create new instances on behalf of the concrete factory. This technique permits variation in product objects at finer levels of granularity than by using distinct concrete factories. Only the classes kept in variables need to be changed.

See Also

Factory Method: Abstract Factories are often implemented using Factory Methods.

STRATEGY — Object Behavioral

Intent

A Strategy objectifies an algorithm or behavior, allowing the algorithm or behavior to be varied independently of its clients.

Motivation

There are many algorithms for breaking a text stream into lines. It is impossible to hard-wire all such algorithms into the classes that require them. Different algorithms might be appropriate at different times.

One way to address this problem is by defining separate classes that encapsulate the different linebreaking algorithms. An algorithm objectified in this way is called a Strategy. InterViews [22] and ET++ [31] use this approach.

Suppose a Composition class is responsible for maintaining and updating the line breaks of text displayed in a text viewer. Linebreaking strategies are not implemented by the class Composition. Instead, they are implemented separately by subclasses of the Compositor class. Compositor subclasses implement different strategies as follows:

- **SimpleCompositor** implements a simple strategy that determines line breaks one at a time.
- **TeXCompositor** implements the TEXalgorithm for finding line breaks. This strategy tries to optimize line breaks globally, that is, one paragraph at a time.
- **ArrayCompositor** implements a strategy that is used not for text but for breaking a collection of icons into rows. It selects breaks so that each row has a fixed number of items.

A Composition maintains a reference to a Compositor object. Whenever a Composition is required to find line breaks, it forwards this responsibility to its current Compositor object. The client of Composition specifies which Compositor should be used by installing the corresponding Compositor into the Composition (see the diagram below).

Applicability

Whenever an algorithm or behavior should be selectable and replaceable at run-time, or when there exist variations in the implementation of the algorithm, reflecting different space-time tradeoffs, for example.

Use a Strategy whenever many related classes differ only in their behavior. Strategies provide a way to configure a single class with one of many behaviors.

Participants

- **Strategy**
  - objectifies and encapsulates an algorithm or behavior.
- **StrategyContext**
  - maintains a reference to a Strategy object.
  - maintains the state manipulated by the Strategy.
  - can be configured by passing it an appropriate Strategy object.

Collaborations

- Strategy manipulates the StrategyContext. The StrategyContext normally passes itself as an argument to the Strategy's methods. This allows the Strategy to call back the StrategyContext as required.
- StrategyContext forwards requests from its clients to the Strategy. Usually clients pass Strategy objects to the StrategyContext. Thereafter clients only interact with the StrategyContext. There is often a family of Strategy classes from which a client can choose.

Diagram

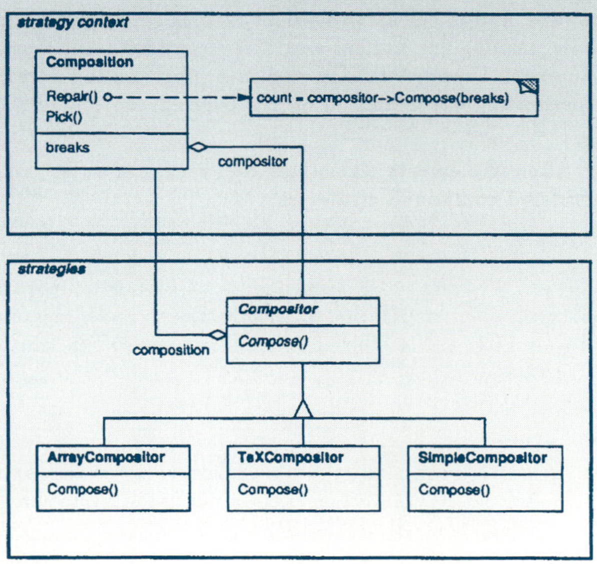

Consequences

Strategies can define a family of policies that a StrategyContext can reuse. Separating a Strategy from its context increases reusability, because the Strategy may vary independently from the StrategyContext.

Variations on an algorithm can also be implemented with inheritance, that is, with an abstract class and subclasses that implement different behaviors. However, this hard-wires the implementation into a specific class; it is not possible to change behaviors dynamically. This results in many related classes that differ only in some behavior. It is often better to break out the variations of behavior into their own classes. The Strategy pattern thus increases modularity by localizing complex behavior. The typical alternative is to scatter conditional statements throughout the code that select the behavior to be performed.

Implementation

The interface of a Strategy and the common functionality among Strategies is often factored out in an abstract class. Strategies should avoid maintaining state across invocations so that they can be used repeatedly and in multiple contexts.

Examples

In the RTL System for compiler code optimization [17], Strategies define different register allocation schemes (RegisterAllocator) and different instruction set scheduling policies (RISCscheduler, CISCscheduler). This gives flexibility in targeting the optimizer for different machine architectures.

The ET++SwapsManager calculation engine framework [10] computes prices for different financial instruments. Its key abstractions are Instrument and YieldCurve. Different instruments are implemented as subclasses of Instrument. The YieldCurve calculates discount factors to present value future cash flows. Both of these classes delegate some behavior to Strategy objects. The framework provides a family of Strategy classes that define algorithms to generate cash flows, to value swaps, and to calculate discount factors. New calculation engines are created by parameterizing Instrument and YieldCurve with appropriate Strategy objects. This approach supports mixing and matching existing Strategy implementations while permitting the definition of new Strategy objects.

See Also

Walker often implements algorithms over recursive object structures. Walkers can be considered compound strategies.

Wrapper

Compound Structural

Intent

A Wrapper attaches additional services, properties, or behavior to objects. Wrappers can be nested recursively to attach multiple properties to objects.

Motivation

Sometimes it is desirable to attach properties to individual objects instead of classes. In a graphical user interface toolkit, for example, properties such as borders or services like scrolling should be freely attachable to any user interface component.

One way to attach properties to components is via inheritance. Inheriting a border from a base class will give all instances of its derived classes a border. This is inflexible because the choice of border is made statically. It is more flexible to let a client decide how and when to decorate the component with a border.

This can be achieved by enclosing the component in another object that adds the border. The enclosing object, which must be transparent to clients of the component, is called a Wrapper. This transparency is the key for nesting Wrappers recursively to construct more complex user interface components. A Wrapper forwards requests to its enclosed user interface component. The Wrapper may perform additional actions before or after forwarding the request, such as drawing a border around a user interface component.

Typical properties or services provided by user interface Wrappers are:

- decorations like borders, shadows, or scroll bars; or
- services like scrolling or zooming.

The following diagram illustrates the composition of a TextView with a BorderWrapper and a ScrollWrapper to produce a bordered, scrollable TextView.

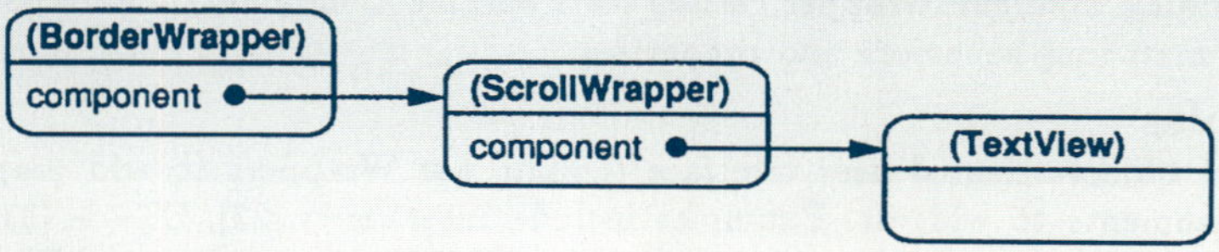

Applicability

When properties or behaviors should be attachable to individual objects dynamically and transparently.

When there is a need to extend classes in an inheritance hierarchy. Rather than modifying their base class, instances are enclosed in a Wrapper that adds the additional behavior and properties. Wrappers thus provide an alternative to extending the base class without requiring its modification. This is of particular concern when the base class comes from a class library that cannot be modified.

Participants

- **Component**
  - the object to which additional properties or behaviors are attached.
- **Wrapper**
  - encapsulates and enhances its Component. It defines an interface that conforms to its Component's.
  - Wrapper maintains a reference to its Component.

Collaborations

- Wrapper forwards requests to its Component. It may optionally perform additional operations before and after forwarding the request.

Diagram

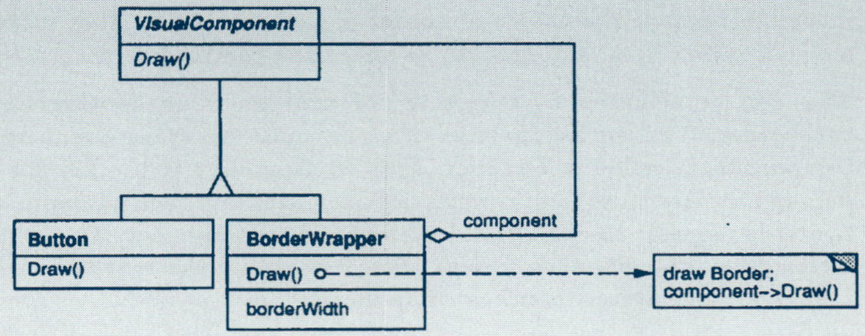

Consequences

Using Wrappers to add properties is more flexible than using inheritance. With Wrappers, properties can be attached and detached at run-time simply by changing the Wrapper. Inheritance would require creating a new class for each property composition (for example, BorderdScrollableTextView, BorderedTextView). This clutters the name space of classes unnecessarily and should be avoided. Moreover, providing different Wrapper classes for a specific Component class allows mixing and matching behaviors and properties.

Examples

Most object-oriented user interface toolkits use Wrappers to add graphical embellishments to widgets. Examples include InterViews [22], ET++ [31], and the ParcPlace Smalltalk class library [28]. More exotic applications of Wrappers are the DebuggingGlyph from InterViews and the PassivityWrapper from ParcPlace Smalltalk. A DebuggingGlyph prints out debugging information before and after it forwards a layout request to its enclosed object. This trace information can be used to analyze and debug the layout behavior of objects in a complex object composition. The PassivityWrapper can enable or disable user interactions with the enclosed object.

Implementation

Implementation of a set of Wrapper classes is simplified by an abstract base class, which forwards all requests to its component. Derived classes can then override only those operations for which they want to add behavior. The abstract base class ensures that all other requests are passed automatically to the Component.

See Also

Adapter: A Wrapper is different from an Adapter, because a Wrapper only changes an object's properties and not its interface; an Adapter will give an object a completely new interface.

Composite: A Wrapper can be considered a degenerate Composite with only one component. However, a Wrapper adds additional services—it is not intented for object aggregation.

Editors
Manfred Broy
Institut für Informatik
Technische Universität München
80290 München, Germany
broy@informatik.tu-muenchen.de

Ernst Denert
sd&m AG
software design & management
Postfach 83 08 51
81708 München, Germany
denert@sdm.de

Library of Congress Cataloging-in-Publication Data

Software pioneers : contributions to software engineering : sd&m conference 2001/
editors, Manfred Broy, Ernst Denert.
p. cm.
Includes bibliographical references and index.
ISBN 3540430814 (alk. paper)
1. Computers--Biography. 2. Electronic data processing personnel--Biography 3.
Software engineering--History. I. Broy, M., 1949- II. Denert, Ernst. III. sd&m AG.

QA76.2.A2 S62 2002
005.1'092'--dc21

ISBN 3-540-43081-4 Springer-Verlag Berlin Heidelberg New York

Springer-Verlag Berlin Heidelberg New York
a member of BertelsmannSpringer Science+Business Media GmbH

http://www.springer.de

Layout: Studio Quitta, München
Typesetting: VerlagsService Hegele, Dossenheim
Printing on acid-free paper SPIN 10861319 45/3142X0 - 543210